FUNDAMENTALS OF FOOD PROCESS ENGINEERING

Second Edition

FUNDAMENTALS OF FOOD PROCESS ENGINEERING

Second Edition

Romeo T. Toledo

CBS Publishers & Distributors Pvt. Ltd.

New Delhi • Bengaluru • Chennai • Kochi • Kolkata • Mumbai
Hyderabad • Nagpur • Patna • Pune • Vijayawada

CBS Pubs ISBN: 978-81-239-1551-7
Chapman ISBN: 0-412-05311-X

First Indian Edition: 1997
Reprint: 2000

This edition has been published in India by arrangement with Chapman & Hall, Inc., New York

Published by **Satish Kumar Jain** and produced by **Varun Jain** for
CBS Publishers & Distributors Pvt. Ltd.,
4819/XI Prahlad Street, 24 Ansari Road, Daryaganj, New Delhi - 110002
delhi@cbspd.com, cbspubs@airtelmail.in • www.cbspd.com
Ph.: 23289259, 23266861, 23266867 • Fax: 011-23243014

Corporate Office: 204 FIE, Industrial Area, Patparganj, Delhi - 110 092
Ph: 49344934 • Fax: 011-49344935
E-mail: publishing@cbspd.com • publicity@cbspd.com

Branches:

- ***Bengaluru:*** 2975, 17th Cross, K.R. Road, Bansankari 2nd Stage, Bengaluru - 70 • Ph: +91-80-26771678/79 • Fax: +91-80-26771680 E-mail: cbsbng@gmail.com, bangalore@cbspd.com
- ***Chennai:*** No. 7, Subbaraya Street, Shenoy Nagar, Chennai - 600030 Ph: +91-44-26681266, 26680620 • Fax: +91-44-42032115 E-mail: chennai@cbspd.com
- ***Kochi:*** Ashana House, 39/1904, A.M. Thomas Road, Valanjambalam, Ernakulum, Kochi • Ph: +91-484-4059061-65 Fax: +91-484-4059065 • E-mail: cochin@cbspd.com
- ***Kolkata:*** 6-B, Ground Floor, Rameshwar Shaw Road, Kolkata - 700014 Ph: +91-33-22891126/7/8 • E-mail: kolkata@cbspd.com
- ***Mumbai:*** 83-C, Dr. E. Moses Road, Worli, Mumbai - 400018 Ph: +91-9833017933, 022-24902340/41 • E-mail: mumbai@cbspd.com

Representatives:

- Hyderabad: 0-9885175004
- Patna: 0-9334159340
- Jharkhand: 0-9811541605
- Nagpur: 0-9021734563
- Pune: 0-9623451994
- Uttarakhand: 0-9716462459

Printed at:

Mudrak, Noida, UP, India

Contents

PREFACE

PREFACE

Ten years after the publication of the first edition of *Fundamentals of Food Process Engineering*, there have been significant changes in both food science education and the food industry itself. Students now in the food science curriculum are generally better prepared mathematically than their counterparts two decades ago. The food science curriculum in most schools in the United States has split into science and business options, with students in the science option following the Institute of Food Technologists' minimum requirements. The minimum requirements include the food engineering course, thus students enrolled in food engineering are generally better than average, and can be challenged with more rigor in the course material.

The food industry itself has changed. Traditionally, the food industry has been primarily involved in the canning and freezing of agricultural commodities, and a company's operations generally remain within a single commodity. Now, the industry is becoming more diversified, with many companies involved in operations involving more than one type of commodity. A number of formulated food products are now made where the commodity connection becomes obscure. The ability to solve problems is a valued asset in a technologist, and often, solving problems involves nothing more than applying principles learned in other areas to the problem at hand. A principle that may have been commonly used with one commodity may also be applied to another commodity to produce unique products. Numerous examples may be cited where processes used in the food industry were adapted from the chemical or textile industries. The food industry is now also involved in sophisticated separation processes, to make higher-value ingredients, remove undesirable components from naturally occurring food sources, or to recover valuable components from food processing byproducts. In addition, federal, state, and local regulations for public health, worker safety, and environmental protection require a knowledge not only of making the product, but also of techniques for the elimination of microbial hazards, for packaging, and for the reduction of point discharge of pollutants to the environment. Present and future consumers are quality and safety conscious, and most are willing to pay for quality, safety, and convenience. Incorporating all of these features into processed food products requires the development of new processes and optimized processing procedures. The old art in food processing and preservation, and the trial-and-error method of making acceptable products, must be replaced by the scientific method, if a company wants priority in product introduction and to generate a marketing edge. These changes in the food industry mean faster career advancement for those who are better prepared technically, and food engineering brings an extra dimension to the training of a food technologist.

Learning food engineering is different from learning other courses in the food science curriculum. Students have to be warned that memorizing a solution to a problem is not recommended. What matters is learning how to recognize principles, and how to utilize these principles in formulating a solution to a problem. To be effective in demonstrating this ability of application, rather than rote memorization, practice and test problems must be made different from solved example problems.

These trends in both the food industry and food science education have been taken into consideration in the revision of *Fundamentals of Food Process Engineering*. New example problems have been added, the calculus is more liberally used in this edition as compared to the first edition, and a number of current developments in the field have been incorporated. A recurring theme in the book is the principle of similitude (i.e., most processes are similar when viewed from the standpoint of mass and energy transport). These show up in the derivation of equations where the same physical principles are generally used as the starting point in the derivation. The description of equipment is included only when it is essential in establishing how parameters for their efficient operation are selected. Quantification of the relationships between operating variables, equipment size, and product quality is the primary objective of food process engineering, as presented in this book.

There is often not enough time in a term to teach all of the subjects covered in this edition. Instructors can choose which subjects should be covered in depth. The coverage of subjects of contemporary interest enables students to connect the relevance of the subject matter to their career development. News media articles, trade publications, and announcements of symposia or short courses may be used as a guide in the selection of subjects of contemporary interest.

A review chapter on mathematics is included. This was viewed as a useful feature in the first edition. It is recommended that this chapter not be taught as a review of mathematics, but rather as assigned reading prior to the first use of a mathematical principle in the solution of a problem in later sections of the textbook. BASIC programming is also included. The use of computers is becoming more common in schools and in the workplace and, in the coming decade, most college-bound high school students should be familiar with writing and using computer programs. Again, this may be treated as assigned reading if computer solutions to some of the examples or end of chapter problems are to be covered in class. The chapter on units and dimensions may be used in the same way, although the author has experienced good results in later chapters when units, conversion, and the assessment of equations for dimensional consistency have been thoroughly discussed.

The chapter on thermal processing has been expanded. The subject of thermobacteriology is one unique area in food processing where chemistry, microbiology, and engineering are simultaneously applied in the solution of a processing problem. With the current emphasis on food safety and product quality, knowing how optimal processes for microbial inactivation are determined is a must for every food technologist. To complement the expansion of the thermal processing chapter, a section on unsteady state heat transfer has been added in the chapter on heat transfer.

The author has been careful, in the preparation of the manuscript for this edition, to eliminate errors. Students and colleagues have been very helpful in catching errors. Special thanks is due to Dr. Richard Hartel for an extremely helpful review. If some errors remain when the book is printed, responsibility for those rests solely on the author.

Romeo T. Toledo

1

Review of Mathematical Principles and Applications in Food Processing

GRAPHING AND FITTING EQUATIONS TO EXPERIMENTAL DATA

Variables and Functions. A variable is a quantity that can assume any value. In algebraic expressions, variables are represented by letters from the beginning and end of the alphabet. In physics and engineering, any letter of the alphabet and Greek letters are used as symbols for physical quantities. Any symbol may represent a variable if the value of the physical quantity it stands for is not fixed in the statement of the problem. In an algebraic expression, the letters from the beginning of the alphabet often represent constants, that is, their values are fixed. Thus, in the expression $ax = 2by$, x and y represent variables and a and b are constants.

A function represents the mathematical relationship between variables. Thus, the temperature in a solid which is being heated in an oven may be expressed as a function of time and position using the mathematical expression $T = F(x, t)$. In an algebraic expression, $y = 2x + 4$, $y = F(x)$, and $F(x) = 2x + 4$.

Variables may be dependent or independent. Unless defined, the dependent variable in a mathematical expression is one which stands alone on one side of an equation. In the expression $y = F(x)$, y is the dependent variable and x is the independent variable. When the expression is rearranged in the form $x = F(y)$, x is the dependent variable and y is the independent variable. In physical or chemical systems, the interdependence of the variables is determined by the design of the experiment. The independent variables are those fixed in the design of the experiment, and the dependent variables are those which are measured. For example, when determining the loss of ascorbic acid in stored canned foods, ascorbic acid concentration is the dependent variable and time is the independent variable. On the other hand, in an experiment where a sample of a food is taken and both moisture content and water activity are measured, either of the two variables may be designated as the dependent or independent variable. In statistical design, the terms *response variable* and *treatment variable* are used for the dependent and independent variables, respectively.

Graphs. Each data point obtained in an experiment is a set of numbers representing the values of the independent and dependent variables. A data point for a response variable which depends on only one independent variable (univariate) will be a number pair, while a data point for a response variable which depends on several independent variables (multivariate) will consist of a value for the response variable and one value each for the treatment variables. Experimental data are often presented as a table of numerical values of the variables or as a graph. The graph traces the path of the dependent variable as the values of the independent variables are changed. For univariate responses, the graph will be two-dimensional; for multivariate responses, the graph will be multidimensional.

When all variables in the function have an exponent of 1, the function is called *first order* and is represented by a straight line. When any variable has an exponent other than 1, the graph is a curve.

The numerical values represented by a data point are called the *coordinates* of that point. When plotting experimental data, the independent variable is plotted on the horizontal axis, or *abscissa*, and the dependent variable is plotted on the vertical axis or *ordinate*. The rectangular, or cartesian, coordinate system is the most common system for graphing data. Both the abscissa and the ordinate are in the arithmetic scale, and the distance from the origin measured along or parallel to the abscissa or ordinate to the point under consideration is directly proportional to the value of the coordinate of that point. Scaling of the abscissa and ordinate is done such that the data points, when plotted, will be symmetrical and centered within the graph. The cartesian coordinate system is divided into four quadrants, with the origin in the center. The upper right quadrant represents points with positive coordinates; the left right quadrant represents negative values of the variable on the abscissa and positive values for the variable on the ordinate; the lower left quadrant represents negative values for both variables; and the lower right quadrant represents positive values for the variable on the abscissa and negative values for the variable on the ordinate.

Equations. An equation is a statement of equality. Equations are useful for presenting experimental data, since they can be mathematically manipulated. Furthermore, if the function is continuous, interpolation between experimentally derived values for a variable may be possible. Experimental data may be fitted to an equation using any of the following techniques:

1. *Linear and polynomial regression*. Statistical methods are employed to determine the coefficients of a linear or polynomial expression involving

the independent and dependent variables. Statistical procedures are based on minimizing the sum of squares for the difference between the experimental values and the values predicted by the equation.

2. *Linearization, data transformation, and linear regression.* The equation to which the data is being fitted is linearized. The data are then transformed in accordance with the linearization equation, and a linear regression determines the appropriate coefficients for the linearized equation.
3. *Graphing.* The raw or transformed data are plotted to form a straight line, and from the slopes and intercept, the coefficients of the variables in the equation are determined.

Linear Equations. Plotting of linear equations can be facilitated by writing the equation in the following forms:

1. *The slope-intercept form*: $y = ax + b$, where a = the slope and b = the y-intercept, the point on the ordinate at $x = 0$. The slope is determined by taking two points on the line with coordinates (x_1, y_1) and (x_2, y_2) and solving for $a = (y_2 - y_1)/(x_2 - x_1)$.
2. *The point-slope form*: $(y - b) = a(x - c)$, where a = slope and b and c represent coordinates of a point (c, b) through which the line must pass.

When linear regression is used on experimental data, the slope and the intercept of the line are calculated. The line must pass through the point which represents the mean of x and the mean of y. A line can then be drawn easily, using either the point-slope or the slope-intercept form of the equation for the line.

The equations for the slope and intercept of a line obtained by regression analysis of N pairs of experimental data are:

$$a = \frac{\Sigma xy - (\Sigma x \Sigma y/N)}{\Sigma x^2 - [(\Sigma x)^2/N]}; \quad b = \frac{\Sigma y \, \Sigma x^2 - \Sigma x \, \Sigma xy}{N(\Sigma x^2 - [(\Sigma x)^2/N])}$$

The process of regression involves minimizing the square of the difference between the value of y calculated by the regression equation and y_i, the experimental value of y. In linear regression, $\Sigma(ax + b - \bar{y})^2$ is called the *explained variation* and $\Sigma(y_i - \bar{y})^2$ is called the *random error* or *unexplained variation*.

The ratio of the explained and unexplained variations is called the *correlation coefficient*. If all the points fall exactly on the regression line, the variation of y from the mean will be due to the regression equation. Therefore the explained variation equals the unexplained variation, and the correlation coefficient is 1.0.

If there is too much data scatter, the random or unexplained variation will be very large and the correlation coefficient will be less than 1.0. Thus, regression analysis not only determines the equation of a line which fits the data points but can also be used to test if a predictable relationship exists between the independent and dependent variables. The formula for the linear correlation coefficient is

$$r = \frac{N\sum xy - \sum x \sum y}{\left[\left[N\sum x^2 - \left(\sum x\right)^2\right]\left[N\sum y^2 - \left(\sum y\right)^2\right]\right]^{0.5}}$$

r will have the same sign as the regression coefficient, a. Values for r which are very different from 1.0 must be tested for significance of the regression. The student is referred to statistics textbooks for procedures to follow in testing the significance of regression from the correlation coefficient.

EXAMPLE: The *protein efficiency ratio (PER)* is defined as the weight gain of an animal fed a diet containing the test protein per unit weight of protein consumed. Data are collected by providing feed and water to the animals so that they can feed at will, determining the amount of feed consumed and weighing each animal at designated time intervals. The PER may be calculated from the slope of the regression line for the weight of the animals (y) against the cumulative weight of protein consumed (x). The following data were collected on five experimental animals. Perform a regression analysis, plot the data, and draw the line represented by the regression equation on the graph.

The sum and sums of squares of x and y are: $\Sigma x = 62.20$; $\Sigma x^2 = 307.00$; $\Sigma y = 408.20$; $\Sigma y^2 = 9138.62$; $\Sigma xy = 1568.28$; $N = 20$. The mean of $x = \Sigma x/N = 62.2/20 = 3.11$. The mean of $y = \Sigma y/N = 408.20/20 = 20.41$. Thus the best-fit line will go through the point (3.11, 20.41).

Table 1.1 Experimental feeding test data for determination of the Protein Efficiency Ratio by linear regression

	Experimental Animal									
	1		2		3		4		5	
Day	x	y	x	y	x	y	x	y	x	y
0	0	11.5	0	12.2	0	14.0	0	13.3	0	12.5
7	2.0	16.8	2.2	16.7	1.8	15.2	2.5	18.4	1.8	16.8
14	3.4	22.8	4.2	22.5	3.7	20.7	4.6	25.3	4.0	23.5
21	6.5	28.0	6.3	29.5	6.8	31.0	5.8	28.5	6.6	29.0

$$a = \frac{408.2 - (62.2)(408.20)/20}{307.00 - (62.20)^2/20} = 2.631$$

$$b = \left(\frac{1}{20}\right)\left[\frac{(408.20)(307.00) - (62.20)(1568.28)}{307.00 - (62.20)^2/20}\right] = 12.23$$

$$r = \frac{20(1568.28) - 62.2(408.20)}{[[20(307.00) - (62.20)^2][20(9138.62) - (408.20)^2]]^{0.5}}$$

$$r = 0.9868$$

The correlation coefficient is very close to 1.0, indicating a very good fit of the data to the regression equation. The plot of the data points and the best-fit line represented by the regression equation are shown in Fig. 1.1. The line is plotted by locating the intercept $y = 12.23$ at $x = 0$, the point for the means of x and y (3.11, 20.41), and connecting the two points. The PER is the slope of the line, 2.631.

Nonlinear Equations: Nonlinear monovariate equations are those in which the exponent of any variable in the equation is a number other than 1. The polynomial $y = a + bx + cx^2 + dx^3$ is often used to represent experimental data. The term with the exponent 1 is the linear term; that with the exponent 2 is the quadratic term; and that with the exponent 3 is the cubic term. Thus b, c, and d are often referred to as the *linear*, *quadratic*, and *cubic* coefficients, respectively.

Nonlinear regression analysis is used to determine the coefficients of a polynomial to which experimental data are being fitted. As in linear regression analysis, the objective is to determine the coefficients of the polynomial such that the sum of the squares of the difference between the experimental and predicted values of the response variable is minimal. Polynomial regression is more difficult to perform manually than linear regression because of the number of coefficients which must be evaluated. Stepwise regression analysis may be performed; that is, additional terms are added to the polynomial, and the contribution of each additional term in reducing the error sum of squares is evaluated. To illustrate the complexity of nonlinear compared to linear regression, the equations which must be solved to determine the coefficients are as follows:

For linear regression, $y = ax + b$:

$$\sum y = aN + b \sum x$$

$$\sum xy = aN \sum x + b \sum^2 x$$

For a second-order polynomial, $y = a + bx + cx^2$:

$$\sum y = aN + b \sum x + c \sum x^2$$
$$\sum xy = aN \sum x + b \sum x^2 + c \sum x^3$$
$$\sum x^2 y = aN \sum x^2 + b \sum x^3 + c \sum x^4$$

Thus, evaluation of coefficients for linear regression is relatively easy, involving the solution of two simultaneous equations. On the other hand, polynomial

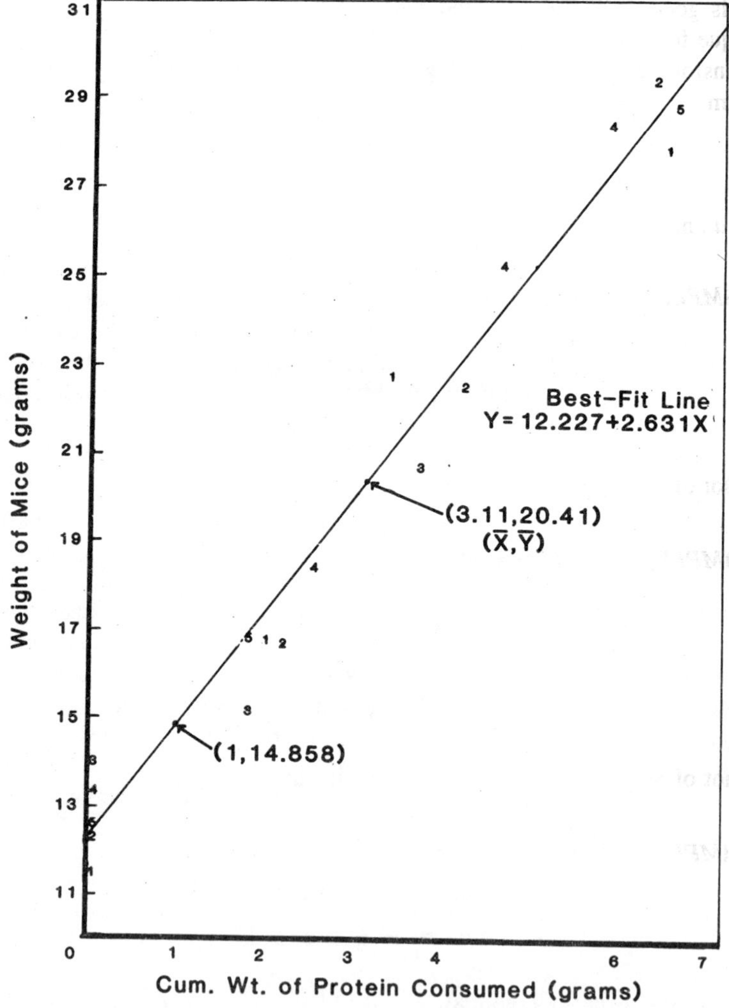

Fig. 1.1. Plot of data points and regression line for PER determination from weight gain data.

regression involves solving $n + 1$ simultaneous equations to evaluate the coefficients of an nth-order polynomial. Determinants (see this section) can be used to determine the constants for an nth-order polynomial. For the second-order polynomial (quadratic) equation, the constants a, b, and c are solved by substituting the values of N, Σx, Σx^2, Σx^3, Σx^4, Σxy, and $\Sigma x^2 y$ into the three equations above and solving them simultaneously.

Linearization of Nonlinear Equations. Nonlinear equations may be linearized by series expansion, but the technique is only an approximation and the result is good only for a limited range of values for the variables. Another technique for linearization involves mathematical manipulation of the function and transformation and/or grouping such that the transformed function assumes the form

$$F(x, y) = aG(x, y) + b,$$

where a and b are constants whose values do not depend on x and y.

EXAMPLE 1: $xy = 5$

$$\text{Linearized form: } y = 5\left(\frac{1}{x}\right)$$

A plot of y against $(1/x)$ will be linear.

EXAMPLE 2: $y = (y^2/x) + 4$.

$$y^2 = xy - 4x$$

$$x = \frac{y^2}{y - 4}$$

A plot of x against $y^2/(y - 4)$ will be linear.

EXAMPLE 3: The hyperbolic function: $y = 1/b + x$.

$$\frac{1}{y} = b + x$$

A plot of $1/y$ against x will be linear.

EXAMPLE 4: The exponential function: $y = ab^x$

$$\log y = \log a + x \log b$$

A plot of log y against x will be linear.

EXAMPLE 5: The geometric function: $y = ax^b$

$$\log y = \log a + b \log x$$

A plot of log y against log x will be linear.

Logarithmic and Semilogarithmic Graphs. Graphing paper is available in which the ordinate and abscissa are in the logarithmic scale. A full logarithmic, or log-log, graphing paper has both the abscissa and the ordinate in the logarithmic scale. A semilogarithmic graphing paper has the ordinate in the logarithmic scale and the abscissa in the arithmetic scale. Full logarithmic graphs are used for geometric functions, as in Example 5 above, and semilogarithmic graphs are used for exponential functions as in Example 4. The distances used in marking the coordinates of points in the logarithmic scale are shown in Fig. 1.2. Each cycle of the logarithmic scale is marked by numbers from 1 to 10. Distances are scaled on the basis of the logarithm of numbers to the base 10. Thus, there is a repeating cycle with multiples of 10. One-cycle semilogarithmic and full logarithmic graphing paper are shown in Fig. 1.3.

When plotting points on the logarithmic scale, label the extreme left and lower coordinates of the graph with the multiple of 10 immediately below the least magnitude of the coordinate to be graphed. Thus, if the least magnitude of the coordinate of the point to be plotted is 0.025, then the extreme left or lower coordinate of the graph should be labeled 0.01.

The number of cycles on the logarithmic scale of the graph to be used must be selected such that the points plotted will occupy most of the graph after plotting. Thus, if the range of numbers to be plotted is from 0.025 to 3.02, three logarithmic cycles will be needed (0.01 to 0.1; 0.1 to 1; 1 to 10). If the range of numbers is from 1.2 to 9.5, only one cycle will be needed (1 to 10).

Numerical values of data points are directly plotted on the logarithmic axis. The scaling of the graph accounts for the logarithmic relationship. Thus, points, when read from the graph, will be in the original rather than the logarithmically transformed data.

Slopes on log-log graphs are determined using the following formula:

$$\text{Slope} = \frac{\log y_2 - \log y_1}{\log x_2 - \log x_1}$$

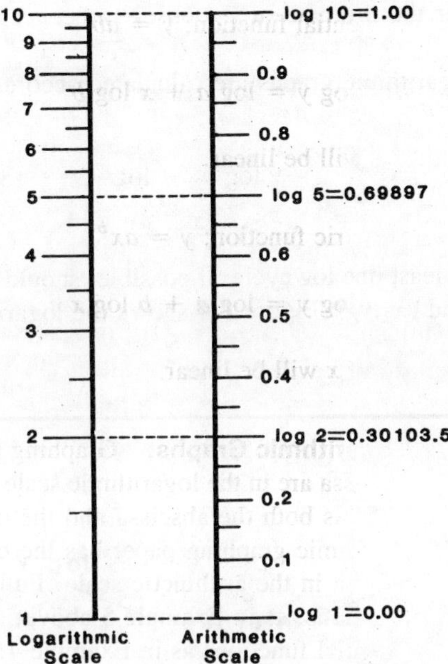

Fig. 1.2. Scaling of a logarithmic scale used on the logarithmic axis of semilogarithmic or full logarithmic graphing paper.

Coordinates of points (x_1,y_1) and (x_2,y_2) which are exactly on the line drawn to fit the data points best are located. Enough separation should be provided between the points to minimize errors. At least one log cycle separation should

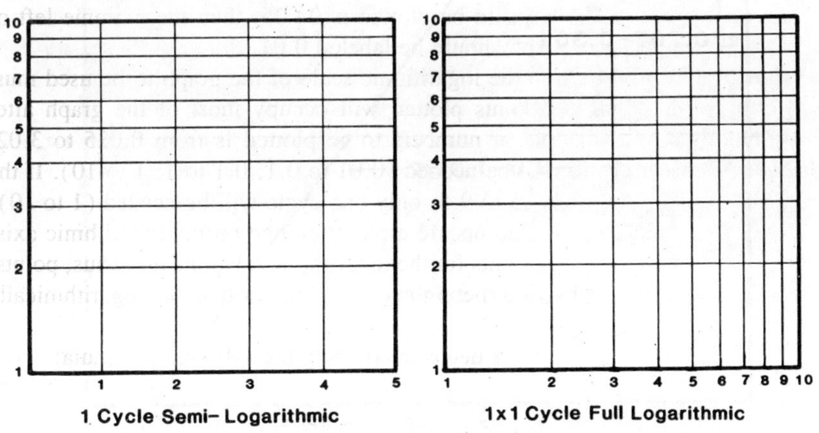

Fig. 1.3. One-cycle semilogarithmic and full logarithmic graphing paper.

be allowed on either the ordinate or the abscissa between the two points selected.

Slopes on semilogarithmic graphs are calculated according to the following formula:

$$\text{Slope} = \frac{\log y_2 - \log y_1}{x_2 - x_1}$$

A separation of at least one log cycle, if possible, should be allowed between the points (x_1,y_1) and (x_2,y_2). Figure 1.4 shows the logarithmic scale relative

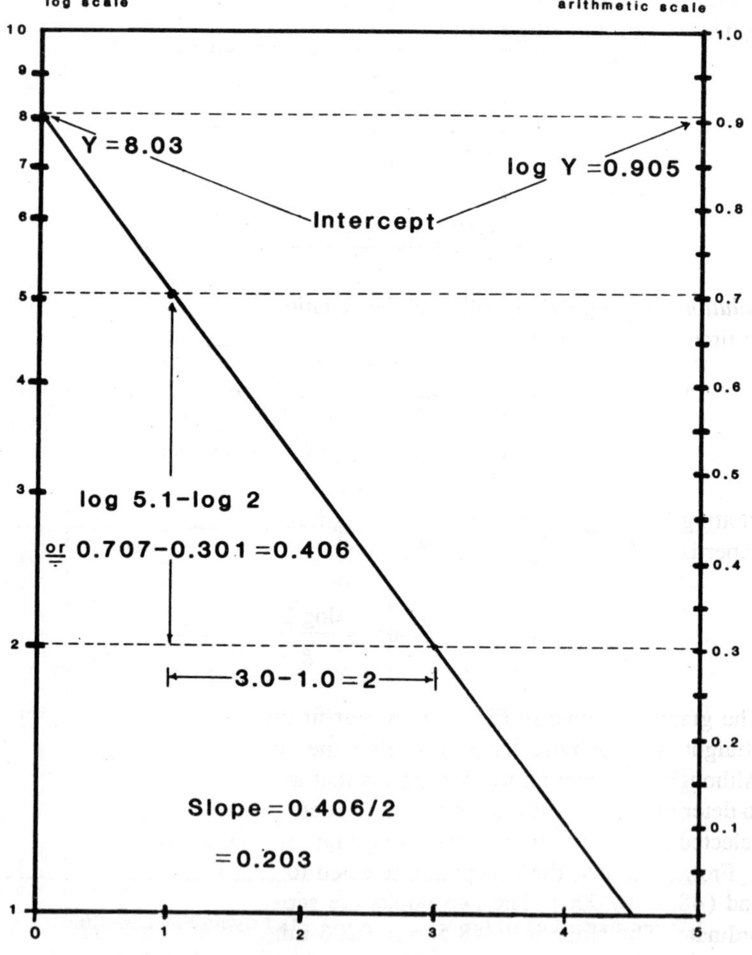

Fig. 1.4. Graph showing the logarithmic scale relative to the arithmetic scale and how the slope and intercept are determined on a semilogarithmic graph.

to the arithmetic scale which would be used if the data are transformed to logarithms prior to plotting. The determination of the slope and intercept is also shown.

The following examples illustrate the use of semilog and log-log graphs:

EXAMPLE 1: An index of the rate of growth of microorganisms is the generation time (g). In the logarithmic phase of microbial growth, the number of organisms (N) changes with time of growth (t) according to

$$N = N_0[2]^{t/g}$$

Find the generation time of a bacterial culture which shows the following numbers with time of growth.

Numbers (N)	Time of growth in minutes (t)
980	0
1700	10
4000	30
6200	40

Solution: Taking the logarithm of the equation for cell numbers as a function of time:

$$\log N = \log N_0 + \left(\frac{t}{g}\right) \log 2$$

Plotting log N against t will give a straight line. A semilogarithmic graphing paper is required for plotting. The slope of the line will be:

$$\text{Slope} = \frac{\log 2}{g}$$

The graph is shown in Fig. 1.5. A best-fit line is drawn by positioning the straight edge such that the points below the line balance those above the line. Although the equation for N suggests that any two data points may be used to determine g, it is advisable to plot the data to make sure that the two points selected lie exactly on the best-fitting line.

From Fig. 1.5, the two points selected to obtain the slope are (0, 1000) and (48.5, 10,000). The two points are separated by one log cycle on the ordinate. The slope is $1/48.5 = 0.0206 \text{ min}^{-1}$. The generation time $g = \log 2/\text{slope} = 14.6$ minutes.

Regression eliminates the guesswork in locating the position of the best-

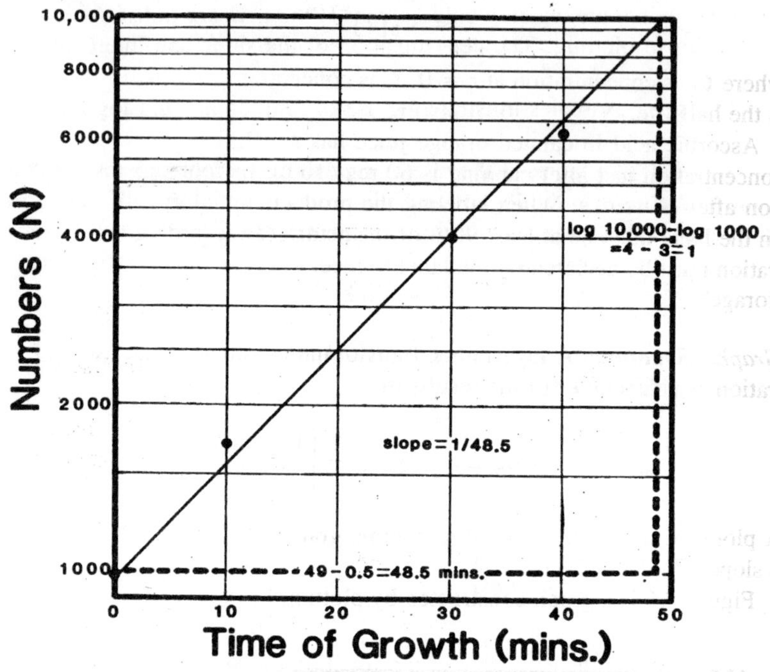

Fig. 1.5. Semilogarithmic plot of microbial growth.

fit line among the data points. Let log $(N) = y$ and $t = x$. The sums are $\Sigma x = 80$, $\Sigma y = 13.616$, $\Sigma x^2 = 2600$, and Σy^2, $\Sigma xy = 292.06$.

$$a = \frac{292.06 - 80/4}{2600 - (80)^2/4} = 0.01974$$

$$b = \frac{13.616(2600) - 80(292.06)}{4[2600 - (80)^2/4]} = 3.0092$$

$$r = \frac{4(292.06) - 80(13.616)}{\{[4(2600) - (80)^2][4(46.740) - (13.616)^2]\}^{0.5}} = 0.9981$$

The correlation coefficient is very close to 1.0, indicating good fit of the data to the regression equation. The slope is 0.01974. $g = \log(2)/0.01974 = 15.2$ minutes.

EXAMPLE 2: The term *half-life* is an index used to express the stability of a compound and is defined as the time required for the concentration to drop to half the original value. In equation form:

$$C = C_0[2]^{-t/t_{0.5}},$$

where C_0 is concentration at $t = 0$, C is concentration at any time t, and $t_{0.5}$ is the half-life.

Ascorbic acid in canned orange juice has a half-life of 30 weeks. If the concentration just after canning is 60 mg/100 mL, calculate the concentration after 10 weeks. When labeling the product, the concentration declared on the label must be at least 90% of the actual concentration. What concentration must be declared on the label to meet this requirement at 10 weeks of storage?

Graphic Solution: A logarithmic transformation of the equation for concentration as a function of time results in:

$$\text{Log } C = \log C_0 - \left[\frac{\log 2}{t_{0.5}}\right] t$$

A plot of C against t on semilogarithmic graphing paper will be linear, with a slope of $-(\log 2)/t_{0.5}$.

Figure 1.6 is a graph constructed by plotting 60 mg/100 mL at $t = 0$ and

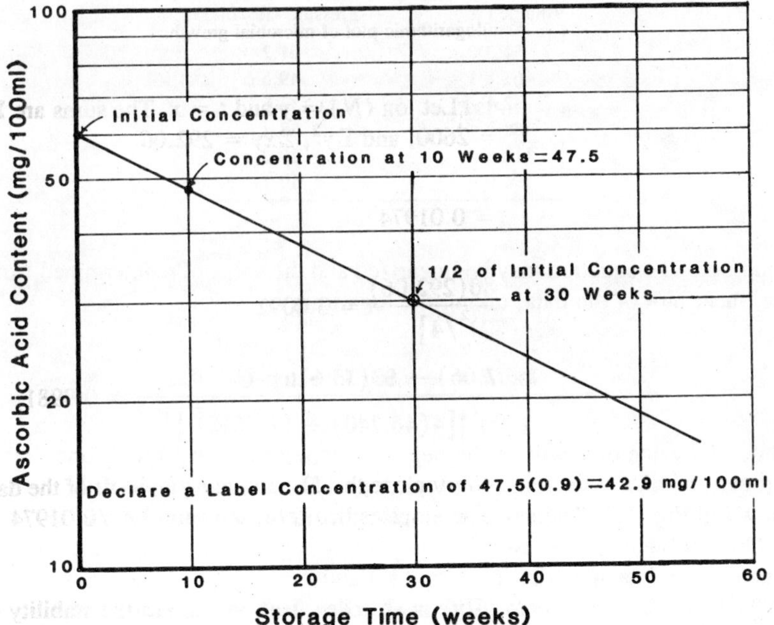

Fig. 1.6. Graphical representation of the half-life as demonstrated by ascorbic acid degradation with time of storage.

half that concentration (30 mg/100 mL) at t = 30 weeks and drawing a line connecting the two points. At t = 10 weeks, a point on the line shows a concentration of 47.5 mg/100 mL. Thus, a concentration of 0.9(47.5), or 42.9 mg/100 mL, would be the maximum that can be declared on the label.

Analytical Solution: Given: $C_0 = 60$, $t_{0.5} = 30$ at $t = 10$, C_{10} = concentration, and the declared concentration on the label, $C_d = 0.9 C_{10}$. Solving for C_d:

$$C_d = 0.9(60)[2]^{-10/30}$$

$$= 0.9(60)(0.7938) = 42.86 \text{ mg/mL}$$

EXAMPLE 3: The pressure–volume relationship which exists during adiabatic compression of a real gas is given by $PV^n = C$, where P = absolute pressure, V = volume, n = adiabatic expansion factor, and C is a constant. Calculate the value of the adiabatic expansion factor, n, for a gas which exhibits the following pressure–volume relationship.

Volume (ft^3)	Absolute Pressure (lb_f/in^2)
54.3	61.2
61.8	49.5
72.4	37.6
88.7	28.4
118.6	19.2
194.0	10.1

Solution: The equation may be linearized and the value of n determined from the linear plot of the data, taking the logarithm:

$$\log P = -n \log V + \log C$$

Thus, the value of n will be the negative slope of the best-fitting line drawn through the data points in a log-log graph. The log-log graph of the data is shown in Fig. 1.7. The slope is -1.39; therefore, the value of n is 1.39.

This problem can be solved using linear regression analysis after transforming the data such that $x = \log(V)$ and $y = \log(P)$. The sums are $\Sigma x = 11.6953$, $\Sigma y = 8.7975$, $\Sigma x^2 = 23.006$, $\Sigma y^2 = 13.3130$, and $\Sigma xy = 16.8544$.

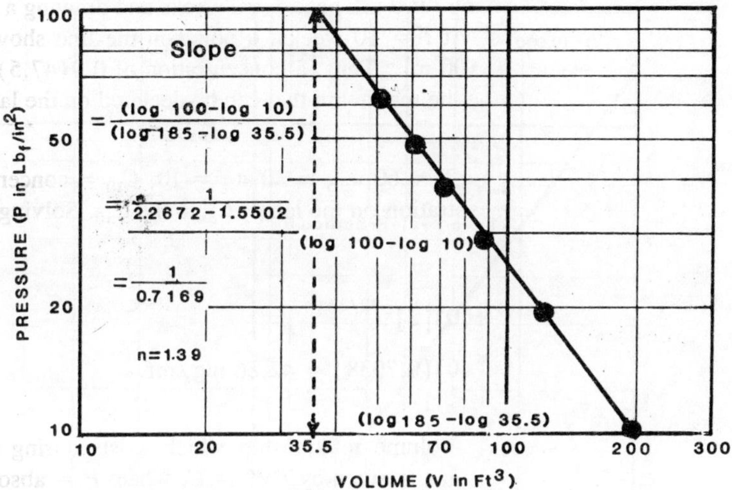

Fig. 1.7. Log-log graph of pressure against volume during adiabatic expansion of a gas to determine the adiabatic expansion coefficient from the slope.

$$a = \frac{16.8544 - 8.7975(11.6953)/6}{23.006 - (11.6953)^2/6} = -1.40$$

$$b = \frac{8.7975(23.006) - (11.6953)(16.8544)}{6[23.006 - (11.6953)^2/6]} = 4.2033$$

$$r = \frac{6(16.8544) - (11.6953)(8.7975)}{[[6(23.006) - (11.6953)^2][6(8.7975) - (13.3130)^2]]^{0.5}}$$

$r = -0.9986$. r is very close to 1.0, indicating a good fit of the data points to the linear regression equation. The value of n equals the slope; therefore, $n = 1.40$

Intercept of Log-Log Graphs. The y-intercept of a log-log graph is determined at a point where $\log x = 0$. On a log-log plot, $\log x = 0$ when $x = 1$. Therefore, the y-intercept is read from the graph at a point where the line crosses $x = 1$. In Example 3 above, $\log c$ is the y-intercept of the line. Figure 1.8 is drawn by extending the graph in Fig. 1.7 to include $V = 1$ in order to show how C may be evaluated from the intercept. The line passes through the point $V = 35.5$ and $P = 100$. The slope, 1.39, is used to generate the other point on the line by reducing V one log cycle to a point with coordinate 3.55 and going up 1.39 log cycles, i.e., from 100 to 1000 ($10^{0.39}$), or 2450. Thus, the coor-

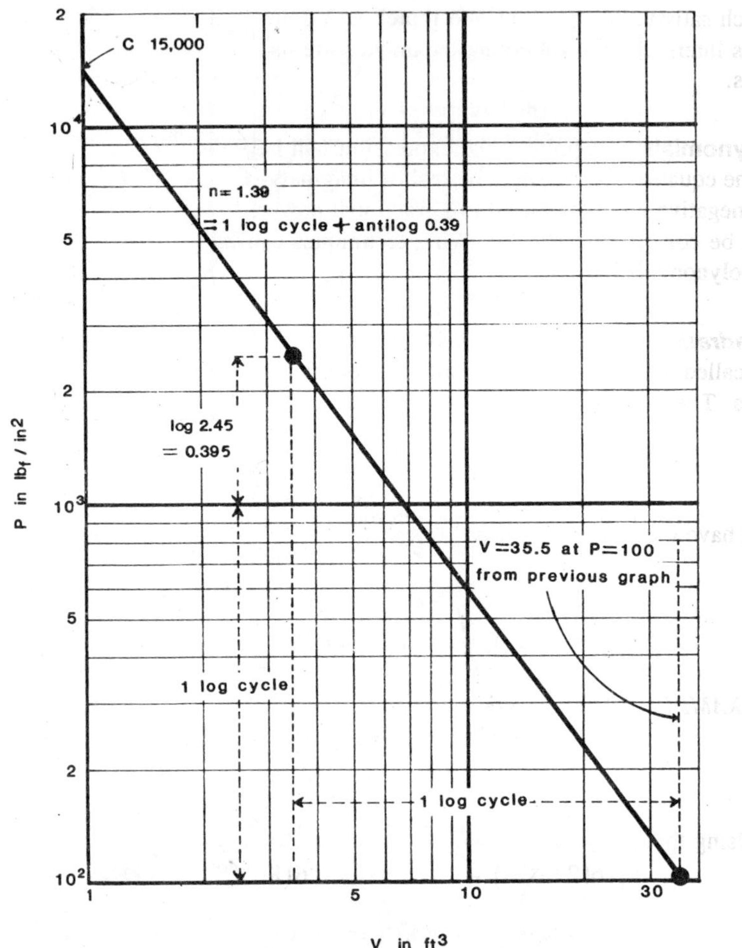

Fig. 1.8. Example of how a linear log-log graph is plotted if one point and the slope are known. The determination of an intercept on logarithmic coordinates is also shown.

dinate of the second point is 3.55, 2480. Joining the two points by a straight line and extrapolating the line to $V = 1$, the value of the intercept, which is 15,000, may be obtained. Thus, $C = 15{,}000$. From the regression, the intercept $= 4.2033$. $C = 10^{4.2033} = 15{,}500$.

ROOTS OF EQUATIONS

The roots of an equation $F(x) = 0$ are the points where the function crosses the abscissa. The roots of a system of equations are values of the variables which satisfy the equations and represent a point in the graph where the equations intersect. The following techniques are used to determine roots of equations.

Polynomials. A polynomial expression will have as many roots as the order of the equation. A root may be real or imaginary. Examples of imaginary roots are negative numbers raised to a fractional power. In this book, only real roots will be considered. The following techniques can be used in evaluating roots of polynomials:

Quadratic Equation. Equations with 2 as the highest power of the variable are called *quadratic equations* and the root is obtained using the quadratic formula. The equation

$$ax^2 + bx + c = 0$$

will have the following roots:

$$x = \frac{-b \pm \sqrt{(b^2 - 4ac)}}{2a}$$

EXAMPLE: Determine the roots of the expression:

$$2x^2 + 3x - 2 = 0$$

Using the quadratic equation, $a = 2$, $b = 3$, and $c = -2$:

$$x = \frac{-3 \pm \sqrt{(3)^2 - (4)(2)(-2)}}{2(2)}$$

$$= 0.686; \quad x = -2.186$$

Factoring. Equations may be factored and the roots of the individual factors calculated. Thus:

$$F(x) = (ax + b)(cx + d)(ex + f) = 0$$

$$x = \frac{-b}{a}; \; x = \frac{-d}{c}; \; x = \frac{-f}{e}$$

All three values of x satisfy the equality $F(x) = 0$.

EXAMPLE: Determine the roots of the equation $2x^3 + 5x^2 - 11x + 4 = 0$. Dividing by $2x - 1$, the quotient is $x^2 + 3x - 4$, which, when further divided by $x - 1$, will give a quotient of $x + 4$. Thus the factors $(2x - 1)(x - 1)(x + 4) = 2x^3 + 5x^2 - 11x + 4 = 0$. The roots are $x = 1/2$, $x - 1$, and $x = -4$

Iteration Techniques. This is a trial-and-error method involving substituting values of the variable into the equation and testing if the equality expressed by the function is satisfied. Inspection will usually identify a range of values of the variable which gives a negative value for the function at one end of the range and a positive value at the other end. Substitution of values within that range and plotting will identify the value of the variable when the function crosses the abscissa.

Another method for iteration involves calculus and is called the *Newton-Raphson iteration procedure*. In this procedure, a value of the variable (x_1) is assumed and the next value (x_2) can be calculated as follows:

$$x_2 = x_1 - \frac{F(x)}{F'(x)}$$

The iteration is continued until $F(x) = 0$. $F'(x)$ is the value of the first derivative of the function evaluated at the assumed value of x. The derivative is discussed in the section on "Differential Calculus."

EXAMPLE: Determine the roots of the function $F(x)$: $2x^3 + 5x^2 - 11x + 4 = 0$.

This is the same function as in the previous example; therefore, the results of the iteration method can be verified. Since this is a cubic expression, three roots are to be expected. Using differential calculus, the derivative of the function is determined.

$$F'(x) = 6x^2 + 10x - 11$$

If this derivative is equated to zero, the values of x where the function exhibits a maximum and a minimum can be determined. The roots of $6x^2 + 10x - 11 = 0$ can be determined using the quadratic equation as follows:

$$x = \frac{-10 \pm \sqrt{100 - 4(6)(-11)}}{2(6)}; \qquad x = -2.423; \qquad x = 0.755$$

Since these two points represent a peak and a valley in the curve, it would be expected that one root might exist at $x < -2.423$, one at $-2.423 < x < 0.755$, and one at $x > 0.755$. To illustrate the Newton-Raphson iterative technique, consider the root at the region $-2.423 < x < 0.755$. First, let $x = -1$; $F(x) = 18$, $F'(x) = -15$; $x_2 = -1 - 18/(-15) = +0.2$. x_2 is assumed to be the new value of x in the next iteration.

Let $x = 0.2$; $F(x) = 2.016$; $F'(x) = -8.76$. $x_2 = 0.2 - (2.086/-8.76) = 0.43$. This is used as the new value of x in the next iteration.

Let $x = 0.43$; $F(x) = 0.353$; $F'(x) = -5.59$.

$$x_2 = 0.43 - \frac{0.353}{-5.59} = 0.493$$

Using the new value of x in the next iteration: Let $x = 0.493$; $F(x) = 0.032$; $F'(x) = -4.612$.

$$x_2 = 0.492 - \frac{0.032}{-4.612} = 0.4999$$

The iteration is terminated when a critical value of $|x - x_2|$ is reached. For example, if the critical $|x - x_2|$ is 0.0001, another iteration is needed with $x = 0.4999$. $F(x) = 0.00045$; $F'(x) = -4.5016$; $x_2 = 0.4999 + 0.00045/4.5016$; $x_2 = 0.5000$. Thus, the critical $|x - x_2|$ of 0.0001 is reached and the iteration is stopped. The value of x is the last one computed, which is 0.5000.

The root in this region of the function, as shown in the previous example, is 0.5. A similar process can be used to determine the other two roots.

Using a Personal Computer by Programming in BASIC. BASIC stands for **B**eginner's **A**ll-purpose **S**ymbolic **I**nstruction **C**ode. It is a program which interprets symbols and codes and converts them into machine language which the computer can process. BASIC is easy to learn since the format of the language is very similar to what would happen if the process were done manually. The computer executes the commands in sequence. When an iterative procedure is used to solve the function $F(x) = 0$, a value of the variable is substituted into the equation, the value of $F(x)$ is calculated, and the process is repeated until $F(x) = 0$. The repeated calculations are done rapidly by the computer. The iteration procedure using the Newton-Raphson iteration technique is particularly easy to do on a personal computer (PC) using the BASIC programs.

BASIC programming or program execution is initiated by loading the software, the BASIC interpreter, into computer memory. The PC is ready to accept BASIC programs when the computer acknowledges that BASIC has been loaded and identifies the amount of memory available for the programs. BASIC can access only 64K of computer memory, although some recent versions enable access to a larger portion of memory. BASIC executes programs only after they have been loaded into computer memory. These programs may be loaded by entering a series of instructions defined by program lines, one line at a time. The program lines are entered into memory by typing the appropriate codes and pressing the return (enter) key. Typed codes, although they appear on the screen, will not be entered into the computer's memory unless the return key is pressed. To clear the screen, type "CLS" and enter. To determine what has been loaded into memory, type "LIST" and enter. Stored programs may also be loaded into memory by typing "LOAD *filename*" and enter. The filename is what the file was called when it was stored.

BASIC executes program lines consecutively; therefore, make sure that the variables have been defined and their values are known before using them in a program line. Data may be entered using the "input" command, by defining values of variables in the opening lines of the program, or by using the "data" statement followed by a "read" statement. The following example illustrates the use of BASIC in an iterative procedure for determining the root of a function:

EXAMPLE: Determine the positive root of the function $x^2 + 109.3x^{1.35} - 20000 = 0$. The solution is based on substitution of various values of x. The BASIC program is:

```
10 Assume x = 40 and increment it by 1 to x = 60
20 for x = 40 to 60
30 fx = x^2 + 109.3 * x^1.35 - 20000
40 print "x = "; x , "fx = "; fx
50 next x
```

Type "run" and enter to run the program. The screen with display values of fx for the different values of x.

The value of fx changes from negative (-620) at $x = 43$ to positive (19.44) at $x = 44$. Thus, the root must be just slightly below and close to 44. To narrow the value of x to the nearest tenth, line 10 in the program is changed to substitute values of x between 43 and 44 in increments of 0.1. Edit line 10 in the program, converting it to 10 for $x = 43$ to 44 step 0.1. Running the revised program will give $F(x) = -44.8$ at $x = 43.9$ and $F(x) = 19.44$ at $x = 44$. Thus x is greater than 43.9 and is closer to 44. The root of the equation to the nearest tenth is $x = 44.0$.

Solution: Using the Newton-Raphson iteration technique, $F'(x) = 2x + 147.555x^{0.35}$. The BASIC program is:

```
10 x = 40
20 fx = x^2 + 109.3 * x^1.35 - 20000
30 fxp = 2 * x + 147.555 * x^0.35
40 x2 = x - (fx / fxp)
50 if abs(fx) < 0.1 then goto 80
```

```
60 x = x2
70 goto 20
80 print "x= "; x2
90 end
```

The program, when run, will output $x = 43.96959$.

Simultaneous Equations. Simultaneous equations are often encountered in problems involving material balances and multistage processes. The following techniques are used in evaluating simultaneous equations.

Substitution. If an expression for one of the variables is fairly simple, substitution is the easiest means of solving simultaneous equations. In a set of equations $x + 2y = 5$ and $2x - 2y = 3$, x may be expressed as a function of y in one equation and substituted into the other to yield an equation with a single unknown. Thus, $x = 5 - 2y$ and $2(5 - 2y) - 2y = 3$, giving values of $y = 7/6$ and $x = 8/3$.

Elimination. Variables may be eliminated either by subtraction or by division. Division is usually used with geometric expressions, and elimination by subtraction is used with linear expressions. When subtraction is used, equations are multiplied by a factor such that the variable to be eliminated will have the same coefficient in the two equations. Subtraction will then yield an equation having one less variable than the original two equations.

EXAMPLE: Calculate the values of Γ and μ which would satisfy the following expressions: $\mu = 3.2(\Gamma)^{0.75}$ and $1.5\mu = 0.35(\Gamma)^{0.35}$.

μ is eliminated by division.

$$\frac{1}{1.5} = \frac{3.2}{0.35}\Gamma^{0.75-0.35}$$

$$\Gamma = \frac{0.35}{(1.5)(3.2)}\frac{1}{0.4}$$

$$= 0.00169;\ \mu = 3.2(0.00169)^{0.75} = 0.02667$$

EXAMPLE: Determine x and y that satisfy the following expressions: $2x + 2y = 32$; $x^2 = y - 2$.

y may be eliminated by multiplying the second equation by 2 and adding the two equations.

$$2x + 2y = 32$$

$$2x^2 - 2y = -4$$

Adding:

$$2x^2 + 2x = 28$$

Solving by the quadratic formula:

$$x = \frac{-2 \pm \sqrt{4 - 4(2)(-28)}}{2(2)}$$

$$x = 3.275; \qquad y = 12.275$$

$$= -4.275; \qquad = 20.275$$

Determinants. Coefficients of a system of linear equations may be set up in an array or matrix and the matrices resolved to determine the values of the variables. Programs are available to perform matrix inversion on a PC. For a system of three equations or more, setting up the matrix to solve the equations using the PC will be the fastest way to determine the values of the variables. In a system of equations:

$$a_1x + b_1y + c_1z = d_1$$

$$a_2x + b_2y + c_2z = d_2$$

$$a_3x + b_3y + c_3z = d_3$$

The array of the coefficients is as follows:

$$\begin{vmatrix} a_1 & b_1 & c_1 \\ a_2 & b_2 & c_3 \\ a_3 & b_2 & c_3 \end{vmatrix} \begin{vmatrix} d_1 \\ d_2 \\ d_3 \end{vmatrix}$$

The values of x, y, and z are determined as follows:

$$x = \frac{\begin{vmatrix} d_1 & b_1 & c_1 \\ d_2 & b_2 & c_2 \\ d_3 & b_3 & c_3 \end{vmatrix}}{\begin{vmatrix} a_1 & b_1 & c_1 \\ a_2 & b_2 & c_3 \\ a_3 & b_3 & c_3 \end{vmatrix}}$$

$$y = \frac{\begin{vmatrix} a_1 & d_1 & c_1 \\ a_2 & d_2 & c_2 \\ a_3 & d_3 & c_3 \end{vmatrix}}{\begin{vmatrix} a_1 & b_1 & c_1 \\ a_2 & b_2 & c_2 \\ a_3 & b_3 & c_3 \end{vmatrix}}$$

$$z = \frac{\begin{vmatrix} a_1 & b_1 & d_1 \\ a_2 & b_2 & d_2 \\ a_3 & b_3 & d_3 \end{vmatrix}}{\begin{vmatrix} a_1 & b_1 & c_1 \\ a_2 & b_2 & c_2 \\ a_3 & b_3 & c_3 \end{vmatrix}}$$

A 2 × 2 matrix is resolved by cross-multiplying the elements in the array and subtracting one from the other. The position of an element in the matrix is designated by the subscript *ij*, with *i* representing the row and *j* representing the column. In order to maintain consistency in the sign of the cross-product, the element in the first column whose subscript adds up to an odd number is assigned a negative value. Thus, the cross-product with that element will have a negative sign. A 2 × 2 matrix and its value are shown below:

$$\begin{vmatrix} a_{11} & a_{12} \\ a_{21} & a_{22} \end{vmatrix} = a_{11}a_{22} - a_{21}a_{12}$$

A 3 × 3 matrix is evaluated by multiplying each of the elements in the first column with the 2 × 2 matrix left-utilizing elements in the second and third columns other than those in the same row as the multiplier. As with the 2 × 2 matrix above, the multiplier whose subscript adds up to an odd number is assigned a negative value. The multiplier and the 2 × 2 matrices as multiplicand are determined as follows:

$$\begin{vmatrix} a_{11} & a_{12} & a_{13} \\ a_{21} & a_{22} & a_{23} \\ a_{31} & a_{32} & a_{33} \end{vmatrix} \begin{vmatrix} a_{11} & a_{12} & a_{13} \\ a_{21} & a_{22} & a_{23} \\ a_{31} & a_{32} & a_{33} \end{vmatrix} \begin{vmatrix} a_{11} & a_{12} & a_{12} \\ a_{21} & a_{22} & a_{23} \\ a_{31} & a_{32} & a_{33} \end{vmatrix}$$

Thus, the 3 × 3 matrix resolves into:

$$a_{11} \begin{vmatrix} a_{22} & a_{23} \\ a_{32} & a_{33} \end{vmatrix} - a_{21} \begin{vmatrix} a_{12} & a_{13} \\ a_{32} & a_{33} \end{vmatrix} + a_{31} \begin{vmatrix} a_{12} & a_{13} \\ a_{22} & a_{23} \end{vmatrix}$$

EXAMPLE: Determine the values of x, y, and z in the following equations:

$$x + y + z = 100$$
$$0.8x + 0.62y + z = 65$$
$$0.89x + 0.14y = 20$$

The array of the coefficients and constants is as follows:

$$\begin{vmatrix} 1 & 1 & 1 \\ 0.8 & 0.62 & 1 \\ 0.89 & 0.14 & 0 \end{vmatrix} \begin{vmatrix} 100 \\ 65 \\ 20 \end{vmatrix}$$

The matrix which consists of the coefficients of the variables will be the denominator of the three equations for x, y, and z.

$$\begin{vmatrix} 1 & 1 & 1 \\ 0.8 & 0.62 & 1 \\ 0.89 & 0.14 & 0 \end{vmatrix} = 1 \begin{vmatrix} 0.62 & 1 \\ 0.14 & 0 \end{vmatrix} - 0.8 \begin{vmatrix} 1 & 1 \\ 0.14 & 0 \end{vmatrix} + 0.89 \begin{vmatrix} 1 & 1 \\ 0.62 & 1 \end{vmatrix}$$

$$= 1(0 - 0.14) - 0.8(0 - 0.14) + 0.89(1 - 0.62)$$
$$= 0.3102$$

The matrix which will be the numerator in the equation for x is:

$$\begin{vmatrix} 100 & 1 & 1 \\ 65 & 0.62 & 1 \\ 20 & 0.14 & 0 \end{vmatrix} = 100 \begin{vmatrix} 0.62 & 1 \\ 0.14 & 0 \end{vmatrix} - 65 \begin{vmatrix} 1 & 1 \\ 0.14 & 0 \end{vmatrix} + 20 \begin{vmatrix} 1 & 1 \\ 0.62 & 1 \end{vmatrix}$$

$$= 100(0 - 0.14) - 65(0 - 0.14) + 20(1 - 0.62)$$
$$= 2.7$$

Thus, $x = 2.7/0.3102 = 8.7$.

The matrix which is the numerator in the equation for y is:

$$\begin{vmatrix} 1 & 100 & 1 \\ 0.8 & 65 & 1 \\ 0.89 & 20 & 0 \end{vmatrix} = 1\begin{vmatrix} 65 & 1 \\ 20 & 0 \end{vmatrix} - 0.8\begin{vmatrix} 100 & 1 \\ 20 & 0 \end{vmatrix} + 0.89\begin{vmatrix} 100 & 1 \\ 65 & 1 \end{vmatrix}$$

$$= 1(0 - 20) - 0.8(0 - 20) + 0.89(100 - 65)$$

$$= 27.15$$

Thus, $y = 27.15/0.3102 = 87.5$.

The matrix which is the numerator of the equation for z is:

$$\begin{vmatrix} 1 & 1 & 100 \\ 0.8 & 0.62 & 65 \\ 0.89 & 0.14 & 20 \end{vmatrix} = 1\begin{vmatrix} 0.62 & 65 \\ 0.14 & 20 \end{vmatrix} - 0.8\begin{vmatrix} 1 & 100 \\ 0.14 & 20 \end{vmatrix} + 0.89\begin{vmatrix} 1 & 100 \\ 0.62 & 65 \end{vmatrix}$$

$$= 1(12.4 - 9.1) - 0.8(20 - 14)$$

$$+ 0.89(65 - 62) = 1.17$$

Thus, $z = 1.17/0.3102 = 3.8$.
Check: $x + y + z = 8.7 + 87.5 + 3.8 = 100$.

Power Functions and Exponential Functions. Power functions and exponential functions consist of a base raised to an exponent. Although the two functions are similar, power functions are those which have numerical exponents, while exponential functions are those which have variable exponents. Both functions are resolved using logarithms or by taking the rth root of both sides of the equation, where r is the exponent of the variable whose value needs to be determined. The following are basic rules when working with exponents:

Multiplication. When the base is the same, add the exponents.

$$x^2(x^3) = x^{2+3} = x^5$$

Division. When the base is the same, subtract the exponents. A negative exponent signifies division.

$$\frac{(x + 2)^3}{(x + 2)^{1.5}} = (x + 2)^{3-1.5} = (x + 2)^{1.5}$$

$$\left(\frac{P}{V}\right)^{3.5} = P^{3.5}V^{-3.5}$$

Exponentiation. Multiply the exponents. Extracting the rth root of a function implies exponentiation to the $1/r$th power.

$$(x^2)^3 = x^{2(3)} = x^6$$
$$\sqrt[4]{(x^2)} = (x^2)^{1/4} = (x)^{2/4} = (x)^{1/2}$$

Logarithmic Functions.

1. Logarithm of a product = sum of the logarithms:

$$\log(xy) = \log x + \log y$$

2. Logarithm of a quotient = difference of the logarithms. A negative logarithm signifies a reciprocal of the terms within the logarithm:

$$\log \frac{x}{y} = \log x - \log y$$

$$-\log x = \log \frac{1}{x}$$

3. Logarithm of a power function = exponent multiplied by the logarithm of the base:

$$\log e^{2x} = 2x \log e$$

EXAMPLE: The Reynolds number of a non-Newtonian fluid is expressed as:

$$\text{Re} = \frac{8V^{2-n}R^n\rho}{k\left(\frac{3n+1}{n}\right)^n}$$

Calculate the value of the velocity V which would result in a Reynolds number of 3000 if $k = 1.5$ Pa.s^n, $n = 0.775$, $\rho = 1030$ kg/m^3, and $R = 0.0178$ m.

Substituting values:

$$2000 = \frac{8(V)^{1.225}(0.0178)^{0.775}(1030)}{1.5\left[\frac{2.325}{0.775}\right]^{0.775}}$$

$$V^{1.225} = \frac{2000(1.5)(2.3429)}{8(0.04406)(1030)} = 25.55$$

$$V = (25.55)^{1/1.225} = 14.08 \text{ m/s}$$

EXAMPLE: The temperature (T, in °C) of a fluid flowing through a tube immersed in a constant-temperature water bath at temperature T_b changes exponentially with position along the length of the tube as follows:

$$T = T_b - (T_b - T_0)(e)^{-3.425L}$$

Calculate the length L such that when fluid enters the tube with an initial temperature $T_0 = 20°\text{C}$, the exit temperature will be 99% of the water bath temperature, T_b, which is 95°C.

$$T = 0.99(95) = 95 - (95 - 20)(e)^{-3.425L}$$

$$(e)^{-3.425L} = \frac{95 - 94.05}{75} = 0.012667$$

Taking the natural logarithm of both sides and noting that $\ln(e) = 1$:

$$-3.425L = \ln(0.012667);\ L = \frac{-4.3687}{-3.425} = 1.276 \text{ m}$$

DIFFERENTIAL CALCULUS

Calculus is a branch of mathematics which deals with infinitesimally small segments of a whole. The concept is analogous to high-speed filming of a moving object. The action can be frozen in an infinitesimal time increment, and it is possible to take measurements on the frozen picture frame. Analysis of a series of frames will define the nature and magnitude of the changes which occur as the object moves. Calculus is particularly useful in predicting point values of variables in a system from global measurements.

Calculus is divided into differential and integral calculus. The former deals with the rate of change of variables or with incremental changes in a variable. The latter originated in earlier studies of areas of plane figures. An application of integral calculus which is particularly useful to engineers is the derivation of a function which defines a variable or processing parameter from a differential expression of the changes in one variable with respect to another variable.

Definition of a Derivative. Any function $F(x)$ may be represented by a graph, as shown in Fig. 1.9. A section of the x axis between x_1 and x_2 is designated Δx. The slope of the function within this section is:

$$\text{Slope} = \frac{\Delta F(x)}{\Delta x} = \frac{F(x + \Delta x) - F(x)}{\Delta x}$$

If Δx is infinitesimal, Δx approaches 0, and $\Delta F(x)/\Delta x$ is the derivative of $F(x)$ with respect to x.

$$\underset{\Delta x \to 0}{\text{Limit}} \frac{\Delta F(x)}{\Delta x} = \frac{dF(x)}{dx} = \underset{\Delta x \to 0}{\text{limit}} \frac{F(x + \Delta x) - F(x)}{\Delta x}$$

If the function $F(x) = x^2$:

$$\frac{dF(x)}{dx} = \underset{\Delta x \to 0}{\text{limit}} \frac{(x + \Delta x)^2 - x^2}{\Delta x}$$

$$= \underset{\Delta x \to 0}{\text{limit}} \frac{x^2 + 2x\Delta x + (\Delta x^2) - x^2}{\Delta x}$$

$$= \underset{\Delta x \to 0}{\text{limit}} (2x + \Delta x) = 2x \qquad \text{when } x = 0$$

$$= 2x$$

The derivative of $F(x) = x^2$ is $dF(x)/dx = 2x$.
The differential form of the above derivative is:

$$dF(x) = 2x\,dx$$

The symbol d is a differential operator, meaning that a differentiation operation has been performed on $F(x)$. Differentiation is the process of obtaining a

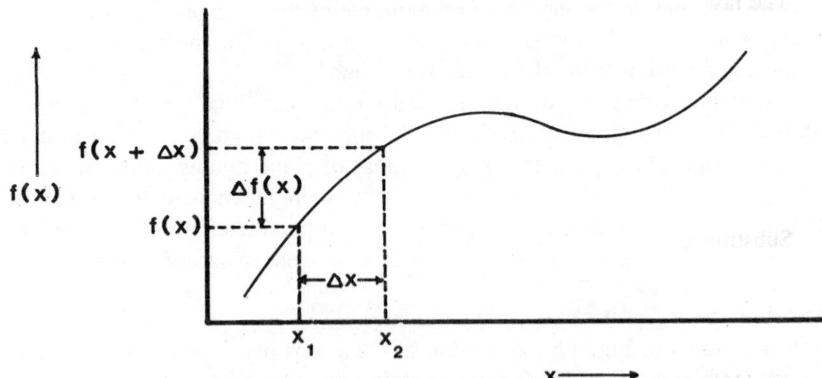

Fig. 1.9. Diagram of a function, its slope, and the derivative.

derivative. The result of a differentiation operation is a differential equation which can then be divided by the differential term of the reference variable to obtain the derivative. The term $dF(x)/dx$ may be written as $F'(x)$.

The derivative is a rate of change, or the slope of a function. Thus, constants will have zero slope and therefore will have a derivative of zero. A linear function will have a constant slope; therefore, the derivative of a variable to the power 1 will be a constant, the coefficient of that variable.

Differentiation Formulas. The following are differentiation formulas which are most often used:

Constant: $d(a) = 0$
Sum: $d[F(x) + G(x)] = dF(x) + dG(x)$
Product: $d[F(x)G(x)] = F(x)\,dG(x) + G(x)\,dF(x)$
Quotient: $d(F(x)/G(x)) = (G(x)\,dF(x) - F(x)\,dG(x))/[G(x)]^2$
Power function: $d[F(x)]^n = n[F(x)]^{n-1}\,dF(x)$
Exponential function: $d(a)^{F(x)} = (a)^{F(x)}[dF(x)] \ln a$
Logarithmic function: $d \ln [F(x)] = dF(x)/F(x)$
$d \log [F(x)] = dF(x)/F(x)(\ln 10)$

EXAMPLE 1: Determine the slope of the function

$$y = 2(x + 2)^3 + 2x^2 + x + 3$$

at $x = 1$.

The slope is the derivative of the function. The slope is obtained by differentiating the function and solving for dy/dx. The sum and power function formulas will be used.

$$dy = d[2(x + 2)^2] + d(2x)^2 + d(x) + d(3)$$

The last term is the derivative of a constant and is 0.

$$dy = 2(3)(x + 2)^2 d(x + 2) + 2(2)(x)\,dx + dx$$
$$= 6(x + 2)^2\,dx + 4x\,dx + dx$$

$$\frac{dy}{dx} = 6(x + 2)^2 + 4x + 1$$

Substituting $x = 1$:

$$\frac{dy}{dx} = 6(3)^2 + 4 + 1 = 54 + 5 = 59$$

EXAMPLE 2: Differentiate the function $H = E + PV$. All terms are variables. Use the sum and the product formulas $dH = dE + PdV + VdP$.

EXAMPLE 3: The expression for the water activity of a sugar solution is given as:

$$\log\left[\frac{a_w}{x}\right] = -k(1 - x)^2$$

For glucose, $k = 0.7$. How much faster will the water activity (a_w) change with a change in water mole fraction (x) at $x = 0.7$ compared to $x = 0.9$?

This problem requires determining da_w/dx at $x = 0.7$ and at $x = 0.9$. The ratio of the two slopes will give the relative effects of small increases in concentration around $x = 0.9$ and $x = 0.7$ on the water activity.

The expression to be differentiated after substituting $k = 0.7$ is:

$$\log a_w - \log x = -0.7(1 - x)^2$$

When differentiated directly in this form, a_w will appear in the denominator of the first term involving $d(\log a_w)$. This can be eliminated by solving first for a_w before differentiation:

$$a_w = x(10)^{-0.7(1-x)^2}$$

Differentiating:

$$d(a_w) = xd[10]^{-0.7(1-x)^2} + [10]^{-0.7(1-x)^2}\,dx$$

$$= x[10]^{-0.7(1-x)^2}(-0.7)(2)(1 - x)[\ln(10)](-dx)$$

$$+ [10]^{-0.7(1-x)^2}\,dx$$

$$\frac{da_w}{dx} = [10]^{-0.7(1-x)^2}[(0.7)(2)(1 - x)(x)[\ln(10)] + 1]$$

Substituting $x = 0.9$,

$$\left[\frac{da_w}{dx}\right] = [(0.7)(2)(1 - 0.9)(0.9)(2.303) + 1][10]^{-0.7(1-0.9)^2}$$

$$= 1.3226(10)^{-0.007} = 1.301$$

Substituting $x = 0.7$,

$$\left[\frac{da_w}{dx}\right] = [(0.7)(2)(1 - 0.7)(0.7)(2.303) + 1][10]^{-0.7(1-0.7)^2}$$

$$= 1.6769(10)^{-0.063} = 1.450$$

a_w will be changing faster as x is incremented at $x = 0.7$ compared to $x = 0.9$.

EXAMPLE 4: Differentiate:

$$y = \frac{3x + 2}{x + 3}$$

Using the formula for the derivative of a quotient:

$$dy = \frac{(x + 3)d(3x + 2) - (3x + 2)d(x + 3)}{(x + 3)^2}$$

$$= \frac{(x + 3)(3\,dx) - (3x + 2)\,dx}{(x + 3)^2} = \frac{(3x + 9 - 3x - 2)\,dx}{(x - 3)^2}$$

$$\frac{dy}{dx} = \frac{7}{(x + 3)^2}$$

EXAMPLE 5: The growth of microorganisms expressed as cell mass is represented by the following:

$$\log\left(\frac{C}{C_0}\right) = kt$$

Determine the rate of increase of cell mass at $t = 10$ h if it took 1.5 h for the cell mass to double and the initial cell mass at time zero (C_0) is 0.10 g/L.

Solution: The value of k is determined from the time required for the cell mass to double. $k = (\log 2)/1.5 = 0.200\ \text{h}^{-1}$. The expression to be differentiated to determine the rate is $\log(C/0.10) = 0.200\,t$. Differentiating using the formula for derivative of a logarithmic function:

$$d \log \frac{C}{0.10} = 0.200\, dt$$

$$\frac{dC/0.1}{(\ln 10)(C/0.1)} = 0.200\, dt; \qquad \frac{dC}{dt} = 0.200\, C \ln(10) = 0.9233\, C$$

C from the original expression is substituted to obtain a rate expression dependent only on t. $C = C_0(10)^{0.200t}$

$$\frac{dC}{dt} = 0.4606\, C_0(10)^{0.200t}$$

At $t = 10$,

$$\frac{dC}{dt} = 0.4606(0.1)(10)^{2.00} = 4.605 \text{ g/L (h)}$$

Maximum and Minimum Values of Functions. One of the most useful attributes of differential calculus is its use in determining the maximum and minimum values of functions. In a previous section, it has been shown that determining the root of a polynomial expression is facilitated by determining where the maximum and minimum points are located. Determination of the maximum and minimum values of a function can be applied in optimizing processes to identify conditions where cost is minimized, profit is maximized, or a product quality attribute is maximized.

When a function has a maximum or minimum point, the slope changes signs upon crossing the crest or valley of the curve. Thus, at the maximum or minimum point, the slope is zero. After a value of the independent variable is determined at a point where the slope of the curve is zero, that point is identified as a maximum or minimum by either of two methods: (1) The second derivative is determined. As the curve approaches a maximum, the slope is positive and decreases with increasing values of the independent variable; therefore, the second derivative is negative. As the curve approaches a minimum, the function has a negative slope which decreases in value; therefore, the second derivative is positive. (2) For functions which have a complex second derivative, substitution of the root of the derivative equation into the original function gives the maximum and minimum values of the function. If the derivative equation has only one root, it is possible to identify this root as the maximum or minimum point by substituting any other value of the independent variable into the original function.

EXAMPLE 1: Plot the curve $y = 2x^3/3 + x^2/2 - 6x$ and determine its maximum and minimum values.

Differentiating:

$$\frac{dy}{dx} = 2x^2 + x - 6$$

At the maximum or minimum, $dy/dx = 0$ and $2x^2 + x - 6 = 0$. The roots of the derivative are:

$$x = \frac{-1 \pm \sqrt{1 - 4(2)(-6)}}{2(2)}; \qquad x = -2; \qquad x = 1.5$$

To determine which of these points is a maximum, take the second derivative:

$$\frac{d^2y}{dx^2} = 4x + 1$$

At $x = -2$, $d^2y/dx^2 = -7$. The point is a maximum. Substituting $x = -2$ in the expression for y:

$$y = \frac{2(-2)^3}{3} + \frac{(-2)^2}{2} - 6(-2) = 8.667$$

At $x = 1.5$, $d^2y/dx^2 = 7$. The point is a minimum. Substitute $x = 1.5$ in the expression for y:

$$y = \frac{2(1.5)^3}{3} + \frac{(1.5)^2}{2} - 6(1.5) = -5.625$$

Even without taking the second derivative, values of y show which roots of the derivative equation represents the maximum and minimum points. A plot of the function is shown in Fig. 1.10. An inflection point in the curve exists where the second derivative is zero at $x = -0.25$.

EXAMPLE 2: Derive an expression for the constants a and b in the equation $y = ax + b$ which is the best-fitting line to a set of experimental data points (x_i, y_i) by minimizing the sum of squares of the error $(y - y_i)$.

$$E = \sum(ax_i + b - y_i)^2$$

where $i = 1$ to n. To evaluate a and b, two independent equations must be formulated. E will be maximized with respect to b at constant a and with

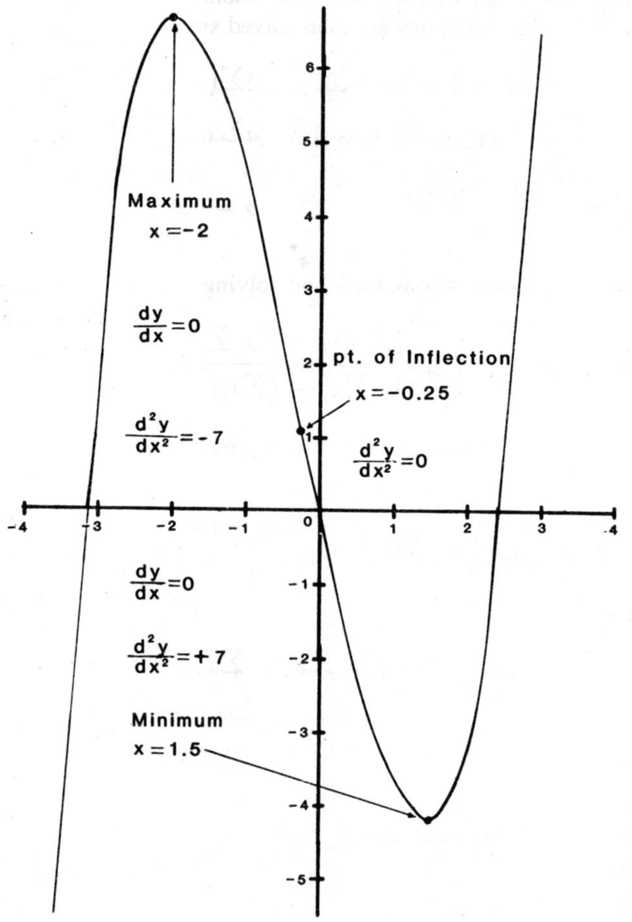

Fig. 1.10. Diagram of a function $y = (2/3)x^3 + (1/2)x^2 - 6x$ showing the maximum and minimum values, the point of inflection, and the values of the second derivative at the maximum and minimum points.

respect to a at constant b. x and y are considered constants during the differentiation process with respect to either a or b. The two equations are:

$$\frac{dE}{da} = 2\Sigma(ax_i + b - y_i)x_i; \qquad \frac{dE}{db} = 2\Sigma(ax_i + b - y_i)$$

The second derivatives of these two expressions will be $2\Sigma x_i^2$ and 2, respectively, both of which are positive quantities for all values of a or b; therefore, the root of the derivative equation will represent a point where E is minimum. The two derivative equations are then solved simultaneously.

$$2\sum x_i(ax_i + b - y_i) = 0; \qquad 2\sum(ax_i + b - y_i) = 0$$

$$a\sum x_i^2 + nb - \sum y_i x_i = 0; \qquad a\sum x_i + nb - \sum y_i = 0$$

$$b = \frac{\sum y_i - a\sum x_i}{n}; \qquad b = \frac{\sum x_i y_i - a\sum x_i^2}{\sum x_i}$$

Equating the two expressions for b and solving for a:

$$a = \frac{\sum x_i y_i - \sum x_i \sum y_i/n}{\sum x_i^2 - (\sum x_i)^2/n}$$

Substituting the expression for a in the expression for b and solving for b:

$$\frac{\sum x_i y_i - \sum x_i \sum y_i/n}{\sum x_i^2 - (\sum x_i)^2/n}\sum x_i + nb - \sum y_i = 0$$

solving for b:

$$b = \frac{1}{n}\left[\frac{\sum y_i \sum x_i^2 - \sum x_i \sum x_i y_i}{\sum x_i^2 - (\sum x_i)^2/n}\right]$$

EXAMPLE 3: Calculate the dimensions of a can that will hold 100 mL of material such that the amount of metal used in its manufacture is a minimum.

Let r = radius and h = height. Two independent equations are needed. One equation must involve the surface area to be minimized. The other equation must involve the volume, since the 100-mL volume requirement must be met. $V = 100 = \pi r^2 h$. $h = 100/(\pi r^2)$.

$A = 2\pi rh + 2\pi r^2$. Substituting h:

$$A = \frac{2\pi r(100)}{\pi r^2} + 2\pi r^2 = \frac{200}{r} + 2\pi r^2$$

$$\frac{dA}{dr} = 4\pi r - \frac{200}{r^2} = 0; \qquad 4\pi r^3 = 200$$

$$r = \left[\frac{50}{\pi}\right]^{0.333} = 2.51 \text{ cm}$$

$$h = \frac{100}{\pi(2.51)^2} = 5.06 \text{ cm}$$

The second derivative, $d^2A/dr^2 = 4\pi + 400/r^3$, will be positive for all positive values of r; therefore, the root of the derivative function represents a minimum point. At $r = 2.51$ cm, $A = 2\pi(2.51)^2 + 200/2.51 = 119.3$ cm^2. At $r = 2.4$ cm, $A = 2\pi(2.4)^2 + 200/2.4 = 119.5$ cm^2. The value of A at $r = 2.51$ cm is less than at $r = 2.4$ cm; therefore, $r = 2.51$ is a minimum point.

INTEGRAL CALCULUS

Integration is the inverse of differentiation. If $dF(x)$ is the differential term, the integral of $dF(x)$ is $F(x)$. Integrals are differential terms preceded by the integral sign $\int$. The terms inside the integral sign consist of a function $F(x)$ and a differential term dx. An integral is represented graphically in Fig. 1.11. The function $F(x)$ traces a curve, and the differential term dx is represented by a small area increment with height $F(x)$ and thickness dx. When the area increments are evaluated consecutively, the sum represents the value of the integral. Limits are needed to place a definite value on the integral. The limit must correspond to the value of the variable in the differential term and is the abscissa of the curve used to plot the function within the integral sign. Graphically, the limits define which region of the curve is covered by the area summation. Fig. 1.11 represents $\int_{x_1}^{x_2} F(x)\,dx$.

When evaluating a definite integral, the limits are substituted into the integrand and the value at the lower limit is subtracted from the value at the upper limit.

When the limits are not specified, the integral is an indefinite integral and the

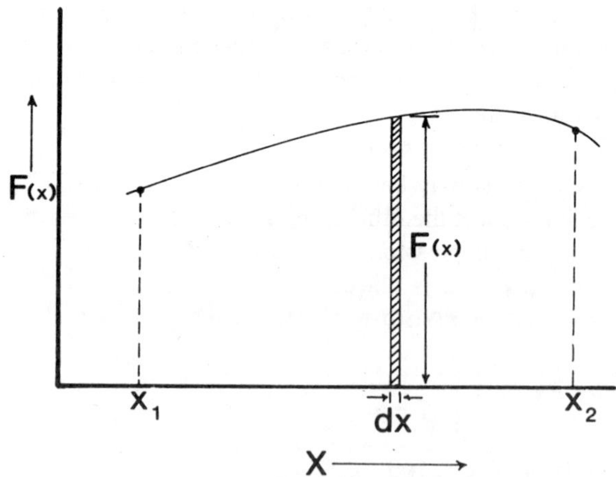

Fig. 1.11. Graphical representation of an integral.

integrand will need a constant of integration. $\int^{\cdot cp3} dF(x) = F(x) + C$. The value of C is determined using boundary or initial conditions which satisfy the function $F(x) + C$.

Integration Formulas. Only simple and most common integrals are presented here. The reader is referred to calculus textbooks for techniques that are used for more complex functions.

Integral of a sum:

$$\int (du + dv + dw) = \int du + \int dv + \int dw$$

Integral of a power function:

$$\int F(x)^n \, dF(x) = \frac{F(x)^{n+1}}{n+1} + C$$

Integral of a quotient yielding a logarithmic function:

$$\int \frac{dF(x)}{F(x)} = \ln F(x) + C$$

Integral of an exponential function:

$$\int e^{F(x)} \, dF(x) = e^{F(x)} + C$$

$$\int 10^{F(x)} \, dF(x) = \frac{10^{F(x)}}{\ln 10} + C$$

Integration Techniques

Constants. Constants may be taken in or out of the integral sign. If only a constant is needed to meet the differential form needed in the above integration formulas, the integral expression may be multiplied and divide by the same constant. The multiplicand is placed within the integral to meet the required differential form, and the divisor is placed outside the integral.

Integration by Parts

$$\int F(x) \, dG(x) = F(x)G(x) - \int G(x) \, dF(x)$$

Partial Fractions

$$\int \frac{F(x)\,dx}{G(x)H(x)} = \int \frac{A\,dx}{G(x)} + \int \frac{B\,dx}{H(x)} + C$$

A and B are constants obtained by clearing the fractions in the above equation and solving for A and B, which would satisfy the equality.

Substitution. A variable is used to substitute for a function. The substitution must result in a simpler expression which is integrable using standard integration formulas or integration by parts.

EXAMPLE 1: $\int (x^2 + 3)^2\,dx$

This does not fit the power function formula, since the differential term dx is not $d(x^2 + 3)$. The function can be expanded and integrated as a sum:

$$\int (x^2 + 3)^2\,dx = \int (x^4 + 6x^2 + 9)\,dx$$

$$= \frac{x^5}{5} + \frac{6x^3}{3} + 9x + C$$

EXAMPLE 2: $\int_{0.1}^{0.3} (2x^2 + 2)^4\,x\,dx$

This function may be integrated using the formula for a power function. $d(2x^2 + 2) = 4x\,dx$. The function will be the same as in the power function formula by multiplying by 4. The whole integral is then divided by 4 to retain the same value as the original expression:

$$\int_{0.1}^{0.3} (2x^2 + 2)^4\,x\,dx = \frac{1}{4}\int_{0.1}^{0.3} (2x^2 + 2)^4\,4x\,dx$$

$$= \frac{1}{4}\frac{(2x^2 + 2)^5}{5}\Bigg|_{0.1}^{0.3}$$

$$= \frac{(2.18)^5 - (2.02)^5}{20}$$

$$= 0.78018$$

EXAMPLE 3: The inactivation of microorganisms under conditions when temperature is changing is given by:

$$\frac{N_0}{N} = \int_0^t \frac{dt}{D_T}$$

N_0 is the initial number of microorganisms, and D_T is the decimal reduction time of the organism. D_T is given by:

$$D_T = D_0[10]^{(250-T)/z}$$

If $D_0 = 1.2$ min, $z = 18°$F, $N_0 = 10{,}000$, and $T = 70 + 1.1t$, where T is in °F and t is time in minutes, calculate N at $t = 250$ min. The integral to be evaluated is:

$$\frac{10{,}000}{N} = \int_0^{250} \frac{dt}{1.2\,[10]^{(250-70-1.1t)/18}}$$

$$= \int_0^{250} \frac{dt}{1.2[10]^{(180-1.1t)/18}}$$

$$= \frac{1}{1.2(10)^{10}} \int_0^{250} \frac{dt}{(10)^{-0.06111t}}$$

$$= 0.833(10)^{-10} \int_0^{250} (10)^{0.06111t}$$

$$= 0.833(10)^{-10} \left[\frac{(10)^{0.06111t}}{0.06111 \ln(10)}\right]_0^{250}$$

$$= \frac{0.833(10)^{-10}}{0.06111 \ln(10)} [10^{15.275} - 1]$$

The 1 in brackets is much smaller than $10^{15.275}$; therefore, it may be neglected. Therefore:

$$\frac{10{,}000}{N} = 5.9209(10)^{-10}(10)^{15.275}$$

$$= 5.9209(10)^{15.275-10}$$

$$= 5.9209(10)^{5.275}$$

Taking the logarithm of both sides:

$$\log\left(\frac{10{,}000}{N}\right) = 5.275 + \log 5.9209 = 6.047$$

$$N = (10)^{4-6.047} = 0.0089$$

EXAMPLE 4: $\int \frac{x\,dx}{(2x^2 + 1)}$

This function will yield a logarithmic function because the derivative of the denominator is $4x\,dx$. The numerator of the integral can be made the same by multiplying by 4 and dividing the whole integral by 4.

$$\int \frac{x\,dx}{(2x^2 + 1)} = \frac{1}{4} \int \frac{4x\,dx}{(2x^2 + 1)} = \frac{1}{4} \ln (2x^2 + 1) + C$$

EXAMPLE 5: Solve for the area under a parabola $y = 4x^2$ bounded by $x =$ 1 and $x = 3$, as given by the following integral:

$$A = \int_1^3 4x^2\,dx = \frac{4x^3}{3} \Bigg|_1^3 = \frac{1}{3} [4(27) - 4] = 34.6667$$

Graphical Integration. Graphical integration is used when functions are so complex that they cannot be integrated analytically. They are also used when numerical data such as experimental results are available and it is not possible to express the data in the form of an equation which can be integrated analytically. Graphical integration is a numerical technique used for evaluation of differential equations by the finite difference method. Three techniques for graphical integration will be shown. Each of these will be used to evaluate the area under a parabola, a problem solved analytically in the preceding example.

Rectangular Rule. The procedure is illustrated in Fig. 1.12A. The domain under consideration is divided into a sequence of bars. The thicknesses of the

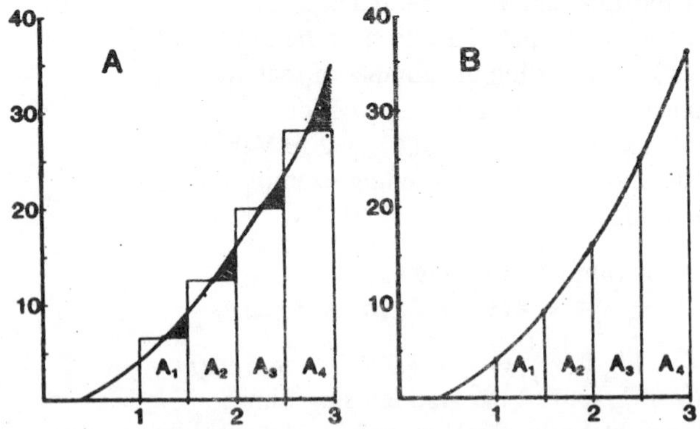

Fig. 1.12. Graphical integration by the rectangular (A) and trapezoidal (B) techniques.

bars represent increments of x. The heights of the bars are set such that the shaded area within the curve which is outside the bar equals the area inside the bar which lies outside the curve. The increments may be unequal, but in Fig. 1.12A equal increments of 0.5 unit are used. The heights of the bars are 6.5, 12.5, 20, and 28 for A_1, A_2, A_3, and A_4, respectively. The sum of the area increments is $0.5(6.5 + 12.5 + 20 + 28) = 33.5$.

Trapezoidal rule. The procedure is illustrated in Fig. 1.12B. The domain under consideration is subdivided into a series of trapezoids. The area of the trapezoid is the width multiplied by the arithmetic mean of the height. For the function $y = 4x^2$, the values (height) are 4, 9, 16, 25, and 36 for $x = 1, 1.5, 2, 2.5$, and 3, respectively. The trapezoids A_1, A_2, A_3, and A_4 will have areas of $0.5(4 + 9)(0.5)$, $0.5(9 + 16)(0.5)$, $0.5(16 + 25)(0.5)$, and $0.5(25 + 36)(0.5)$. The sum is $3.25 + 6.25 + 10.25 + 15.25 = 35$.

Simpson's Rule. This procedure assumes a parabolic curve between area increments. The curve is divided into an even number of increments; therefore, including the values of the functions at the lower and upper limits, there will be i values of the height of the function and $(i - 1)$ area increments. The number of increments $(i - 1)$ must be an even integer. The thickness of the area increments must be uniform, therefore for the limits $x = a$ to $x = b$, $\delta x = (b - a)/(i - 1)$. Simpson's rule is as follows:

$$A = \frac{\delta x}{3}\left[F(a) + 4F(1) + 2F(2) + 4F(3) + 2F(4) + \cdots 2F(i-2) + 4F(i-1) + F(b)\right]$$

Note the repeating 4, 2, 4, 2, 4 multiplier of the value of the function as succeeding increments are considered. If the index $i = 0$ at $x = a$, the lower limit, all values of the function when the index is even are multiplied by 2 and the multiplier is 4 when the index is odd. For the example of the parabola $y = 4x^2$ from $x = 1$ to $x = 3$, setting the number of increments at four gives five values of y which must be successively evaluated for the area increment. The thickness of the increments $\delta x = (3 - 1)/(4) = 0.5$. Values of $F(x)$ at 1, 1.5, 2, 2.5, and 3 were determined in the preceding examples. Substituting in the formula for Simpson's rule:

$$A = \frac{0.5}{3}\left[4 + 4(9) + 2(16) + 4(25) + 36\right] = 34.6667$$

The accuracy of graphical integration using Simpson's rule is greater than that using the trapezoidal rule. Both the trapezoidal rule and Simpson's rule can be programmed easily in BASIC. The following program in BASIC illustrates how the area of a curve drawn to fit numerical data is determined using Simpson's rule. The area under the parabola $y = 4x^2$ is being analyzed. A smaller thickness of the area increment is used than in the previous example. Ten area increments are chosen; therefore, 11 values of the function will be needed to include $x = 1$ and $x = 3$, the upper limit. These values are entered as a data statement in line 10.

```
10 Dim Y(10)
20 DATA 4,5.76,7.84,10.24,12.96,16,19.36,23.04,27.04,31.36,36
30 FOR I = 1 to 11
40 READ Y(I)
50 Next I
60 DX = 0.2
70 FOR N = 2 to I STEP 2
80 SA = SA + Y(N)
90 NEXT N
100 FOR N = 3 TO I - 1 STEP 2
110 SB = SB + Y(N)
120 NEXT N
130 PRINT "A = "; (DX / 3) * (4 * SA + 2 * SB + Y(1) + Y(11))
```

The program, when run, will display $A = 34.66667$ on the screen.

Differential Equations. Differential equations are used to determine the functionality of variables from information on rate of change. The simplest differential equations are those in which the variables are separable. These equations take the following form:

$$\frac{dy}{dx} = F(x); \qquad dy = F(x)\,dx$$

The solution to the differential equation then involves integration of both sides of the equation. Constants of integration are determined using initial or boundary conditions.

EXAMPLE 1: If the rate of dehydration is constant at 2 kg/h from a material which weighs 20 kg and contains 80% moisture at time = 0, derive an expression for moisture content as a function of time.

Let W = weight at any time, t = time, and x = moisture content expressed as a mass fraction of water. From the rate data, the differential equation is:

$$\frac{dW}{dt} = -2$$

The negative sign occurs because the rate is expressed as a weight loss; therefore, W will be decreasing with time. Separating variables: $dW = -2dt$. Integrating: $W = -2t + C$. The constant of integration is determined by substituting $W = 20$ at $t = 0$, and the expression for W becomes $W = 20 - 2t$. The moisture content x is determined from W as follows:

$$W = \frac{\text{weight dry solids}}{\text{mass fraction dry solids}} = \frac{20(1 - 0.8)}{1 - x}$$

Substituting W and simplifying:

$$\frac{1}{1 - x} = 5 - 0.5t$$

EXAMPLE 2: The venting of air from a retort by steam displacement is analogous to dilution in which the air is continuously diluted with steam and the diluted mixture is continuously vented. Let C = concentration of air at any time, t = time, V = the volume of the retort, and R = rate of addition of steam. The differential equation for C with respect to time is:

$$V\frac{-dC}{dt} = RC; \qquad \frac{dC}{C} = \frac{-R}{V}\,dt$$

The integration constant will be evaluated using, as an initial condition, $C = C_0$ at $t = 0$. Integrating: $\ln C = (-R/V)\,t + \ln C_0$.

Finite Difference Approximation of Differential Equations. Differential equations may be approximated by finite differences in the same manner as integrals are solved by numerical methods using graphical integration. Finite increments are substituted for the differential terms, and the equation is solved within the specified boundaries. Finite difference approximations can be easily done using BASIC.

EXAMPLE: A stirred tank having a volume V, in liters, contains a number of cells of microorganisms, N. The tank is continuously fed with cell-free medium at the rate of R L/h, and the volume is maintained constant by allowing the medium to overflow at the same rate as the feed. The number of cells in the tank increases due to reproduction, and some of the cells are washed out in the overflow. (a) Derive an equation for the cell number at any time as a function of the generation time, g, the feed rate, R, and the volume, V. (b) Calculate N after 5 hours if $g = 0.568$ h, $N_0 = 10{,}000$, $V = 1.3$ L, and $R = 1.5$ L/h.

Solution: Microorganisms in the most rapid phase of growth will increase in cell numbers with time of growth according to the formula $N = N_0(2)^{t/g}$, where g = generation time.

$$\text{Cell growth} = \left(\frac{dN}{dt}\right) g$$

$$= \frac{d}{dt}(N_0)(2)^{t/g}$$

$$= N_0(2)^{2/g} \ln 2$$

$$= N\frac{\ln 2}{g}$$

Washout = $N(R)$; net accumulation = $dN/dt(V)$. A balance of cell numbers will give the number of cells generated, which equals washout plus accumulation. The differential equation is:

$$N\frac{\ln 2}{g} = NR + V\frac{dN}{dt}$$

Rearranging and separating variables:

$$\frac{dN}{N} = \left(\frac{\ln 2}{g} - \frac{R}{V}\right)dt$$

Integrating and using as an initial condition $N = N_0$ at $t = 0$:

$$\ln\frac{N}{N_0} = \left(\frac{\ln 2}{g} - \frac{R}{V}\right)t$$

To solve the problem by finite difference techniques, values of V, R, N_0, and g must be defined. If g = 0.568 h, N_0 = 10,000, R = 1.5 L/h, and V = 1.5 L, the number of organisms in the tank at t = 5 h is as follows: From the analytical solution:

$$\ln N = \ln 10{,}000 + \left[\frac{\ln 2}{0.568} - \frac{1.5}{1.5}\right](5)$$

$$= 9.21 + 1.101 = 10.311$$

$$N = e^{10.311} = 30{,}060$$

The differential equation can be solved using a finite difference technique utilizing BASIC. The program in BASIC for the differential equation is:

```
10 NO = 10000: R = 1.5: V = 1.5: G = 0.368: dt = 0.1
20 N = NO
30 DN = N * ( LOG(2) / G - (R/V) ) * DT
40 N = N + DN
50 TIME = TIME + DT
60 IF (TIME) <5 THEN GOTO 30
70 PRINT "TIME = "; (TIME-DT), "N = "
80 END
```

When the program is run, "Time = 5 N = 30120" will be displayed on the screen. The value of N calculated by finite difference is 30,120 compared to 30,060 determined analytically.

Problems

1. Determine the maximum and minimum values of the following function and prove that the points are maximum or minimum.

$$C = \frac{315 + 52.5T}{(0.21T - 0.76)^{0.5} - 1.61}$$

2. A warehouse having a volume of 10,000 ft^3 and a floor area of 1000 ft^2 is to be built. The cost of constructing the floor is \$6/ft^2; the roof, \$10/ft^2; and the walls, \$20/ft^2. If W is the width, L is the length, and H is the height of the building, what should the dimensions be such that the cost is minimum?
3. Calculate the maximum or minimum value of the following expression. Show that the calculated value is maximum or minimum.

$$q = \frac{240R}{-0.02 + 20 \ln (0.5/R) + 0.02R}$$

4. Calculate the maximum (or minimum value) of y.

$$y = 2X^2 + 0.5R + X + 3$$

$$X = 0.5R$$

5. Calculate the maximum velocity of a fluid flowing inside a pipe expressed in terms of the average velocity. The point velocity (velocity at any point r measured from the center) is:

$$V = \left[\frac{\Delta P}{2\ Lk}\right]^{1/n} \left[1 - \left[\frac{r}{R}\right]^{(n+1)/n}\right] \left[\frac{n}{n+1}\right] [R]^{(n+1)/n}$$

The average velocity is:

$$\bar{V} = \left[\frac{\Delta P}{2\,Lk}\right]^{1/n} \left[\frac{n}{3n + 1}\right] [R]^{(n+1)/n}$$

6. The rate of evaporation of component A divided by the rate of evaporation of component B in a mixture containing A moles of A and B moles of B is directly proportional to A/B. If a mixture originally contained 5 moles of A and 3 moles of B and the rate of evaporation of A is 5 moles/h and that of B is 2.6 moles/h, derive an equation for the concentration of A relative to that of B.
7. The rate of production of ethanol in a fermenter is directly proportional to the number of cells of yeast present. At a cell mass of 14 g/L yeast, the ethanol production rate is 18 g ethanol/(L · h). If a batch fermenter is inoculated with 0.1 g/L of yeast, and the generation time of the yeast is 1.5 h, what will be the ethanol production after 10 h of fermentation? Assume that there is no alcohol-induced inhibition of yeast growth and that yeast doubles in cell mass with each generation time.
8. The work required to compress a gas is given by the following integral:

$$W = \int_{V_1}^{V_2} P\,dV$$

The Van der'Waals equation of the state for a gas is as follows:

$$P = \frac{nRT}{V - nb} - \frac{n^2 a}{V^2}$$

$R = 82.06\ (\text{cm}^3 \cdot \text{atm})/(\text{gmole} \cdot \text{K})$, $b = 36.6\ \text{cm}^3/\text{gmole}$, and $a = 1.33 \times 10^6\ (\text{atm} \cdot \text{cm}^6)/(\text{gmole})^2$ when P is expressed in atm, T in °K, and V in cm^3. Calculate the work done when the pressure of the gas is increased from 1 to 10 atm. There was originally 1 L of gas at 293°K.
9. Write a computer program in BASIC which can be used to calculate a value for the water activity of a solution containing two sugars having percentage composition by weight of $x1\%$ and $x2\%$. $x1$ is sucrose (molecular weight 342; $k = 2.7$); $x2$ is glucose (molecular weight 180; $k = 0.5$). The water activity of the mixture is expressed as:

$$a_w = (a_{w1})^\circ (a_{w2})^\circ$$

where $(a_{w1})^\circ$ = water activity of a solution containing all the water in the mixture and component 1 and $(a_{w2})^\circ$ = water activity of a solution containing all the water in the mixture and component 2.

$$\ln\left[\frac{(a_{wi})^\circ}{f_i}\right] = k_i(1 - f_i)^2,$$

where f_i = mole fraction of component i in a solution containing only component i and all the water present in the mixture.

$$f_i = 1 - \left[\frac{x_i/M_i}{(x_i/M_i) + (1 - x_1 - x_2)/18}\right]$$

10. Determine the slope of the following functions:

$$f(x) = 2x^4 - 3x^2 + 7x \qquad \text{at } x = 2$$

$$xy = (0.5x + 3)(x + 2) \qquad \text{at } x = 1$$

11. Determine the maximum and minimum values of the functions in problem 10.

12. Determine the slope of the following function at the indicated point:

$$x = (0.5\,xy + 3)(3 + x^2) \qquad \text{at } x = 1$$

13. Write a computer program in BASIC which can be used to determine the boiling temperature of a liquid in an evaporator as it is being concentrated. The boiling point of a liquid is the temperature at which the vapor pressure equals the atmospheric pressure. A solution will exhibit a boiling point rise because the solute will lower the water activity, resulting in a higher temperature to be reached before boiling occurs. The vapor pressure of water ($P°$) as a function of temperature is expressed by the following equation:

$$\ln(P°) = \left(\frac{H}{R}\right)\left(\frac{1}{T}\right) + C$$

where $P°$ is the vapor pressure in kiloPascals. H/R is the ratio of the latent heat of vaporization and the gas constant, C is a constant, and T is the absolute temperature in °K. The value for H/R is 4950, and C is 17.86.

Atmospheric pressure is 101 kiloPascals. The vapor pressure of a solution is $P = a_w P°$. The water activity (a_w) is given by:

$$\log\left(\frac{a_w}{f_1}\right) = -k(1 - f_1)^2$$

f_1 is the mole fraction of water and is calculated by:

$$f_1 = \frac{(1 - x_1)/18}{(1 - x_1)/18 + x_1/M_1}$$

Assume that the solute is only sucrose, with a molecular weight (M_1) of 342 and a k value of 2.7. Have the program display the value of the boiling point when the sucrose concentration is 20% and at concentrations in 5% intervals to a final concentration of 60%.

14. Calculate a value of x which would give the minimum value for s in the following expression:

$$S = 50(9.522 \times 10^{-6}) - \frac{50}{(50X + 1)}(9.522 \times 10^{-6}) - \frac{60(0.22)X}{30(24)}$$

Show that the function is a minimum or maximum.

15. In the dehydration of diced potatoes, the following weights were recorded at various times in the process when the dehydration rate was slowest:

t = 6 h from start of drying; weight = 2350 g
t = 8 h from start of drying; weight = 2275 g; moisture content = 15.6%

In this range of moisture content, the drying rate is proportional to the moisture content, expressed in mathematical form as follows:

$$\frac{dW}{dt} = kW$$

W is the moisture content in g water/g dry matter, and k is a constant.

(a) Derive an equation for the moisture content, W, as a function of time, t, which satisfies the experimental conditions given above.

(b) How long would it take for a product to be dehydrated from 22.5% to 12.5% moisture?

16. Write a computer program in BASIC which will solve the following integral:

$$\frac{N_0}{N} = \frac{R^2\overline{V}}{2}\left[\frac{1}{\int_0^R rV_r[10]^{-L/(V_rD)}\,dr}\right]$$

where

$$V_r = \overline{V}\left[\frac{3n + 1}{n + 1}\right]\left[1 - \left[\frac{r}{R}\right]^{(n+1)/n}\right]$$

given $\overline{V} = 2$, $R = 0.02$, $n = 0.6$, $L = 10$, and $D = 1$. Use the trapezoidal rule in evaluating the integral, and use increments in r of 0.0001.

17. Linearize the following equations. In each case, indicate which function should be plotted as the independent and dependent variables to obtain a linear plot. What is the slope and intercept of each plot?

(a) $\log(a_w/X_i) = -k(1 - X_i)^2$, where a_w and X_i are variables.

(b) $(1/x) = (a + b)/2y + c/4y$, where x and y are variables and a, b, and c are constants.

(c) $N = N_0(10)^{-(t/D)}$, where N and t are variables and N_0 and D are constants.

18. The velocity of an enzyme-catalyzed reaction (V) expressed as a function of the substrate concentration (S) is given by the following equation:

$$V = \frac{V_{max}(S)}{K_m + S}$$

where V_{max} is the maximum reaction rate and K_m is a constant for the reaction. Linearize the equation and determine the independent and dependent variables to be plotted to obtain V_{max} and K_m from the slope and intercept.

19. The temperature (T) of a refrigerant during compression in a refrigeration system increases with pressure (P), as shown in the following equation. P_1 and T_1 are reference temperature and pressure and are considered constants.

$$\frac{T}{T_1} = \left[\frac{P}{P_1}\right]^{(k-1)/k}$$

The following data are available on the temperature and pressure of a refrigerant during compression:

T(°R)	450	510	550	608	640
P(lb$_f$/in.^{2})	4	6	10	20	30

Linearize the above equation, plot the data, and determine the constant k for this refrigerant.

20. The heat of respiration of leafy greens as a function of time when the temperature is changing during a cooling operation is as follows:

$$q = \int_0^t 0.009854\,[e]^{0.073T}\,dt$$

where q is in BTU/lb, T is in °F, and t is time in hours. The temperature of a box of spinach containing 50 lb, originally at 110°F, will cool down exponentially when placed in a refrigerated room at 35°F according to the formula

$$T = 35 + (75)\exp(-t/5)$$

Write a computer program in BASIC which can be used to determine the total heat generated by the spinach as the box cools down from 110° to 40°F. Use the trapezoidal rule and Simpson's rule in determining the value of the integral and compare your results.

21. An immobilized enzyme reactor must have the enzyme regenerated periodically because of the decay in the activity of the enzyme. The enzyme will convert 87% of the substrate to product within the first day of operation, and this conversion changes to $0.87/(t)^{0.82}$ after the first day. If the feed rate is 150 lb/day of substrate, the amount of product formed after t days of operation will be:

$$P = 150(0.87) + \left[\frac{150(0.87)}{(t)^{0.82}}\right](t - 1)$$

At a substrate cost of \$2.50/lb and an operating cost of \$600/day, a product cost of \$14/lb, and an enzyme replacement cost of \$600, the return from the operation will be:

$$S = 14P - 600(t) - 2.5(P) - \frac{600}{t}$$

Calculate the number of days, t, which the reactor must be operated before recharging in order that the return, S, will be maximum. Prove that S is maximum at the value of t identified.

SUGGESTED READING

Burrows, W. H. 1965. *Graphical Techniques for Engineering Computation*. Chemical Publishing Co., New York.

Cumming, H. G., and Anson, C. J. 1967. *Mathematics and Statistics for Technologists*. Chemical Publishing Co., New York.

Dull, R. W., and Dull, R. 1951. *Mathematics for Engineers*. McGraw-Hill Book Co., New York.

Greenberg, D. A. 1965. *Mathematics for Introductory Science Courses. Calculus and Vectors with a Review of Basic Algebra*. W. A. Benjamin Co., New York.

Kelly, F. H. C. 1963. *Practical Mathematics for Chemists*. Butterworths, London.

Mackie, R. K., Shephard, T. M., and Vincent, C. A. 1972. *Mathematical Methods for Chemists*. English University Press, London.

Perry, R. H., Chilton, C. H., and Kirkpatrick, S. D. 1963. *Chemical Engineers' Handbook*, 4th ed. McGraw-Hill Book Co., New York.

2

Units and Dimensions

The units used to designate magnitude of a dimension have evolved based on common usage and the instruments available for measurement. Two major systems for measurement have been used: the English system, used primarily in industry, and the metric system, used in the sciences. The confusion that resulted from the use of various terms to represent the same dimension led to the development of a common system of units that is proposed for use in both science and industry. The international system of units and the official international designation SI were adopted in 1960 by the General Conference on Weights and Measures. This body consists of delegates from member countries of the Meter Convention and meets at least once every six years. At least 44 countries are represented in this convention, one of which is the United States.

The use of SI units is gaining momentum. However, it is still often necessary to convert data from one system to another, since tables in handbooks may be in a different unit from what is needed in the calculations, or experimental data may be obtained using instruments calibrated in a different unit from what is desired in reporting the results.

In this chapter, the various SI units are discussed and techniques for conversion of units using the dimensional equation are presented. Also emphasized in this chapter is the concept of dimensional consistency of mathematical equations involving physical quantities, and how units of variables in an equation are determined to ensure dimensional consistency.

DEFINITION OF TERMS

Dimension, used to designate a physical quantity under consideration, (e.g., time, distance, weight).

Unit, used to designate the magnitude or size of the dimension under consideration (e.g., m for length, kg for weight).

Base unit, units that are dimensionally independent. They are used to designate only one dimension (e.g., units of length, mass, and time).

Derived units, a combination of various dimensions. An example is the unit of force, which includes the dimensions of mass, length, and time.

Precision, synonymous with reproducibility, the degree of deviation of the measurements from the mean. This is often expressed as a plus or minus term or as the smallest value of a unit that can be consistently read in all determinations.

Accuracy, how a measured quantity relates to a known standard. To test the accuracy of a measurement, the mean of a number of determination is compared with a known standard. Accuracy depends upon proper calibration of an instrument.

SYSTEMS OF MEASUREMENT

The various systems in use are shown in Table 2.1. These systems vary in the base units used. Under the English system, variations exist in expressing the unit of force. The chemical and food industries in the United States use the American Engineering System, although SI is the preferred system in scientific articles and textbooks.

The metric system uses prefixes on the base unit to indicate magnitude. Industry has adopted the mks system, while the sciences have adopted the cgs system. SI is designed to meet the needs of both science and industry.

Table 2.1. Systems of Measurement

System	Dimension						
	Use	Length	Mass	Time	Temperature	Force	Energy
English							
English absolute	Scientific	Foot	Pound mass	Second	°F	Poundal	BTU ft (poundal)
British Engineering	Industrial	Foot	Slug	Second	°F	Pound force	BTU ft (pound force)
American Engineering	U.S. industrial	Foot	Pound mass	Second	°F	Pound force	BTU ft (pound force)
Metric							
cgs	Scientific	Centimeter	Gram	Second	°C	Dyne	Calorie, erg
mks	Industrial	Meter	Kilogram	Second	°C	Kilogram force	Kilocalorie joule
SI	Universal	Meter	Kilogram	Second	°K	Newton	Joule

THE SI SYSTEM

The following discussion of the SI system and the convention followed in rounding after conversion are based on the American National Standard, Metric Practice, adopted by the American National Standards Institute, the American Society for Testing and Materials, and the Institute of Electrical and Electronics Engineers.

Units in SI and Their Symbols. SI uses base units and prefixes to indicate magnitude. All dimensions can be expressed in either a base unit or combinations of base units. The latter are called *derived units*, and some have specific names. The base units and the derived units with assigned names are shown in Table 2.2.

Table 2.2. Base Units of SI and Derived Units with Assigned Names and Symbols

Quantity	Unit	Symbol[a]	Formula
Length	meter	m	—
Mass	kilogram	kg	—
Electric current	ampere	A	—
Temperature	kelvin	K	—
Amount of substance	mole	mol	—
Luminous intensity	candela	cd	—
Time	second	s	—
Frequency (of a periodic phenomenon)	hertz	Hz	1/s
Force	newton	N	$kg\ m/s^2$
Pressure, stress	pascal	Pa	N/m^2
Energy, work, quantity of heat	joule	J	N · m
Power, radient flux	watt	W	J/s
Quantity of electricity, electric charge	coulomb	C	A · s
Electric potential, potential difference, electromotive force	volt	V	W/A
Capacitance	farad	F	C/V
Electric resistance	ohm		V · A
Conductance	siemens	S	A/V
Magnetic flux	weber	Wb	V · s
Magnetic flux density	tesla	T	Wb/m^2
Inductance	henry	H	Wb/A
Luminous flux	lumen	lm	cd · sr[b]
Illuminance	lux	lx	lm/m^2
Activity (of radionuclides)	becquerel	Bq	1/s
Absorbed dose	gray	Gy	J/kg

Source: American National Standard, 1976. Metric Practice. IIEE Std. 268-1976. Institute of Electrical and Electronics Engineers, New York.

[a]Symbols are written in lowercase letters unless they are from the name of a person.

[b]sr stands for steradian, a supplementary unit used to represent solid angles.

Table 2.3. Prefixes Recommended for Use in SI

Prefix	Multiple	Symbol[a]
tera	10^{12}	T
giga	10^{9}	G
mega	10^{6}	M
kilo	1000	k
milli	10^{-3}	m
micro	10^{-6}	μ
nano	10^{-9}	n
pico	10^{-12}	p
femto	10^{-15}	f

[a]Symbols for the prefixes are written in capital letters when the multiplying factor is 10^6 and larger. Prefixes designating multiplying factors less than 10^6 are written in lower case letters.

Prefixes Recommended for Use in SI. Prefixes are placed before the base multiples of 10. Prefixes recommended for general use are shown in Table 2.3.

A dimension expressed as a numerical quantity and a unit must be such that the numerical quantity is between 0.1 and 1000. Prefixes should be used only on base units or named derived units. Double prefixes should not be used.

Examples: (1) 10,000 cm should be 100 m, not 10 kcm.
(2) 0.0000001 m should be 1 μm.
(3) 3000 m^3 should not be written as 3 km^3.
(4) 10,000 N/m can be written as 10 kPa but not as 10 kN/m^2.

CONVERSION OF UNITS

Precision, Rounding-Off Rule, Significant Digits. Conversion from one system of units to another should be done without gain or loss of precision. Results of measurements must be reported such that a reader can determine the degree of precision. The easiest way to convey precision is by the number of significant figures.

Significant figures include all nonzero digits and nonterminal zeroes in a number. Terminal zeroes are significant in decimals and may be significant in whole numbers when specified. Zeroes preceding nonzero digits in decimal fractions are not significant.

Examples: (1) 123 has three significant figures.
(2) 103 has three significant figures.
(3) 103.03 has five significant figures.
(4) 10.030 has five significant figures.
(5) 0.00230 has three significant figures.
(6) 1500 has two significant figures unless the two terminal zeroes are specified as significant.

When the precision of numbers used in mathematical operations is known, the answer should be rounded off following these rules recommended by the American National Standards Institute. All conversion factors are assumed to be exact.

1. In addition or substraction, the answer should not contain a significant digit to the right of the least precise number. Example: 1.030 + 1.3 + 1.4564 = 3.8. The least precise number is 1.3. Any digit to the right of the tenth digit is not significant.
2. In multiplication or division, the number of significant digits in the answer should not exceed that of the least precise original number. Example 123 × 120 = 15,000 if the terminal zero in 120 is not significant. The least precise number has two significant figures, and the answer should also have two significant figures.
3. When rounding off, raise the terminal significant figure retained by one if the discarded digits start with 5 or larger; otherwise, the terminal significant figure retained is unchanged. If the start of the digits discarded is 5, followed by zeroes, make the terminal significant digit retained even.

If the precision of numbers containing terminal zeroes is not known, assume that the numbers are exact. When performing a string of mathematical operations, round off only after the last operation is performed.

Examples: 1253 rounded off to two significant figures = 1300.
1230 rounded off to two significant figures = 1200 (2 is even).
1350 rounded off to two significant figures = 1400 (3 is odd).
1253 rounded off to three significant figures = 1250.
1256 rounded off to three significant figures = 1260.

The Dimensional Equation. The magnitude of a numerical quantity is uncertain unless the unit is written along with the number. To eliminate this ambiguity, make a habit of writing both the number and its unit.

An equation that contains both numerals and their units is called a *dimensional equation*. The units in a dimensional equation are treated just like algebraic terms. All mathematical operations done on the numerals must also be done on their corresponding units. The numeral may be considered as a coefficient of an algebraic symbol represented by the unit. Thus,

$$(4\text{m})^2 = 16\ \text{m}^2 \quad \text{and} \quad \left[5\,\frac{\text{J}}{\text{kg}\cdot\text{K}}\right](10\ \text{kg})(5\ \text{K}) = 5(10)(5)\left[\frac{\text{J}\cdot\text{kg}\cdot\text{K}}{\text{kg}\cdot\text{K}}\right]$$

= 250 J. Addition and subtraction of numerals and their units also follow the rules of algebra; that is, only like terms can be added or subtracted. Thus, 5 m

− 3 m = (5 − 3)m, but 5 m − 3 cm cannot be simplified unless their units are expressed in like units.

Conversion of Units Using the Dimensional Equation Determining the appropriate conversion factors to use in conversion of units is facilitated by a dimensional equation. The following procedure may be used to set up the dimensional equation for conversion.

1. Place the units of the final answer on the left-hand side of the equation.
2. The number being converted and its unit are the first entry on the right-hand side of the equation.
3. Set up the conversion factors as a ratio, using Appendix Table A.1.
4. Sequentially multiply the conversion factors such that the original units are systematically eliminated by cancellation and replacement with the desired units.

EXAMPLE 1: Convert BTU/(lb · °F) to J/(g · K).

$$\frac{\text{J}}{\text{g} \cdot \text{K}} = \frac{\text{BTU}}{\text{lb} \cdot {}^\circ\text{F}} \times \text{appropriate conversion factors}$$

The numerator J on the left corresponds to BTU on the right-hand side of the equation. The conversion factor is 1054.8 J/BTU. Since the desired unit has J in the numerator, the conversion factor must have J in the numerator. The factor 9.48×10^4 BTU/J may be obtained from the table, but it should be entered as J/9.48×10^4 BTU in the dimensional equation. The other factors needed are 2.2046×10^3 lb/g or lb/453.6 g and 1.8°F/K.
The dimensional equation is:

$$\frac{\text{J}}{\text{g} \cdot \text{K}} = \frac{\text{BTU}}{\text{lb} \cdot {}^\circ\text{F}} \, \frac{1054.8\ \text{J}}{\text{BTU}} \, \frac{2.2046 \times 10^3\ \text{lb}}{\text{g}} \, \frac{1.8^\circ\text{F}}{\text{K}}$$

Another form of the dimensional equation is:

$$\frac{\text{J}}{\text{g} \cdot \text{K}} = \frac{\text{BTU}}{\text{lb} \cdot {}^\circ\text{F}} \, \frac{\text{J}}{9.48 \times 10^{-4}\ \text{BTU}} \, \frac{\text{lb}}{453.6\ \text{g}} \, \frac{1.8^\circ\text{F}}{\text{K}}$$

Canceling out units and carrying out the arithmetic operations:

$$\frac{\text{J}}{\text{g K}} = \frac{\text{BTU}}{\text{lb }{}^\circ\text{F}} \times 4.185$$

EXAMPLE 2: The heat loss through the walls of an electric oven is 6500 BTU/h. If the oven is operated for 2 h, how many kilowatt-hours of electricity will be used just to maintain the oven temperature (heat input = heat loss)?

To solve this problem, rephrase the question. In order to supply 6500 BTU/h for 2 h, how many kilowatt-hours are needed? Note that power is energy/time; therefore, the product of power and time is the amount of energy. Energy in BTU is to be converted to J. The dimensional equation is:

$$\mathrm{J} = \frac{6500\ \mathrm{BTU}}{\mathrm{h}}\ \frac{2\ \mathrm{h}}{}\ \frac{1054.8\ \mathrm{J}}{\mathrm{BTU}}$$

$$\mathrm{W} = \mathrm{J/s};\ \text{therefore, } \mathrm{W \cdot s} = \mathrm{J},\ \mathrm{kW \cdot h} = \frac{\mathrm{W \cdot s}}{}\ \frac{1\ \mathrm{kW}}{1000\ \mathrm{W}}\ \frac{1\ \mathrm{h}}{3600\ \mathrm{s}}$$

$$\mathrm{kW \cdot h} = \frac{6500\ \mathrm{BTU}}{\mathrm{h}}\ \frac{2\ \mathrm{h}}{}\ \frac{1054.8\ \mathrm{J}}{\mathrm{BTU}}\ \frac{\mathrm{kW}}{1000\ \mathrm{W}}\ \frac{\mathrm{h}}{3600}$$

$$\mathrm{kW \cdot h} = 3.809$$

An alternative dimensional equation is:

$$\mathrm{kW \cdot h} = \frac{6500\ \mathrm{BTU}}{\mathrm{h}}\ \frac{2\ \mathrm{h}}{}\ \frac{\mathrm{h}}{60\ \mathrm{min}}\ \frac{1.757 \times 10^{-2}\ \mathrm{kW}}{\mathrm{BTU/min}} = 3.809$$

The conversion factors in Appendix Table A.1 express the factors as a ratio and are most convenient to use in a dimensional equation. Although column I of this table is labeled "Denominator" and columns II and III are labeled "Numerators," the reciprocal may be used. Just make sure that the numerals in column II go with the units in column III.

THE DIMENSIONAL CONSTANT (g_c)

In the American Engineering System of measurement, the units of force and the units of mass are both expressed in pounds. It is necessary to differentiate between the two types of units, since they have different physical significance. To eliminate confusion between the two types of units, the pound force is usually written as lb_f and the pound mass is written as lb_m.

Since the force of gravity on the surface of the earth is the weight of a given mass, in the American Engineering System the weight of 1 lb m is exactly 1 lb_f. Thus, the system makes it easy to conceptualize the magnitude of a 1-lb_f.

The problem with the use of lb_f and lb_m units is that an additional factor is introduced into an equation to make the units dimensionally consistent if units of force and mass are both in the same equation. This factor is the dimensional constant g_c.

The dimensional constant g_c is derived from the basic definition of force: Force = mass × acceleration.

If lb_f is the unit of force and lb_m is the unit of mass, substituting ft/s^2 for acceleration results in a dimensional equation for force that is not dimensionally consistent.

$$lb_f = lb_m \times \frac{ft}{s^2}$$

The American Engineering System of measurement is based on the principle that the weight, or the force of gravity on a pound mass on the surface of the earth, is a pound force. Introducing a dimensional constant, the equation for force becomes:

$$lb_f = lb_m \times \frac{ft}{s^2} \times g_c$$

To make lb_f numerically equal to lb_m when acceleration due to gravity is 32.174 ft/s^2, it is obvious that g_c would have a numerical value of 32.174 and would be a denominator in the force equation. 32.174 is g_c and has the following units:

$$lb_f = lb_m \times \frac{ft}{s^2} \times \frac{1}{g_c}$$

$$g_c = \frac{ft \cdot lb_m}{lb_f \cdot s^2}$$

The dimensional constant, g_c, should not be confused with the acceleration of gravity, g. The two quantities have different units. In SI the unit of force can be expressed in the base units following the relationship between force, mass, and acceleration; therefore, g_c is not needed. Use of the conversion factors in Appendix Table A.1 eliminates the need for g_c in conversion from the American Engineering System of units to SI.

DETERMINATION OF APPROPRIATE SI UNITS

A key feature of SI is expression of any dimension in terms of the base units of meter, kilogram, and second. Some physical quantities having assigned names, as shown in Table 2.2, can also be expressed in terms of the base units when used in a dimensional equation. When properly used, the coherent nature of SI ensures dimensional consistency when all quantities used for substitution into an equation are in SI units. The following examples illustrate selection of appropriate SI units for a given quantity.

EXAMPLE 1: A table for viscosity of water at different temperatures lists viscosity in units of $lb_m/(ft \cdot h)$. Determine the appropriate SI unit and calculate a conversion factor.

The original units have units of mass (lb_m), distance (ft), and time (h). The corresponding SI base units should be $kg/(m \cdot s)$. The dimensional equation for the conversion is:

$$\frac{kg}{m \cdot s} = \frac{lb_m}{ft \cdot h} \times \text{conversion factor} = \frac{lb_m}{ft \cdot h} \frac{1\ kg}{2.2046\ lb_m}$$

$$\frac{1\ m}{3.281\ ft} \frac{1\ h}{3600\ s} \frac{kg}{m \cdot s} = \frac{lb_m}{ft \cdot h} \times 3.84027 \times 10^{-5}$$

Viscosity in SI is also expressed in $Pa \cdot s$. Show that this has the same base units as in the preceding example.

$$Pa \cdot s = \frac{N}{m^2}\ s = \frac{kg \cdot m}{s^2} \frac{s}{m^2} = \frac{kg}{m \cdot s}$$

EXAMPLE 2: Calculate the power available in a fluid which flows down the raceway of a reservoir at a rate of 525 lb_m/min from a height of 12.3 ft. The potential energy (*PE*) is:

$$PE = mgh$$

where m = mass, g = the acceleration due to gravity (32.2 ft/s^2), and h = height.

From Table 2.2, the unit of power in SI is the watt (W), and the formula is J/s. Expressing in base units:

$$P = W = \frac{N \cdot m}{s} = \frac{kg \cdot m}{s^2} \frac{m}{s} = \frac{kg \cdot m^2}{s^3}$$

$$\frac{kg \cdot m^2}{s^3} = \frac{525\ lb_m}{min} \frac{32.2\ ft}{s^2} \frac{12.3\ ft}{} \frac{kg}{2.2046\ lb_m}$$

$$\frac{1\ min}{60\ s} \frac{m^2}{(3.281)^2\ ft^2} = 1.53\ W$$

DIMENSIONAL CONSISTENCY OF EQUATIONS

All equations must have the same units on both sides of the equation. Equations should be tested for dimensional consistency before substitution of values of variables. A dimensional equation would easily verify dimensional consistency. Consistent use of SI for units of variables substituted into an equation ensures dimensional consistency.

EXAMPLE 1: The heat transfer equation expresses the rate of heat transfer (q = energy/time) in terms of the heat transfer coefficient h, the area A, and the temperature difference δT.

Test the equation for dimensional consistency and determine the units of h.

$$q = hA\Delta T$$

The dimensional equation will be set up using SI units of the variables:

$$\frac{\mathrm{J}}{\mathrm{s}} = \mathrm{W} = (\text{--})(\mathrm{m}^2)(\mathrm{K})$$

The equation will be dimensionally consistent if h has the units $\mathrm{W/m^2 \cdot K}$. $\mathrm{m^2}$ and K will cancel out on the right-hand side of the equation, leaving W on both sides, signifying dimensional consistency.

EXAMPLE 2: The Van der Waal's equation of state is as follows:

$$\left(P + \frac{n^2 a}{V^2}\right)(V - nb) = nRT$$

where P = pressure, n = moles, V = volume, R = gas constant, and T = temperature. Determine the units of the constants a, b, and R and test for dimensional consistency.

Since the two terms on the left-hand side of the equation involve subtraction, both terms in the first set of parentheses will have units of pressure and both terms in the second set of parentheses will have units of volume. The product of pressure times volume will be the same units on the right-hand side of the equation.

Using SI units for pressure of $\mathrm{N/m^2}$ and volume, $\mathrm{m^3}$, the dimensional equation is:

$$\frac{\mathrm{N}}{\mathrm{m}^2}(\mathrm{m}^3) = (\mathrm{kgmole})(\text{---})(^\circ\mathrm{K})$$

The equation will be dimensionally consistent if R has units of N · m/kgmole · °K, since that will give the same units, Nm, on both sides of the equation. The units of a and b are determined as follows: The first term, which contains a, has units of pressure:

$$\frac{\text{N}}{\text{m}^2} = \frac{(\text{kgmole})^2(\text{----})}{(\text{m}^3)^2}$$

To make the right-hand side have the same units as the left, a will have units of $\text{N}(\text{m})^4/(\text{kgmole})^2$. The second term, which contains b, has units of volume.

$$\text{m}^3 = (\text{kgmole})(\text{----})$$

Thus, b will have units of m^3/kgmole.

CONVERSION OF DIMENSIONAL EQUATIONS

Equations may be dimensionless, i.e., dimensionless groups are used in the equation. Examples of dimensionless groups are the Reynolds number (Re), Nusselt number (Nu), Prandtl number (Pr), Fourier number (Fo), and Biot number (Bi). These numbers are defined below, and the dimensional equations for their units show that each group is dimensionless. The variables are V = velocity, D = diameter, ρ = density, μ = viscosity, k = thermal conductivity, h = heat transfer coefficient, α = thermal diffusivity, C_p = specific heat, L = thickness, t = time.

$$\text{Re} = \frac{DV\rho}{\mu} = \text{m}\,\frac{\text{m}}{\text{s}}\,\frac{\text{kg}}{\text{m}^3}\,\frac{1}{\text{kg}/(\text{m}\cdot\text{s})}$$

$$\text{Nu} = \frac{hD}{k} = \frac{\text{W}}{\text{m}^2\cdot\text{K}}\,\text{m}\,\frac{1}{\text{W}/(\text{m}\cdot\text{K})}$$

$$\text{Pr} = \frac{C_p\mu}{k} = \frac{\text{J}}{\text{kg}\cdot\text{K}}\,\frac{\text{kg}}{\text{m}\cdot\text{s}}\,\frac{1}{\text{J}/(\text{s}\cdot\text{m}\cdot\text{K})}$$

$$\text{Fo} = \frac{\alpha t}{L^2} = \frac{\text{m}^2}{\text{s}}\,\text{s}\,\frac{1}{\text{m}^2}$$

$$\text{Bi} = \frac{hL}{k} = \frac{\text{W}}{\text{m}^2\cdot\text{K}}\,\text{m}\,\frac{1}{\text{W}/(\text{m}\cdot\text{K})}$$

An example of a dimensionless equation is the Dittus-Boelter equation for heat transfer coefficients in fluids flowing inside tubes:

$$\mathrm{Nu} = 0.023\,(\mathrm{Re})^{0.8}\,(\mathrm{Pr})^{0.3}$$

Since all terms are dimensionless, there is no need to change the equation, regardless of what system of units is used in determining the values of the dimensionless groups.

Dimensional equations result from empirical correlations, e.g., statistical analysis of experimental data. With dimensional equations, the units must correspond to those used on the original data and substitution of a different system of units will require a transformation of the equation.

When converting dimensional equations, the following rules are useful:

1. When variables appear in the exponent, the whole exponent must be dimensionless; otherwise, it will not be possible to achieve dimensional consistency for the equation.
2. When variables occur in arguments of logarithmic functions, the whole argument must be dimensionless.
3. Constants in an equation may not be dimensionless. The units of these constants provide for dimensional consistency.

EXAMPLE 1: The number of surviving microorganisms in a sterilization experiment is linear in a semilogarithmic graph; therefore, the equation assumes the form:

$$\log N = -at + b$$

where N = number of survivors and t = time. a and b are constants. The equation will be dimensionally consistent if b is expressed as log N at time 0, designated log N_0, and a will have units of reciprocal time. Thus, the correct form of the equation which is dimensionally consistent is:

$$\log \frac{N}{N_0} = -at$$

In this equation, both sides are dimensionless.

EXAMPLE 2: An equation for a heat transfer coefficient between air flowing through a bed of solids and the solids is:

$$h = 0.0128\, G^{0.8}$$

where G = mass flux of air in lb/(ft^2 · h) and h = heat transfer coefficient in BTU/(h · ft^2 · F). Derive an equivalent equation in SI.

The dimensional equation is:

$$\frac{\text{BTU}}{h \cdot \text{ft}^2 \cdot \text{F}} = (\text{---})\left(\frac{\text{lb}}{\text{ft}^2 \cdot h}\right)^{0.8}$$

An equation must be dimensionally consistent; therefore, the constant 0.0128 in the above equation must have units of

$$\frac{\text{BTU}}{\text{h} \cdot \text{ft}^2 \cdot \text{F}}\left(\frac{\text{ft}^{1.6}\text{h}^{0.8}}{\text{lb}^{0.8}}\right) = \frac{\text{BTU}}{h^{0.2}\text{ft}^{0.4}\text{lb}^{0.8}\text{F}}$$

Converting the equation involves conversion of the coefficient. The equivalent SI unit for h = W/(m^2 · K) and G = kg/(m^2 · s). Thus, the coefficient will have units of J/(s$^{0.2}$m$^{0.4}$kg$^{0.8}$K). The dimensional equation for the conversion is:

$$\frac{\text{J}}{\text{s}^{0.2}\text{m}^{0.4}\text{kg}^{0.8}\text{K}} = \frac{0.0128\ \text{BTU}}{\text{h}^{0.2}\text{ft}^{0.4}\text{lb}^{0.8}\text{F}}\ \frac{1054.8\ \text{J}}{\text{BTU}}\ \frac{\text{ft}^{0.4}}{(0.3048)^{0.4}\text{m}^{0.4}}$$

$$\frac{\text{h}^{0.2}}{(3600)^{0.2}\text{s}^{0.2}}\ \frac{(2.2048)^{0.8}\ \text{lb}^{0.8}}{\text{kg}^{0.8}}\ \frac{1.8\ \text{F}}{\text{K}}$$

$$= 14.305$$

The converted equation is:

$$h = 14.305\ G^{0.8}$$

where both h and G are in SI units. To check: If G = 100 lb/(ft^2 · h) and $h = 0.0128(100)^{0.8}$, h = 0.5096 BTU/(h · ft^2 · °F). The equivalent SI value is:

$$\frac{0.5096\ \text{BTU}}{\text{h} \cdot \text{ft}^2\text{F}}\ \frac{5.678263\ \text{W}/(\text{m}^2 \cdot \text{K})}{\text{BTU}/(\text{h} \cdot \text{ft}^2 \cdot {}^\circ\text{F})} = 2.893\ \frac{\text{W}}{(\text{m}^2 \cdot \text{K})}$$

In SI, G = [100 lb/(ft^2 · h)] [ft^2/(0.3048)^{2}m^2] (h/3600 s) (kg/2.2046 lb) = 0.1356. Using the converted equation, $h = 14.305\ (0.1356)^{0.8}$ = 2.893 W/(m^2 · K).

PROBLEMS

1. Set up dimensional equations and determine the appropriate conversion factor to use in each of the following:

$$\frac{\text{lb}}{\text{ft}^3} = \frac{\text{lb}}{\text{gal}} \times \text{conversion factor}$$

$$\frac{\text{lb}}{\text{in}^2} = \frac{\text{lb}}{\text{ft}^2} \times \text{conversion factor}$$

$$\text{W} = \frac{\text{cal}}{\text{s}} \times \text{conversion factor}$$

2. The amount of heat required to change the temperature of a material from T_1 to T_2 is given by:

$$q = mC_p(T_2 - T_1)$$

where q = BTU, m = mass of material in lb, C_p = specific heat of material in BTU/lb · °F, and T_1 and T_2 = initial and final temperatures in °F.
   (a) How many BTUs of heat are required to cook a roast weighing 10 lb from 40° to 130°F? C_p = 0.8 BTU/(lb · °F).
   (b) Convert the number of BTUs of heat in (a) into watt-hours.
   (c) If this roast is heated in a microwave oven having an output of 200 watts, how long will it take to cook the roast?
3. How many kilowatt-hours of electricity will be required to heat 100 gal of water (8.33 lb/gal) from 60° to 100°F? C_p of water is 1 BTU/(lb · °F).
4. Calculate the power requirements for an electric heater necessary to heat 10 gal of water from 70° to 212°F in 10 min. Express this in J/min and in watts. Use the following conversion factors in your calculations:

   Specific heat of water = 1 BTU/(lb · °F)
   3.414 BTU/(W · h)
   60 min/h
   8.33 lb water/gal
   1.054×10^3 J/BTU

5. One ton of refrigeration is defined as the rate of heat withdrawal from a system necessary to freeze 1 t (2000 lb) of water at 32°F in 24 h. Express this in watts and in BTU/h. Heat of fusion of water = 80 cal/g.
6. (a) In the equation $T = \mu(\alpha)$, what would be the units of T in the equation if μ is expressed in dyne · s/cm^2 and α is in sec^{-1}?
   (b) If μ is to be expressed in lb$_m$/(ft · sec), T in lb$_f$/ft^2, and α in sec^{-1}, what is needed in the equation to make it dimensionally consistent?
7. In the equation

$$\overline{V} = \frac{1000(\rho_1 - \rho_2)}{m\rho_1\rho_2} + \frac{M}{\rho_2}$$

what units should be used for the density, ρ, such that $\overline{V}$ would have the units ml/mole?

m = moles/1000 g
M = g/mole

8. Express the following in SI units. Follow the rounding-off rule in your answer.
   (a) The pressure at the base of a column of fluid 8.325 in. high when the acceleration due to gravity is 32.2 ft/s^2 and the fluid density is 1.013 g/cm. P = density × height × acceleration due to gravity
   (b) The compressive stress (same units as pressure) on a specimen having a diameter of 0.525 in. when the applied force is 5.62 lb_f. Stress = force/area.
   (c) The force needed to restrain a piston having a diameter of 2.532 in. when a pressure of 1500 (exact) lb_f/in^2 is in the cylinder behind the piston. Force = pressure × area
9. An empirical equation for heat transfer coefficient in a heat exchanger is:

$$h = a(V)^{0.8}(1 + 0.011T)$$

where h = BTU/(h · ft^2 · °F), V = ft/s, and T = °F. In one experimental system, a had a value of 150. What would be the form of the equation and the value of a if h, V, and T are in SI units?
10. A correlation equation for the density of a liquid as a function of temperature and pressure is:

$$d = (1.096 \times 0.0086T)(e)^{0.000953P}$$

where d is density in g/cm^3, T = temperature in °K, P = pressure in atm, and e is the base of natural logarithms. A normal atmosphere is 101.3 kPa. Determine the form of the equation if all variables are to be expressed in SI.

SUGGESTED READING

American National Standards Institute. 1976. *American National Standard, Metric Practice*. American National Standards Institute IIEE Standard 268-1976. Institute of Electrical and Electronics Engineers, New York.

American Society of Agricultural Engineers. 1978. Use of customary and SI (metric) units. *American Society for Agricultural Engineers Yearbook—1978*. ASAE, St. Joseph, Mich.

Benson, S. W. 1971. *Chemical Calculations*, 3rd ed. John Wiley & Sons, New York.

Watson, E. L., and Harper, J. C. 1988. *Elements of Food Engineering*, 2nd ed. Van Nostrand Reinhold Co., New York.

Himmelblau, D. M. 1967. *Basic Principles and Calculations in Chemical Engineering*, 2nd ed. Prentice-Hall, Englewood Cliffs, N.J.

Kelly, F. H. C. 1963. *Practical Mathematics for Chemists*. Butterworths, London.

McCabe, W. L., Smith, J. C., and Harriott, P. 1985. *Unit Operations of Chemical Engineering*, 4th ed. McGraw-Hill Book Co., New York.

Obert, E., and Young, R. L. 1962. *Elements of Thermodynamics and Heat Transfer*. McGraw-Hill Book Co., New York.

3

Material Balances

Material balance calculations are employed in tracing the inflow and outflow of material in a process, and thus establish the quantities of various materials in each process stream. The procedure is useful in making formulations, evaluating final compositions after blending, evaluating yields, and evaluating separation efficiencies in mechanical separation systems.

BASIC PRINCIPLES

Law of Conservation of Mass. Material balances are based on the principles that matter is neither created nor destroyed. Thus, in any process, a mass balance can be made as follows:

$$\text{Inflow} = \text{outflow} + \text{accumulation}$$

Inflow may include formation of material by chemical reaction or microbial growth processes, and outflow may include material depletion by chemical or biological reactions.

If accumulation is 0, inflow = outflow and the process is at steady state. If the accumulation term is not 0, then the quantity and concentration of components in the system could change with time and the process is at unsteady state.

Process Flow Diagrams. Before formulating a material balance equation, visualize the process and determine the boundaries of the system for which the material balance is to be made. It is essential that everything about the process that affects the distribution of components be known. The problem statement should be adequate to enable the reader to draw a flow diagram. However, in some cases, certain physical principles which could affect the distribution of components in the system may not be stated in the problem. It is essential that a student remembers the physical principles that apply in the various processes presented in the example problems. Knowing these principles not only allows the student to solve similar material balance problems but also provides a reservoir of information that may be used later as a basis for the design of a new process or for evaluation of parameters affecting the efficiency of a process.

As an example of these principles, consider a problem in crystallization. The problem may be simply stated: Determine the amount of sugar (water-free basis) that can be produced from 100 kg of sugar solution that contains 20% by weight of sugar and 1% of a water-soluble, uncrystallizable impurity. The solution is concentrated to 75% sugar, cooled to 20°C, and centrifuged, and the crystals are dried.

The problem statement is adequate to draw a process flow diagram. This is shown in Fig. 3.1. However, this flow diagram does not give a complete picture of where various streams separate and leave the system.

Figure 3.2 shows the same process flow diagram, but after taking into consideration the characteristics of the various steps in the process, additional streams leaving the system are drawn. To concentrate a 20% solution to 75% requires the removal of water. Thus, water leaves the system at the evaporator. The process of cooling does not alter the mass; therefore, the same process stream enters and leaves the crystallizer. Centrifugation separates most of the liquid phase from the solid phase, and the crystals, the solid phase containing a small amount of retained solution, enter the drier. A liquid phase leaves the system at the centrifuge. Water leaves the system at the drier.

There are three physical principles involved in this problem that are not stated. (1) Crystals will crystallize out of a solution when the solute concentration exceeds the saturation concentration. The solute concentration in the liquid phase is forced towards the saturation concentration as crystals are formed. Given enough time to reach equilibrium, the liquid phase that leaves the system at the centrifuge is a saturated sugar solution. (2) The crystals consist of pure solute, and the only impurities are those adhering to the crystals from the solution. (3) It is not possible to completely eliminate the liquid from the solid phase by centrifugation. The amount of impurities that will be retained with the sugar crystals depends upon how efficiently the centrifuge separates the solid from the liquid phase. This principle of solids purity being dependent upon the degree of separation of the solid from the liquid phase applies not only in crystallization but also in solvent extraction.

In order to solve this problem, the saturation concentration of sugar in water at 20°C and the water content of the crystal fraction after centrifugation must be known.

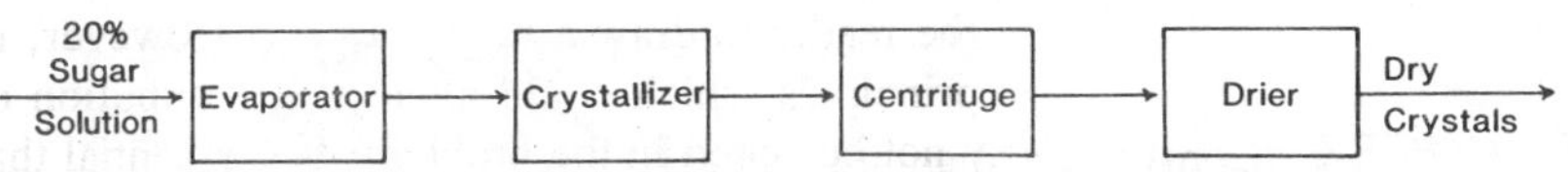

Fig. 3.1. Process flow diagram for a crystallization problem.

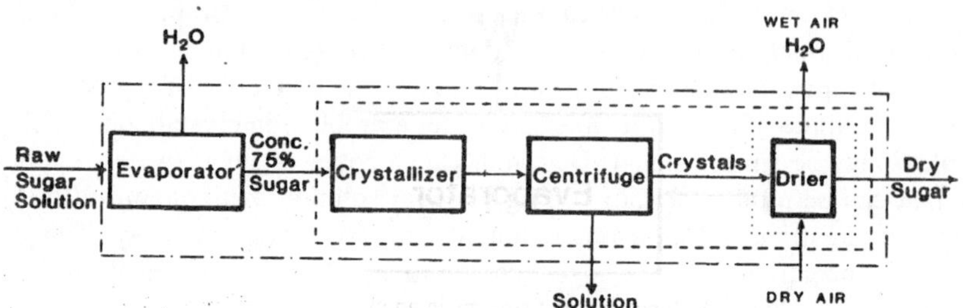

Fig. 3.2. Process flow diagram for a crystallization problem showing input and exit streams and boundaries enclosing sub-systems for analyzing sections of the process.

System Boundaries. Figure 3.2 shows how the boundaries of the system can be moved to facilitate solution of the problem. If the boundary completely encloses the whole process, one stream will be entering and four streams leaving the system. The boundary can also be set just around the evaporator, in which case there will be one stream entering and two streams leaving. The boundary can also be set around the centrifuge or around the drier. A material balance can be carried out around any of these subsystems or around the whole system. The material balance equation may be a total mass balance or a component balance.

Total Mass Balance. The equation in the section "Law of Conservation of Mass," when used on the total weight of each stream entering or leaving a system represents a total mass balance. The following examples illustrate how total mass balance equations are formulated for systems and subsystems.

EXAMPLE 1: In an evaporator, dilute material enters and concentrated material leaves the system. Water is evaporated during the process. If I is the weight of the dilute material entering the system, W is the weight of water vaporized, and C is the weight of the concentrate, write an equation that represents the total mass balance for the system. Assume that a steady state exists.

Solution: The problem statement describes a system depicted in Fig. 3.3. The total mass balance is:

$$\text{Inflow} = \text{outflow} + \text{accumulation}$$

$$I = W + C \text{ (Accumulation is 0 in a steady-state system)}$$

EXAMPLE 2: Construct a diagram and set up a total mass balance for a dehydrator. Air enters at the rate of A lb/min, and wet material enters at W lb/min. Dry material leaves the system at D lb/min. Assume a steady state.

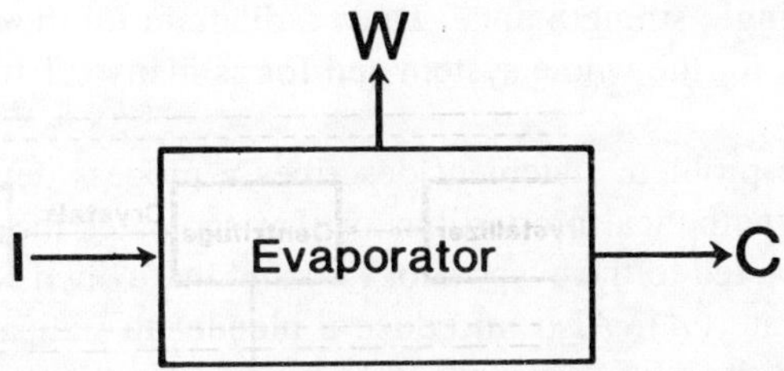

Fig. 3.3. Input and exit streams in an evaporation process.

Solution: The problem statement describes a system (dehydrator) in which air and wet material enter and dry material leaves. Obviously, air and water must also leave the system. A characteristic of a dehydrator not written into the problem statement is that the water removed from the solids is transferred to air and leaves the system with the air stream. Figure 3.4 shows the dehydrator system and its boundaries. Also shown are two separate subsystems—one for the solids and the other for air—with their corresponding boundaries. Considering the whole dehydrator system, the total mass balance is:

$$W + A = \text{wet air} + D$$

Considering the air subsystem:

$$A + \text{water} = \text{wet air}$$

The mass balance for the solids subsystem is:

$$W = \text{water} + D$$

EXAMPLE 3: Orange juice concentrate is made by concentrating single-strength juice to 65% solids followed by dilution of the concentration to 45%

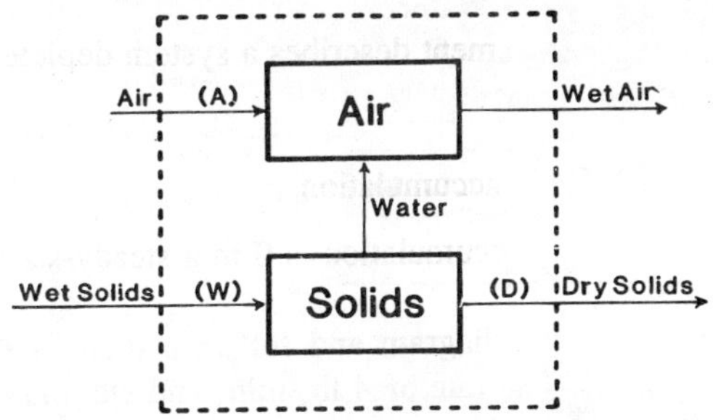

Fig. 3.4. Diagram of material flow in a dehydration process.

solids using single-strength juice. Draw a diagram for the system, and set up mass balances for the whole system and for as many subsystems as possible.

Solution: The problem statement describes a process depicted in Fig. 3.5. Consider a hypothetical proportionator that separates the original juice (S) to that which is fed to the evaporator (F) and that which is used to dilute the 65% concentrate (A). Also introduce a blender to indicate that part of the process where the 65% concentrate (C_{65}) and the single-strength juice are mixed to produce the 45% concentrate (C_{45}). The material balance equations for the whole system and the various subsystems are:

$$\text{Overall: } S = W + C_{45}$$

$$\text{Proportionator: } S = F + A$$

$$\text{Evaporator: } F = W + C_{65}$$

$$\text{Blender: } C_{65} + A = C_{45}$$

Component Mass Balance. The same principles apply as in the total mass balance, except that the components are considered individually. If there are n components, n independent equations can be formulated—one equation for total mass balance and $n - 1$ component balance equations.

Since the object of a material balance problem is to identify the weight and composition of various streams entering and leaving a system, it is often necessary to establish several equations and solve these equations simultaneously to evaluate the unknowns.

It is helpful to include the known quantities of process streams and concentrations of components in the process diagram so that all streams where a component may be present can be easily accounted for. In a material balance, use mass units and concentration in mass fraction or mass percentage. If the quantities are expressed in volume units, convert them to mass units using density.

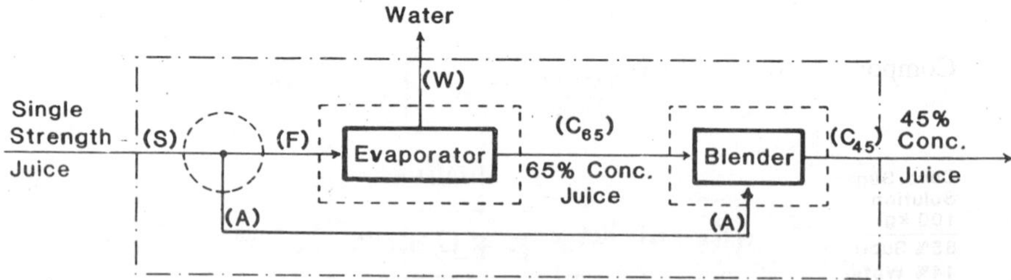

Fig. 3.5. Diagram of an orange juice concentrate process involving evaporation and blending of concentrate with freshly squeezed juice.

A form of a component balance equation that is particularly useful in problems involving concentration or dilution is the expression for the mass fraction or weight percentage.

$$\text{Mass fraction } A = \frac{\text{mass of component } A}{\text{total mass of mixture containing } A}$$

Rearrange the equation:

$$\text{Total mass of mixture containing } A = \frac{\text{mass of component } A}{\text{mass fraction of } A}$$

Thus, if the weight of component A in a mixture is known, and its mass fraction in that mixture is known, the mass of the mixture can be easily calculated.

EXAMPLE 1: Draw a diagram and set up a total mass and component balance equation for a crystallizer where 100 kg of a concentrated sugar solution containing 85% sucrose and 1% inert, water-soluble impurities (balance, water) enters. Upon cooling, the sugar crystallizes from solution. A centrifuge then separates the crystals from a liquid fraction called the *mother liquor*. The crystal slurry fraction has, for 20% of its weight, a liquid with the same composition as the mother liquor. The mother liquor contains 60% sucrose by weight.

Solution: The diagram for the process is shown in Fig. 3.6.

Based on a system boundary enclosing the whole process of crystallization and centrifugation, the material balance equations are:
Total mass balance:

$$S = C + M$$

Component balance of sucrose:

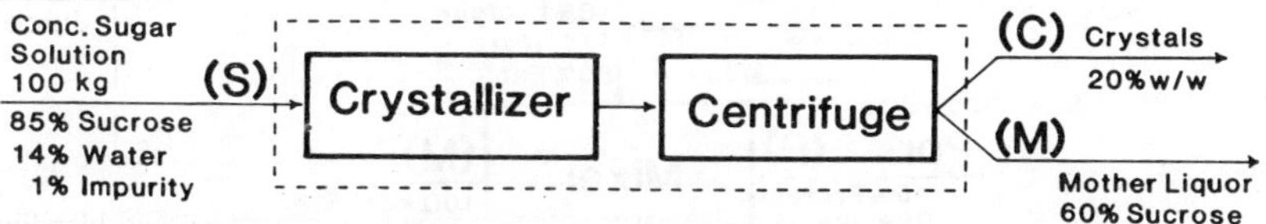

Fig. 3.6. Diagram showing composition and material flow in a crystallization process.

$$S(0.85) = M(0.6) + C(0.2)(0.6) + C(0.8)$$

Sucrose in inlet stream	=	sucrose in mother liquor	+	sucrose in mother liquor carried by crystals	+	sucrose in crystals

Component balance on water:
Let x = mass fraction of impurity in the mother liquor

$$S(0.14) = M(0.4 - x) + C(0.2)(0.4 - x)$$

Water in inlet stream	=	water in mother liquor	+	water in mother liquor adhering to the crystals

Component balance on impurity:

$$S(0.01) = M(x) + C(0.2)(x)$$

Note that a total of four equations can be formulated, but there are only three unknown quantities (C, M, and x). One of the equations is redundant.

EXAMPLE 2: Draw a diagram and set up equations representing the total mass balance and the component mass balance for a system involving the mixing of pork (15% protein, 20% fat, and 63% water) and backfat (15% water, 80% fat, and 30% protein) to make 100 kg of a mixture containing 25% fat.

Solution: The diagram is shown in Fig. 3.7.
Total mass balance:

$$P + B = 100$$

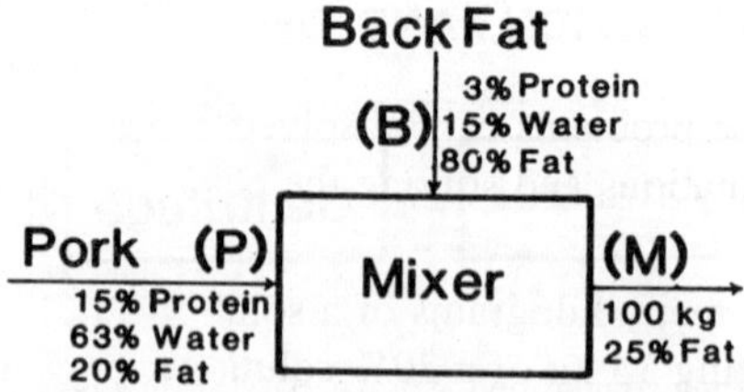

Fig. 3.7. Composition and material flow in a blending process.

Fat balance:

$$0.2P + 0.8B = 0.25(100)$$

These two equations are solved simultaneously by substituting $P = 100 - B$ into the second equation.

$$0.2(100 - B) + 0.8B = 25$$

$$B = \frac{25 - 20}{0.8 - 0.2} = 8.33 \text{ kg}$$

$$P = 100 - 8.33 = 91.67 \text{ kg}$$

Basis and Tie Material. A *tie material* is a component used to relate the quantity of one process stream to the quantity of another. It is usually the component that does not change during a process. Examples of tie material are solids in dehydration or evaporation processes and nitrogen in combustion processes. Although it is not essential that these tie materials be identified, the calculations are often simplified if they are identified and included in one of the component balance equations. This is illustrated in Example 1 of the section "steady state," where the problem is solved readily using a component mass balance in the solid (the tie material in this system) compared to when the mass balance was made on water. In a number of cases, the tie material need not be identified, as illustrated in examples in the section "Blending of Food Ingredients."

A *basis* is useful in problems where no initial quantities are given and the answer required is a ratio or a percentage. It is also useful in continuous flow systems. Material balance in a continuous flow system is obtained by assuming as a basis a fixed time of operation. A material balance problem can be solved on any assumed basis. After all the quantities of process streams are identified, the specific quantity asked for in the problem can be found using ratio and proportion. It is possible to change the basis when considering each subsystem within a defined boundary inside the total system.

MATERIAL BALANCE PROBLEMS INVOLVED IN DILUTION, CONCENTRATION, AND DEHYDRATION

Steady State. These problems can be solved by formulating total mass and component balance equations and solving the equations simultaneously..

EXAMPLE 1: How many kilograms of a solution containing 10% NaCl can be obtained by diluting 15 kg of a 20% solution with water?

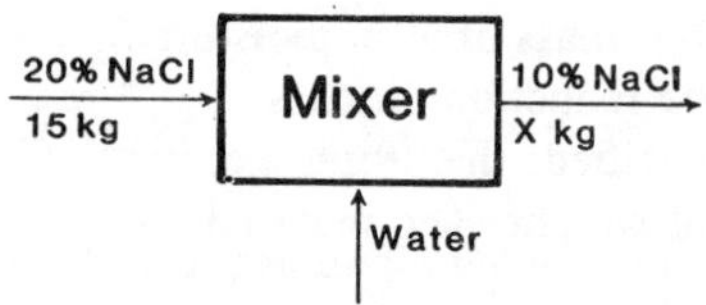

Fig. 3.8. Composition and material flow in a dilution process.

Solution: The process diagram in Fig. 3.8 shows that all NaCl enters the mixer with the 20% NaCl solution and leaves in the diluted solution. Let x = kg 10% NaCl solution and y = kg water. The material balance equations are:

$$\text{Total mass: } 15 = x - y$$

$$\text{Component: } 15(0.20) = x(0.10); \qquad x = \frac{3}{0.1} = 30 \text{ kg}$$

The total mass balance equation is redundant, since the component balance equation alone can be used to solve the problem.

The mass fraction equation can also be used in this problem. Here 15 kg of a 20% NaCl solution contains 3 kg NaCl. Dilution would not change the quantity of NaCl, so that 3 kg of NaCl is in the diluted mixture. The diluted mixture contains 10% NaCl; therefore:

$$x = \frac{3 \text{ kg NaCl}}{\text{mass fraction NaCl}} = \frac{3}{0.1} = 30 \text{ kg}$$

EXAMPLE 2: How much weight reduction would result when a material is dried from 80% moisture to 50% moisture?

Solution: The process diagram is shown in Fig. 3.9. Dehydration involves removal of water; the mass of solids remains constant. There are two components, solids and water, and a decrease in the concentration of water, indicating a loss, will increase the solids concentration. Let W = mass of 80%

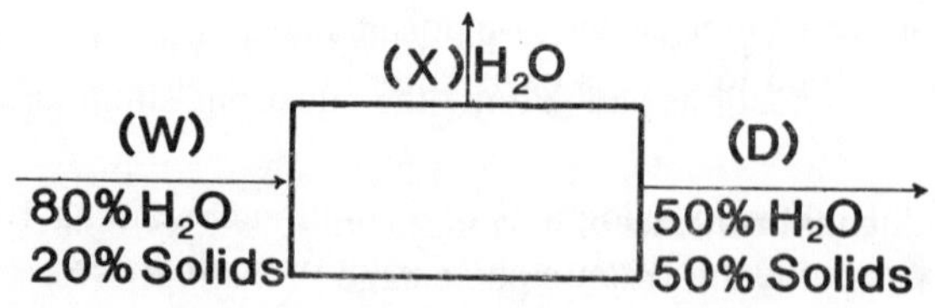

Fig. 3.9. Composition and material flow in a dehydration process.

moisture material, D = mass of 50% moisture material, and W = mass water lost, which is also the reduction in mass.

No weights are specified; therefore, express the reduction in weight as a ratio of final to initial weight. The material balance equations are:

$$\text{Total mass: } W = X + D$$

$$\text{Water: } 0.8W = 0.5D + X$$

Solving simultaneously:

$$W = X + D; \qquad X = W - D$$

$$0.8W = 0.5D + (W - D)$$

$$0.5D = 0.2W; \qquad \frac{D}{W} = \frac{0.2}{0.5} = 0.4$$

$$\% \text{ wt. reduction} = \frac{W - D}{W}(100) = \left(1 - \frac{D}{W}\right)100$$

$$= (1 - 0.4)(100) = 60\%$$

The problem can also be solved by using as a basis 100 kg of 80% moisture material. Let D = 100.

$$\text{Total mass: } 100 = X + D$$

$$\text{Water: } 0.8(100) = X + 0.5D$$

Solving simultaneously, by subtracting one equation from the other: $100 - 80 = D(1 - 0.5)$; $D = 20/0.5 = 40$ kg.

$$\% \text{ wt. reduction} = \frac{100 - 40}{100}(100) = 60\%$$

Volume Changes on Mixing. When two liquids are mixed, the volumes are not always additive. This is true with most solutions and miscible liquids. NaCl solution, sugar solutions, and ethanol solution all exhibit volume changes on mixing. Because of volume changes, material balances must be based on the mass rather than the volume of the components. Concentrations on a volume basis must be converted to a mass basis before the material balance equations are formulated.

EXAMPLE: Alcohol contents of beverages are reported as percentage by volume. A *proof* is twice the volume percentage of alcohol. The density of absolute ethanol is 0.7893 g/cm^3. The density of a solution containing 60% by weight of ethanol is 0.8911 g/cm^3. Calculate the volume of absolute ethanol which must be diluted with water to produce 1 L of 60% by weight ethanol solution. Calculate the proof of a 60% ethanol solution.

Solution: Use as a basis 1 L of 60% w/w ethanol. Let X = volume of absolute ethanol in liters. The component balance equation on ethanol is:

$$X(1000)(0.7983) = 1(1000)(0.8911)$$

$$X = 1.1162 \text{ L}$$

This is a dilution problem, and a component balance on ethanol was adequate to solve it. Note that a mass balance was made using the densities given. There is obviously a volume loss on mixing, since more than 1 L of absolute ethanol is needed to make 1 L to 60% w/w ethanol solution.

To calculate the proof of 60% w/w ethanol use as a basis 100 g of solution.

$$\text{Volume of solution} = \frac{100}{0.8911} = 112.22 \text{ cm}^3$$

$$\text{Volume of ethanol} = \frac{100(0.6)}{0.7893} = 76.016 \text{ cm}^3$$

$$\text{Volume percent} = \left(\frac{76.016}{112.22}\right)(100) = 67.74\%$$

$$\text{Proof} = 2(\text{volume }\%) = 135.5 \text{ proof}$$

Continuous vs. Batch. Material balance calculations are the same regardless of whether a batch or a continuous process is being evaluated. In a batch system, the total mass considered includes what entered or left the system at one time. In a continuous system, a basis of a unit time of operation may be used, and the material balance will be made on what entered or left the system during that period of time. The previous examples were batch operations. If the process is continuous, the quantities given will all be mass/time, e.g., kg/h. If the basis used is 1 h of operation, the problem is reduced to the same form as a batch process.

EXAMPLE 1: An evaporator has a rated evaporation capacity of 500 kg/h of water. Calculate the rate of production of juice concentrate containing 45% total solids from raw juice containing 12% solids.

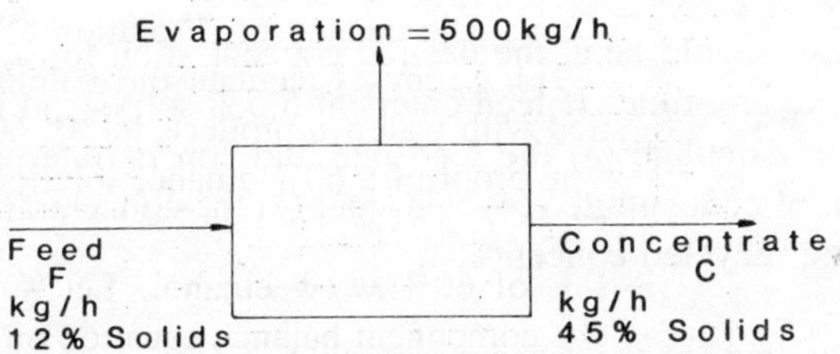

Fig. 3.10. Flow diagram of an evaporation process.

Solution: The diagram of the process is shown in Fig. 3.10. Use as a basis 1 h of operation. Here 500 kg of water leaves the system. A component balance on solids and a total mass balance will be needed to solve the problem. Let F = the feed, 12% solids juices, and C = concentrate containing 45% solids. The material balance equations are:

$$\text{Total mass: } F = C + 500$$

$$\text{Solids: } 0.12F = 0.45C; \qquad F = \frac{0.45C}{0.12} = 3.75C$$

Substituting and solving for C: $3.75C = C + 500$.

$$C = \frac{500}{3.75 - 1} = 181.8 \text{ kg}$$

Since the basis is 1 h of operation, the answer will be: Rate of production of concentration = 181.8 kg/h.

Recycle. Recycle is evaluated as in the previous examples, but the boundaries of the subsystems analyzed are moved around to isolate the process streams being evaluated. A system is defined which has a boundary surrounding the total system, which reduces the problem to a simple material balance problem without recycle.

EXAMPLE: A pilot plant model of a falling film evaporator has an evaporation capacity of 10 kg/h of water. The system consists of a heater through which the fluid flows down in a thin film, and the heated fluid discharges into a collecting vessel maintained under a vacuum in which flash evaporation reduces the temperature of the heated fluid to the boiling point. In a continuous operation, a recirculating pump draws part of the concentrate from the reservoir, mixes this concentrate with feed, and pumps the mixture through the heater. The recirculating pump moves 20 kg/h of fluid. The fluid in the collecting vessel should be at the desired concentration for withdrawal from the evaporator at any time. If feed enters at 5.5% solids and a 25% concentrate is desired, calculate (a) the feed rate and concentrate production rate, (b) the amount of concentrate recycled, and (c) the concentration of the mixture of feed and recycled concentrate.

Solution: A diagram of the system is shown in Fig. 3.11. The basis is 1 h of operation.

A mass and solids balance over the whole system will establish the quantity of feed and concentrate produced per hour.

$$\text{Total mass: } F = C + V; \qquad F = C + 10$$

$$\text{Solids: } F(0.055) = C(0.25); \qquad F = C\left(\frac{0.25}{0.055}\right) = 4.545C$$

Substituting F:

$$4.545C = C + 10; \; C = \frac{10}{4.545 - 1} = 2.82 \text{ kg}$$

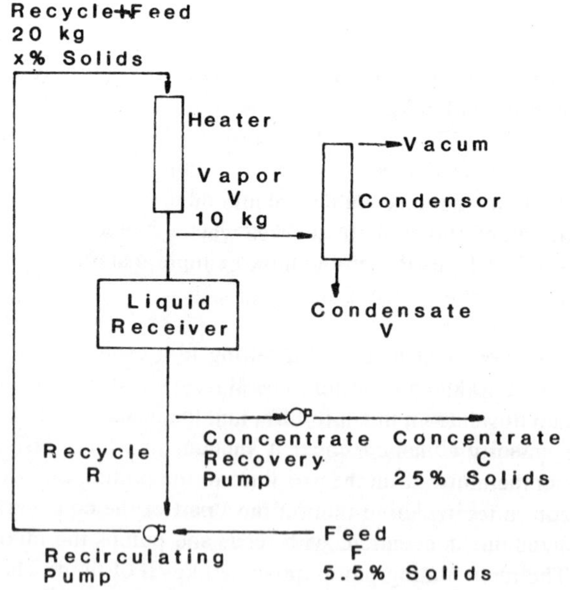

Fig. 3.11. Diagram of material flow in a falling film evaporator with product recycle.

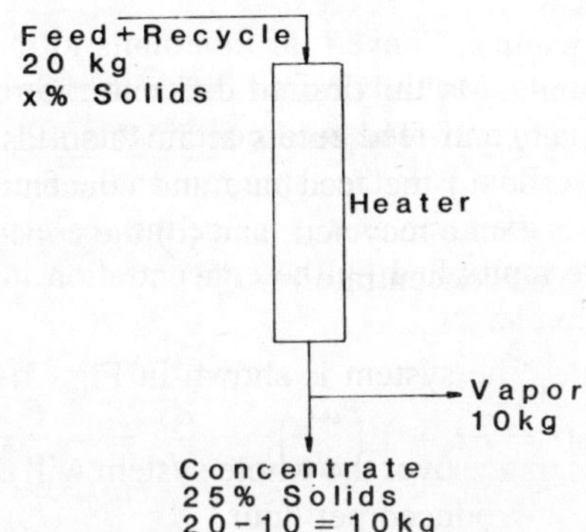

Fig. 3.12. Diagram of material balance around the heater of a falling film evaporator.

(a) Solving for F: $F = 4.545(2.82) = 12.82$ kg/h. The concentrate production rate is 2.82 kg/h.
(b) Material balance around the recirculating pump: $R + F = 20$; $R = 20 - 12.82 = 7.18$ kg. Recycle rate = 7.18 kg/h.
(c) A material balance can be made around the part of the system where the vapor separates from the heated fluid as shown in Fig. 3.12.

Solids balance: $20(x) = 10(0.25)$; $x = 2.5/20 = 0.125$. The fluid entering the heater contains 12.5% solids.

Unsteady State. Unsteady-state material balance equations involve an accumulation term in the equation. Accumulation is expressed as a differential term of the rate of change of a variable with respect to time. A mass balance is made in the same manner as in steady state problems. Since a differential term is involved, the differential equation must be integrated to obtain an equation for the value of the dependent variable as a function of time.

EXAMPLE 1: A stirred tank with a volume of 10 L contains a salt solution at a concentration of 100 g/L. If salt-free water is continuously fed into this tank at a rate of 12 L/h, and the volume is maintained constant by continuously overflowing the excess fluid, what will be the concentration of salt after 90 min?

Solution: This is similar to a dilution problem, except that the continuous overflow removes salt from the tank, thus reducing not only the concentration but also the quantity of salt present. A mass balance will be made on the mass of salt in the tank and in the streams entering and leaving the system.

Let x = the concentration of salt in the vessel at any time. Representing time by the symbol t, the accumulation term will be dx/dt, which will have units of g salt/(L · min). Convert all time units to minutes in order to be consistent with the units. Multiplying the differential term by the volume will result in units of g salt/min. The feed contains no salt; therefore, the input term is zero. The overflow is at the same rate as the feed, and the concentration in the overflow is the same as it is inside the vessel; therefore, the output term is the feed rate multiplied by the concentration inside the vessel. The mass balance equation is:

$$0 = Fx + V\left(\frac{dx}{dt}\right); \qquad \frac{dx}{dt} = -\frac{F}{V}x$$

The negative sign indicates that x will be decreasing with time. Separating variables:

$$\frac{dx}{x} = -\frac{F}{V}dt$$

Integrating:

$$\ln x = -\left(\frac{F}{V}\right)t + C$$

The constant of integration is obtained by substituting $x = 100$ at $t = 0$; $C = \ln(100)$; $F = \text{L/h} = 0.2\ \text{L/min}$; $V = 10$ L.

$$\ln\left(\frac{x}{100}\right) = -\frac{0.2\,t}{10}$$

At 90 min, $\ln(0.01\,x) = -1.800$.

$$x = 100(e^{-1.80}) = 16.53\ \text{g/L}$$

EXAMPLE 2: The generation time is the time required for a cell mass to double. The generation time of yeast in a culture broth has been determined from turbidimetric measurements to be 1.5 h. (a) If this yeast is used in a continuous fermentor, which is a well-stirred vessel having a volume of 1.5 L, and the inoculum is 10,000 cells/mL, at what rate can cell-free substrate be fed into this fermentor so the yeast cell concentration will remain constant? The fermentor volume is maintained constant by continuous overflow. (b) If the feed rate is 80% of what is needed for a steady-state cell mass, calculate the cell mass after 10 h of operation.

Solution: This problem is a combination of dilution with continuous cell removal, but with the added factor of cell generation by growth inside the

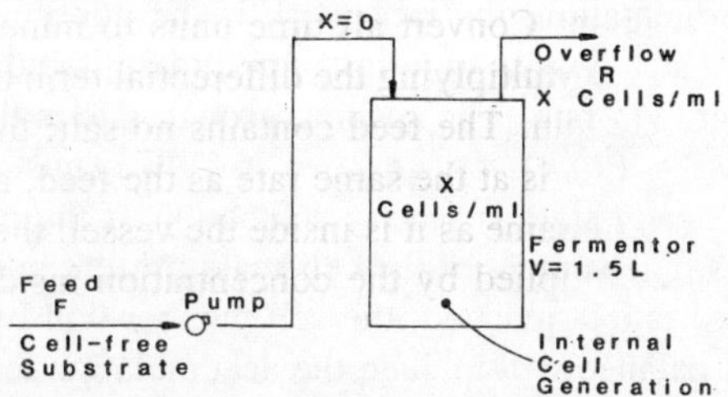

Fig. 3.13. Diagram of material balance around a fermentor.

vessel. The diagram shown in Fig. 3.13 represents the material balance on cell mass around a fermentor. The balance on cell mass or cell numbers is as follows:

Input (from feed + cell growth) = output + accumulation

(a) The substrate entering the fermentor is cell free; therefore, this term in the material balance equation is zero. Let R = substrate feed rate = overflow rate, since the fluid volume in the fermentor is maintained constant. Let x = cells/mL. The material balance on cell numbers with the appropriate time derivative for rate of increase in cell numbers, $(dx/dt)_{gen}$, and accumulation $(dx/dt)_{acc}$, is:

$$V\left[\frac{dx}{dt}\right]_{gen} = Rx + V\left[\frac{dx}{dt}\right]_{acc}$$

If the cell mass is constant, then the last term on the right is zero. Cell growth is often expressed in terms of a generation time, G, which is the average time for doubling of cell numbers. Let t = time in hours. Then:

$$x = x_0(2)^{(t/G)}$$

Differentiating to obtain the generation term in the material balance equation:

$$\left[\frac{dx}{dt}\right]_{gen} = \frac{d}{dt}\left[x_0(2)^{(t/G)}\right] = x_0(G^{-1})(2)^{(t/G)} \ln 2$$

$$= xG^{-1} \ln 2$$

Substituting in the material balance equation and dropping out zero terms: $VxG^{-1} \ln 2 = Rx$; x cancels out on both sides.

$$R = V(\ln 2)G^{-1}$$

Substituting known quantities:

$$R = \frac{1.5 \ln 2}{1.5} = \frac{0.693 \text{ L}}{\text{h}}$$

In continuous fermentation, the ratio R/V is the dilution rate, and when the fermentor is at a steady state in terms of cell mass, the dilution rate must equal the specific growth rate. The specific growth rate is the quotient: rate of growth/cell mass $= (1/x)(dx/dt)_{gen} = \ln 2/G$. Thus, the dilution rate needed to achieve steady state is strictly a function of the rate of growth of the organism and is independent of the cell mass present in the fermentor.

(b) When the feed rate is reduced, the cell number will be in an unsteady state. The material balance will include the accumulation term. Substituting the expression for cell generation into the original material balance equation with accumulation:

$$V\frac{x \ln 2}{G} = Rx + V\frac{dx}{dt}$$

$$x\left[\frac{V \ln 2}{G} - R\right] = V\frac{dx}{dt}$$

Separating variables and integrating:

$$\frac{dx}{x} = \left[\frac{\ln 2}{G} - \frac{R}{V}\right] dt$$

$$\ln x = \left[\frac{\ln 2}{G} - \frac{R}{V}\right] t + C$$

At $t = 0$, $x = 10{,}000$ cells/mL; $C = \ln(10{,}000)$.

$$\ln\left(\frac{x}{10{,}000}\right) = \left[\frac{\ln 2}{G} - \frac{R}{V}\right] t$$

Substituting: $V = 1.5$ L; $R = 0.8(0.693$ L/h$) = 0.554$ L/h; $G = 1.5$ h; $t = 10$ h.

$$\ln\frac{x}{10{,}000} = 0.925; \qquad x = 10{,}000(2.5217)$$

$$x = 25{,}217 \text{ cells/mL}$$

BLENDING OF FOOD INGREDIENTS

Total Mass and Component Balances. These problems involve setting up total mass and component balances and solving several equations simultaneously.

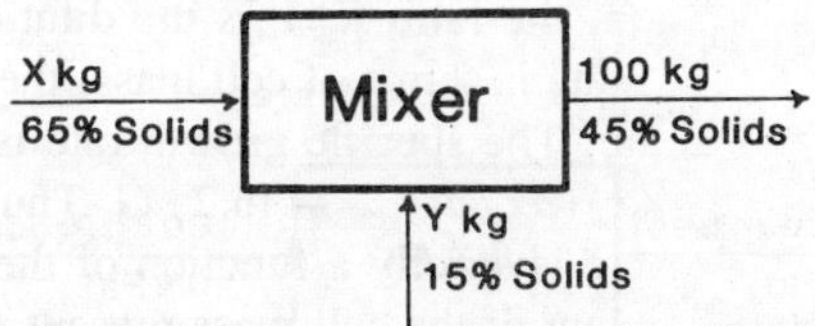

Fig. 3.14. Material flow and composition in a process for blending juice concentrates.

EXAMPLE 1: Determine the amount of a juice concentrate containing 65% solids and single-strength juice containing 15% solids that must be mixed to produce 100 kg of a concentrate containing 45% solids.

Solution: The diagram for the process is shown in Fig. 3.14.

$$\text{Total mass balance: } X + Y = 100; \qquad X = 100 - Y$$

$$\text{Solid balance: } 0.65X + 0.15Y = 100(0.45) = 45$$

Substituting $(100 - Y)$ for X:

$$0.65(100 - Y) + 0.15Y = 45$$

$$65 - 0.65Y + 0.15Y = 45$$

$$65 - 45 = 0.65Y - 0.15Y$$

$$20 = 0.5Y$$

$$Y = 40 \text{ kg single-strength juice}$$

$$X = 60 \text{ kg } 65\% \text{ concentrate}$$

EXAMPLE 2: Determine the amounts of lean beef, pork fat, and water that must be used to make 100 kg of a frankfurter formulation. The compositions of the raw materials and the frankfurter are:

Lean beef—14% fat, 67% water, 19% protein.
Pork fat—89% fat, 8% water, 3% protein.
Frankfurter—20% fat, 15% protein, 65% water.

Solution: The diagram representing the various mixtures being blended is shown in Fig. 3.15.

$$\text{Total mass balance: } Z + X + Y = 100$$

$$\text{Fat balance: } 0.14X + 20$$

$$\text{Protein balance: } 0.19X + 0.03Y = 15$$

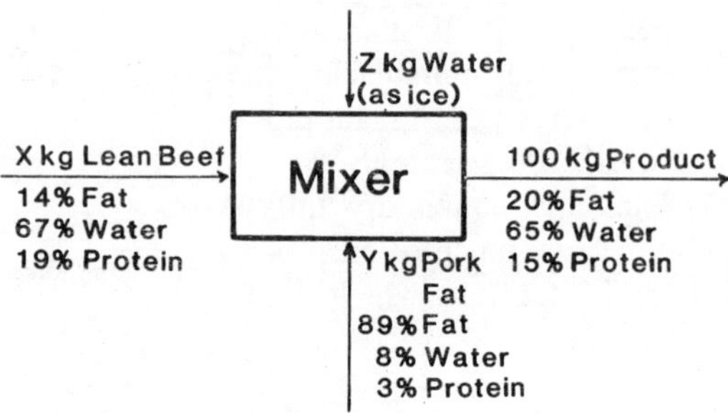

Fig. 3.15. Composition and material flow in blending of meats for a frankfurter formulation.

These equations will be solved by determinants. Note that the fat and protein balance equations involve only X and Y; therefore, solving these two equations first will give values for X and Y. Z will be solved using the total mass balance equation. See the section "Determinants" in Chapter 1 for a discussion of how this is done.

The matrices for the fat and protein balance equations are:

$$\begin{vmatrix} 0.14 & 0.89 \\ 0.19 & 0.03 \end{vmatrix} \begin{vmatrix} X \\ Y \end{vmatrix} = \begin{vmatrix} 20 \\ 15 \end{vmatrix}$$

$$X = \frac{\begin{vmatrix} 20 & 0.89 \\ 15 & 0.03 \end{vmatrix}}{\begin{vmatrix} 0.14 & 0.89 \\ 0.19 & 0.03 \end{vmatrix}} = \frac{20(0.03) - 15(0.89)}{0.14(0.03) - 0.19(0.89)}$$

$$X = \frac{0.6 - 13.35}{0.0042 - 0.1691} = 77.3 \text{ kg}$$

$$Y = \frac{\begin{vmatrix} 0.14 & 20 \\ 0.19 & 15 \end{vmatrix}}{\begin{vmatrix} 0.14 & 0.89 \\ 0.19 & 0.03 \end{vmatrix}} = \frac{0.14(15) - 0.19(20)}{0.0042 - 0.1691}$$

$$Y = \frac{-1.7}{-0.1649} = 10.3 \text{ kg}$$

Total mass balance: $Z = 100 - 77.3 - 10.3 = 12.4$ kg

EXAMPLE 3: A food mix is to be made which would balance the amount of methionine (MET), a limiting amino acid in terms of food protein nutritional value, by blending several types of plant proteins. Corn which contains 15% protein has 1.2 g MET/100 g protein; soy flour with 55% protein has 1.7 g MET/100 g protein; and nonfat dry milk with 36% protein has 3.2 g MET/100 g protein. How much of each of these ingredients must be used to produce 100 kg of formula which contains 30% protein and 2.2 g MET/100 g protein?

Solution: This problem is solved by setting up a component balance on protein and MET. Let C = kg corn, S = kg soy flour, and M = kg nonfat dry milk. The material balance equations are:

$$\text{Total mass: } C + S + M = 100$$

$$\text{Protein: } 0.15C + 0.55S + 0.36M = 30$$

$$\text{MET: } \frac{(1.2)(0.15)}{100} C + \frac{(1.7)(0.55)}{100} S + \frac{(3.2)(0.36)}{100} M = \frac{2.2}{100}(30)$$

$$0.18C + 0.935S + 1.152M = 66$$

The matrices representing the three simultaneous equations are:

$$\begin{vmatrix} 1 & 1 & 1 \\ 0.15 & 0.55 & 0.36 \\ 0.18 & 0.935 & 1.152 \end{vmatrix} \begin{vmatrix} C \\ S \\ M \end{vmatrix} = \begin{vmatrix} 100 \\ 30 \\ 66 \end{vmatrix}$$

Thus:

$$C = \frac{\begin{vmatrix} 100 & 1 & 1 \\ 30 & 0.55 & 0.36 \\ 66 & 0.935 & 1.152 \end{vmatrix}}{\begin{vmatrix} 1 & 1 & 1 \\ 0.15 & 0.55 & 0.36 \\ 0.18 & 0.935 & 1.152 \end{vmatrix}}$$

The denominator matrix is resolved to:

$$1\begin{vmatrix} 0.55 & 0.36 \\ 0.935 & 1.152 \end{vmatrix} - 0.15\begin{vmatrix} 1 & 1 \\ 0.935 & 1.152 \end{vmatrix} + 0.18\begin{vmatrix} 1 & 1 \\ 0.55 & 0.36 \end{vmatrix}$$

$$= 1[0.55(1.152) - 0.935(0.36)] - 0.15[1(1.152) - 1(0.935)]$$
$$+ 0.18[(1)(0.36) - (0.55)(1)]$$
$$= 0.297 - 0.03255 - 0.0342 = 0.23025$$

The numerator matrix resolves as follows:

$$100\begin{vmatrix} 0.55 & 0.36 \\ 0.935 & 1.152 \end{vmatrix} - 30\begin{vmatrix} 1 & 1 \\ 0.935 & 1.152 \end{vmatrix} + 66\begin{vmatrix} 1 & 1 \\ 0.55 & 0.36 \end{vmatrix}$$

$$= 100[0.55(1.152) - 0.935(0.36)] - 30[1(1.152) - 0.935(1)]$$
$$+ 66[1(0.36) - 0.55(1)]$$
$$= 100(0.297) - 30(0.217) + 66(-0.19) = 10.65$$

$$C = \frac{10.65}{0.23025} = 46.25 \text{ kg}$$

$$S = \frac{\begin{vmatrix} 1 & 100 & 1 \\ 0.15 & 30 & 0.36 \\ 0.18 & 66 & 1.152 \end{vmatrix}}{0.23025}$$

The numerator matrix resolves as follows:

$$1\begin{vmatrix} 30 & 0.36 \\ 66 & 1.152 \end{vmatrix} - 0.15\begin{vmatrix} 100 & 1 \\ 66 & 1.152 \end{vmatrix} + 0.18\begin{vmatrix} 100 & 1 \\ 30 & 0.36 \end{vmatrix}$$

$$= 1[30(1.152) - 66(0.36)] - 0.15[100(1.152) - 66(1)]$$
$$+ 0.18[100(0.36) - 30(1)]$$
$$= 1(10.9) - 0.15(49.2) + 0.18(6) = 4.5$$

$$S = \frac{4.5}{0.23025} = 19.54 \text{ kg}$$

The total mass balance may be used to solve for M, but as a means of checking the calculations the matrix for M will be formulated and resolved.

$$M = \frac{\begin{vmatrix} 1 & 1 & 100 \\ 0.15 & 0.55 & 30 \\ 0.18 & 0.935 & 66 \end{vmatrix}}{0.23025}$$

The numerator matrix is resolved as follows:

$$1\begin{vmatrix} 0.55 & 30 \\ 0.935 & 66 \end{vmatrix} - 0.15\begin{vmatrix} 1 & 100 \\ 0.935 & 66 \end{vmatrix} + 0.18\begin{vmatrix} 1 & 100 \\ 0.55 & 30 \end{vmatrix}$$

$$= 1[0.55(66) - 0.935(30)] - 0.15[1(66) - 0.935(100)]$$
$$+ 0.18[1(30) - 0.55(100)]$$
$$= 1(8.25) - 0.15(-27.5) + 0.18(-25) = 7.875$$

$$M = \frac{7.875}{0.23025} = 34.2 \text{ kg}$$

Check: $C + S + M = 46.25 + 19.54 + 34.2 = 99.99$

Use of Specified Constraints in Equations. When blending ingredients, constraints may be imposed either by the functional properties of the ingredients or its components or by regulations. Examples of constraints are limits on the amount of nonmeat proteins which can be used in meat products like frankfurters, moisture content in meat products, or functional properties such as fat or water binding properties. Another constraint may be cost. It is necessary that the constraints specified be included as one of the equations to be resolved.

In some cases, a number of ingredients may be available for making a food formulation. To determine effectively the quantity of each ingredient using the procedures discussed in the previous sections, the same number of independent equations must be formulated as there are unknown quantities to be determined. If three components can be used to formulate component balance equations, only three independent equations can be formulated. Thus a unique solution for components of a blend, using the solution to simultaneous equations, will be possible only if there are three ingredients to be used.

When there are more possible ingredients for use in a formulation than there are independent equations which can be formulated using the total mass and component balances, constraints will have to be used to allow consideration of all possible ingredients. The primary objectives are minimizing cost and maximizing quality. Cost and quality factors will form the basis for specifying constraints.

The constraints may not be specific, i.e., they may only define a boundary rather than a specific value. Thus, a constraint cannot be used as a basis for an equation which must be solved simultaneously with other equations.

One example of a system of constraints that includes as many ingredients as possible in a formulation is the *least-cost formulation* concept used in the meat industry. Several software companies market least-cost formulation strategies; the algorithm in these computer programs may vary from one company to the other. However, the basis for these calculations is basically the same: (1) The composition requirement must be met, usually 30% fat and water and protein contents must satisfy the U.S. Department of Argiculture's requirement of water not to exceed four times the protein content plus 10% ($4P + 10$) or some other specified moisture range allowable for certain category of products. (2) The bind values for fat must be adequate for emulsifying the fat contained in the formulation. Software companies that market least-cost sausage formulation strategies vary in their use of these fat-holding capacities. (3) Water-holding capacity must be maximized, i.e., the water and protein content relationship should stay within the $4P + 10$ tolerance in the finished product. This requirement is not usually followed because of the absence of reliable water-holding capacity data, and processors simply add an excess of water during formulation to compensate for the water loss that occurs during cooking.

To illustrate these concepts, a simple least-cost formulation will be set-up for frankfurters utilizing a choice of five different meat types. Note that as in the previous examples, a total mass balance and only two component balance equations can be formulated (water can be determined from the total mass balance). Thus, only three independent equations can be formulated. Product textural properties may be used as a basis for a constraint. If previous experience has shown that a minimum of the protein present in the final blend must come from a specific type of meat (e.g. pork) to achieve a characteristic flavor and texture, then this constraint will be one factor to be included in the analysis. Other constraints are that no negative values of the components are acceptable.

The solution will involve an iterative technique. Component balance equations are formulated. Then constraints which can be expressed in equation form are used to formulate equations. Values are then assumed for some of the ingredients such that the number of unknown quantities in the simultaneous equations equals the number of independent equations. The system of equations is then solved iteratively, using constraints to discard combinations which are out of bounds. This technique reduces the solution to a few combinations of ingredients, and the combination having the least cost is selected.

EXAMPLE: Derive a least-cost formulation involving four meat types from the five types shown. The composition and pertinent functional data, such as bind values, for each of the ingredients are shown in Table 3.1. The fat content in the blend is 30% and the protein content is 15%.

Table 3.1. Data for the Least-Cost Formulation Problems

Meat type	Composition Fat	Protein	Water	Bind constant kg fat/100·kg	Cost ($/kg)
Bull meat	11.8	19.1	67.9	30.01	1.80
Pork trm.	25.0	15.9	57.4	19.25	1.41
Beef chk.	14.2	17.3	68.0	13.96	1.17
Pork chk.	14.1	17.3	67.0	8.61	1.54
Back fat	89.8	1.9	8.3	1.13	0.22

The constraints are as follows: (1) for product color and textural considerations, at least 20% of the total protein must come from pork trim and pork cheeks; (2) the amount of beef cheek meat should not exceed 15% of the total mixture; (3) the bind value (sum of the product of the mass of each ingredient and its bind constant) must be greater than 15. These constraints are established by experience as the factors needed to impart the desired texture and flavor in the product.

The following defines the variables used in the equations:

A = Bull meat

B = pork trim

C = beef cheek

D = pork cheek

E = back fat

PA, PB, PC, PD, and PE = mass fraction protein in A, B, C, D, and E, respectively

FA, FB, FC, FD, and FE = mass fraction fat in A, B, C, D, and E, respectively

BA, BB, BC, BD, and BE = bind constants for A, B, C, D, and E, respectively

To solve the problem, 100 kg of the blend is used as a basis, and the following algorithm is utilized:

1. Values for C and D are assumed.
2. Constraint 1 will result in the following equation:

$$B(PB) + D(PD) = 15(0.20)$$

$$B = \left(\frac{1}{PB}\right)[15(0.20) - D(PD)]$$

3. Fat balance:

$$A(FA) + B(FB) + C(FC) + D(FD) + E(FE) = 30$$

$$A(FA) + E(FE) = 30 - B(FB) - C(FC) - D(FD)$$

$$\text{Let } F = 30 - B(FB) - C(FC) - D(FD)$$

$$A(FA) + E(FE) = F \qquad \text{(a.3)}$$

4. Protein balance:

$$A(PA) + B(PB) + C(PC) + D(PD) + E(PE) = 15$$
$$A(PA) + E(PE) = 15 - B(PB) - C(PC) - D(PD)$$
$$\text{Let } P = 15 - B(PB) - C(PC) - D(PD)$$
$$A(PA) + E(PE) = P \qquad \text{(a.4)}$$

5. Since B, D and C have values at this point from steps 1 and 2, solving equations a.3 and a.4 will give values for A and E. The determinants for solving these two equations are:

$$A = \frac{\begin{vmatrix} P & PE \\ F & FE \end{vmatrix}}{\begin{vmatrix} PA & PE \\ FA & FE \end{vmatrix}} = \frac{F(PE) - P(FE)}{FA(PE) - PA(FE)}$$

$$B = \frac{\begin{vmatrix} PA & P \\ FA & F \end{vmatrix}}{FA(PE) - PA(FE)} = \frac{FA(P) - PA(F)}{FA(PE) - PA(FE)}$$

6. The bind constants are then used to determine the bind: $B = A(BA) + B(BB) + C(BC) + D(BD) + E(BE)$. B must exceed 15.
7. The other constraints are as follows: none of the meat values must exceed 100; no negative values are acceptable, and the sum of the meat weights must not be greater than 100.
8. The cost of each combination which satisfies the constraints is calculated, and the least costly formulation is selected.

A computer program in BASIC is used to solve this problem. The program is as follows. REM statements will indicate the step in the algorithm just discussed that is being performed in the next series of lines until the next REM statement. The REM statements need not be keyed in when using this program.

```
10    REM: PROGRAM FOR LEAST COST FORMULATION
15    REM: DATA ENTRY
```

```
20 DATA .118,.25,.142,.141,.898,.191,.159,.173,.173,
   .019,.3001,.1925,.1396,.0861,.0113,1.80,
 1.41,1.17,1.54,.22
25     REM: DEFINE VARIABLES AND ASSIGN VALUES
30 READ FA,FB,FC,FD,FE,PA,PB,PC,PD,PE,
   BA,BB,BC,BD,BE,CTA,CTB,CTC,CTD,CTE
35      REM: STEP 1 IN ALGORITHM, ASSUMING VALUES FOR C
             AND D
40 FOR D = 0 TO 20 STEP 2
50 FOR C = 0 TO 15 STEPS 2
55      REM: STEP 2 IN ALGORITHM, USE OF CONSTRAINT 1
        AND SOLVING FOR B.
60 B = (15 * .20 - PD * D) / PB
65      REM: STEPS 3 AND 4 IN ALGORITHM, SOLVING FOR
        PROTEIN (PCD) AND FAT (FCD), IN C AND D.
70 PCD = PC * C + PD * D
80 FCD = FC * C + FC * D
85      REM: EQN. A.4
90 P = 15 - PCD -PB * B
95      REM:EQN. A.3
100 F = 30 - FCD -FB * B
105     REM: STEPS 5 AND 6 IN THE ALGORITHM, SOLVING FOR
             THE DENOMINATOR OF THE DETERMINANTS OF A
             AND B
110 DENUM = FA * PE - PA * FE
115     REM: STEPS 5 AND 6 IN THE ALGORITHM, SOLVING FOR
             A AND E BY DETERMINANTS.
120 A = (F * PE - P * FE) / DENUM
130 E = (FA * P - PA * F) / DENUM
135      REM: STEP 6 IN THE ALGORITHM, SOLVING FOR BIND
         VALUE
140 BIND = A * BA + BB * B + BC * C + BD * D + BE * E
145      REM: CONSTRAINT ON BIND
150 IF BIND < 15 THEN GOTO 230
155      REM: CONSTRAINT ON INDIVIDUAL INGREDIENT NOT TO
              EXCEED 100, AND NO NEGATIVE VALUES.
160 IF A > 100 THEN GOTO 230
170 IF B > 100 THEN GOTO 230
180 IF E > 100 THEN GOTO 230
190 IF A < 0 THEN GOTO 230
200 IF B < 0 THEN GOTO 230
210 IF E < 0 THEN GOTO 230
212       REM: CONSTRAINT ON TOTAL MIXTURE NOT GREATER
          THAN 100
215 IF (A + B + C + D + E) > 100 THEN GOTO 230
220 GOSUB 260
230 NEXT C
240 NEXT D
250 END
260 REM: PRINT SUBROUTINE
270 COST = A * CTA + B * CTB + C * CTC + D * CTD + E * CTE
```

```
290 PRINT "A = ";A; "B = ";B;"C = ";C;"D = ";D;"E = ";E
300 PRINT "COST = ";COST, "BIND = ";BIND
310 RETURN
```

The program when run will print:

```
A = 60.82  B = 18.86  C = 0  D = 0  E =   .16
Cost = 136.08          Bind = 22.11
A = 59.02  B = 18.87  C = 2  D = 0  E = 20.08
Cost = 135.17          Bind = 21.84
A = 60.79  B = 16.69  C = 0  D = 2  E = 20.46
Cost = 136.04          Bind = 21.86
```

The least-cost formulation is the one with 59.02 kg bull meat, 18.87 kg pork trimmings, 2 kg beef cheeks, 0 kg pork cheeks, and 20.08 kg back fat. The total mass of meats is 99.97 kg, and the cost of the formulation is \$135.17/100 kg.

The least-cost formulation will change as the cost of the ingredients changes; therefore, processors need to update cost information regularly and determine the current least-cost formula.

MULTISTAGE PROCESSES

Problems of this type require drawing a process flow diagram and moving the system boundaries for the material balance formulations around parts of the process as needed to the unknown quantities. It is always good to define the basis used in each stage of the calculations. It is possible to change the basis as the calculations proceed from an analysis of one subsystem to the next. A tie material will also be helpful in relating one part of the process to another.

EXAMPLE 1: The standard of identity for jams and preserves specifies that the ratio of fruit to added sugar in the formulation is 45 parts fruit to 55 parts sugar. A jam also must have a soluble content of at least 65% to produce a satisfactory gel. The standard of identity requires soluble solids of at least 65% for fruit preserves from apricot, peach, pear, cranberry, guava, nectarine, plum, gooseberry, figs, quince, and currants. The process of making fruit preserves involves mixing the fruit and sugar in the required ratio, adding pectin, and concentrating the mixture by boiling in a vacuum, steam-jacketed kettle until the soluble solids content is at least 65%. The amount of pectin added is determined by the amount of sugar used in the formulation and by the grade of the pectin (a 100 grade pectin is one that will form a satisfactory gel at a ratio of 1 kg pectin to 100 kg sugar).

If the fruit contains 10% soluble solids and 100 grade pectin is used, cal-

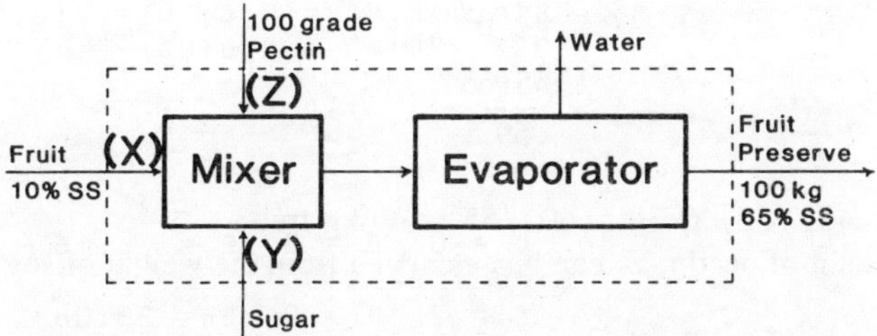

Fig. 3.16. Process and material flow for manufacturing fruit preserves.

culate the weight of the fruit, sugar, and pectin necessary to produce 100 kg of fruit preserve. For quality control purposes, soluble solids are those which change the refractive index and can be measured on a refractometer. Thus only the fruit soluble solids and sugar are considered soluble solids in this context; pectin is excluded.

Solution: The process flow diagram is shown in Fig. 3.16. The boundary used for the system on which the material balance is made encloses the whole process.

The information given in the problem, 100 kg fruit preserve, is used as a basis. The tie material is the soluble solids, since pectin is unidentifiable in the finished product. The designation "fruit" in the finished product is meaningless because the water that evaporates comes from the fruit, and the soluble solids in the fruit mixes with the rest of the system.

$$\text{Soluble solids balance: } 0.1(X) + Y = 100(0.65)$$

Since the ratio 45 parts fruits to 55 parts sugar is a requirement, the other equation is:

$$\frac{X}{Y} = \frac{45}{55}; \qquad X = \frac{45}{55}Y$$

Substituting the expression for X in the equation representing the soluble solids balance and solving for Y:

$$0.1\left(\frac{45}{55}Y\right) + Y = 65$$

$$0.1(45)(Y) + 55Y = 55(65)$$

$$4.5Y + 55Y = 55(65)$$

$$59.5Y = 55(65)$$

$$Y = \frac{55(65)}{59.5} = 60 \text{ kg sugar}$$

Since $X = 45Y/55$, $X = 45(60)/55 = 49$ kg fruit.

The amount of pectin, Z, can be calculated from the weight of sugar used:

$$Z = \frac{\text{kg sugar}}{\text{grade of pectin}} = \frac{60}{100} = 0.6 \text{ kg pectin}$$

The problem can also be solved by selecting a different variable as a basis: 100 kg fruit. Equation based on the required ratio of fruit to sugar:

$$\text{Sugar} = 100 \text{ kg fruit} \times \frac{55 \text{ kg sugar}}{45 \text{ kg fruit}}$$

$$= 122 \text{ kg}$$

Let X = kg of jam produced.

$$\text{Soluble solids balance: } 100(0.1) + 122 = X(0.65)$$

$$X = \frac{132}{0.65} = 203 \text{ kg}$$

Since 100 kg fruit will produce 203 kg of jam, the quantity of fruit required to produce 100 kg of jam can be determined by ratio and proportion:

$$100:102 = X:100$$

$$X = \frac{100(100)}{203} = 49 \text{ kg fruit}$$

$$\text{Sugar} = 49 \times \frac{45}{55} = 60 \text{ kg}$$

$$\text{Pectin} = \frac{1}{100}(60) = 0.6 \text{ kg}$$

EXAMPLE 2: In solvent extractions, the material to be extracted is thoroughly mixed with a solvent. An ideal system is one in which the component to be extracted dissolves in the solvent and the ratio of solute to solvent in the liquid phase equals the ratio of solute to solvent in the liquid absorbed in the solid phase. This condition occurs with thorough mixing until equilibrium is reached and if sufficient solvent is present such that the solubility of solute in the solvent is not exceeded.

Meat (15% protein, 20% fat, 64% water, and 1% inert insoluble solids) is extracted with five times its weight of a fat solvent that is miscible in all proportions with water. At equilibrium, the solvent mixes with water and fat dissolves in this mixture. Assume that there is sufficient solvent to allow all the fat to dissolve.

After thorough mixing, the solid is separated from the liquid phase by filtration and is dried until all the volatile material is removed. The weight of the dry cake is only 50% of the weight of the cake leaving the filter. Assume that none of the fat, protein, and inerts are removed from the filter cake by drying and that no nonfat solids are in the liquid phase leaving the filter. Calculate the fat content in the dried solids.

Solution: The diagram of the process is shown in Fig. 3.17. Using kg of meat as a basis, the ratio of solvent to meat of 5 would require 500 kg of solvent. Consider as a system one which has a boundary that encloses the filter and the drier. Also, draw a subsystem with a boundary that encloses only the drier. Write the known weights of the process streams and the weights of the components in each stream in the diagram. Entering the system at the filter will be 600 kg of material containing 15 kg of protein, 1 kg of inerts, 64 kg of water, 20 kg of fat and 500 kg of solvent. Since all nonfat solids leave at the drier, the weight of the dried solids, D, includes 16 kg of the protein and inerts. From the condition given in the problem that the weight of the dried solids is 50% of the weight of the filter cake, the material entering the drier should be $D/0.5$.

All fat dissolves in the solvent-water mixture. Fat, however, also appears in the solids fraction because some of the fat-water-solvent solution is re-

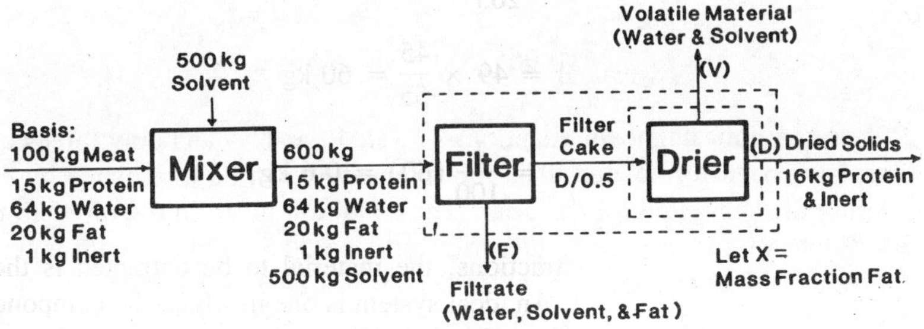

Fig. 3.17. Process and material flow for a solvent extraction process involving simultaneous removal of fat and moisture.

tained by the solids after filtration. Although the filtrate does not constitute all of the fat-water-solvent solution, the mass fraction of fat in the filtrate will be the same as in the whole solution, and this condition can be used to calculate the mass fraction of fat in the filtrate.

$$\text{Mass fraction fat in filtrate} = \frac{\text{wt fat}}{\text{wt fat} + \text{wt solv} + \text{wt } H_2O}$$

$$= \frac{20}{20 + 64 + 500} = 0.034246$$

Consider the system of the filter and drier. Let x = mass fraction of fat in D. The following component balance can be made.

$$\text{Fat balance: } F(0.034246) + Dx = 20 \quad (1)$$

Protein and inerts balance:

Sixteen kilograms of protein and inerts enters the system, and all of it leaves the system with the dried solids, D. Since D consists only of fat + protein + inerts, and x is the mass fraction of fat, then $1 - x$ would be the mass fraction of protein and inerts.

$$D(1 - x) = 16; \qquad D = \frac{16}{1 - x} \quad (2)$$

Solvent and water balance:

Five hundred kilograms of solvent and 64 kg of water enter the system. These components leave the system with the volatile material V, at the drier and with the filtrate, F, at the filter. The mass fraction of fat in the filtrate is 0.034246. The mass fraction of water and solvent = 1 − 0.034246 = 0.965754.

$$F(0.965754) + V = 564 \quad (3)$$

There are four unknown quantities—F, D, V, and x—and only three equations have been formulated. The fourth equation can be formulated by considering the subsystem of the drier. The condition given in the problem that the weight of solids leaving the drier must be 50% of the weight entering gives the following total mass balance equation around the drier:

$$\frac{D}{0.5} = D + V; \qquad D = V \quad (4)$$

The above four equations can be solved simultaneously. Substituting D for V in equation 3:

$$F(0.965754) + D = 564 \qquad (5)$$

Substituting equation 2 in equation 5:

$$F(0.965654) + \frac{16}{1 - x} = 564 \qquad (6)$$

Substituting equation 2 in equation 1:

$$F(0.034246) + \frac{16}{1 - x}(x) = 20 \qquad (7)$$

Solving for F in equations 6 and 7 and equating:

$$\frac{20 - 36x}{(0.034246)(1 - x)} = \frac{548 - 564x}{(0.965754)(1 - x)}$$

Simplifying and solving for x:

$$0.965754(20 - 36x) = 0.034246(548 - 564x)$$

$$19.31508 - 34.767144x = 18.7868 - 19.314744x$$

$$x(34.76144 - 19.314744) = 19.31508 - 18.7868$$

$$x = \frac{0.548272}{15.446696} = 0.03549$$

The percentage fat in the dried solids = 3.549%.

The solution can be shortened and considerably simplified if it is realized that the fat/(solvent + water) ratio is the same in the filtrate as it is in the liquid that adheres to the filter cake entering the drier. Since the volatile material leaving the drier is only solvent + water, it is possible to calculate the amount of fat carried with it from the fat/(solvent + water) ratio.

$$\frac{\text{fat}}{\text{solvent + water}} = \frac{20}{500 + 64} = 0.03546$$

The amount of fat entering the drier = $0.03546V$.
Since $D = 16 + \text{fat}$, $D = 16 + 0.03546V$.

Total mass balance around the drier gives:

$$D = V.$$

Substituting D for V gives:

$$D = 16 + 0.03546D$$

$$= \frac{16}{1 - 0.03546} = 16.5882$$

$$\text{Wt. fat in } D = 16.5882 - 16 = 0.5882$$

$$\% \text{ fat in } D = \frac{0.5882}{16.5882}(100) = 3.546\%$$

A principle presented in an example at the beginning of this chapter was that in solvent extraction or crystallization, the purity of the product depends strongly on the efficiency of separation of the liquid from the solid phase. This is demonstrated here; the amount of fat entering the drier is a direct function of the amount of volatile matter present in the wet material. Thus, the more efficient the solid-liquid separation process before drying the solids, the less fat will be carried over to the finished product.

EXAMPLE 3: Determine the quantity of sucrose crystals that will crystallize out of 100 kg of a 75% sucrose solution after cooling to 15°C. The mother liquor contains 66% sucrose.

Solution: The diagram for crystallization is shown in Fig. 3.18. A 75% sucrose solution enters the crystallization, and the material separates into a solution containing 66% sucrose and crystals of 100% sucrose.

Figure 3.18 shows sucrose appearing in the crystals and saturated solution

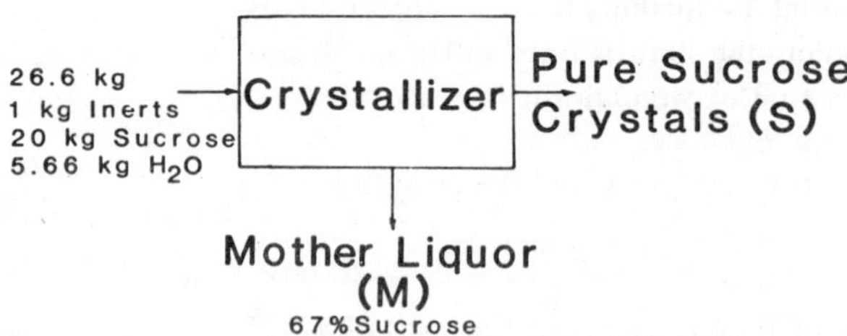

Fig. 3.18. Composition and material flow in a crystallization process.

fractions; water appears only in the saturated solution fraction. Water can be used as a tie material in this process.

The mother liquor contains 66% sucrose. The balance (34%) is water. Using the mass fraction principle, 25 kg of water enters the system and all of it leaves with the mother liquor, in which the mass fraction of water is 0.34. Thus, the weight of the mother liquor is 25/0.34 = 73.52 kg.

$$\text{Total mass balance: Wt crystals} = 100 - 73.52 = 26.48 \text{ kg}$$

EXAMPLE 4: For the problem statement and the process represented by the diagram in Fig. 3.2, calculate the yield and the purity of the sugar crystals. The mother liquor contains 67% sucrose w/w. The crystal fraction from the centrifuge loses 15% of its weight in the drier and emerges moisture free.

Solution: The problem can be separated into two parts. First, a material balance around the evaporator will be used to determine the weight and composition of the material entering the crystallizer. The second part of the problem, which involves the crystallizer, centrifuge, and drier, can be solved utilizing the principles demonstrated in the two preceding examples.

Material balance around the evaporator:

Twenty kilograms of sucrose enters, and this represents 75% of the weight of the concentrate.

$$\text{Wt concentrate} = 20/0.75 = 26.66 \text{ kg}$$

The concentrate entering the crystallizer consists of 20 kg of sucrose, 1 kg of inerts, and 5.66 kg of water.

Consider the centrifuge as a subsystem in which the cooled concentrate is separated into a pure crystals fraction and a mother liquor fraction. The weights of these two fractions and the composition of the saturated solution fraction can be determined using a procedure similar to the one used in the preceding example. In Fig. 3.19, the known quantities are indicated on the diagram.

The composition of the mother liquor fraction is determined by making a material balance around the crystallizer. The mass and composition of the streams entering and leaving the crystallizer are shown in Fig. 3.20.

All the water and inert material entering the centrifuge go into the mother liquor fraction. The weights of the water and inert material in the mother liquor are 5.66 and 1 kg, respectively.

The material balances around the crystallizer are:

$$\text{Total mass: } M + S = 26.66$$

$$\text{Sucrose balance: } M(0.67) + S = 20$$

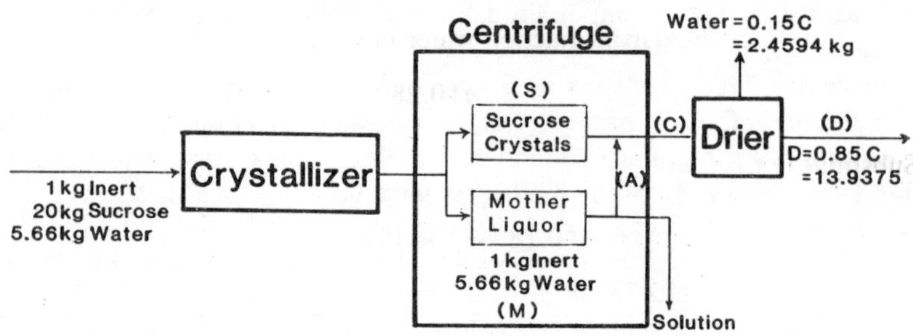

Fig. 3.19. Material flow and system boundaries used for analysis of crystal purity from a crystallizer.

Eliminating S by subtracting the two equations:

$$M(1 - 0.67) = 26.66 - 20; \qquad M = \frac{6.66}{0.33} = 20.18 \text{ kg}$$

From the known weights of the components, the composition of the mother liquor, in mass fraction, is as follows:

Sucrose = 0.67, as defined

Inert: 1.00 kg; mass fraction = 1/20.18 = 0.04955

Water: 5.66 kg; mass fraction = 5.66/20.18 = 0.28048

The solids fraction, C, leaving the centrifuge is a mixture of the mother liquor adhering to the crystals, A, and the pure sucrose crystals, S. Since 26.66 kg of material entered the crystallizer and 20.18 kg is in the mother liquor, the pure sucrose crystals, S, = 26.66 = 20.18 = 6.48 kg.

$$C = 6.48 + A$$

Water balance around the drier:

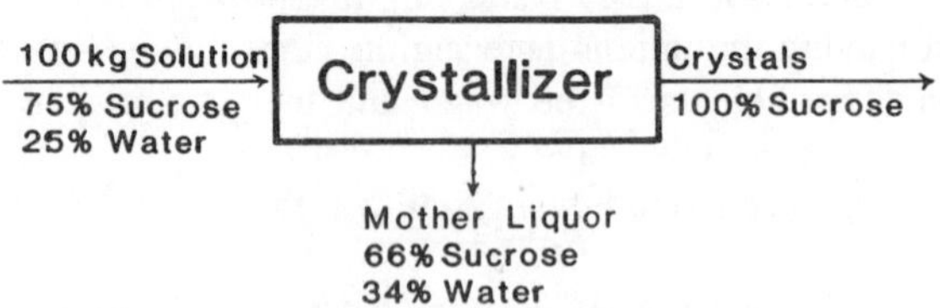

Fig. 3.20. Material balance around the crystallizer to determine the quantities of mother liquor and pure crystals which separate on crystallization.

$$A(\text{mass fraction water in } A) = 0.15C$$

$$A(0.28048) = 0.15C$$

Substituting C:

$$A(0.28948) = 0.15(6.48 + A)$$

$$A(0.28948 - 0.15) = 0.15(6.48)$$

$$A = 7.4494 \text{ kg}$$

$$\text{Wt dry sugar, } D = 0.85(6.48 + 7.4494)$$

$$= 11.84 \text{ kg}$$

$$\text{Wt inert material} = A(\text{mass fraction inert in } A)$$

$$= 7.4494(0.04955) = 0.3691 \text{ kg}$$

$$\% \text{ inert in dry sugar} = \frac{0.3691}{11.84}(100) = 3.117\%$$

The dry sugar is only 96.88% sucrose.

EXAMPLE 5: An ultrafiltration system has a membrane area of 0.75 m^2 and a water permeability of 180 kg water/h(m^2) under the conditions used in concentrating whey from 7% to 25% total solids at a pressure of 1.033 MPa. The system is fed by a pump which delivers 230 kg/h, and the appropriate concentration of solids in the product is obtained by recycling some of the product through the membrane. The concentrate contained 11% lactose and the unprocessed whey contained 5.3% lactose. There is no protein in the permeate.

Calculate: (1) the production rate of 25% concentrate through the system; (2) the amount of product recycled/h; (3) the amount of lactose removed in the permeate/h; (4) the concentration of lactose in the mixture of fresh and recycled whey entering the membrane unit; and (5) the average rejection factor by the membrane for lactose based on the average lactose concentrations entering and leaving the unit. The rejection factor (F_r) of solute through a membrane is defined by:

$$F_r = \frac{X_f - X_p}{X_f}$$

where X_f = solute concentration on the feed side of the membrane, which may be considered as the mean of the solute concentrations in the fluid en-

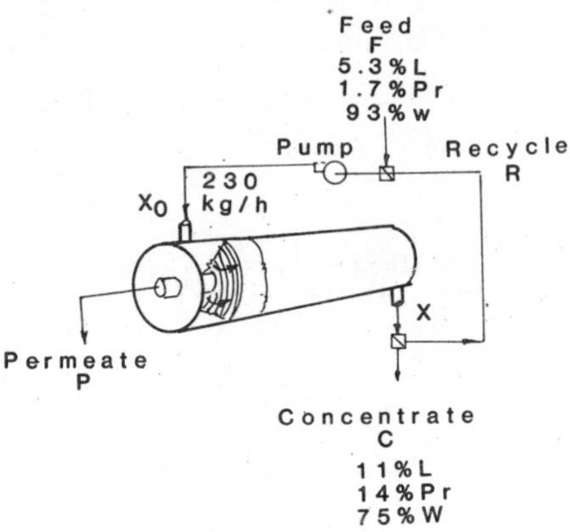

Fig. 3.21. Diagram of a spiral-wound membrane module in an ultrafiltration system for cheese whey.

tering and leaving the membrane unit, and X_p = solute concentration in the permeate.

Solution: The process flow diagram is shown in Fig. 3.21.
Let L, Pr, and W represent lactose, protein, and water, respectively. Let F = feed, P = permeate or fluid passing through the membrane, and R = recycled stream which has the same composition as the concentrate, C. The fluid stream leaving the membrane unit, which is the sum of the concentrate and recycle streams, is also referred to as the *retentate*.
Whole system material balance: Basis: 1 h of operation.

$$\text{Water in permeate} = \frac{180\ \text{kg}}{\text{m}^2\ \text{h}}(0.75\ \text{m}^2)(1\ \text{h})$$

$$= 180(0.75) = 135\ \text{kg}$$

Water balance:

$$F(0.93) = 135 + C(0.75); \qquad 0.93F - 0.75C = 135$$

Protein balance:

$$F(0.017) = C(0.14); \qquad 0.017F - 0.14C = 0$$

Solving for F and C by determinants:

$$\begin{vmatrix} 0.93 & -0.75 \\ 0.017 & -0.14 \end{vmatrix} \begin{vmatrix} F \\ C \end{vmatrix} = \begin{vmatrix} 135 \\ 0 \end{vmatrix}$$

$$F = \frac{135(-0.14) - 0}{0.93(-0.14) - 0.017(-0.75)} = \frac{-18.9}{-0.1302 + 0.01275}$$

$$F = 160.9 \text{ kg/h}$$

$$C = \frac{0 - 0.017(135)}{-0.1302 + 0.01275} = \frac{-2.295}{-0.11745} = 19.45 \text{ kg/h}$$

1. Amount of product = 19.54 kg/h.
2. Amount recycled is determined by a material balance around the pump. The diagram is shown in Fig. 3.22.

$$R = 230 - 160.9 = 69.1 \text{ kg/h}$$

3. Lactose in permeate: Lactose balance around whole system:

$$L_p = \text{lactose in permeate} = F(0.053) - C(0.011)$$
$$= 160.9(0.053) - 19.54(0.011) = 6.379$$
$$X_p = \text{mass fraction lactose in permeate}$$
$$= \frac{6.379}{135 + 6.379} = 0.04511$$

4. Lactose balance around pump:

$$F(0.053) + R(0.011) = 230(X_{fl})$$

X_{fl} = mass fraction lactose in fluid leaving the membrane unit

$$X_{fl} = \frac{160.9(0.053) + 69.1(0.11)}{230} = 0.0702$$

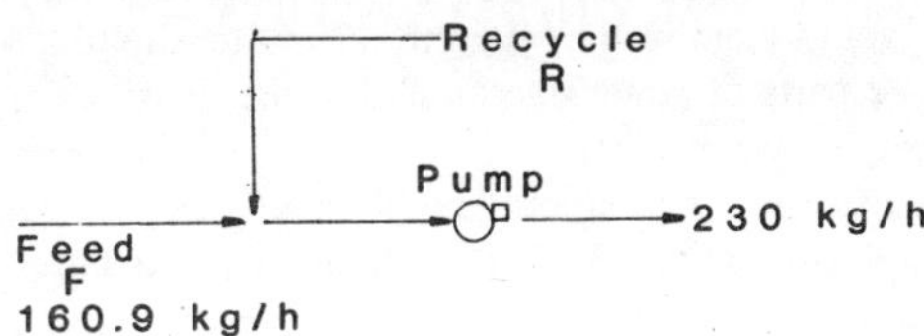

Fig. 3.22. Diagram of material balance during mixing of feed and recycle streams in an ultrafiltration system.

X_f = mean mass fraction lactose on the retentate side of the membrane unit

$$X_f = 0.5(0.0702 + 0.11) = 0.09$$

5. Rejection factor for lactose:

$$F_r = \frac{0.09 - 0.045}{0.09} = 0.498$$

PROBLEMS

1. A frankfurter formulation is to be made from the following ingredients:

   Lean beef—14% fat, 67% water, 19% protein
   Pork fat—89% fat, 8% water, 3% protein
   Soy protein isolate—90% protein, 8% water

   Water needs to be added (usually in the form of ice) to achieve the desired moisture content. The protein isolate added is 3% of the total weight of the mixture. How much lean beef, pork fat, water, and soy isolate must be used to obtain 100 kg of a formulation having the following composition?

   protein—15%; moisture—65%; fat—20%

2. If 100 kg of raw sugar containing 95% sucrose, 3% water, and 2% soluble uncrystallizable inert solids is dissolved in 30 kg of hot water and cooled to 20°C, calculate:
   (a) The amount of sucrose (in kg) that remains in solution,
   (b) The amount of crystalline sucrose,
   (c) The purity of the sucrose (in %) obtained after centrifugation and dehydration to 0% moisture. The solid phase contained 20% water after separation from the liquid phase in the centrifuge.
   A saturated solution of sucrose at 20°C contains 67% sucrose (w/w).
3. Tomato juice flowing through a pipe at a rate of 100 kg/min is salted by adding saturated salt solution (26% salt) to the pipeline at a constant rate. At what rate would the saturated salt solution be added to provide 2% salt in the product?
4. If fresh apple juice contains 10% solids, what would be the solids content of a concentrate that would yield single-strength juice after diluting one part of the concentrate with three parts of water? Assume that the densities are constant and are equal to the density of water.
5. In a dehydration process, the product, which was at 80% moisture initially, has lost half of its weight during processing. What is the final moisture content?
6. Calculate the quantity of dry air that must be introduced into an air drier that dries 100 kg/h of food from 80% to 5% moisture. Air enters with a moisture content of 0.002 kg water/kg of dry air and leaves with a moisture content of 0.2 kg H_2O/kg of dry air.

7. How much water is required to raise the moisture content of 100 kg of a material from 30% to 75%?
8. In the section "Multistage Processes," Example 2, solve the problem if the meat: solvent ratio is 1:1. The solubility of fat in the water-solvent mixture is such that the maximum fat content in the solution is 10%.
9. How many kg of peaches would be required to produce 100 kg of peach preserves? The standard formula of 45 parts fruit to 55 parts sugar is used, the soluble solids content of the finished product is 65%, and the peaches have 12% initial soluble solids content. Calculate the weight of 100 grade pectin required and the amount of water removed by evaporation.
10. The peaches in Problem 9 come in a frozen form to which sugar has been added in the ratio three parts fruit to one part sugar. How much peach preserves can be produced from 100 kg of this frozen raw materials?
11. Yeast has a proximate analysis of 47% carbon, 6.5% hydrogen, 31% oxygen, 7.5% nitrogen, and 8% ash on a dry weight basis. Based on a factor of 6.25 for converting protein nitrogen to protein, the protein content of yeast on a dry basis is 46.9%. In a typical yeast culture process, the growth medium is aerated to convert substrate primarily to cell mass. The dry cell mass yield is 50% of a sugar substrate. Nitrogen is supplied as ammonium phosphate.

    The cowpea is a high-protein, low-fat legume which is a valuable protein source in the diet of many third world nations. The proximate analysis of the legume is 30% protein, 50% starch, 6% oligosaccharides, 6% fat, 2% fiber, 5% water, and 1% ash. It is desired to produce a protein concentrate by fermenting the legume with yeast. Inorganic ammonium phosphate is added to provide the nitrogen source. The starch in cowpea is first hydrolyzed with amylase, and yeast is grown on the hydrolyzate.

    (a) Calculate the amount of inorganic nitrogen added as ammonium phosphate to provide the stoichiometric amount of nitrogen needed to convert all the starch present to yeast mass. Assume that none of the cowpea protein is utilized by the yeast.
    (b) If the starch is 80% converted to cell mass, calculate the proximate analysis of the fermented cowpea on a dry basis.
12. Cottage cheese whey contains 1.8 g/L of protein, 5.2 g/L of lactose, and 0.5 g/L of other solids. This whey is spray-dried to a final moisture content of 3%, and the dry whey is used in an experimental batch of summer sausage.

    In summer sausage, the chopped meat is inoculated with a bacterial culture which converts sugars to lactic acid as the meat ferments prior to cooking in a smokehouse. The level of acid produced is controlled by the amount of sugar in the formulation. The lactic acid level in the summer sausage is 0.5 g/100 g dry solids. Four moles of lactic acid is produced from one mole of lactose. The following formula is used for the summer sausage:

    3.18 kg lean beef (16% fat, 16% protein, 67.1% water, 0.9% ash)
    1.36 kg pork (25% fat, 12% protein, 62.4% water, 0.6% ash)
    0.91 kg ice
    0.18 kg soy protein isolate (5% water, 1% ash, 94% protein)

Calculate the amount of dried whey protein which can be added to this formulation so that when the lactose is 80% converted to lactic acid the desired acidity will be obtained.

13. Osmotic dehydration of blueberries was accomplished by contacting the berries with an equal weight of a corn syrup solution containing 60% soluble solids for 6 h and draining the syrup from the solids. The solid fraction left on the screen after draining the syrup was 90% of the original weight of the berries. The berries originally contained 12% soluble solids, 86.5% water, and 1.5% insoluble solids. The sugar in the syrup penetrated the berries so that the berries remaining on the screen, when washed free of the adhering solution, showed a soluble solids gain of 1.5% based on the original dry solids content. Calculate:
    (a) The moisture content of the berries and adhering solution remaining on the screen after draining the syrup.
    (b) The soluble solids content of the berries after drying to a final moisture content of 10%.
    (c) The percentage of soluble solids in the syrup drained from the mixture. Assume that none of the insoluble solids are lost in the syrup.
14. The process for producing dried mashed potato flakes involves mixing wet mashed potatoes with dried flakes in a 95:5 weight ratio and passing the mixture through a granulator before drying it on a drum dryer. The cooked potatoes, after mashing, contain 82% water and the dried flakes contain 3% water. Calculate:
    (a) The amount of water that must be removed by the dryer for every 100 kg of dried flakes produced.
    (b) The moisture content of the granulated paste fed to the dryer.
    (c) The amount of raw potatoes needed to produce 100 kg of dried flakes; 8.5% of the raw potato weight is lost on peeling.
    (d) Potatoes should be purchased on a dry matter basis. If the base moisture content is 82% and potatoes at this moisture content cost $200/ton, what is the purchase price for potatoes containing 85% moisture?
15. Diafiltration is a process used to reduce the lactose content of whey recovered using an ultrafiltration membrane. The whey is first passed through the membrane and concentrated to twice the initial solids content, rediluted, and passed through the membrane a second time. Two membrane modules in series, each with a membrane surface area of 0.5 m^2, are to be used for concentrating and removing lactose from acid whey, which contains 7.01% total solids, 5.32% lactose, and 1.69% protein. The first module accomplished the initial concentration, and the retentate is diluted with water to 9.8% total solids and reconcentrated in the second module to a solids content of 14.02%. Under the conditions of the process, each module has an average water permeation rate of 254 kg/(h · m^2). The rejection factor for lactose by the membrane, based on the arithmetic mean of the feed and retentate lactose concentrations and the mean permeate lactose concentration, is 0.2. The protein rejection factor is 1. The rejection factor is defined as $F_r = (C_r - C_p)/C_r$, where C_r = concentration on the retentate side and C_p = concentration on the permeate side of the membrane. Calculate:
    (a) The amount of 14.02% solids delactosed whey concentrate produced from the second module per hour.
    (b) The lactose content of the delactosed whey concentrate.

16. An orange juice blend containing 42% soluble solids is to be produced by blending stored orange juice concentrate with the current crop of freshly squeezed juice. The following are the constraints: The soluble solids : acid ratio must equal 18, and the currently produced juice may be concentrated before blending, if necessary. The currently produced juice contains 14.5% soluble solids, 15.3% total solids, and 0.72% acid. The stored concentrate contains 60% soluble solids, 62% total solids, and 4.3% acid. Calculate:

(a) The amount of water which must be removed or added to adjust the concentration of the soluble solids in order to meet the specified constraints.

(b) The amounts of currently processed juice and stored concentrate needed to produce 100 kg of the blend containing 42% soluble solids.

17. The process for extraction of sorghum juice from sweet sorghum for production of sorghum molasses, which is still used in some areas in the rural southern United States, involves passing the cane through a three-roll mill to squeeze the juice out. Under the best conditions, the squeezed cane (bagasse) still contains 50% water.

(a) If the cane originally contained 13.4% sugar, 65.6% water, and 21% fiber, calculate the amount of juice squeezed from the cane per 100 kg of raw cane, the concentration of sugar in the juice, and the percentage of sugar originally in the cane which is left unrecovered in the bagasse.

(b) If the cane is not immediately processed after cutting, moisture and sugar loss occurs. Loss of sugar has been estimated to be as much as 1.5% within a 24-h holding period, and total weight loss for the cane during this period is 5.5%. Assume that sugar is lost by conversion to CO_2; therefore, the weight loss is attributable to water and sugar loss. Calculate the juice yield based on the freshly harvested cane weight of 100 kg, the sugar content in the juice, and the amount of sugar remaining in the bagasse.

18. In a continuous fermentation process for ethanol from a sugar substrate, the sugar is converted to ethanol and part of it is converted to a yeast cell mass. Consider a 1000-L continuous fermentor operating in a steady state. Cell-free substrate containing 12% glucose enters the fermentor. The yeast has a generation time of 1.5 h, and the concentration of yeast cells within the fermentor is 1×10^7/mL. Under these conditions, a dilution rate (F/V, where F is the rate of feeding of cell-free substrate and V is the volume of the fermentor) which causes the cell mass to stabilize at a steady state results in a residual sugar content in the overflow of 1.2%. The stoichiometric sugar : dry cell mass ratio is 1 : 0.5 on a weight basis, and the sugar : ethanol ratio is based on 2 moles of ethanol produced per mole of glucose. A dry cell mass of 4.5 g/L is equivalent to a cell count of 1.6×10^7/mL. Calculate the ethanol productivity of the fermentor in g ethanol/(L · h).

19. A protein solution is to be dimineralized by dialysis. The solution is placed in a dialysis tube immersed in continuously flowing water. For all practical purposes, the concentration of salt in the water is zero and the dialysis rate is proportional to the concentration of salt in the solution in the tube. The contents of the tube may be considered to be well mixed. A solution which initially contained 500 μg/mL of salt and 15 mg/mL of protein, contains 400 μg/mL of salt and 13 mg/mL of protein at the end of 2 h. Assume that there is no permeation of protein through the membrane and the solution was originally 1 L. The rate of permeation of water into the membrane and the rate of permeation of salt out of the membrane are directly proportional to the concentration of salt in the solution inside the membrane. Calculate the time of dialysis needed to reduce the salt concentration inside the membrane to 10 $\mu g/mL$. What will be the protein concentration inside the membrane at this time?

SUGGESTED READING

Charm, S. E. 1971. *Fundamentals of Food Engineering*, 2nd ed. AVI Publishing Co., Westport, Conn.

Watson, E. L., and Harper, J. C. 1989. *Elements of Food Engineering*, 2nd ed. Van Nostrand Reinhold Co., New York.

Himmelblau, D. M. 1967. *Basic Principles and Calculations in Chemical Engineering*, 2nd ed. Prentice-Hall, Englewood Cliffs, N.J.

Hougen, O. A., and Watson, K. M. 1946. Chemical process principles. Part I. *Material and Energy Balances*. John Wiley & Sons, New York.

4

Gases and Vapors

Gases and vapors are naturally associated with foods and food processing systems. The equilibrium between food and water vapor determines the temperatures achieved during processing. Dissolved gases in foods such as oxygen affect shelf life. Gases are used to flush packages in order to eliminate oxygen and prolong shelf life. Modified atmospheres in packages are used to prolong the shelf life of packaged foods. Air is used for dehydration. Gases are used as propellants in aerosol cans and as refrigerants.

The distinction between gases and vapors is very imprecise, since theoretically all vapors are gases. The term *vapor* is generally used for the gaseous phase of a substance which exists as a liquid or a solid at ambient conditions.

EQUATIONS OF STATE FOR IDEAL AND REAL GASES

Equations of state are expressions for the relationships between the pressure, volume, temperature, and quantity of gases within a given system. The simplest equation of state, the ideal gas equation, closely approximates the actual behavior of gases at near ambient temperature and pressure where the effect of molecular interactions is minimal. At high pressures and temperatures, however, most gases deviate from ideal behavior, and several equations of state have been proposed to fit experimental data. In this section, two equations of state will be discussed: the ideal gas equation and one of the most often used equations of state for real gases, Van der Waal's equation.

The Kinetic Theory of Gases. The fundamental theory governing the behavior of gases, the kinetic theory, was first proposed by Bernoulli in 1738 and was tested and extended by Clausius, Maxwell, Boltzman, Van der Waal, and Jeans. The postulates of the kinetic theory are as follows:

1. Gases are composed of a discrete particles called *molecules* which are in constant random motion, colliding with each other and with the walls of the surrounding vessel.
2. The force resulting from the collision between the molecules and the walls of the surrounding vessel is responsible for the *pressure* of the gas.
3. The lower the pressure, the farther apart the molecules are. Thus, attractive forces between molecules have a reduced influence on the overall properties of the gas.
4. The average kinetic energy of the molecules is directly proportional to the absolute temperature.

Absolute Temperature and Pressure. The pressure (P) of a gas is the force of collisions of gas molecules against a surface in contact with the gas. Since pressure is force per unit area, pressure is proportional to the number of gas molecules and their velocity. This pressure is the absolute pressure.

Pressure is often expressed as *gage pressure* when the measured quantity is greater than atmospheric pressure and as *vacuum* when the measured quantity is below atmospheric pressure. Measurement of gage and atmospheric pressures is shown in Fig. 4.1. Two types of pressure-measuring devices are depicted, a manometer and a Bourdon tube pressure gage. On the left side of the diagram, the pressure of the gas is counteracted by atmospheric pressure such that if pressure is atmospheric, the gages will read zero. The reading given by these type of gages is the *gage pressure*. In the American Engineering System of measurement, gage pressure is expressed as pounds of force per square inch gage (psig). In the SI system, gage pressure is expressed as kPa above atmospheric pressure.

The diagram on the right side in Fig. 4.1 shows the measuring element completely isolated from the atmosphere. The pressure reading from these gages represents the actual pressure or force of collision of gas molecules, the *absolute*

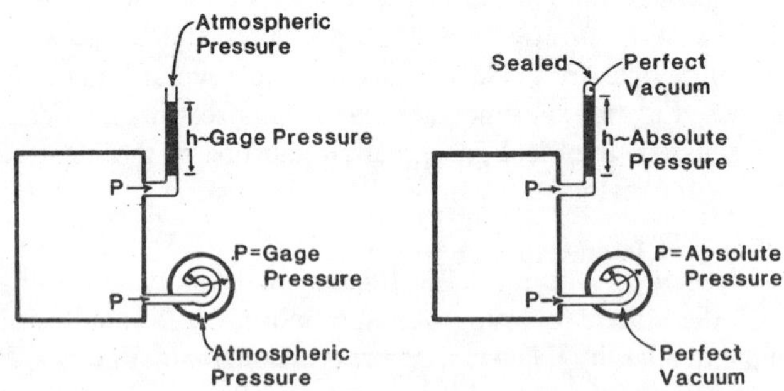

Fig. 4.1. Diagram showing differences in measurement of gage and absolute pressure.

pressure. In the American Engineering System, absolute pressure is expressed as pounds of force per square inch absolute (psia). In the SI system, it is expressed as kPa absolute.

Conversion from gage to absolute pressure is done using the following equation:

$$P_{gage} = P_{absolute} - P_{atmospheric} \tag{1}$$

The term vacuum after a unit of pressure indicates how much the pressure is below atmospheric pressure. Thus, vacuum may be interpreted as a negative gage pressure and is related to the atmospheric and absolute pressures as follows:

$$P_{absolute} = P_{atmospheric} - P_{vacuum} \tag{2}$$

Pressure is sometimes expressed in *atmospheres*(*atm*) instead of the traditional force/area units. This has led to some confusion between pressure specified in atmospheres and atmospheric pressure. Unless otherwise specified, the pressure term *atm* refers to a *standard atmosphere*, the mean atmospheric pressure at sea level, equivalent to 760 mm Hg, 29.921 in. Hg. 101.325 kPa, or 14.696 lb_f/in^2. A *technical atmosphere* is a pressure of 1 kg_f/cm^2. Some technical articles express pressure in *Bar*, which is equivalent to 100 kPa. *Atmospheric pressure* is actual pressure exerted by the atmosphere in a particular location, and varies with time and location. Atmospheric pressure must be specified if it is different from a standard atmosphere.

EXAMPLE: Calculate the absolute pressure inside an evaporator operating under 20 in. Hg vacuum. Atmospheric pressure is 30 in. Hg. Express this pressure in SI and in the American Engineering System of units.

From the table of conversion factors, Appendix Table A.1, the following conversion factors are obtained:

$$\frac{0.4912\ lb_f/in.^2}{in.\ Hg};\qquad \frac{3.386338 \times 10^3\ Pa}{in.\ Hg}$$

$$P_{abs.} = P_{atm} - P_{vac.}$$

$$= (30 - 20)\ in.\ Hg = 10\ in.\ Hg$$

$$= 10\ in.\ Hg\ \frac{0.4912\ lb_f/in.^2}{in.\ Hg} = 4.912\ psia$$

$$= 10\ in.\ Hg\ \frac{3386.338\ Pa}{in.\ Hg} = 33.863\ kPa\ absolute$$

Temperature (T) is a thermodynamic quantity related to the velocity of motion of molecules. The temperature scales are based on the boiling (liquid-gas equilibrium) and freezing points (solid, liquid, gas equilibrium) of pure substances such as water at 1 standard atm (101.325 kPa) pressure. The Kelvin (K) is defined as the fraction, 1/273.16, of the thermodynamic temperature of the triple point of water. The absolute temperature scales are based on a value on the scale which is zero at the temperature when the internal energy of molecules is zero. These are the Kelvin in SI and the Rankine (R) in the American Engineering System. Conversion from the commonly used Celsius and Fahrenheit temperature scales to the absolute or thermodynamic scales is as follows:

$$K = {}^\circ C + 273.16$$

$$R = {}^\circ F + 460$$

In common usage, the conversion to Kelvin is rounded off to:

$$K = {}^\circ C + 273$$

The absolute temperature is used in equations of state for gases.

Quantity of Gases. The quantity of gases is often expressed as volume at a specific temperature and pressure. Since molecules of a gas have a very weak attraction toward each other, they will disperse and occupy all the space within a confining vessel. Thus, the volume (V) of a gas is the volume of the confining vessel and is not definitive of the quantity of the gas unless the pressure and temperature are also specified.

The most definitive quantification of gases is by mass or as the number of moles, n, which is the quotient of mass and the molecular weight. The unit of mass is prefixed to mole (e.g., kgmole) to indicate the ratio of mass in kg and the molecular weight. The number of molecules in 1 gmole is the *Avogadro Number*, 6.023×10^{23} molecules/gmole.

At 273°K and 760 mm Hg pressure (101.325 kPa), 1 gmole, kgmole, and lbmole occupy 22.4 L, 22.4 m^3, and 359 ft^3, respectively.

The Ideal Gas Equation. The ideal gas equation is the simplest equation of state and was originally derived from the experimental work of Boyle, Charles, and Guy-Lussac. The equation is derived based on the kinetic theory of gases.

Pressure, the force of collision between gas molecules and a surface, is directly proportional to the temperature and the number of molecules per unit volume. Expressing the proportionality as an equation, and using a proportionality constant, R:

$$P \alpha \frac{n}{V} T; \qquad P = R \frac{n}{V} T$$

Rearranging:

$$PV = nRT \tag{3}$$

Equation 3 is the *ideal gas equation*. R is the gas constant and has values of 0.08206 lit(atm)/(gmole · K) or 8315 N(m)/(kgmole · K) or 1545 ft(lb_f)/(lbmole · °R).

P-V-T Relationships for Ideal Gases. When a fixed quantity of a gas that follows the ideal gas equation (equation 3) undergoes a process in which the volume, temperature, or pressure is allowed to change, the product of the number of moles, n, and the gas constant R, is a constant and:

$$\frac{PV}{T} = \text{constant}$$

If the initial temperature, pressure, and volume are known and are designated by subscript 1, these properties at another point in the process will be expressed by:

$$\frac{PV}{T} = \frac{P_1 V_1}{T_1} \tag{4}$$

Ideality of gases predicted by equations 3 and 4 occurs only under conditions close to ambient or during processes which span a relatively narrow range of temperature and pressure. Thus, these equations are useful in solving problems usually encountered when using gases in food processing or packaging.

EXAMPLE 1: Calculate the quantity of oxygen entering a package in 24 h if the packaging material has a surface area of 3000 cm^2, an oxygen permeability of 100 $cm^3/(m^2)(24\ h)$ S.T.P. (standard temperature and pressure = 0°C and 1 standard atm of 101.325 kPa).

Solution: Solving for the volume of oxygen permeating through the package in 24 h:

$$V = \frac{100\ \text{cm}^3}{\text{m}^2\ (24\ \text{h})} (24\ \text{h}) \frac{1\ \text{m}^2}{(100)^2\ \text{cm}^2} (3000\ \text{cm}^2)$$

$$= 30\ \text{cm}^3$$

Using equation 3: $n = PV/RT$.
Use $R = 0.08206$ L(atm)/(gmole · K).

$$n = \frac{1 \text{ atm } (30 \text{ cm}^3)}{[0.08206 \text{ L(atm)/(gmole} \cdot \text{K)}](273 \text{ K})} \frac{1 \text{ L}}{1000 \text{ cm}^3}$$
$$= 0.001339 \text{ gmole}$$

EXAMPLE 2: Calculate the volume of CO_2 in ft^3 at 70°F and 1 atm which would be produced by vaporization of 1 lb of dry ice.

Solution: Dry ice is solid CO_2 (mol. wt. = M = 44 lb/lbmole). T = 70°F + 460 = 530°R.

$$n = \frac{W}{M} = \frac{1 \text{ lb}}{44 \text{ lb/lbmole}} = 0.02273 \text{ lbmole}$$

$$P = 1 \text{ atm} = \frac{14.7 \text{ lb}_f}{\text{in}^2} \frac{144 \text{ in}^2}{\text{ft}^2} = 2116.8 \text{ lb}_f/\text{ft}^2$$

Substituting in the ideal gas equation: $V = nRT/P$.

$$V = 0.02273 \text{ lbmole} \frac{1545 \text{ ft lb}_f}{\text{lbmole (°R)}} \frac{530\text{°R}}{2116 \text{ lb}_f/\text{ft}^2}$$
$$= 8.791 \text{ ft}^3$$

EXAMPLE 3: Calculate the density of air (M = 29) at 70°F and 1 atm in (a) American Engineering System and (b) SI units.

Solution: Density is mass/volume = W/V. Using the ideal gas equation:

$$\text{Density} = \frac{W}{V} = \frac{PM}{RT}$$

(a) In the American Engineering System of units, P = 2116.8 lb_f/ft^2; M = 29 lb/lbmole; R = 1545 ft lb_f/lbmole (°R); and T = 70 + 460 = 530°R.

$$\text{Density} = \frac{2116.8 \text{ lb}_f/\text{ft}^2}{1545 \text{ ft lb}_f/(\text{lbmole} \cdot \text{°R})} \frac{29 \text{ lb/lbmole}}{530\text{°R}}$$
$$= 0.07498 \text{ lb/ft}^3$$

(b) In SI, $P = 101{,}325\ \text{N/m}^2$; $T = (70 - 32)/1.8 = 21.1°\text{C} = 21.1 + 273 = 294.1°\text{K}$; $V = 1\ \text{m}^3$; $M = 29$ kg/kgmole; and $R = 8315$ Nm/(kgmole · K).

$$\text{Density} = \frac{101{,}325\ \text{N/m}^2}{8315\ \text{Nm/(kgmole} \cdot \text{K)}} \frac{29\ \text{kg/kgmole}}{294.1\ \text{K}}$$

$$= 1.202\ \text{kg/m}^3$$

EXAMPLE 4: A process requires $10\ \text{m}^3/\text{s}$ at 2 atm absolute pressure and 20°C. Determine the rating of a compressor in m^3/s at S.T.P. (0°C and 101.325 kPa) which must be used to supply air for this process.

Solution: Unless otherwise specified, the pressure term *atm* means a standard atmosphere, or 101.325 kPa.
Use the *P-V-T* equation, equation 4. $V_1 = 10\ \text{m}^3$; $T_1 = 293$ K; $P_1 = 202{,}650\ \text{N/m}^2$; $P = 101{,}325\ \text{N/m}^2$; $T = 273$ K. Solving for V:

$$V = \frac{P_1 V_1}{T_1} \frac{T}{P} = \frac{(202{,}650\ \text{N/m}^2)(2\ \text{m}^3)}{293\ \text{K}} \frac{373\ \text{K}}{101{,}325\ \text{N/m}^2}$$

$$= 18.625\ \text{m}^3/\text{s}$$

EXAMPLE 5: An empty can was sealed in a room at 80°C and 1 atm pressure. Assuming that only air is inside the sealed can, what will be the vacuum after the can and contents cool to 20°C?

Solution: The quantity of gas remains constant; therefore, the *P-V-T* equation can be used. The volume does not change; therefore, it will cancel out of equation 4. $P_1 = 101{,}325\ \text{N/m}^2$; $T_1 = 353$ K; $T = 293$ K. The pressure to be used in equation 4 will be absolute.

$$P = \frac{101{,}325\ \text{N/m}^2}{353\ \text{K}} (293\ \text{K}) = 84{,}103\ \text{Pa absolute.}$$

The vacuum in Pa will be 101,325 − 84,103 = 17,222 Pa.

$$\text{Vacuum} = 17{,}222\ \text{Pa} \left(\frac{1\ \text{cm Hg}}{1333.33\ \text{Pa}}\right) = 12.91\ \text{cm Hg vacuum}$$

Van der Waal's Equation of State. The ideal gas equation is based on unhindered movement of gas molecules within a confined space; therefore, at constant temperature when molecular energy is constant, the product of pressure and volume is constant. However, as pressure is increased, the molecules are drawn closer together, and attractive and repulsive forces between the molecules affect molecular motion. When molecules are far apart, an attractive force exists. The magnitude of this force is inversely proportional to the square of the distance between the molecules. When molecules collide, they approach a limiting distance of separation when the repulsive force becomes effective, preventing them from contacting each other directly. Molecular contact may cause a chemical reaction; this occurs only at very high molecular energy levels which exceed the repulsive force. The separation distance between molecules at which the repulsive force is effective defines an exclusion zone which reduces the total volume available for molecules to move randomly. The attractive force between molecules also restricts molecular motion; this has the effect of reducing the quantity and magnitude of the impact against the walls of the confining vessel. This attractive force is referred to as an *internal pressure*. At low pressures, molecules are separated by large distances, therefore the internal pressure is small, the excluded volume is small compared to the total volume, and the gas obeys the ideal gas equation. However, at high pressures, the pressure-volume-temperature relationship deviates from ideality. Gases which deviate from ideal behavior are considered real gases. One of the commonly used equations of state for real gases is Van der Waal's equation.

Van der Waal proposed corrections to the ideal gas equation based on the excluded volume, nb, and a factor, $n^2 a/V^2$, the internal pressure. The fit between the experimental P-V-T relationship and calculated values using Van der Waal's equation of state is very good except in the region of temperature and pressure near the critical point of the gas. For n moles of gas, Van der Waal's equation of state is:

$$\left(P + \frac{n^2 a}{V^2}\right)(V - nb) = nRT \tag{5}$$

Values of the constants a and b in SI units are given in Table 4-1.

EXAMPLE 1. Calculate the density of air at 150°C and 5 atm pressure, using Van der Waal's equation of state and the ideal gas equation.

Solution: Using the ideal gas equation:

$$\text{Density} = \frac{W}{M} = \frac{PVM}{RT}; \qquad V = 1\ \text{m}^3$$

$$= \frac{5(101{,}325)(29)}{8315\,(150 + 273)} = 4.1766\ \text{kg/m}^3$$

Table 4.1. Values of Van der Waal's Constants for Different Gases

Gas	a Pa(m^3/kgmole)^2	b m^3/kgmole
Air	1.348×10^5	0.0366
Ammonia	4.246×10^5	0.0373
Carbon dioxide	3.648×10^5	0.0428
Hydrogen	0.248×10^5	0.0266
Methane	2.279×10^5	0.0428
Nitrogen	1.365×10^5	0.0386
Oxygen	1.378×10^5	0.0319
Water vapor	5.553×10^5	0.0306

Source: Calculated from values in the International Critical Tables.

Using Van der Waal's equation of state, $V = 1$ m^3.

$$\left(P + \frac{n^2 a}{V^2}\right)(V - nb) = nRT; \qquad n = \frac{W}{M}; \qquad V = 1$$

$$\left(P + \frac{W^2 a}{M^2 V^2}\right)\left(V - \frac{W}{M} b\right) = \frac{W}{M} RT$$

Expanding and collecting like terms:

$$W^3\left(\frac{ab}{M^3 V^2}\right) - W^2\left(\frac{a}{M^2 V}\right) - W\left(\frac{Pb + RT}{M}\right) + PV = 0$$

Substituting $a = 1.348 \times 10^5$; $b = 0.0366$; $R = 8315$; $M = 29$; $P = 5(101{,}325) = 506{,}625$ Pa; $T = 150 + 273 = 423$°K; $V = 1$ m^3.

$$0.2023\,W^3 - 160.3\,W^2 - 121{,}426.8\,W + 500{,}625 = 0$$

Dividing through by 0.2023:

$$W^3 - 792.3875\,W^2 - 600{,}231.3\,W + 2{,}474{,}666 = 0$$

Solve using the Newton-Raphson iteration technique:

$$f = W^3 - 792.3875\,W^2 - 600{,}231.3\,W + 2{,}474{,}666$$

$$f' = 3W^2 - 1584.775\,W - 600{,}231.3$$

Let $W = 4; f = 61{,}126.5; f\ = -606{,}522.3$

$$W_2 = W - \frac{f}{f'} = \frac{4 - 61{,}126.5}{(-606{,}522.3)} = 4.100782 \text{ kg}$$

Let $W = 4.100782; f = -8; f' = -6{,}006{,}679.6$

$$W_2 = \frac{4.100782 - (-8)}{(-6{,}006{,}679.6)} = 4.1007 \text{ kg}$$

Thus, the mass of 1 m^3 of air at 150°C and 5 atm is 4.1007 kg = the density in kg/m^3. There is just a slight discrepancy between the values calculated using the ideal gas equation and Van der Waal's equation of state.

Critical Conditions. When the pressure of a gas is increased by compression, the molecules are drawn closer together and attractive forces between them becomes strong enough to restrict their movement. At a certain temperature and pressure, a saturation point of the gas is reached and an equilibrium condition between gas and liquid can exist. If the energy level in the gas is reduced by removal of the latent heat of vaporization, it will condense into a liquid. The higher the pressure, the higher the saturation temperature of the gas.

If the pressure of the gas is increased as the temperature is maintained following the saturation temperature curve, a point will be reached where gas and liquid become indistinguishable. This particular combination of temperature and pressure is called the *critical point*. The properties of the gas at the critical point are very similar to those of a liquid in terms of dissolving certain solutes, and this property is put into practical use in a process called *supercritical fluid extraction*, which will be discussed in more detail in Chapter 14.

Gas Mixture. In this section, the concept of partial pressures and partial volumes will be used to elucidate *P-V-T* relationships of components in a gas mixture.

If the components of a gas mixture at constant volume are removed one after the other, the drop in pressure accompanying complete removal of one component is the *partial pressure* of that component.

If P_t is the total pressure and P_a, P_b, $P_c \cdots P_n$ are partial pressures of the components $a, b, c, \cdots n$, then:

$$P_t = P_a + P_b + P_c + \cdots P_n \tag{6}$$

Equation 6 is *Dalton's law of partial pressures*. Since the same volume is occupied by all components, the ideal gas equation may be used on each component to determine the number of moles of that component from the partial pressure.

$$P_a V = n_a RT \tag{7}$$

The *partial volume* is the change in volume of a gas mixture when each component is removed separately at constant pressure. If V_t is the total volume and $V_a, V_b, V_c, \cdots V_n$ are the partial volumes of the components, then:

$$V_t = V_a + V_b + V_c + \cdots V_n \tag{8}$$

Equation 8 is *Amagat's law of partial volumes*. The ideal gas equation may also be used on the partial volume of each component to determine the number of moles of that component in the mixture.

$$PV_a = n_a RT \tag{9}$$

Compositions of gases are expressed as a percentage of each component by volume. Equation 9 indicates that the volume percentage of a gas is numerically the same as the mole percentage.

EXAMPLE 1: Calculate the quantity of air in the headspace of a can at 20°C when the vacuum in the can is 10 in. Hg. Atmospheric pressure is 30 in. Hg. The headspace has a volume of 16.4 cm^3. It is saturated with water vapor.

Solution: The steam tables (Appendix Table A.4) give the vapor pressure of water at 20°C of 2336.6 Pa. Let P_t be the absolute pressure inside the can.

$$P_t = (30 - 10) \text{ in. Hg} \left(\frac{3386.38 \text{ Pa}}{\text{in. Hg}}\right) = 67{,}728 \text{ Pa}$$

Using the ideal gas equation: $V = 16.4 \text{ cm}^3 (10^{-6}) \text{ m}^3/\text{cm}^3 = 3 \times 10^{-5}$ m^3; $T = 20 + 273 = 293$ K.

$$P_{air} = P_t - P_{water}$$

$$= 67{,}728 - 2336.6 = 65{,}392.4 \text{ Pa}$$

$$n_{air} = \frac{P_{air} V}{RT} = \frac{(65{,}392.4 \text{ N/m}^2)(1.64 \times 10^{-5} \text{ m}^3)}{8315/(293)}$$

$$= 4.40 \times 10^{-7} \text{ kgmoles}$$

EXAMPLE 2: A gas mixture used for controlled atmosphere storage of vegetables contains 5% CO_2, 5% O_2, and 90% N_2. The mixture is generated by mixing appropriate quantities of air and N_2 and CO_2 gases. Of this mixture, 100 m^3 at 20°C and 1 atm is needed per hour. Air contains 21% O_2 and 79% N_2. Calculate the rate at which the component gases must be metered into the system in m^3/h at 20°C and 1 atm.

Solution: All percentages are by volume. No volume changes occur on mixing of ideal gases. Since volume percentage in gases is the same as mole percentage, material balance equations may be made on the basis of volume and volume percentages. Let X = volume O_2, Y = volume CO_2, and Z = volume, N_2, fed into the system per hour.

Oxygen balance: $0.21(X) = 100(0.05)$; $X = 23.8\ m^3$

CO_2 balance: $Y = 0.05(100)$; $Y = 5\ m^3$

Total volumetric balance: $X + Y + Z = 100$

$$Z = 100 - 23.8 - 5 = 71.2\ m^3$$

THERMODYNAMICS

Thermodynamics is a branch of science which deals with energy exchange between components within a system or between a system and its surroundings. A *system* is any matter enclosed in a boundary. The boundary may be real or imaginary and depends solely upon the part of the process under study. Anything outside the boundary is the *surroundings*. The *properties* of a system determine its state, just as a state may imply certain properties. Properties may be *extrinsic* (i.e., measurable) or intrinsic (i.e., no measurable external manifestations of those properties exist). Changes in the intrinsic properties, however, may be measured through changes in energy associated with the changes in the properties.

Equilibrium is a fundamental requirement in thermodynamic transformations. When a system reaches a condition where its properties remain constant, a state of equilibrium is attained. The science of thermodynamics deals with a system in equilibrium. Thermodynamics cannot predict how long it will take for equilibrium to occur within a system; it can only predict the final properties at equilibrium.

Thermodynamic Variables. The energy involved in thermodynamic transformations is expressed in terms of *heat* (Q), the energy which crosses a system's boundaries due to a difference in temperature, and *work* (W), the energy associated with force displacement. The term *internal energy* (E) is used to define an intrinsic property which is energy not associated with either work or heat. Internal energy is not measurable, but changes in internal energy can be measured. An intrinsic property of *entropy* (S) is also possessed by a system and is a measure of the disorder which exists within the system. The first and second laws of thermodynamics are based on the relationships of the main thermodynamic variables, discussed above.

If δQ and δW are small increments of heat and work energy which cross a system's boundaries, the accompanying differential change in the internal energy of the system is:

$$dE = \delta Q - \delta W \tag{10}$$

Equation 10 is also expressed as:

$$\Delta E = Q - W \tag{11}$$

Equations 10 and 11 are expressions for the *first law of thermodynamics*, a fundamental relationship also known as the *law of conservation of energy*. The symbol δ, called *del*, operating on W and Q is not a true differential but rather a finite difference. Although it is considered in mathematical operations as a differential operator, integration produces an increment change in a function rather than an absolute value. This principle is used when integrating functions such as E and S, which do not have absolute values but for which increment change associated with a process can be calculated.

Entropy cannot be measured, but the change in entropy is defined as the ratio of the reversible energy (Q_{rev}) which crosses a system's boundaries and the absolute temperature:

$$dS = \frac{\delta Q_{rev}}{T}; \qquad \Delta S = \frac{Q_{rev}}{T} \tag{12}$$

The *second law of thermodymanics* states that any process which occurs is accompanied by a positive entropy change, which approaches zero for reversible processes. If the process is reversible, any change in entropy in a system is compensated for by a change in the entropy of the surroundings such that the net entropy change for the system and its surroundings is zero.

The concept of the entropy, or state of disorder, of a system has several interpretations. One interpretation is that no truly reversible process is possible, since any process produces a more disordered system than the one that existed before the change. This change in the state of order requires the addition or loss of energy. Another interpretation of the concept of entropy involves the spontaneity of processes. Spontaneous processes will occur only when enough energy is available initially to overcome the requirement for increasing the entropy. Thus, process spontaneity always requires a change from a higher to a lower energy state.

Another intrinsic thermodynamic variable is *enthalpy* (H), defined as:

$$H = E + PV \tag{13}$$

In differential form:

$$dH = dE + PdV + V\,dP \tag{14}$$

Since $PdV = \delta W$, $dE + \delta W = \delta Q$ and:

$$dH = \delta Q + V\,dP \tag{15}$$

For a constant pressure process, $dP = 0$ and:

$$\delta Q = dH; \qquad \delta H = Q \tag{16}$$

A specific heat at constant pressure may be defined as:

$$C_p = \left.\frac{dQ}{dT}\right|_P \tag{17}$$

and as:

$$\Delta H = C_p\,dT \tag{18}$$

Equation 18 shows why the enthalpy is referred to as the heat content. In a constant-volume process, work is zero and equation 10 becomes:

$$dE = \delta Q; \qquad \Delta E = Q \tag{19}$$

A specific heat at constant volume may be defined as:

$$C_v = \left.\frac{dQ}{dT}\right|_V \tag{20}$$

and as:

$$\Delta E = C_v\,dT \tag{21}$$

The Relationship Between C_p and C_v for Gases. The relationship between C_p and C_v for gases is derived as follows. Substituting $W = PdV$ for work in equation 10 for a constant-pressure process:

$$dE = dQ - PdV \tag{22}$$

Taking the derivative with respect to temperature:

$$\frac{dE}{dT} = \left.\frac{dQ}{dT}\right|_P - P\frac{dV}{dT} \tag{23}$$

Equation 17 gives $dQ/dT = C_p$, and equation 21 gives $dE/dT = C_v$. From the ideal gas equation, for 1 mole of gas, $PV = RT$ and dV/dT at constant pressure is R/P. Substituting in equation 23:

$$C_v = C_p - R \tag{24}$$

A useful property in calculating *P-V-T* and other thermodynamic variables involved in expansion and compression of a gas is the specific heat ratio, C_p/C_v, designated by the symbol γ. The ratio C_p/R is expressed in terms of γ as follows:

$$\frac{C_p}{R} = \frac{\gamma}{\gamma - 1} \tag{25}$$

***P-V-T* Relationships for Ideal Gases in Thermodynamic Processes.** *Adiabatic processes* are those in which no heat is added or removed from the system; therefore, $\delta Q = 0$. Equation 15 then becomes $dH = V\,dP$. Since for 1 mole of an ideal gas $V = RT/P$, and since $dH = C_p\,dT$,

$$C_p\,dT = RT\left(\frac{dP}{P}\right)$$

$$\frac{C_p}{R}\int_{T_1}^{T_2}\frac{dT}{T} = \int_{P_1}^{P_2}\frac{dP}{P} \tag{26}$$

Integrating and substituting $C_p/R = \gamma/(\gamma - 1)$:

$$\ln\left[\frac{P_2}{P_1}\right] = \frac{\gamma}{\gamma - 1}\ln\left[\frac{T_2}{T_1}\right] \tag{27}$$

$$\frac{P_2}{P_1} = \left[\frac{T_2}{T_1}\right]^{\gamma/\gamma - 1} \tag{28}$$

The ideal gas equation may be used to substitute for P in equation 27 to obtain an expression for V as a function of T in an adiabatic process. Substituting $P_1 = nRT_1/V_1$ and $P_2 = nRT_2/V_2$ in equation 27 and simplifying:

$$\frac{V_1}{V_2} = \left[\frac{T_2}{T_1}\right]^{(1/\gamma - 1)} \tag{29}$$

Equations 28 and 29 can be used to derive:

$$\left[\frac{V_1}{V_2}\right]^{\gamma} = \left[\frac{P_2}{P_1}\right] \tag{30}$$

Isothermal processes are those in which the temperature is maintained constant. Thus, $P_1V_1 = P_2V_2$.

Isobaric processes are those in which the pressure is maintained constant. Thus, $V_1/T_1 = V_2/T_2$.

Isocratic processes are those in which the volume is maintained constant. Thus, $P_1/T_1 = P_2/T_2$.

Changes in Thermodynamic Properties, Work, and Heat Associated with Thermodynamic Processes

Adiabatic: $\Delta Q = 0; \quad \Delta S = 0; \quad \Delta E = W = \int P\,dV; \; \Delta H = \int V\,dP$

Isothermal: $Q = W = \int P\,dV; \quad \Delta S = \frac{Q}{T}; \; \Delta E = 0; \quad \Delta H = 0$

Isobaric: $Q = \int C_p\,dT = \Delta H; \quad \Delta S = \int C_p \frac{dT}{T}; \; W = P\Delta V; \quad \Delta E = Q -$

Isocratic: $Q = \Delta E = \int C_v\,dT; \quad \Delta S = \int C_v \frac{dT}{T}; \; W = 0; \quad \Delta E = Q$

Work and Enthalpy Change on Adiabatic Expansion or Compression of an Ideal Gas. Work and enthalpy changes are important in determining the power input into compressors. Adiabatic compression is what occurs during the compression of a refrigerant in a refrigeration cycle.

In adiabatic compression or expansion:

$$W = \int P\,dV; \quad \Delta H = \int V\,dP$$

Equation 30 is rearranged, substituting P and V for P_2 and V_2. Solving for V:

$$V = [(P_1)^{1/\gamma} V_1](P)^{-1/\gamma}$$

Differentiating with respect to P:

$$dV = \frac{-1}{\gamma}[(P_1)^{1/\gamma} V_1][P]^{-1/\gamma - 1}$$

Substituting dV in the expression for work and integrating:

$$W = \int_{P_2}^{P_2} -\frac{1}{\gamma}[(P_1)^{1/\gamma} V_1][P]^{-1/\gamma - 1 + 1}$$

$$= -\frac{(P_1)^{1/\gamma} V_1}{\gamma} \int_{P_1}^{P_2} (P)^{-1/\gamma}\, dP$$

The integral is:

$$-\frac{1}{1/\gamma + 1}\left[[P_2]^{(1/\gamma)+1} - [P_1]^{(1/\gamma)+1}\right]$$

Combining the integral with the multiplier and simplifying:

$$W = \frac{P_1 V_1}{1 - \gamma}\left[\left[\frac{P_2}{P_1}\right]^{(\gamma - 1)/\gamma} - 1\right] \tag{31}$$

The enthalpy change is determined by evaluating the following integral:

$$\Delta H = V_1 \int_{P_1}^{P_2} \left[\frac{P_1}{P}\right]^{(1/\gamma)} dP$$

$$= \left[\frac{\gamma}{\gamma - 1}\right](P_1 V_1)\left[\left[\frac{P_2}{P_1}\right]^{(\gamma - 1)/\gamma} - 1\right] \tag{32}$$

Work and Enthalpy Change on Isothermal Expansion or Compression of an Ideal Gas. When an ideal gas is subjected to an isothermal process.

$$Q = W = \int P\, dV \text{ and } \Delta H = \int P\, dV + \int V\, dP$$

Since T is constant, $P = nRT/V$ and $dP = -(nRTV^{-2})\, dV$.

$$Q = W \int_{V_1}^{V_2} nRT \frac{dV}{V}$$

$$= nRT \ln \left[\frac{V_2}{V_1} \right] \quad (33)$$

$$\Delta H = W + \int_{V_1}^{V_2} - \left[nRT \frac{dV}{V} \right]$$

The second term in the above expression is exactly the same as the expression for W but has a negative sign; therefore, in an isothermal process, $\Delta H = 0$.

EXAMPLE 1: N_2 gas trapped inside a cylinder with a movable piston at 80 atm pressure is then allowed to expand adiabatically (no heat added or removed) until the final pressure is 1 atm. The gas is initially at 303°K, and the initial volume is 1 L. Calculate (a) the work performed by the gas, (b) the entropy change, (c) the enthalpy change, and (d) the change in internal energy. The gas has a specific heat ratio, γ, of 1.41.

Solution: (a) Equation 31 will be used to calculate the work. $V_1 = 0.001$ m^3; $P_1 = 80(101{,}325) = 8{,}106{,}000$ Pa; and $P_2 = 101{,}325$ Pa. From equation 31:

$$W = \frac{(8{,}106{,}000 \text{ Pa})(0.001 \text{ m}^3)}{-0.41} \left[\left(\frac{101{,}325}{8{,}106{,}000} \right)^{0.41/1.41} - 1 \right]$$

$$= -14241.8 \text{ J}$$

(b) $\Delta S = 0$

(c) From equation 32:

$$\Delta H = \left[\frac{1.41}{0.41} (8{,}106{,}000 \text{ Pa})(0.001 \text{ m}^3) \right]$$

$$\left[\left[\frac{101{,}325}{8{,}106{,}000} \right]^{(0.41/1.41)} - 1 \right]$$

$$= -20{,}080.9 \text{ J}$$

(d) $\Delta E = W = -14{,}241.7$ J

VAPOR-LIQUID EQUILIBRIUM

The molecular attraction which holds liquid molecules together is not strong enough to prevent some molecules from escaping; therefore, some molecules will escape in gaseous form. If the volume of the system containing the liquid and vapor is held constant, eventually equilibrium will be attained when the rate of escape of molecules from the liquid phase to become vapor equals the rate at which the vapor molecules are recaptured by the liquid phase. This condition of equilibrium exists in all liquids. A liquid will always maintain an envelope of vapor around its surface. The pressure exerted by that vapor is known as the *vapor pressure*. The pressure exerted by the molecules of vapor in equilibrium with a liquid is a function only of temperature. In the absence of any other gas exerting pressure on the liquid surface (e.g., when the liquid is introduced into a container which is under a perfect vacuum), the pressure of the vapor at equilibrium is the vapor pressure. If the total pressure above a liquid is maintained constant at the vapor pressure, heating will not increase the temperature but will cause more liquid molecules to enter the vapor phase. The temperature at a pressure which corresponds to the vapor pressure of a liquid is the *boiling point*, a condition in which the whole atmosphere over the liquid surface consists only of gaseous molecules of that liquid. The heat added to the system to generate a unit mass of vapor from liquid at the boiling point is the *heat of vaporization*. When the pressure above the liquid is higher than the vapor pressure, some other gas molecule (e.g., air) would be exerting that pressure in addition to the vapor molecules; therefore, the vapor pressure will be the partial pressure of vapor in the gas mixture surrounding the liquid.

The Clausius-Clapeyron Equation. The temperature dependence of the vapor pressure is expressed by equation 34, the Clausius-Clapeyron equation:

$$\ln\left[\frac{P_2}{P_1}\right] = \left[\frac{\Delta H_v}{R}\right]\left[\frac{T - T_1}{TT_1}\right] \tag{34}$$

where ΔH_v = heat of vaporization, P = vapor pressure at temperature T, R = gas constant, and P_1 = vapor pressure at temperature T_1. Equation 34 assumes constant heat of vaporization, but in reality, this quantity changes with temperature. Thus, the most useful function of equation 34 is in interpolating between values of vapor pressure listed in abbreviated tables where the temperature interval between entries in the table is quite large. The steam tables (Appendix Tables A.3 and A.4) give values of the vapor pressure for water at different temperatures. Since the temperature intervals in these tables are rather small, a linear interpolation may be used instead of a logarithmic function interpolation, as suggested by equation 34.

Liquid Condensation from Gas Mixtures. When the partial pressure of a component in a gas mixture exceeds the vapor pressure of that component, condensation will occur. If the mixture is in a closed container where the volume is constant, condensation will result in a decrease in total pressure. This principle is responsible for the vacuum that results when a hot liquid is placed in a can and the can is sealed. Replacement of air in the headspace by vapors from the product itself or by steam in a system where steam is flushed over the can headspace prior to sealing results in a reduced pressure after the can cools adequately to condense the vapors in the headspace.

EXAMPLE 1: A canned food at the time of sealing is at a temperature of 80°C, and the atmospheric pressure is 758 mm Hg. Calculate the vacuum (in mm Hg) formed inside the can when the contents cool down to 20°C.

Solution: Assume that there are no dissolved gases in the product at the time of sealing; therefore, the only gases in the headspace are air and water vapor. From Appendix Table A.3, the vapor pressures of water at 20° and 80°C are 2.3366 and 47.3601 kPa, respectively. In the gas mixture in the headspace, air is assumed to remain at the same quantity in the gaseous phase, while water condenses on cooling.

$$P_{air} = P_t - P_{water}$$

At 80°C, P_t = 758 mm Hg (1 cm Hg/10 mm Hg)(1333.33 Pa/cm Hg) = 101,064 Pa. Let P = pressure in the headspace at 20°C.

$$n_{air} = \frac{PV}{RT} = \frac{(101{,}064 - 47{,}360.1)V}{8315(80 + 273)} = 0.018296\,V \text{ kgmole}$$

at 20°C, $n_{air} = PV/8315(293) = 4.1014 \times 10^{-7}\,PV$ kgmole. Since the number of moles of air trapped in the headspace is constant:

$$4.1014 \times 10^{-7}\,PV = 0.018294\,V$$

$$P = 44{,}575 \text{ Pa absolute} = 332 \text{ mm Hg absolute}$$

$$\text{Vacuum} = 758 - 332 = 426 \text{ mm Hg vacuum}$$

EXAMPLE 2: Air at 5 atm pressure is saturated with water vapor at 50°C. If this air is allowed to expand to 1 atm pressure and the temperature is reduced to 20°C, calculate the amount of water that will be condensed per m^3 of high-pressure air at 50°C.

Solution: The vapor pressures of water at 50° and 20°C are 12.3354 and 2.3366 kPa, respectively. Basis: 1 m^3 air at 5 atm pressure and 50°C. The number of moles of air will remain the same on cooling.

$$n_{air} = \frac{[5(101{,}325) - 12{,}335.4]\,1}{8315(50 + 273)} = 0.1840 \text{ kgmole}$$

At 20°C, n_{air} = (101,325 − 2336.6)V/8315(20 + 273) = 0.04063V. Equating: 0.04063V = 0.1840; V = 4.529 m³ at 20°C and 1 atm. At 50°C, n_{water} = 12,335.4(1)/8315(323) = 0.004593. At 20°C, n_{water} = 2336.6(4.529)/8315(293) = 0.004344.

$$\text{Moles water condensed} = 0.004593 - 0.004344$$
$$= 0.000249 \text{ kg moles}$$

EXAMPLE 3: The partial pressure of water in air at 25°C and 1 atm is 2.520 kPa. If this air is compressed to 5 atm total pressure to a temperature of 35°C, calculate the partial pressure of water in the compressed air.

Solution: Increasing the total pressure of a gas mixture will proportionately increase the partial pressure of each component.

Using the ideal gas equation (equation 3) for the mixture and for the water vapor, let V_1 = the volume of the gas mixture at 25°C and 1 atm, P_t = total pressure and, P_w = partial pressure of water vapor.

The total number of moles of air and water vapor is:

$$n_t = \frac{P_1V_1}{RT_1} = \frac{P_2V_2}{RT_2}$$

The number of moles of water vapor is:

$$n_w = \frac{P_{w1}V_1}{RT_1} = \frac{P_{w2}V_2}{RT_2}$$

Assuming no condensation, the ratio n_t/n_w will be the same in low-pressure and high-pressure air; therefore:

$$\frac{P_1}{P_{w1}} = \frac{P_2}{P_{w2}}; \quad P_{w2} = P_{w1}\frac{P_1}{P_2}$$

$$P_{w2} = \frac{5 \text{ atm}}{1 \text{ atm}}(2.520 \text{ kPa}) = 12.60 \text{ kPa}$$

The temperature was not used in calculating the final partial pressure of water. The temperature is used in verifying the assumption of no condensation, comparing the calculated final partial pressure with the vapor pressure of water at 35°C. From the steam tables, Appendix Table A.4, the vapor pressure of water at 35°C is the pressure of saturated steam at 35°C, which is 5.6238 kPa. Since the calculated partial pressure of water in the compressed air is greater than the vapor pressure, condensation of water must have occurred and the correct partial pressure of water will be the vapor pressure at 35°C. Thus, P_{w2} = 5.6238 kPa.

PROBLEMS

1. Air used for dehydration is heated by burning natural gas and mixing the combustion products directly with air. The gas has a heating value of 1050 BTU/ft^3 at 70°F and 1 atm pressure. Assume that the gas is 98% methane and 2% nitrogen.
   (a) Calculate the quantity of natural gas in ft^3 at 70°F and 1 atm needed to supply the heating requirements for a dryer which uses 1500 lb/h dry air at 170°F and 1 atm. Assume that the products of combustion will have the same specific heat as dry air: 0.24 BTU/(lb + °F).
   (b) If the air used to mix with the combustion gases is completely dry, what will be the humidity of the air mixture entering the dryer?
2. A package having a void volume of 1500 cm^3 is to be flushed with nitrogen to displace oxygen prior to sealing. The process used involved drawing a vacuum of 700 mm Hg on the package, breaking the vacuum with nitrogen gas, and drawing another 700 mm Hg vacuum before sealing. The solids in the package prevent total collapse of the package as the vacuum is drawn; therefore, the volume of gases in the package may be assumed to remain constant during the process. If the temperature is maintained constant at 25°C during the process, calculate the number of gmoles of oxygen left in the package at the completion of the process. Atmospheric pressure is 760 mm Hg.
3. Compression of air in a compressor is an adiabatic process. If air at 20°C and 1 atm pressure is compressed to 10 atm pressure, calculate (a) the temperature of the air leaving the compressor and (b) the theoretical compressor horsepower required to compress 100 kg of air. The specific heat ratio of air is 1.40; the molecular weight is 29.
4. Air at 25°C and 1 atm containing water vapor at a partial pressure which is 50% of the vapor pressure at 25°C (50% relative humidity) is required for a process. This air is generated by saturating room air by passing through water sprays, compressing this saturated air to a certain pressure, P, and cooling the compressed air to 25°C. The partial pressure of water in the cooled saturated air which leaves the compressor is the vapor pressure of water at 25°C. This air is allowed to expand to 1 atm pressure isothermally. Calculate P such that, after expansion, the air will have 50% relative humidity.
5. An experiment requires a gas mixture containing 20% CO_2, 0.5% O_2, and 79.5% N_2 at 1 atm and 20°C. This gas mixture is purchased premixed and comes in a 10-L tank at a pressure of 130 atm gage. The gas will be used to displace air from packages using a packaging machine which operates by drawing a vacuum completely inside a chamber where the packages are placed, displacing the vacuum with the gas mixture, and sealing the packages. The chamber can hold four packages at a time, and the total void volume in the chamber with the packages in place is 2500 cm^3. How many packages can be treated in this manner before the pressure in the gas tank drops to 1 atm gage?

6. A vacuum pump operates by compressing gases from a closed chamber to atmospheric pressure so that these gases can be ejected to the atmosphere. A vacuum dryer operating at 700 mm Hg vacuum (atmospheric pressure is 760 mm Hg) and 50°C generates 500 g of water vapor per minute by evaporation from a wet material in the dryer. In addition, the leakage rate for ambient air infiltrating the dryer is estimated to be 1 L/h at 1 atm and 20°C.
   (a) Calculate the total volume of gases which must be removed by the vacuum pump per minute.
   (b) If the pump compresses the gas in an adiabatic process, calculate the theoretical horsepower required for the pump. The specific heat ratio for water is 1.30, and that for air is 1.40.
7. The mass rate of flow of air (G) used in correlation equations for heat transfer in a dryer is expressed in kg air/m^2 (h). Use the ideal gas equation to solve for G as a function of the velocity of flow (V, in m/h) of air at temperature T and 1 atm pressure.
8. Use Van der Waal's equation of state to calculate the work done on isothermal expansion of a gas from a volume of 10 to 300 m^3 at 80°C. The initial pressure was 10 atm. Calculate the entropy change associated with the process.

SUGGESTED READING

Himmelblau, D. M. 1967. *Basic Principles and Calculations in Chemical Engineering*, 2nd ed. Prentice-Hall, Englewood Cliffs, N.J.

Hougen, O. A., and Watson, K. W. 1946. *Chemical Process Principles*, Part I. *Material and Energy Balances*. John Wiley & Sons, New York.

Martin, M. C. 1986. *Elements of Thermodynamics*. Prentice-Hall, Englewood Cliffs, N.J.

Peters, M. S. 1954. *Elementary Chemical Engineering*. McGraw-Hill Book Co., New York.

5

Energy Balances

Energy used to be a term everybody took for granted. Now, to a layman, energy has been added to the list of the basic necessities of life. Increasing energy costs have forced people to recognize and appreciate the value of energy more than ever before. Energy conservation is being stressed not only in industrial operations but also in almost all aspects of an individual's daily activities.

Energy is not static; it is always in flux. Even under steady-state conditions, an object absorbs energy from its surroundings and at the same time emits energy to its surroundings at the same rate. When there is an imbalance between the energy absorbed and emitted, the steady state is altered, molecular energy of some parts of the system may increase, new compounds may be formed, or work may be performed. Energy balance calculations can be used to account for the various forms of energy involved in a system.

Energy audits are essential in identifying the effectiveness of energy conservation measures and pinpointing areas where energy conservation can be done. The technique is also useful in the design of processing systems involving heating or cooling to ensure that fluids used for heat exchange are adequately provided and that the equipment is sized adequately to achieve the processing objective at the desired capacity. When energy exchange involves a change in mass due to evaporation or condensation, energy balances can be used during formulation such that, after processing, the product will have the desired composition.

GENERAL PRINCIPLES

An energy balance around a system is based on the first law of thermodynamics, the law of conservation of energy. Mechanical (work), electrical, and thermal energies can all be reduced to the same units. Mechanical input into a system to overcome friction, electrical energy, or electromagnetic energy such as microwaves will be manifested by an increase in the heat content of the system. Defining the surroundings of the system on which to make an energy balance is a process similar to the material balance calculations described in Chapter 3. The basic energy balance equation is:

$$\text{Energy in} = \text{energy out} + \text{accumulation} \qquad (1)$$

If the system is in a steady state, the accumulation term is zero; an unsteady state system will have the accumulation term as a differential expression. When equation 1 is used, all energy terms known to change within the system must be accounted for. The heat contents are expressed as enthalpy based on an increase in enthalpy from a set reference temperature. Mechanical, electrical, or electromagnetic inputs must all be accounted for if their effects on the total heat content are significant. If the system involves only an exchange of energy between two components, the energy balance will be:

$$\text{Energy gain by component 1} = \text{Energy loss by component 2} \qquad (2)$$

Either equation 1 or 2 will give similar results if equation 2 is applicable. Equation 1 is a general form of the energy balance equation.

ENERGY TERMS

The unit of energy in SI is the joule. Conversion factors in Appendix Table A.1 can be used to convert mechanical and electrical energy units to their SI equivalents. Microwave and radiant energies absorbed by a material are usually expressed as a rate of energy flow, energy/time. Energy from ionizing radiation is expressed as an absorbed dose, the Gray (Gy), which has the base unit J/kg. Another accepted form for reporting absorbed ionizing radiation is the rad; 100 rad = 1 Gy.

HEAT

Sensible heat is defined as the energy transferred between two bodies at different temperatures or the energy present in a body by virtue of its temperature. Latent heat is the energy associated with phase transitions, heat of fusion from solid to liquid, and heat of vaporization from liquid to vapor.

Heat Content, Enthalpy. Enthalpy, as defined in Chapter 4, is an intrinsic property, the absolute value of which cannot be measured directly. However, if a reference state is chosen for all components which enter and leave a system such that at this state the enthalpy is considered to be zero, then the change in enthalpy from the reference state to the current state of a component can be considered as the value of the absolute enthalpy for the system under consideration. The reference temperature (T_{ref}) for determining the enthalpy of water in the steam tables (Appendix Tables A.3 and A.4) is 32.018°F or 0.01°C. The enthalpy of any component of a system which would be equivalent to the enthalpy of water obtained from the steam tables at any temperature T is given by:

$$H = C_p(T - T_{ref}) \tag{3}$$

In equation 3, C_p is the specific heat at constant pressure, defined in Chapter 4.

Specific Heat of Solids and Liquids. The specific heat (C_p) is the amount of heat that accompanies a unit change in temperature for a unit mass. The specific heat which varies with temperature is more variable for gases than for liquids or solids. Most solids and liquids have a constant specific heat over a fairly wide temperature range. The enthalpy change of a material with mass m is:

$$q = m \int_{T_1}^{T_2} C_p \, dt \tag{4}$$

Handbook tables give specific heats averaged over a range of temperature. When average specific heats are given, equation 4 becomes:

$$q = mC_{avg}(T_2 - T_1) \tag{5}$$

For solids and liquids, equations 3 and 5 are valid over the range of temperatures encountered in food processing systems.

Table 5.1 shows the average specific heats of various solids and liquids.

For fat-free fruits and vegetables, purees, and concentrates of plant origin, Siebel (1918) observed that the specific heat varies with moisture content and that the specific heat can be determined as the weighted mean of the specific heat of water and the specific heat of the solids.

For a fat-free plant material with a mass fraction of water M, the specific heat of water above freezing is 1 BTU/(lb · °F) or 4186.8 J/(kg · K), and that of nonfat solids is 0.2 BTU/(lb · °F) or 837.36 J/(kg · K). Since the mass fraction of nonfat solids is $(1 - M)$, the weighted average specific heat for unit mass of material above freezing is:

$$\begin{aligned} C_{avg} &= 1(M) + 0.2(1 - M) = (1 - 0.8)M + 0.2 \\ &= 0.8M + 0.2 \quad \text{in BTU/(lb} \cdot {}^\circ\text{F)} \end{aligned} \tag{6}$$

$$\text{In SI: } C_{avg} = 3349M + 837.36 \qquad \text{in J/(kg} \cdot \text{K)} \tag{7}$$

Table 5.1. Specific Heat of Food Products

Product	% H_2O	C_{pm}
Dairy products		
Butter	14	2050
Cream, sour	65	2930
Milk, skim	91	4000
Fresh meat, fish, poultry, and eggs		
Codfish	80	3520
Chicken	74	3310
Egg white	87	3850
Egg yolk	48	2810
Pork	60	2850
Fresh fruits, vegetables, and juices		
Apples	75	3370
Apple juice	88	3850
Apple sauce	—	3730
Beans, fresh	90	3935
Cabbage, white	91	3890
Carrots	88	3890
Corn, sweet, kernels	—	3320
Cucumber	97	4103
Mango	93	3770
Orange juice, fresh	87	3890
Plums, fresh	76.5	3500
Spinach	87	3800
Strawberries	91	3805
Other products		
Bread, white	44	2720
Bread, whole wheat	48.5	2850
Flour	13	1800
Flour	13	1800

Source: Adapted from Polley, S. L., Snyder, O. P., and Kotnour, P. A Compilation of thermal properties of foods. *Food Technol.* 36 (11):76, 1980.

The same principle is used to derive the specific heat below freezing. The specific heat of ice is 0.5 BTU/(lb · °F) and the specific heat of solids nonfat is the same as it is at above freezing.

$$C'_{avg} = 0.3M + 0.2 \quad \text{in BTU/(lb} \cdot \text{°F), frozen} \qquad (8)$$

$$\text{In SI: } C_{avg} = 1256M + 837.36 \quad \text{in J/(kg} \cdot \text{K), frozen} \qquad (9)$$

Equations 6 to 9 are different forms of Siebel's equation, which has been used by ASHRAE (1965) in tabulated values for specific heat of fruits and vegetables.

When fat is present, the specific heat may be estimated from the mass fraction fat (F), mass fraction solids nonfat (SNF), and mass fraction moisture (M) as follows:

Above freezing:

$$C_{avg} = 0.4F + 0.2SNF + M \quad \text{in BTU/(lb} \cdot \text{°F)} \tag{10}$$

$$C'_{avg} = 1674.72F + 837.36SNF + 4186.8M \quad \text{in J/(kg} \cdot \text{K)} \tag{11}$$

Below freezing:

$$C_{avg} = 0.4F + 0.2SNF + 0.5M \quad \text{in BTU/(lb} \cdot \text{°F)} \tag{12}$$

$$C'_{avg} = 1674.72F + 837.36SNF + 2093.4M \quad \text{in J/(kg} \cdot \text{K)} \tag{13}$$

Equations 10 to 13 are more general and can be used instead of equations 6 to 9.

EXAMPLE 1: Calculate the specific heat of beef roast containing 15% protein, 20% fat, and 65% water.

Solution:

$$C'_{avg} = 0.15(0.2) + 0.20(0.4) + 0.65(1)$$
$$= 0.76 \text{ BTU/(lb} \cdot \text{°F)}$$
$$C_{avg} = 0.15(837.36) + 0.2(1674.72) + 0.65(4186.8)$$
$$= 3182 \text{ J/(kg} \cdot \text{K)}$$

EXAMPLE 2: Calculate the specific heat of orange juice concentrate having a solids content of 45%.

Solution: Using Siebel's equation: Calculating a weighted average specific heat:

$$C'_{avg} = 0.2(0.45) + 1(0.55) = 0.64 \text{ BTU/(lb} \cdot \text{°F)}$$

$$C_{avg} = 837.36(0.45) + 4186.8(0.55) = 2679 \text{ J/(kg} \cdot \text{K)}$$

EXAMPLE 3: Calculate the heat required to raise the temperature of a 4.535-kg (10-lb) roast containing 15% protein, 20% fat, and 65% water from 4.44°C (40°F) to 65.55°C (150°F). Express this energy in (a) BTUs, (b) joules, and (c) watt-hours.

Solution: C'_{avg} from Example 1 = 0.76 BTU/(lb · °F).

(a) $q = mC_{avg}(T_2 - T_1)$

$$= 10\ \text{lb}\ \frac{0.76\ \text{BTU}}{\text{lb} \cdot °\text{F}}(150 - 40)°\text{F} = 836\ \text{BTU}$$

(b) Use C_{avg} from Example 1 = 3182 J/(kg · K).

$q = 4.535$ kg [3182 J/(kg · K)](65.55 − 4.44)°K = 0.882 MJ

(c) $$q = 0.882\ \text{MJ} \cdot \frac{10^6\ \text{J}}{\text{MJ}} \cdot \frac{1\ \text{Ws}}{\text{J}} \cdot \frac{1\ \text{h}}{3600\ \text{s}} = 245\ \text{W} \cdot \text{h}$$

The specific heat calculated by Siebel's equation is used by the American Society for Heating, Refrigerating, and Air Conditioning Engineers in one of the most comprehensive tabulated values for specific heat of foodstuffs. However, it is oversimplified, and the assumption that all types of nonfat solids have the same specific heat may not always be correct. Furthermore, Siebel's equation for specific heat below the freezing point assumes that all the water is frozen, and this is most inaccurate.

Specific heats of solids and liquids may also be estimated using correlations obtained from Choi and Okos (1987). The procedure is quite unwieldy. However, a simple BASIC program can be written to calculate the values of specific heat if a number of these calculations have to be made. The program is presented in Chapter 7 in the section on thermophysical properties of foods. The specific heats, in J/(kg · K), as a function of T, (°C), for various components of foods are as follows:

Protein: $C_{pp} = 2008.2 + 1208.9 \times 10^{-3}\,T - 1312.9 \times 10^{-6}\,T^2$

Fat: $C_{pf} = 1984.2 + 1473.3 \times 10^{-3}\,T - 4800.8 \times 10^{-6}\,T^2$

Carbohydrate: $C_{pc} = 1548.8 + 1962.5 \times 10^{-3}\,T - 5939.9 \times 10^{-6}\,T^2$

Fiber: $C_{pfi} = 1845.9 + 1930.6 \times 10^{-3}\,T - 4650.9 \times 10^{-6}\,T^2$

Ash: $C_{pa} = 1092.6 + 1889.6 \times 10^{-3} - 3681.7 \times 10^{-6}\,T^2$

Water above freezing:

$$C_{waf} = 4176.2 - 9.0862 \times 10^{-5}\,T + 5473.1 \times 10^{-6}\,T^2$$

The specific heat of the mixture above freezing is:

$$C_{avg} = P(C_{pp}) + F(C_{pf}) + C(C_{pc}) + Fi(C_{pfi}) + A(C_{pa}) + M(C_{waf}) \quad (14)$$

where P, F, Fi, A, and M represent the mass fraction of protein, fat, fiber, ash, and moisture, respectively.

EXAMPLE 4: Calculate the specific heat of a formulated food product which contains 15% protein, 20% starch, 1% fiber, 0.5% ash, 20% fat, and 43.5% water at 25°C.

Solution: The calculated values, in J/(kg · K), are: $C_{pp} = 2037.6$; $C_{pf} = 2018.0$; $C_{pc} = 1593.8$; $C_{pfi} = 1891.3$; $C_{pa} = 1137.5$; and $C_{waf} = 4184.9$. Substituting in equation 14:

$$\begin{aligned} C_{pavg} &= 0.15(2037.6) + 0.2(1593.8) + 0.01(1891.3) \\ &\quad + 0.005(1137.5) + 0.2(2018) + 0.435(4184.9) \\ &= 2865.06 \text{ J}/(\text{kg} \cdot \text{K}) \end{aligned}$$

By comparison, using Siebel's equation:

$$\begin{aligned} C_p &= 1674.72(0.2) + 837.36(0.15 + 0.01 + 0.005 + 0.2) \\ &\quad + 4186.8(0.435) = 2462 \text{ J}/(\text{kg} \cdot \text{K}) \end{aligned}$$

Values for C_p calculated using Choi and Okos' (1987) correlations, are generally higher than those calculated using Siebel's equations at high moisture contents. Siebel's equations have been found to agree closely with experimental values when $M > 0.7$ and when no fat is present. Choi and Okos' correlation is more accurate at low moisture contents and for a wider range of product composition, since it is based on published literature values for a wide variety of foods. The simplicity of Siebel's equations, however, appeals to most users, particularly when tolerance for error is not too low.

For enthalpy change calculations, Choi and Okos' equations for specific heat must be expressed as an average over the range of temperatures under consideration. The mean specific heat, C^*, over a temperature range T_1 to T_2, where $(T_2 - T_1) = \delta$, $T_2^2 - T_1^2 = \delta^2$ and $T_2^3 - T_1^3 = \delta^3$ is:

$$C^* = \frac{1}{\delta} \int_{T_1}^{T_2} C_p \, dT$$

Thus the equations for the mean specific heats of the various components over the temperature range δ become:

$$\text{Protein: } C_{pp}^* = \left[\frac{1}{\delta}\right][2008.2(\delta) + 0.6045(\delta^2) - 437.6 \times 10^{-6}(\delta^3)]$$

Fat: $C^*_{pf} = \left[\frac{1}{\delta}\right][1984.2(\delta) + 0.7367(\delta^2) - 1600 \times 10^{-6}(\delta^3)]$

Carbohydrate: $C^*_{pc} = \left[\frac{1}{\delta}\right][1548.8(\delta) + 0.9812(\delta^2) - 1980 \times 10^{-6}(\delta^3)]$

Fiber: $C^{*p}_{fi} = \left[\frac{1}{\delta}\right][1845.9(\delta) + 0.9653(\delta^2) - 1500 \times 10^{-6}(\delta^3)]$

Ash: $C^*_{pa} = \left[\frac{1}{\delta}\right][1092.6(\delta) + 0.9448(\delta^2) - 1227 \times 10^{-6}(\delta^3)]$

Water: $C^*_{waf} = \left[\frac{1}{\delta}\right][4176.2(\delta) - 4.543 \times 10^{-5}(\delta^2) + 1824 \times 10^{-6}(\delta^3)]$

$$C^*_{\text{avg}} = P(C^*_{pp}) + F(C^*_{pf}) + C(C^*_{pc}) + Fi(C^*_{pfi}) + A(C^*_{pa}) + M(C^*_{waf})$$

EXAMPLE 5: Calculate the mean specific heat of the formulated food product in Example 4 in the temperature range 25° to 100°C.

Solution: $\delta = 75°C$, $\delta^2 = 9375°C^2$, and $\delta^3 = 984{,}375°C^3$. The mean specific heats of the components, in J/(kg · K), are: $C^*_{pp} = 2078$; $C^*_{pf} = 2055$; $C^*_{pc} = 1645$; $C^*_{pfi} = 1946$; $C^*_{pa} = 1195$; $C^*_{waf} = 4200$.
The mean specific heat is: $C^*_{pavg} = 2881$ J/(kg · K).

Enthalpy Change with a Change in Phase. When considering the heat to be removed during freezing of a food product, a change in phase is involved and the latent heat of fusion must be considered. Not all water in a food changes into ice at the freezing point. Some unfrozen water exists below the freezing point; therefore, Siebel's equations for specific heat below the freezing point are very inaccurate. The best method for determining the amount of heat which must be removed during freezing, or the heat input for thawing, is to calculate the enthalpy change. One method for calculating the enthalpy change below the freezing point (good only for moisture contents between 73 and 94%) is the procedure of Chang and Tao (1981). In this correlation, it is assumed that all water is frozen at 227.6°K (−50°F).

A reduced temperature (T_r) is defined as:

$$T_r = \frac{T - 227.6}{T_f - 227.6}$$

T_f = the freezing point temperature and T = the temperature at which the enthalpy is being determined. Two parameters, a and b, have been calculated for different products as a function of the mass fraction of moisture in the product, M. The correlation equations are:

Meats:

$$a = 0.316 - 0.247(M - 0.73) - 0.688(M - 0.73)^2 \tag{15}$$

$$b = 22.95 + 54.68(a - 0.28) - 5589.03(a - 0.28)^2 \tag{16}$$

Vegetables, fruits, juices:

$$a = 0.362 + 0.0498(M - 0.73) - 3.465(M - 0.73)^2 \tag{17}$$

$$b = 27.2 - 129.04(a - 0.23) - 481.46(a - 0.23)^2 \tag{18}$$

The freezing point (T_f), in °K, is:

$$\text{Meats: } T_f = 271.18 \times 1.47M \tag{19}$$

$$\text{Fruits and vegetables: } T_f = 287.56 - 49.19M + 37.07M^2 \tag{20}$$

$$\text{Juices: } T_f = 120.47 + 327.35M - 176.49M^2 \tag{21}$$

The enthalpy at the freezing point, H_f, in J/kg, relative to 227.6°K, is:

$$H_f = 9792.46 + 405{,}096M \tag{22}$$

The enthalpy at temperature T relative to 227.6°K is determined by:

$$H = H_f\left[aT_r + (1 - a)T_r^b\right] \tag{23}$$

EXAMPLE 1: Calculate the freezing point and the amount of heat which must be removed in order to freeze 1 kg of grape juice containing 25% solids from the freezing point to −30°C.

Solution: $Y = 0.75$
Using equation 21 for juices:

$$T_f = 120.47 + 327.35(0.75) - 176.49(0.75)^2 = 266.7°\text{K}$$

Using equation 22:

$$H_f = 9792.46 + 405{,}096(0.75) = 313{,}614$$

Using equation 17:

$$a = 0.362 + 0.0498(0.02) - 3.465(0.02)^2$$
$$= 0.3616$$

Using equation 18:

$$b = 27.2 - 129.04(0.1316) - 481.46(0.1316)^2$$
$$= 1.879$$

$$T_r = \frac{-30 + 272 - 227.6}{266.7 - 227.6} = 0.394$$

Using equation 23:

$$H = 313{,}614\left[(0.3616)0.394 + (1 - 0.3616)(0.394)^{1.879}\right]$$
$$= 79{,}457 \text{ J/kg}$$

The enthalpy change from T_f to $-30°C$ is:

$$\Delta H = 313{,}614 - 79{,}457 = 234{,}157 \text{ J/kg}$$

Specific Heats of Gases and Vapors. As stated in Chapter 4, the specific heat of gases depends upon whether the process is carried out at constant pressure or at constant volume. If a gas is heated by blowing air across heating elements, the process is a constant-pressure process. The specific heat is designated by C_p, the specific heat at constant pressure. The heat required to raise the temperature of a gas with mass m at constant pressure equals the change in enthalpy, ΔH. The enthalpy change associated with a change in temperature from a reference temperature T_0 to T_2 is:

$$\Delta H = m \int_{T_r}^{T_2} C_p \, dT = mC_{pm}(T_2 - T_0) \tag{24}$$

C_{pm} is the mean specific heat over the temperature range T_0 to T_2, and C_p is the expression for specific heat as a function of T.

If a gas is heated from any temperature T_1 to a final temperature T_2, the change in enthalpy accompanying the process must be calculated as follows:

$$\Delta H = q = mC_{pm}(T_2 - T_0) - mC'_{pm}(T_1 - T_0) \tag{25}$$

where C'_{pm} = mean specific heat from the reference temperature T_0 to T_1. Tabulated values for the mean specific heats of gases are based on an ambient temperature of 77°F or 25°C as the reference temperature. Table 5.2 lists the mean specific heats of common gases in the American Engineering System of units, and Table 5.3 lists the specific heats of the same gases in SI. The values for C_{pm} in Tables 5.2 and 5.3 change very little at the temperatures ordinarily

Table 5.2. Mean Heat Capacities of Various Gases from 77°F to Indicated Temperature

Temperature (°F)	Mean specific heat in BTU/lb(°F)				
	O_2	N_2	CO_2	H_2O vapor	Air
77	0.219	0.248	0.202	0.446	0.241
100	0.219	0.248	0.203	0.446	0.241
200	0.221	0.249	0.209	0.448	0.241
300	0.223	0.249	0.216	0.451	0.242
400	0.224	0.250	0.221	0.454	0.242
500	0.226	0.250	0.226	0.457	0.243
600	0.228	0.251	0.231	0.461	0.245
700	0.230	0.253	0.236	0.456	0.246

Source: Calculated from Mean molal heat capacity of gases at constant pressure. *In* Harper, J. C. 1976. *Elements of Food Engineering.* AVI Publishing Co., Westport, Conn.

used in food processes. Therefore, C_{pm} based on the temperature range from T_0 to T_2 can be used for ΔH between T_1 and T_2, with very little error in comparison with the use of equation 25.

EXAMPLE 1: Calculate the heating requirement for an air drier that uses 2000 ft^3/min of air at 1 atm and 170°F if ambient air at 70°F is heated to 170°F for use in the process.

Solution:

$$q = mC_{pm}(170 - 70)$$

Table 5.3. Mean Heat Capacities of Various Gases from 25°C to Indicated Temperature

Temperature (°C)	Mean specific heat J/kgK				
	O^2	N_2	CO_2	H_2O vapor	Air
25	916	1037	847	1863	1007
50	919	1039	858	1868	1008
100	926	1041	879	1880	1011
125	930	1043	889	1886	1012
150	934	1044	899	1891	1014
175	936	1045	910	1897	1015
200	938	1047	920	1903	1017
250	945	1049	941	1914	1020
300	952	1052	962	1925	1023
350	960	1054	983	1937	1026

Source: Calculated from Mean molal heat capacity of gases at constant pressure. *In* Harper, J. C. 1976. *Elements of Food Engineering.* AVI Publishing Co., Westport, Conn.

m, the mass of air/min going through the drier, in pounds, will be calculated using the ideal gas equation.

$$PV = \frac{m}{M} RT; \qquad M = \text{mol. wt. air} = 29 \text{ lb/lb mole}$$

$$m = \frac{PVM}{RT} = \frac{14.7(144)\text{ lb/ft}^2[2000\text{ ft}^3/\text{min}][29\text{ lb/lbmole}]}{[1545\text{ ft lb}_f/\text{lbmole}°\text{R}](460 + 170)°\text{R}}$$

$$= 126.14 \text{ lb/min}$$

From Table 5.2, C_{pm} is constant at 0.241 BTU/(lb · °F) up to 200°F; therefore, it will be used for the temperature range 70° to 170°F.

$$q = 126.13 \frac{\text{lb}}{\text{min}} \times \frac{0.241\text{ BTU}}{(\text{lb} \cdot °\text{F})(170 - 70)°\text{F}}$$

$$= 3039.8 \text{ BTU/min}$$

EXAMPLE 2: How much heat would be required to raise the temperature of 10 m³/s of air at 50°C to 120°C at 1 atm?

Solution: Using equation 25:

$$C'_{pm} \text{ at } T_1 = 50°\text{C} = \frac{1008\text{ J}}{(\text{kg} \cdot \text{K})}$$

$$C'_{pm} \text{ at } T_2 = 120°\text{C} = \frac{1012\text{ J}}{\text{kg} \cdot \text{K}}$$

The mean specific heat from T_0 to 50°C and from T_0 to 120°C are different enough to require the use of equation 25 to determine ΔH. The reference temperature, T_0 for C_{pm}, in Table 5.3 is 25°C. The mass is determined using the ideal gas equation.

$$m = \frac{PVM}{RT}$$

$$= \frac{1\text{ atm}(10\text{ m}^3)[29\text{ kg/kg mole}]}{8.206[\text{m}^3\text{ atm/kg mole K}](50 + 273)(\text{K})}$$

$$= 0.1094 \text{ kg/s}$$

$$q = 0.1094(1012)(120 - 25) - 0.1094(1008)(50 - 25)$$

$$= 7762 \text{ J/s}$$

PROPERTIES OF SATURATED AND SUPERHEATED STEAM

Steam and water are the two most frequently utilized heat transfer media in food processing. Water is also a major component of food products. The steam tables, which list the properties of steam, are a very useful reference when determining heat exchange involving a food product and steam or water. At temperatures above the freezing point, water can exist in any of the following forms:

Saturated Liquid. Liquid water in equilibrium with its vapor. The total pressure above the liquid must be equal to or higher than the vapor pressure. If the total pressure above the liquid exceeds the vapor pressure, some other gas is present in the atmosphere above the liquid. If the total pressure above a liquid equals the vapor pressure, the liquid is at the boiling point.

Saturated Vapor. This is also known as *saturated steam* and is vapor at the boiling temperature of the liquid. Lowering the temperature of saturated steam at constant pressure by a small increment will cause the vapor to condense to liquid. The phase change is accompanied by a release of heat. If heat is removed from the system, temperature and pressure will remain constant until all vapor is converted to liquid. Adding heat to the system will change either the temperature or the pressure or both.

Vapor-Liquid Mixtures. Steam with less than 100% quality. Temperature and pressure correspond to the boiling point; therefore, water can exist as either saturated liquid or saturated vapor. Addition of heat will not change the temperature and pressure until all saturated liquid is converted to vapor. Removing heat from the system will also not change temperature and pressure until all vapor is converted to liquid.

Steam Quality. The percentage of a vapor-liquid mixture that is in the form of saturated vapor.

Superheated Steam. Water vapor at a temperature higher than the boiling point. The number of degrees the temperature exceeds the boiling temperature is the *degrees superheat.* Addition of heat to superheated steam may increase the superheat at constant pressure or change both the pressure and the temperature at constant volume. Removing heat will allow the temperature to drop to the boiling temperature, where it will remain constant until all the vapor has condensed.

The Steam Tables. The steam tables are tabulated values for the properties of saturated and superheated steam.

The Saturated Steam Table. The saturated steam table consists of entries under the headings of temperature, absolute pressure, specific volume, and enthalpy. A saturated steam table is in the Appendix (Tables A.3 and A.4).

The temperature and absolute pressure correspond to the boiling point, or the temperature and pressure under which steam can be saturated. The absolute pressure at a given temperature is also the vapor pressure.

The rest of the table consists of three general headings, each of which is subdivided into saturated liquid, evaporation, and saturated vapor. The entries under saturated liquid give the properties of liquid water at the indicated temperature. The entries under saturated vapor give the properties of steam at the boiling point. The entries under evaporation are changes due to the phase transformation and are the difference between the properties of saturated vapor and saturated liquid.

Specific volume is the reciprocal of the density. It is the volume in cubic feet occupied by 1 lb of water or steam under the conditions given.

Enthalpy is the heat content of a unit mass of steam or water at the indicated temperature and pressure. Enthalpy values in the steam tables are calculated from a base temperature of 0°C.

The energy change associated with a change in the temperature or pressure of steam is the difference in the initial and final enthalpies. The following examples illustrate the use of the steam tables. In these examples, atmospheric pressure is specified and may not be a standard atmosphere. Please refer to the section "Absolute Temperature and Pressure" in Chapter 4 for a discussion of the difference between atmospheric pressure and a standard atmosphere.

EXAMPLE 1: At what vacuum would water boil at 80°F? Express this in (a) inches of mercury vacuum (given: atm pressure = 30 in. Hg) and (b) absolute pressure in pascals.

Solution: From steam tables, Appendix Table A.4: The boiling pressure for water at 80°F is 0.50683 psia.

$$\text{(a) Inches Hg absolute pressure} = 0.5068\ \frac{\text{lb}}{\text{in.}^2}\left[\frac{2.35\ \text{in. Hg}}{\text{lb/in.}^2}\right]$$

$$= 1.03\ \text{in. Hg absolute}$$

$$\text{Vacuum} = 30\ \text{in. Hg} - 1.03\ \text{in. Hg}$$

$$= 29.97\ \text{in. Hg vacuum}$$

$$\text{(b) Pressure} = \frac{0.50683\ \text{lb}_f}{\text{in.}^2}\left[\frac{6894.757\ \text{Pa}}{\text{lb/in.}^2}\right]$$

$$= 3.494\ \text{kPa absolute pressure}$$

EXAMPLE 2: If 1 lb of water at 100 psig and 252°F is allowed to expand to 14.7 psia, calculate (a) the resulting temperature after expansion and (b) the quantity of vapor produced.

Solution: The absolute pressure = 100 + 14.7 = 114.7 psia. At 252°F water will not boil until the pressure is reduced to 30.9 psia. The water therefore is at a temperature much below the boiling point at 114.7 psia, and it would have the properties of liquid water at 252°F.

(a) After expansion to 14.7 psia, the boiling point at 14.7 psia is 212°F. Part of the water will flash to water vapor at 212°F, and the remaining liquid will also be at 212°F.

(b) The enthalpy of water at 252°F is (h_f at 252°F) 220.62 BTU/lb. Basis: 1 lb H_2O. Heat content = 220.62 BTU. When pressure is reduced to 14.7 psia, some vapor will be formed, but the total heat content of both vapor and liquid at 212°F and 14.7 psia will still be 220.62 BTU.

If x = wt vapor produced, $1 - x$ = wt water at 212°F and 14.7 psia:

$$x(h_g) + (1 - x)(h_f) = 220.62$$

$$h_g = 1150.5 \text{ BTU/lb}; \; h_f = 180.17 \text{ BTU/lb}$$

$$x(1150.5) + (1 - x)(180.17) = 210.62$$

$$x = \frac{220.62 - 180.17}{1150.5 - 180.17} = \frac{40.45}{970.33} = 0.0417 \text{ lb } H_2O$$

EXAMPLE 3: If water at 70°F is added to an evacuated vessel, initially at 0 psia, what will be the pressure inside the vessel when equilibrium is finally attained at 70°F?

Solution: Since the vessel is completely evacuated, the gaseous phase after introduction of water will be 100% water vapor. Upon introduction, the water will vaporize until the pressure of water in the space above the liquid equals the vapor pressure.

Pressure = vapor pressure of water at 70°F

From steam tables, pressure at 70°F = 0.36292 psia

EXAMPLE 4: If the vessel in Example 3 had initially 14.7 psia absolute pressure and contained completely dry air, what would be the absolute pressure after introducing the water, assuming that none of the original air had escaped during the process?

Solution:

$$\text{Pressure} = \text{partial pressure of air} + \text{partial pressure of water (volume is constant)}$$

$$= \text{original pressure of dry air} + \text{vapor pressure of water}$$

$$= 14.7 + 0.36292$$

$$= 15.063 \text{ psia}$$

EXAMPLE 5: How much heat would be given off by cooling steam at 252°F and 30.883 psia to 248°F at the same pressure?

Solution: First, check the state of water at 30.883 psia and 252°F and 248°F.

From the steam tables, the boiling point of water at 30.883 psia is 252°F. Therefore, steam at 252°F and 30.883 psia is saturated vapor. At 30.883 psia and 248°F, water will be in the liquid state, since 248°F is below the boiling temperature at 30.883 psia.

$$\text{Heat given off} = q = h_g \text{ at } 252°\text{F} - h_f \text{ at } 248°\text{F}$$

From the steam tables, Appendix Table A.3:

$$h_g \text{ at } 252°\text{F} = 1164.78 \text{ BTU/lb}$$

$$h_f \text{ at } 248°\text{F} = 216.56 \text{ BTU/lb}$$

$$q = 1164.78 - 216.56 = 948.22 \text{ BTU/lb}$$

Saturated steam is a very efficient heat transfer medium. Note that with only a 2°F change in temperature, 948 BTU/lb of steam is given off. The heat content of saturated vapors come primarily from the latent heat of vaporization, and it is possible to extract this heat simply by causing a phase change at constant temperature and pressure.

The Superheated Steam Tables. A superheated steam table is presented in Appendix Table A.2. Both temperature and absolute pressure must be specified to define the degree of superheat accurately. From the temperature and absolute pressure, the specific volume v in ft^3/lb and the enthalpy h in BTU/lb can be read from the table. Example problems on the use of the superheated steam tables are as follows.

EXAMPLE 1: How much heat is required to convert 1 lb of water at 70°F to steam at 14.696 psia and 250°F?

Solution: First, determine the state of steam at 14.696 psia and 250°F At 14.696 psia, the boiling point is 212°F. Steam at 250°F and 14.696 psia is superheated steam. From the superheated steam table, h at 250°F is 1168.8 BTU/lb.

$$\text{Heat required} = h_g \text{ at } 250°\text{F and } 14.696 \text{ psia} - h_f \text{ at } 70°\text{F}$$
$$= 1168.8 \text{ BTU/lb} - 38.05 \text{ BTU/lb}$$
$$= 1130.75 \text{ BTU/lb}$$

EXAMPLE 2: How much heat would be given off by cooling superheated steam at 14.696 psia and 500°F to 250°F at the same pressure?

Solution: Basis: 1 lb of steam.

$$\text{Heat given off} = q = h \text{ at } 14.696 \text{ psia and } 500°\text{F} - h_g \text{ at } 14.696 \text{ psia and } 250°\text{F}$$
$$= 1287.4 - 1168.8$$
$$= 118.6 \text{ BTU/lb}$$

Superheated steam is not a very efficient heating medium. Note that a 250°F change in temperature is accompanied by the extraction of only 118.6 BTUs of heat.

Double Interpolation from Superheated Steam Tables. Since the entries in the steam tables are not close enough to cover all conditions, it may be necessary to interpolate between entries to obtain the properties under a given set of conditions. In the case of superheated steam where both temperature and pressure are necessary to define the state of the system, double interpolation is sometimes necessary. The following example shows how the double interpolation is carried out.

EXAMPLE: Calculate the enthalpy of superheated steam at 320°F and 17 psia.

Solution: Entries in Appendix Table A.2 show enthalpies at 15 and 20 psia and 300° and 350°F. The tabular entries and the need for interpolation are as follows:

P (psia)	Enthalpy	
	300°F	350°F
15	1192.5	1216.2
17	?	?
20	1191.4	1215.4

Enthalpies at 300° and 350°F at 17 psia are obtained by interpolating between 15 and 20 psia at each temperature.

At 300°F

$P = 15$, $h = 1192.5$; $P = 20$, $h = 1191.4$; at $P = 17$, $h = ?$

$$h_{(300°\text{F and } 17\text{ psia})} = 1191.4 + \frac{(1192.5 - 1191.4)}{(20 - 15)}(20 - 17)$$

$$= 1192.06 \text{ BTU/lb}$$

At 350°F

$P = 15$, $h = 1216.2$; $P = 20$, $h = 1215.4$; at $P = 17$, $h = ?$

$$h_{(350°\text{F}, 17\text{ psia})} = 1215.4 + \frac{(1215.4 - 1216.2)}{(20 - 15)}(20 - 17)$$

$$h = 1215.88 \text{ BTU/lb}$$

Now it is possible to interpolate between 300° and 350°F to obtain the enthalpy at 17 psia and 320°F.

$$h = 1215.88 - \frac{(1192.06 - 1215.88)}{(350 - 300)}(350 - 320)$$

$$= 1215.88 - 14.29$$

$$= 1201.59 \text{ BTU/lb}$$

Properties of Steam Having Less Than 100% Quality. If steam is not 100% vapor, the properties can be determined on the basis of the individual properties of the component.

If x = % quality:

$$v = xv_g + (1 - x)v_f$$

$$h = xh_g + (1 - x)h_f$$

EXAMPLE 1: Calculate the enthalpy of steam at 252°F having 80% quality.

Solution: From the saturated steam tables, Appendix Table A.4: At 252°F saturated steam or water has the following properties:

$$h_f = 220.62 \text{ BTU/lb}$$

$$h_g = 1164.78 \text{ BTU/lb}$$

$$h = 1164.78(0.8) + 220.62(0.2)$$

$$= 931.82 + 44.12$$

$$= 975.94 \text{ BTU/lb}$$

Note that only the temperature of steam is given in this problem. If either the temperature or the pressure is given, but not both, steam is at the boiling point.

HEAT BALANCES

Heat balance calculations are treated in the same manner as material balances. The amount of heat entering a system must equal the amount of heat leaving the system, or: Heat in = heat out + accumulation. At a steady state, the accumulation term is zero and heat entering the system must equal that leaving the system.

Heat balance problems are facilitated by using diagrams which show process streams bringing heat in and taking heat out of a system.

EXAMPLE 1: Calculate the amount of water that must be supplied to a heat exchanger that cools 100 kg/h of tomato paste from 90° to 20°C. The tomato paste contains 40% solids. The increase in water temperature should not exceed 10°C while passing through the heat exchanger. There is no mixing of water and tomato paste in the heat exchanger.

Solution: The diagram of the system is shown in Fig. 5.1. This problem may be solved by assuming a datum temperature from which enthalpy calculations are made. This datum temperature is the lowest of the process stream temperatures. Let T_1 = inlet water temperature = 20°C and T_2 = exit water temperature = 20 + 10 = 30°C. Let 20°C be the reference temperature for enthalpy calculations. The specific heat of water = 4187 J/(kg · K), and that of tomato paste is obtained, using equation 7.

$$C_{avg} = 3349(0.6) + 837.36 = 2846.76 \text{ J/(kg} \cdot \text{K)}$$

Heat content of entering tomato paste:

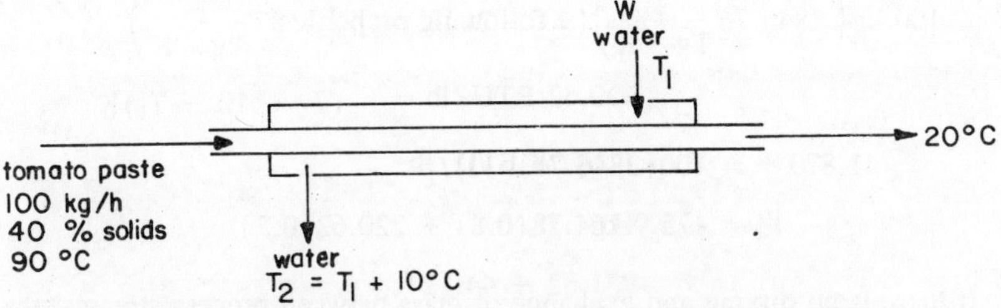

Fig. 5.1. Diagram of a heat exchange system for cooling tomato paste.

$$q_1 = (100 \text{ kg})[2846.76 \text{ J}/(\text{kg} \cdot \text{K})](90 - 20)\,°\text{K}$$

$$= 19.927 \text{ MJ}$$

Heat content of tomato paste leaving system:

$$q_2 = 100 \text{ kg}[2846.76 \text{ J}/(\text{kg} \cdot \text{K})](20 - 20)\text{K} = 0$$

Let W = kg water entering the system

$$q_3 = W \text{ kg}[4187 \text{ J}/(\text{kg} \cdot \text{K})](20 - 20)\text{K} = 0$$

$$q_4 = \text{heat content of water leaving the system}$$

$$= W \text{ kg}\left(\frac{4187 \text{ J}}{\text{kg} \cdot \text{K}}\right)(30 - 20)\text{K} = 41{,}870(W)\text{J}$$

The heat balance is:

$$q_1 + q_3 = q_2 + q_4$$

Since q_2 and $q_3 = 0$, $q_1 = q_4$ and:

$$(19.927 \text{ MJ})\frac{10^6 \text{ J}}{\text{MJ}} = 41{,}870(W)\text{J}$$

$$W = \frac{19.927 \times 10^5}{4187} = 475.9 \text{ kg}$$

The heat balance may also be expressed as follows:

Heat gain by water = Heat loss by the tomato paste

$$100\ \text{kg}\left(2846.76\ \frac{\text{J}}{\text{kg}\cdot\text{K}}\right)(90-20)\text{K} = W\left(4187\ \frac{\text{J}}{\text{kg}\cdot\text{K}}\right)(T_1 + 10 - T_1)\text{K}$$

$$41{,}870\,W = 100(2846.76)(70)$$

$$W = 475.9\ \text{kg}$$

If there is no mixing and exchange of mass between process streams, the calculation is simplified by equating heat gain by one process stream to heat loss by the other. However, when mixing and material transfer occur, a heat balance based on the heat content of each stream entering and leaving the system will simplify analysis of the problem.

EXAMPLE 2: Calculate the amount of saturated steam at 121.1°C that must be supplied to a dehydrator per hour. Steam condenses in the heater which heats the drying air from steam to water at 121.1°C. The dehydrator is operated as follows: Apples at 21.1°C enter the dehydrator with 80% moisture and leave the dehydrator at 37.7°C and 10% moisture. One hundred pounds per hour of fresh apples enter the drier. Fresh air at 21.1°C and a humidity of 0.002 kg H_2O/kg dry air enters the drier, mixes with recycled hot air until the humidity is 0.026 kg H_2O/kg dry air, and is heated to 76.7°C using steam in a finned heat exchanger. Hot air leaves the heater at 43.3°C and a humidity of 0.04 kg H_2O/kg dry air.

Solution: The diagram for the process is shown in Fig. 5.2. Use as a basis 1 h of operation.

The system looks complicated with the hot air recycling, but if the bound-

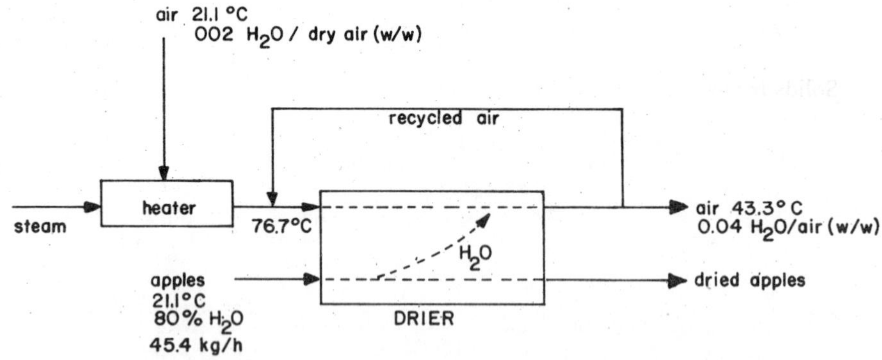

Fig. 5.2. Diagram of material and energy balance during dehydration of apples.

ary of the system is enlarged as shown in Fig. 5.2, solving the problem is considerably simplified.

The problem is solved by separating air into dry air and water vapor, and apples into dry matter and water. The heat contents of each component entering and leaving the system will be calculated. The lowest temperature, 70°F, may be used as the base temperature for heat content calculation, as in the preceding example, but since the steam tables will be used to determine the enthalpy of water and steam, it will be more consistent to use the reference temperature for the steam tables of 0°C.

The following enthalpies are to be calculated:

q_1 = enthalpy of H_2O in entering air (vapor at 21.1°C)

q_2 = enthalpy of dry air entering at 21.1°C

q_3 = enthalpy of H_2O in apples entering (liquid at 21.1°C)

q_4 = enthalpy of dry matter in apples entering at 21.1°C

Heat input = q

q_5 = enthalpy of H_2O in exit air (vapor at 43.3°C)

q_6 = enthalpy of dry air leaving at 43.3°C

q_7 = enthalpy of H_2O in apples leaving (liquid at 37.7°C)

q_8 = enthalpy of dry matter in apples leaving at 37.3°C

The heat balance is:

$$q + q_1 + q_2 + q_3 + q_4 = q_5 + q_6 + q_7 + q_8$$

q_2 and q_6 are calculated using the quantity of dry air that enters the system per hour. The latter is obtained by performing a material balance on water:

Amount of water lost by apples = amount of water gained by air

Solids balance for apples:

$$45.4(0.2) = x(0.9); \qquad x = \text{wt dried apples}$$

$$x = \frac{45.4(0.2)}{0.9} = 10.09 \text{ kg/h}$$

$$\text{Water lost by apples} = 45.4 - 10.09 = 35.51 \text{ kg/h}$$

$$\frac{\text{Water gained by air}}{\text{kg dry air}} = (0.04 - 0.002) = 0.038$$

Let w = mass of dry air:

Total water gained by air = $0.038\,w$

Material balance on water: $0.038\,w = 35.31$

$$w = 929.21 \text{ kg dry air/h}$$

The mean specific heat of air is obtained from Table 5.3. These values are:

$$25°\text{C}; \quad C_{pm} = 1008 \text{ J/(kg} \cdot \text{K)}$$

$$50°\text{C}; \quad C_{pm} = 1007 \text{ J/(kg} \cdot \text{K)}$$

These mean specific heats will be used for determining enthalpies of air at 21.10 and 43.3°C. From the section "Specific Heat of Solids and Liquids":

$$C_p \text{ for solids nonfat in apples} = 837.36 \text{ J/(kg} \cdot \text{K)}$$

Use steam tables for the heat contents of water and steam. Calculating the enthalpies: $q = mC_{pm}(T - 0)$, since 0°C is used as the reference temperature.

$$q_1 = (929.21 \text{ kg dry air}) \frac{0.002 \text{ kg } H_2O}{\text{kg dry air}} (h_g \text{ at } 21.1°\text{C})$$

From Appendix Table A.3, h_g at 21.1°C by interpolation = 2.54017 MJ/kg.

$$q_1 = 020.21(0.002)(2.54017 \times 10^6) = 4.7207 \text{ MJ}$$

$$q_2 = (929.21) \text{ kg dry air } (1008) \text{ J/kg} \cdot \text{K } (21.1 - 0)$$

$$q_2 = 19.7632 \text{ MJ}$$

$$q_3 = 45.4(0.8)(h_f \text{ at } 70°\text{F}).$$

From Appendix Table A.3, h_f at 21.1°C = 0.08999 MJ/kg.

$$q_3 = 45.5(0.8)(0.08999 \times 10^6) = 3.2684 \text{ MJ}$$

$$q_4 = (45.4)(0.2) \text{ lb dry solids} \left(\frac{837.36 \text{ J}}{\text{kg} \cdot \text{K}}\right)(21.1 - 0)°\text{C}$$

$$= 0.16043 \text{ MJ}$$

$$q_5 = (929.21 \text{ kg dry air})\left(0.04 \frac{\text{kg } H_2O}{\text{kg dry air}}\right)(h_g \text{ at } 43.3°\text{C})$$

From Appendix Table A.3, h_g at 43.3°C = 2.5802 M/kg.

$$q_5 = (929.21)(0.04)(2.5802 \times 10^6) = 95.9019 \text{ MJ}$$

$$q_6 = (929.21 \text{ kg dry air}) \frac{1007 \text{ J}}{\text{kg K}} (43.3 - 0)$$

$$= (929.21)(1007)(43.3) = 40.5164 \text{ MJ}$$

$$q_7 = (10.09)(0.1) \text{ lb } H_2O \ (h_f \text{ at } 37.7°\text{C})$$

From Appendix Table A.3, h_f at 37.7°C = 0.15845 MJ/kg.

$$q_7 = (10.09)(0.1)(0.15845) = 0.15987 \text{ MJ}$$

$$q_8 = (10.09)(0.9) \text{ lb dry matter} \left(\frac{837.36 \text{ J}}{\text{kg F}}\right)(37.7 - 0)$$

$$= (10.09)(0.9)(837.36)(37.7) = 0.28667 \text{ MJ}$$

The heat balance with q in MJ is:

$$q + (4.7207 + 19.7632 + 3.2684 + 0.16043)$$

$$= (95.9019 + 40.5164 + 0.15988 + 0.28667)$$

$$q = 136.8648 - 27.91273 = 108.952 \text{ MJ}$$

Since the basic is 1 h of operation, the heating requirement for the process is 108.952 MJ/h. From Appendix Table A.3, the heat of vaporization of steam at 121.1°C is 2.199144 MJ/kg.

$$\text{The amount of steam supplied/h} = \frac{108.952 \text{ MJ/h}}{2.199144 \text{ J/kg}}$$

$$= 49.543 \text{ kg/h}$$

EXAMPLE 3: Calculate the amount of steam at 121.1°C (250°F) that must be added to 100 kg of a food product with a specific heat of 3559 J/(kg · K) to heat the product from 4.44°C (40°F) to 82.2°C (180°F) by direct steam injection.

Solution: The diagram of the system is shown in Fig. 5.3.

Let x = kg of steam required. From the steam table, the enthalpies of water and steam are as follows:

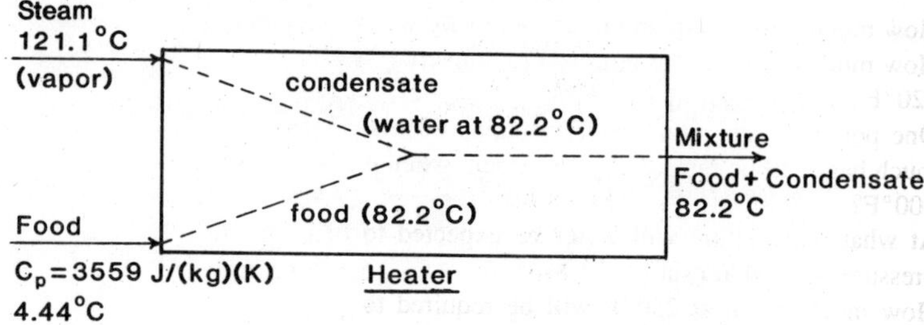

Fig. 5.3. Diagram of a process involving direct steam injection heating of a food product.

	h From Table A.4 BTU/lb	h Converted to MJ/kg
Steam at 121.1°C (250°F)	$h_g = 1164.1$	2.70705
Water at 82.2°C (180°F)	$h_f = 148.00$	0.34417

$$\frac{\text{Heat loss from steam}}{\text{kg}} = h_g \text{ at } 121.1°\text{C} - h_f \text{ at } 82.2°\text{C}$$

$$= 2.70705 - 0.34417 = 2.36288 \text{ MJ/kg}$$

$$\text{Total heat loss by steam} = x(2.36288) \text{ MJ}$$

$$\text{Heat gain by product} = 100 \text{ kg} \left(\frac{3559 \text{ J}}{\text{kg} \cdot \text{K}}\right)(82.2 - 4.44)\text{K}$$

$$= 27.67478 \text{ MJ}$$

$$x(2.36288) = 27.67478$$

$$x = 11.71 \text{ kg steam required}$$

PROBLEMS

1. What pressure is generated when milk is heated to 135°C in a closed system? If the system is not pressured, can this temperature be attained?
2. A process for heating food with saturated steam at temperatures below the boiling point of water is performed under a vacuum. At what vacuum should a system be operated to heat a material with saturated steam at 150°F?
3. If a retort indicates a pressure of 15 psig but the temperature indicator registers only 248°F, what does this indicate?
4. An evaporator is operated at 15 in. Hg vacuum. What is the temperature of the product inside the evaporator?

5. How much heat is required to convert 1 kg water at 20°C to steam at 120°C?
6. How much heat must be removed to convert 1 lb of steam at 220°F to (a) water at 220°F and (b) water at 120°F?
7. One pound of steam at 260°F contains 80% steam and 20% liquid water. How much heat will be released by this steam when it is allowed to condense to water at 200°F?
8. At what temperature will water be expected to boil at 10 in. Hg vacuum? Atm pressure = 14.696 psia.
9. How much steam at 250°F will be required to heat 10 lb of water from 70° to 210°F in a direct steam injection heater?
10. How much heat will be required to convert steam at 14.696 psig to superheated steam at 600°F at the same pressure?
11. Ten pounds of water at a pressure of 20 psig is heated to a temperature of 250°F. If this water is allowed to empty into an open vessel at atmospheric pressure, how much of it will remain in the liquid phase?
12. (a) If water at 70°F is introduced into an evacuated vessel where the original pressure is 0 psia, what will be the pressure inside the vessel at equilibrium? Assume no change in the temperature of water.
    (b) If the original pressure is 14.696 psia, what will be the final pressure?
13. Determine the heat content in BTU/lb for water (it could be liquid, saturated steam, or superheated steam) under the following conditions:
    (a) 180°F and 14.696 psia pressure.
    (b) 300°F and 14.696 psia pressure.
    (c) 212.01°F and 14.696 psia pressure.
14. In the formulation of a pudding mix, it is desired that the solids content of the product will be 20%. The product leaving the batch tank has a temperature of 26.67°C (80°F) and is preheated to 90.56°C (195°F) by direct steam injection, using culinary steam (saturated) at 104.4°C (220°F) followed by heating in a closed system to sterilizing temperatures. There is no further loss or gain of moisture in the rest of the process. What should be the solids content of the formulation in the batch tank such that, after the direct steam injection heating, the final solids content of the product will be 20%. Use Siebel's equation for calculating the specific heat of the product.
15. A fruit juice at 190°F is allowed to flash into an essence recovery system maintained at a vacuum of 29 in. of mercury. Atmospheric pressure is 29.9 in. The vapors that flash off are rectified to produce an essence concentrate, and the juice, after being stripped of its aromatic constituents, is sent to an evaporator for concentration. Assuming sufficient resident time for the juice in the system to allow equilibrium in temperature between the liquid and the vapor, calculate:
    (a) The temperature of the juice leaving the essence recovery system.
    (b) The solids content of the juice leaving the system if the original solids content is 10%. Assume that there is no additional heat input and that the latent heat of vaporization is derived from the loss in sensible heat of the liquid. The specific heat of the solids is 0.2 BTU/(lb · °F).
    (c) The quantity of water vaporized per 100 lb of juice entering the system.
16. An evaporator has a heat transfer surface area that allows the transfer of heat at the rate of 100,000 BTU/h. If this evaporator is concentrating apple juice from 10 to

45% solids under a vacuum of 25 in. Hg (atmospheric pressure is 30 in. Hg), how much apple juice can be processed per hour?

17. Orange juice concentrate at 45% total solids leaves the evaporator at 50°C. This is frozen into slush in swept surface heat exchangers until half of the water is in the form of ice crystals prior to filling the cans, and the cans are frozen at −25°C. Assume that the sugars are all hexose sugars (mol. wt. 180) and that the freezing point reduction can be determined using $\Delta T_f = K_f\, m$, where K_f = the cryoscopic constant = 1.86 and m = the molality. Calculate:
    (a) The total heat which must be removed from the concentrate in the swept surface heat exchangers per kg of concentrate processed.
    (b) The amount of heat which must be further removed from the concentrate in frozen storage.
    (c) The amount of water still in the liquid phase at −25°C.

    Note: The moisture content is beyond the range where Chang and Tao's correlation is applicable. Determine the freezing point by calculating the freezing point depression: $\Delta T_b = K_f m$. The specific heat of the solids is the same below and above freezing. The specific heat of ice = 2093.4 J/(kg · K). The heat of fusion of ice = 334860 J/kg. The juice contains 42.75% soluble solids.

18. In a falling film evaporator, fluid is pumped to the top of a column and falls down as a film along the heated wall of the column, increasing in temperature as it drops. When the fluid emerges from the column it is discharged into a vacuum chamber, where it drops in temperature by flash evaporation until it reaches the boiling temperature at the vacuum employed. If juice containing 15% solids is being concentrated to 16% solids in one pass through the heated column and the vacuum in the receiving vessel is maintained at 25 in. Hg, calculate the temperature of the fluid as it leaves the column such that, when flashing occurs, the desired solids content will be obtained.

19. When sterilizing foods containing particulate solids in the Jupiter system, solids are heated separately from the fluid components by tumbling the solids in a double-cone processing vessel with saturated steam contacting the solids. The fluid component of the food is heated, held until sterile, and cooled using conventional fluid heating the cooling equipment. The cooled sterile liquid is pumped into the double-cone processing vessel containing the hot solids, which cools the latter and drops the pressure to atmospheric. After being allowed to cool by cooling the walls of the processing vessel, the sterile mixture is transferred aseptically to sterile containers.
    (a) Meat and gravy sauce is being prepared. Beef cubes containing 15% SNF, 22% fat, and 63% water are heated from 4° to 135°C, during which time condensate accumulates within the processing vessel with the meat. Calculate the total amount of meat and condensate at 135°C.
    (b) The gravy mix has the same weight as the raw meat processed, and consists of 85% water and 15% solids nonfat. Calculate the temperature of the mixture after equilibration if the gravy mix is at 20°C when it is pumped into the processing chamber containing the meat at 135°C.

20. The chillers in a poultry processing plant cool broilers by contacting the broilers with a mixture of water and ice. Broilers enter the chillers at 38°C and leave at 4°C. The U.S. Department of Agriculture requires an overflow of 0.5 gal of water per broiler processed, and this must be replaced with fresh water to maintain the

liquid level in the chiller. Melted ice is part of this overflow requirement. If a plant processes 7000 broilers per hour and the broilers average 0.98 kg, with a composition of 17% fat, 18% solids nonfat, and 65% water, calculate the ratio by weight of ice to fresh water which must be added to the chiller to meet the overflow requirement and the cooling load. Fresh water is at 15°C, and the overflow is at 1.5°C. The latent heat of fusion of ice is 334,860 J/kg.

21. Saturated steam at 280°F is allowed to expand to a pressure of 14.696 psia without a loss of enthalpy. Calculate:
(a) The temperature.
(b) The weight of high-pressure steam needed to produce 100 m_3/min of low pressure steam at 14.696 psia and the temperature calculated in (a).

22. In one system for ultra-high-temperature sterilization, milk enters a chamber maintained at 60 psia and 800°F in an atmosphere of superheated steam. Here it discharges from a plemum into vertical tubes, where it falls down in a thin film while exposed to the steam. The milk will be at the boiling temperature at 60 psia on reaching the bottom of the heating chamber. Following a sterilizing hold time at constant temperature, the milk is discharged into a vacuum chamber for rapid cooling. If the vacuum chamber is at 15 in. Hg vacuum, calculate: (a) the temperature of the milk leaving the flash chamber and (b) the total solids content. Raw milk enters the heater at 2°C and contains 89% water, 2% fat, and 9% solids nonfat.

SUGGESTED READING

American Society for Heating, Refrigerating and Air Conditioning Engineers. 1967. *Guide and Data Book, Applications for 1966 and 1967*. ASHRAE, New York.

American Society of Mechanical Engineers. 1967. American Society of Mechanical Engineers 1967 steam tables. Properties of saturated and superheated steam from 0.08865 to 15,500 lb per sq in. absolute pressure. Combustion Engineering Inc., Windsor, Conn.

Chang, H. D., and Tao, L. C. 1981. Correlation of enthalpy of food systems. *J. Food Sci.* 46:1493.

Charm, S. E. 1971. *Fundamentals of Food Engineering*, 2nd ed. AVI Publishing Co., Westport, Conn.

Choi, Y., and Okos, M. R. 1987. Effects of temperature and composition on thermal properties of foods. In: *Food Engineering and Process Applications*, M. Le Maguer and P. Jelen (eds.). Vol. I. Elsevier Applied Science Publishers, New York, pp. 93–102.

Himmelblau, D. M. 1967. *Basic Principles and Calculations in Chemical Engineering*, 2nd ed. Prentice-Hall, Englewood, Cliffs, N.J.

Hougen, O. A., and Watson, K. M. 1946. *Chemical Process Principles*. Part I. *Material and Energy Balances*. John Wiley & Sons, New York.

McCabe, W. L., Smith, J. C., and Harriott, P. 1985. *Unit Operations of Chemical Engineering*, 4th ed. McGraw-Hill Book Co., New York.

Siebel, J. E. 1918. *Compend of Mechanical Refrigeration and Engineering*, 9th ed. Nickerson and Collins, Chicago.

Watson, E. L., and Harper, J. C. 1988. *Elements of Food Engineering*, 2nd ed. Van Nostrand Reinhold, New York.

6

Flow of Fluids

Fluids are substances that flow without disintegration when pressure is applied. This definition of a fluid includes gases, liquids, and certain solids. A number of foods are fluids. In addition, gases such as compressed air and steam are also used in food processing, and they exhibit resistance to flow just as liquids do. In this chapter, the subject of fluid flow will be discussed from two standpoints: (1) the resistance to flow and its implications for the design of a fluid-handling system and (2) evaluation of the rheological properties of fluid foods.

THE CONCEPT OF VISCOSITY

Viscosity is a measure of resistance to flow of a fluid. Although molecules of a fluid are in constant random motion, the net velocity in a particular direction is zero unless some force is applied to cause the fluid to flow. The magnitude of the force needed to induce flow at a certain velocity is related to the viscosity of a fluid. Flow occurs when fluid molecules slip past one another in a particular direction on any given plane. Thus, there must be a difference in velocity, *a velocity gradient,* between adjacent molecules. In any particular plane parallel to the direction of flow, molecules above and below that plane exert a resistance to the force which propels one molecule to move faster than the other. This resistance of a material to flow or deformation is known as *stress*. *Shear stress* (τ) is the term given to the stress induced when molecules slip past one another along a defined plane. The velocity gradient ($-dV/dr$ or γ) is a measure of how rapidly one molecule is slipping past another; therefore, it is also referred to as the *rate of shear*. The position from which distance is measured in determining the shear rate is the point in the flow stream where velocity is maximum; therefore, as distance r increases from this point of reference, V decreases and the velocity gradient is a negative quantity. Since shear stress is always positive, expressing the shear rate as $-dV/dr$ satisfies the equality in the equation of

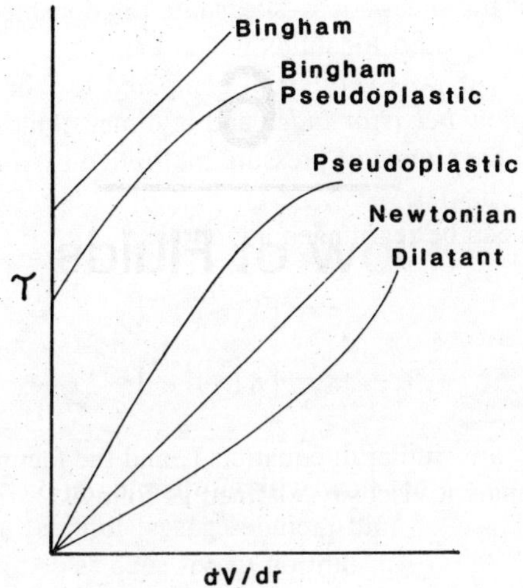

Fig. 6.1. A plot showing the relationship between shear stress and shear rate for different types of fluids.

shear stress as a function of shear rate. A plot of shear stress against shear rate for various fluids is shown in Fig. 6.1.

Fluids which exhibit a linear increase in the shear stress with the rate of shear (equation 1) are called *Newtonian fluids*. The proportionality constant (μ) is called the viscosity.

$$\tau = \mu\left(-\frac{dV}{dr}\right) \quad (1)$$

Newtonian fluids are those which exhibit a linear relationship between the shear stress and the rate of shear. The slope μ is constant; the viscosity of a Newtonian fluid is independent of the rate of shear. The term *viscosity* is appropriate for use only with Newtonian fluids.

Fluids with characteristics deviating from those of equation 1 are called *non-Newtonian* fluids. These fluids exhibit either shear thinning or shear thickening behavior, and some exhibit a yield stress, i.e., a threshold stress which must be overcome before the fluid starts to flow. The two most commonly used equations for characterizing non-Newtonian fluids are the power law model (equation 2) and the Herschel-Bulkley model for fluids (equation 3):

$$\tau = K(\gamma)^n \quad (2)$$

$$\tau = \tau_0 + K(\gamma)^n \quad (3)$$

Equation 2 can fit the shear stress–shear rate relationships of a wide variety of foods. It can also be used for fluids which exhibit a yield stress, as represented by equation 3, if τ_0 is very small compared to τ or if the shear rate is very high. n is the *flow behavior index* and is dimensionless. K is the *consistency index* and will have units of pressure multiplied by time raised to the nth power.

Equation 2 and 3 can be rearranged as follows:

$$\tau = \left[K(\gamma)^{n-1}\right]\gamma \tag{4}$$

$$\tau - \tau_0 = \left[K(\gamma)^{n-1}\right]\gamma \tag{5}$$

Equations 4 and 5 are similar to equation 1, and the factor which is the multiplier of γ is the apparent viscosity, defined as follows:

$$\mu_{app} = K(\gamma)^n \tag{6}$$

For fluids having characteristics which fit equations 4 and 5, the apparent viscosity can also be expressed as:

$$\mu_{app} = \frac{\tau}{\gamma} \tag{7}$$

$$\mu_{app} = \frac{(\tau - \tau_0)}{\gamma} \tag{8}$$

Substitution of equation 2 and τ in equation 7 also yields equation 6.

The apparent viscosity has the same units as viscosity, but the value varies with the rate of shear. Thus, when reporting an apparent viscosity, the shear rate under which it was determined must also be specified.

Equation 6 shows that the apparent viscosity will be decreasing with increasing shear rate if the flow behavior index is less than 1. These types of fluids exhibit "shear-thinning" behavior and are referred to as *shear-thinning fluids* or *pseudoplastic fluids*. When the flow behavior index of a fluid is greater than 1, μ_{app} increases with increasing rates of shear and the fluid is referred to as *shear-thickening* or *dilatant* fluids. Shear-thinning behavior is exhibited by emulsions and suspensions, where the dispersed phase has a tendency to aggregate following the path of least resistance in the flow stream or the particles align themselves in a position which presents the least resistance to flow. Shear-thickening behavior will be exhibited when the dispersed phase swells or changes shape when subjected to a shearing action or when the molecules are so long that they tend to cross-link with each other, trapping molecules of the dispersion medium. Shear-thickening behavior has not been observed in foods, although it has been reported in suspensions of clay and high molecular weight organic polymers.

Fluids which exhibit a yield stress and a flow behavior index of 1 are referred to as *Bingham plastics*. Some foods exhibit a yield stress and a flow behavior index of less than 1. There is no general term used for these fluids, although the terms *Herschel-Bulkley plastic* and *Bingham pseudoplastic* have been used.

Rheology. Rheology is the science of flow and deformation. When a material is stressed it deforms, and the rate and nature of the deformation which occurs characterize its rheological properties. The science of rheology can be applied to solids or fluids. Some food materials exhibit both fluid and solid behavior and are called *viscoelastic*. In general, *flow* indicates the existence of a velocity gradient within the material; this characteristic is exhibited only by fluids. The discussion of rheology in this section covers only the flow properties of fluids. In foods, rheology is useful in defining a set of parameters which can be used to correlate with a quality attribute. These parameters can also be used to predict how the fluid will behave in a process and in determining the energy requirements for transporting the fluid from one point in a processing plant to another.

Viscometry. Instruments used for measuring flow properties of fluids are called *viscometers*. Newtonian viscosity can be easily measured, since only one shear rate need be used. Therefore, viscometers for this purpose are relatively simple compared to those for evaluating non-Newtonian fluids. Viscometers require a mechanism for inducing flow which should be measurable and a mechanism for measuring the applied force. In addition, the geometry of the system in which flow is occurring must be simple in design so that the force and the flow can be translated easily into a shear stress and shear rate.

Viscometers Based on Fluid Flow Through a Cylinder. These viscometers are called *capillary* or *tube viscometers*, depending upon the inside diameter. The principle of operation is based on the Poiseuille equation if the fluid is Newtonian. The Rabinowitsch-Mooney equation applies when the fluid is non-Newtonian. The Poiseuille equation will be derived based on the Newtonian flow equation (equation 1), and the Rabinowitsch-Mooney equation will be derived based on the power law equation (equation 2).

Derivation of the Poiseuille Equation. Figure 6.2 shows a tube of length L and radius R. A pressure P_1 exists at the entrance and a pressure of P_2 exists at the end of the tube. At a distance r from the center, a ring of thickness dr is isolated; at this point, the fluid velocity is V, the point velocity. The shear stress at this point is the force resisting flow per area of the fluid undergoing shear. The force resisting flow on the fluid occupying the area of the ring is the pres-

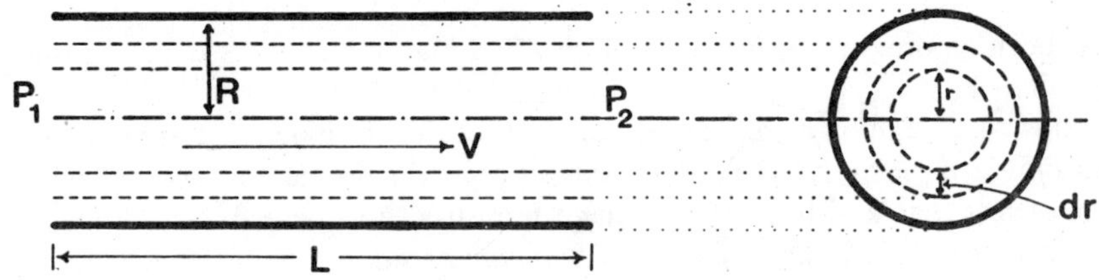

Fig. 6.2. Differential control element for analysis of fluid flow through a tube.

sure drop times the area of the ring, and the area of the cylinder formed when the ring is projected through length L.

$$\tau = \frac{(P_1 - P_2)(\pi r^2)}{2\pi r L} = \frac{(P_1 - P_2)r}{2L} \tag{9}$$

Substituting equation 9 in equation 1:

$$\frac{(P_1 - P_2)r}{2L} = \mu\left(-\frac{dV}{dr}\right)$$

Separating variables and integrating:

$$\int dV = \frac{(P_1 - P_2)}{2L\mu}\int - r\,dr$$

$$V = \frac{(P_1 - P_2)}{2L\mu}\left(\frac{-r^2}{2}\right) + C$$

The constant of integration can be determined by applying the boundary condition: at $r = R$, $V = 0$. $C = (P_1 - P_2)(R^2)/4L$.

$$V = \frac{(P_1 - P_2)}{4L\mu}(R^2 - r^2) = \frac{\Delta P}{4L\mu}(R^2 - r^2) \tag{10}$$

Equation 10 gives the point velocity in a flow stream for any fluid flowing within a cylindrical tube. Point velocity is not easily measured. However, an average velocity, defined as the volumetric flow rate per area, can be easily measured. Equation 10 will be expressed in terms of the average velocity.

Consider a pipe of radius R and a control volume which is the walls of a hollow cylinder of thickness dr within the pipe, shown in Fig. 6.3. The cross-sectional area of the ring of thickness dr is $dA = \pi[(r + dr)^2 - r^2)] = \pi(r^2$

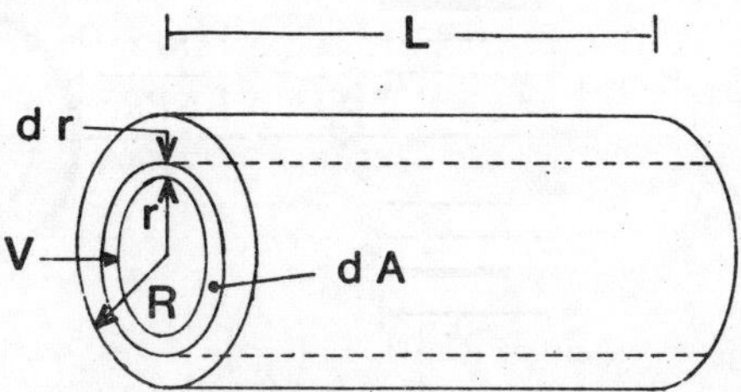

Fig. 6.3. Pipe showing ring thickness dr used as a control element for analysis of fluid flow.

$+ 2r\,dr + (dr)^2 - r^2) = 2\pi r\,dr + (dr)^2$. Since dr is small, $(dr)^2$ is negligible; therefore, $dA = 2\pi r\,dr$. The volumetric rate of flow (volume/time) through the control volume is $VdA = 2\pi rV\,dr$. The total volume going through the pipe will be the integral of the volumetric rate of flow through the control volume from $r = 0$ to $r = R$. Substituting equation 10 for V:

$$\overline{V}(\pi R^2) = \frac{(P_1 - P_2)}{4L\mu}(2\pi)\int_0^R (R^2 - r^2)r\,dr$$

Rearranging and integrating:

$$\overline{V} = \frac{(P_1 - P_2)}{2L\mu R^2}\left[\frac{R^2 r^2}{2} - \frac{r^4}{4}\right]_0^R$$

Substituting limits, combining terms, and substituting ΔP for $(P_1 - P_2)$:

$$\overline{V} = \frac{(P_1 - P_2)R^2}{8L\mu} = \frac{\Delta P R^2}{8L\mu} \qquad (11)$$

Equation 11 is the Poiseuille equation and can be used to determine the viscosity of a Newtonian fluid from pressure drop data when the fluid is allowed to flow through a tube or a capillary.

When equation 11 is used to determine the viscosity of non-Newtonian fluids, the viscosity obtained will be an apparent viscosity. Thus a viscosity obtained from measurements using a single rate of flow is:

$$\mu_{app} = \frac{\Delta P R^2}{8L\overline{V}}$$

Equations 10 and 11 may be combined to obtain an expression for $V/\overline{V}$ which can be differentiated to obtain a rate of shear as a function of the average velocity:

$$V = \bar{V}\left[\frac{\Delta P}{4L\mu}\right](R^2 - r^2)\left[\frac{8L\mu}{\Delta P R^2}\right]$$

Simplifying:

$$V = 2\bar{V}\left[1 - \left(\frac{r}{R}\right)^2\right] \tag{12}$$

Equation 12 represents the velocity profile of a Newtonian fluid flowing through a tube expressed in terms of the average velocity. The equation represents a parabola where the maximum velocity is $2\bar{V}$ at the center of the tube ($r = 0$). Differentiating equation 12:

$$\frac{dV}{dr} = 2\bar{V}\left[\frac{2r}{R^2}\right]$$

The shear rate at the wall ($r = R$) for a Newtonian fluid is:

$$-\left.\frac{dV}{dr}\right|_w = \frac{4\bar{V}}{R} \quad , \tag{13}$$

Velocity Profile and Shear Rate for a Power Law Fluid. Equations 2 and 9 can be combined to give:

$$\frac{(P_1 - P_2)r}{2L} = K\left(-\frac{dV}{dr}\right)^n$$

Rearranging and using ΔP for $(P_1 - P_2)$:

$$\frac{dV}{dr} = \left[\frac{\Delta P}{2LK}\right]^{1/n}(r)^{1/n}$$

Integrating and substituting the boundary condition, $V = 0$ at $r = R$:

$$V = \left[\frac{\Delta P}{2LK}\right]^{1/n}\left[\frac{1}{(1/n) + 1}\right]\left[R^{(1/n)+1} - r^{(1/n)+1}\right] \tag{14}$$

Equation 14 represents the velocity profile of a power law fluid. The velocity profile equation will be more convenient to use if it is expressed in terms of the average velocity. Using a procedure similar to that used in section "Derivation of the Poiseuille Equation," the following expression for the average velocity can be derived:

$$\overline{V}(\pi R^2) = \int_0^R 2\pi r \left[\frac{\Delta P}{2LK}\right]^{1/n} \left[\frac{n}{n+1}\right] \left[R^{(1/n)+1} - r^{(1/n)+1}\right] dr$$

Integrating and substituting limits:

$$\overline{V} = \left[\frac{\Delta P}{2LK}\right]^{1/n} [R]^{(n+1)/n} \left[\frac{n}{3n+1}\right] \tag{15}$$

The velocity profile in terms of the average velocity is:

$$V = \overline{V}\left[\frac{3n+1}{n+1}\right]\left[1 - \left[\frac{r}{R}\right]^{(n+1)/n}\right] \tag{16}$$

Figure 6.4 shows $V/\overline{V}$ as a function of r/R for various values of n. When $n = 1$, the velocity profile is parabolic and equations 12 and 16 give the same results. The velocity profile flattens out near the center of the pipe as n decreases.

Differentiating equation 16 and substituting $r = R$ for the shear rate at the wall:

$$-\left.\frac{dV}{dr}\right|_w = \overline{V}\left[\frac{3n+1}{n+1}\right]\left[\frac{n+1}{n}\right][R]^{-(n+1)/n}[R]^{1/n}$$

Simplifying:

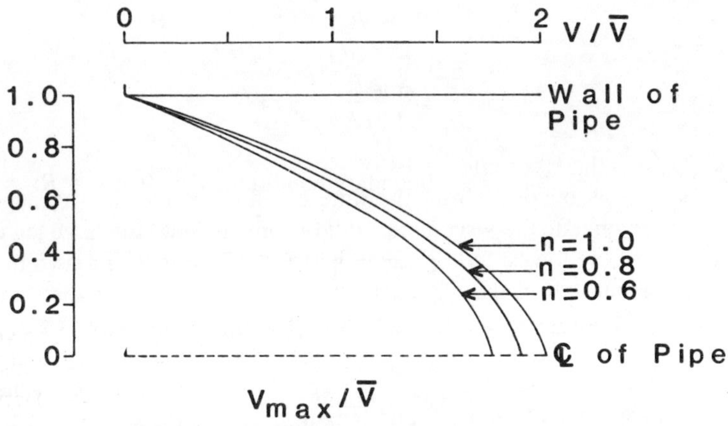

Fig. 6.4. Plot of $V/\overline{V}$ as a function of position in the pipe for fluids with different values of n.

$$-\left.\frac{dV}{dr}\right|_w = \bar{V}\left(\frac{3n+1}{n}\right)\left(\frac{1}{R}\right)$$

Equation 17 is a form of the Rabinowitsch-Mooney equation used to calculate shear rates for non-Newtonian fluids flowing through tubes. Converting into a form similar to equation 13:

$$-\left.\frac{dV}{dr}\right|_w = \frac{4\bar{V}}{R}\left[\frac{3}{4} + \frac{1}{4n}\right] \tag{18}$$

Equation 18 shows that the shear rate at the wall for a non-Newtonian fluid is similar to that for a Newtonian fluid except for the multiplying factor $(0.75 + 0.25/n)$.

Glass Capillary Viscometers. The simplest viscometers operate on gravity-induced flow and are commercially available in glass. Figure 6.5 shows a Cannon-Fenske-type viscometer. These viscometers are available with varying capillary sizes. The capillary size is chosen to minimize the time of efflux for viscous fluids. The Poiseuille equation (equation 11) is used to determine the viscosity in these viscometers. The pressure drop needed to induce flow is:

$$\Delta P \doteq \rho g h$$

where ρ is the density of the fluid and h is the height available for free fall in the viscometer (Fig. 6.5). The viscosity is then calculated, using the above expression for ΔP in equation 13 as follows:

$$\frac{\mu}{\rho} = \frac{ghR^3}{8L\bar{V}} \tag{19}$$

The ratio μ/ρ is the *kinematic viscosity*.

The viscosity is obtained from the time of efflux of the fluid through the viscometer. To operate the viscometer, fluid is pipetted into the large leg of the viscometer until the lowest bulb is about half full. Fluid is then drawn into the small leg to about half of the highest bulb. When the suction on the small leg is released, fluid will flow; timing is started when the fluid meniscus passes the first mark. When the fluid meniscus passes the second mark, timing is stopped and the efflux time is recorded. The lower portion of the small leg of the viscometer is a capillary of radius R and length L. If t_e is the time of efflux, the average velocity $\bar{V}$ will be L/t_e. Substituting in equation 18:

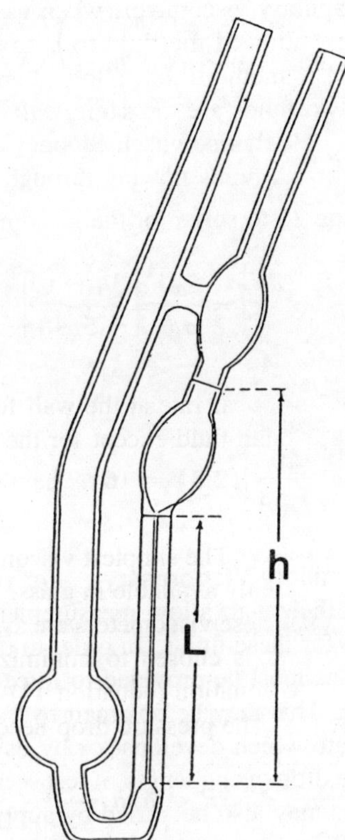

Fig. 6.5. Diagram of a glass capillary viscometer showing the components and parameters used in fitting viscometer data to the Poiseuille equation.

$$\frac{\mu}{\rho} = \frac{ghR^2}{8L} t_e \tag{20}$$

For each viscometer, the length and diameter of the capillary and the height available for free fall are specific; therefore, these factors can be grouped into a constant, k_v, for a particular viscometer. The kinematic viscosity can then be expressed as:

$$\frac{\mu}{\rho} = k_v t_e \tag{21}$$

The viscometer constant is determined from the efflux time of a fluid of known viscosity and density.

EXAMPLE: A glass capillary viscometer, when used on a fluid with a viscosity of 10 centipoises, allowed the fluid to efflux in 1.5 min. This same viscometer, when used on another fluid, allowed an efflux time of 2.5 min. If the densities of the two fluids are the same, calculate the viscosity of the second fluid.

Solution: Using equation 21 to solve for the viscometer constant:

$$k_v = \frac{\mu}{\rho t_e} = \frac{10}{1.5\rho}$$

For the second fluid:

$$\mu = \rho\left[\frac{10}{1.5\rho}\right](2.5) = 16.67 \text{ centipoises}$$

Forced Flow Tube or Capillary Viscometry. For very viscous fluids, gravity-induced flow is not sufficient to allow measurement of viscosity. Forced-flow viscometers are used for these fluids. In order to obtain varying flow rates in viscometers, some means must be provided to force a fluid through the viscometer at a constant rate. This may be obtained by using a constant pressure and measuring the flow rate which develops, or by using a constant flow rate and measuring the pressure drop over a length of test section. Flow rates through glass capillary viscometers may also be varied by applying pressure instead of relying solely on gravity flow.

The flow properties of a fluid are the constants in equations 1 to 3 which can be used to characterize the relationship between the shear stress and the rate of shear. These are the viscosity, if the fluid is Newtonian, the flow behavior index, the consistency index, and the yield stress. Equation 3 is most general. If a fluid has a yield stress, the test is carried out at very high rates of shear so that τ_0 is very small in comparison to τ and equation 3 simplifies to equation 2. A fluid with properties which fit equation 1 is a special case of a general class of fluids which fit equation 2 when $n = 1$. Evaluation of τ_0 in equation 3 from data at low shear rates is simple once n is established from data at high shear rates.

The shear rate at the wall may be determined using equations 13 or 18. These equations show that the shear rate is the product of a factor which is independent of the rate of flow and another factor which is a function of the rate of flow, i.e., the average velocity, $\overline{V}$, the volumetric rate of flow Q, or even the speed of a piston V_p which delivers fluid to the tube. In equation form:

$$\gamma_w = F_1\overline{V} = F_2 Q = F_3 V_p$$

The shear stress at the wall is determined by substituting R for r in equation 9.

$$\tau_w = \frac{\Delta PR}{2L} \tag{22}$$

Equation 22 shows that τ_w is a product of a factor which is independent of the rate of flow and the pressure drop. In equation form: $\tau_w = F_{p1}(\Delta P) = F'_{p2}$ (height of manometer fluid) $= F_{p3}$ (transducer output). Substituting any of the above expressions for τ_w and Δ_w into equation 2:

$$PF_{p1} = [\overline{V}F_1]^n$$

P and $\overline{V}$ are, respectively, measurements from which pressure or velocity can be calculated. Taking logarithms of both sides:

$$\log P + \log F_{p1} = n \log \overline{V} + n \log F_1$$

Rearranging:

$$\log P = n \log \overline{V} + (n \log F_1 - \log F_{p1})$$

Thus, a log-log plot of any measure of pressure against any measure of velocity can give the flow behavior index n for a slope.

The shear rate can then be calculated using either equation 13 or equation 18, and the shear stress can be calculated using equation 22. Taking the logarithm of equation 2: $\log \tau_w = \log K + n \log \gamma_w$. The intercept of a log-log plot of τ_w against γ_w will give K.

EXAMPLE 1: A tube viscometer having an inside diameter of 1.27 cm and a length of 1.219 m is used to determine the flow properties of a fluid having a density of 1.09 g/cm^3. The following data were collected for the pressure drop at various flow rates measured as the weight of fluid discharged from the tube. Calculate the flow behavior and consistency indices for the fluid.

Data for Pressure Drop (in kPa)

$(P_1 - P_2)$	Flow Rate (g/s)
19.197	17.53
23.497	26.29
27.144	35.05
30.350	43.81
42.925	87.65

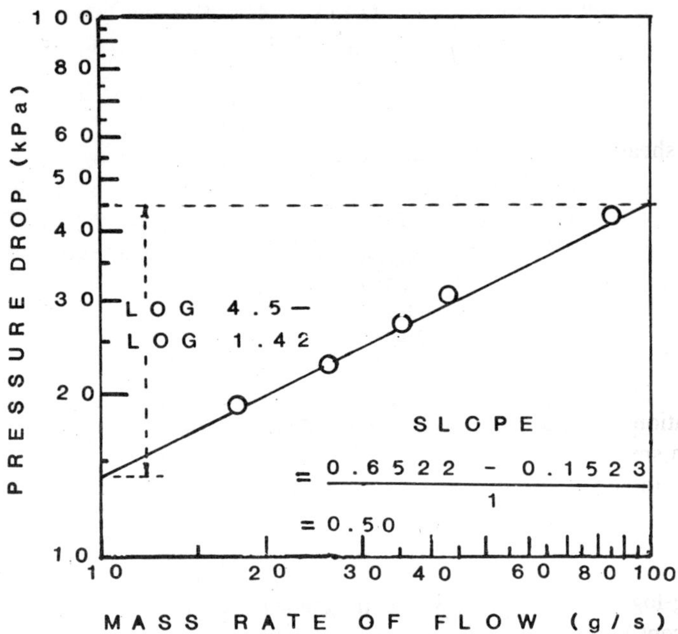

Fig. 6.6. Log-log plot of pressure drop against mass rate of flow to obtain the flow behavior index, n, from the slope.

Solution: Figure 6.6 shows a plot of log ΔP against log (mass rate of flow). The slope, $n = 0.5$, indicates that the fluid is non-Newtonian. Equation 18 will be used to solve for γ_w and equation 22 for τ_w.
Solving for τ_w: $R = 0.5(1.27 \text{ cm})(0.01 \text{ m/cm}) = 0.00635$ m; $L = 1.219$ m.

$$\tau_w = [0.00635(0.5)/1.219]\ \Delta P = 0.002605\ \Delta P \text{ Pa}.$$

The average velocity $\overline{V}$ in m/s can be calculated by dividing the mass rate of flow by the density and the cross-sectional area of the tube. Let q = mass rate of flow in g/s.

$$\overline{V} = q \frac{\text{g}}{\text{s}} \frac{\text{cm}^3}{1.09 \text{ g}} \frac{\text{m}^3}{(100)^3 \text{ cm}^3} \frac{1}{\pi(0.00635)^2 \text{ m}^2}$$

$$= 0.007242\ q \quad \text{m/s}$$

Equation 18 is used to calculate the shear rate at the wall. $n = 0.5$.

$$\gamma_w = \frac{4\bar{V}}{R}\left(0.75 + \frac{0.25}{n}\right) = \frac{4(0.007242\ q)}{0.00635}(1.25)$$

$$= 5.7047\ q$$

The shear stress and shear rates are:

τ_w (Pa)	γ_w (1/s)
50.008	99.966
61.209	149.92
70.710	199.87
79.062	249.83
111.82	499.83

Equation 2 will be used to determine K. The yield stress τ_0 is assumed to be much smaller than the smallest value for τ_w measured. Thus, equation 3 reduces to equation 2, the logarithm of which is as follows:

$$\log \tau_w = \log K + n \log \gamma_w$$

A log-log plot of shear stress against the shear rate is shown in Fig. 6.7. The intercept is the consistency index; $k = 5$ Pa · s^n.

EXAMPLE 2: A fluid induces a pressure drop of 700 Pa as it flows through a tube having an inside diameter of 0.75 cm and a length of 30 cm at a flow rate of 50 cm^3/s.

(a) Calculate the apparent viscosity, defined as the viscosity of a Newtonian fluid which would exhibit the same pressure drop as the fluid at the same rate of flow through this tube.

(b) This same fluid, flowing at the rate of 100 cm^3/s through a tube 20 cm long and 0.75 cm in inside diameter, induces a pressure drop of 800 Pa. Calculate the flow behavior and consistency indices for this fluid.

(c) What would be the shear rates at 50 and 100 cm^3/s?

Solution: Equation 11 will be used to calculate the apparent viscosity from the pressure drop of a fluid flowing through a tube.

(a) $R = (0.0075)(0.5) = 0.00375$ m; $L = 0.3$ m; $\Delta P = 700$ Pa.

$$\bar{V} = 50\ \frac{\text{cm}^3}{\text{s}}\ \frac{1\ \text{m}^3}{(100)^3\ \text{cm}^3}\ \frac{1}{\pi(0.00375)^2} = 1.1318\ \text{m/s}$$

$$\mu_{app} = \frac{\Delta P R^2}{8L\bar{V}}$$

$$= \frac{700(0.00375)^2}{8(0.3)(1.1318)} = 0.003624\ \text{Pa} \cdot \text{s}$$

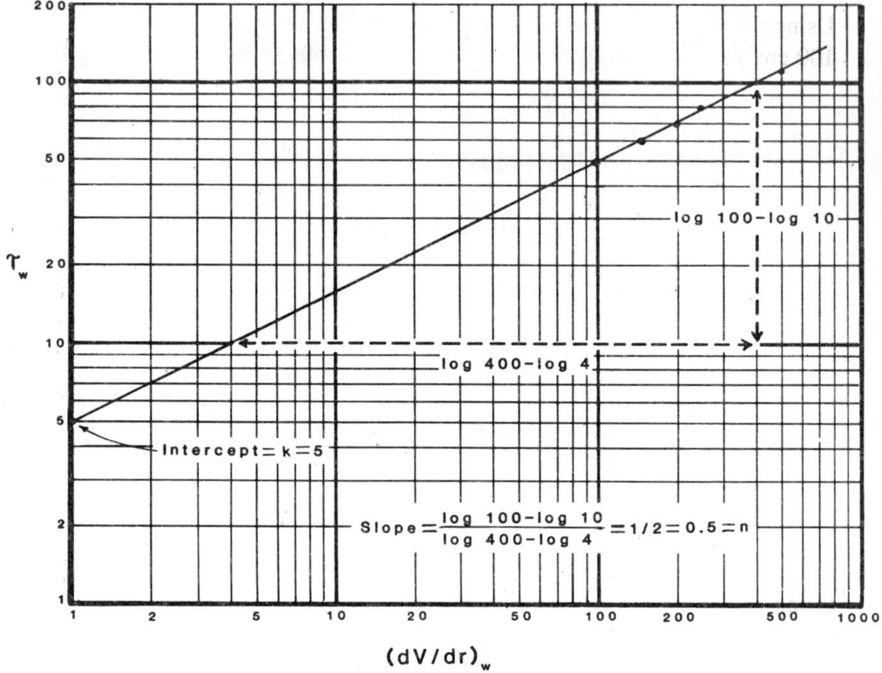

Fig. 6.7. Log-log plot of shear stress against shear rate to obtain the consistency index from the intercept.

(b)
$$\bar{V} = 100 \frac{\text{cm}^3}{\text{s}} \frac{1 \text{ m}^3}{(100)^3 \text{ cm}^3} \frac{1}{\pi(0.00375)^2}$$

$$= 2.2635 \text{ m/s}$$

$$\mu_{app} = \frac{800(0.00375)^2}{8(0.2)(2.2635)} = 0.003106 \text{ Pa} \cdot \text{s}$$

Equation 6 will be used to solve for K and n.
Let $\phi = 0.75 + 0.25/n$.

$$\text{Equation 18: } \gamma_w = \frac{4\bar{V}}{R}(\phi)$$

$$\text{Equation 6: } \mu_{app} = K(\gamma)^{n-1}$$

$$\text{Substituting } \phi\text{: } \mu_{app} = K\left[\frac{4\phi}{R}\right]^{n-1} \bar{V}^{n-1}$$

Using the subscripts 1 and 2 for the velocity and apparent viscosity at 50 and 100 cm^3/s, respectively:

$$\frac{\mu_{app1}}{\mu_{app2}} = \frac{(K(4\phi/R)^{n-1}\overline{V}_1)^{n-1}}{(K(4\phi/R)^{n-1}\overline{V}_2)^{n-1}} = \left(\frac{V_1}{V_2}\right)^{n-1}$$

$$\log\left(\frac{0.003624}{0.003106}\right) = (n-1)\log\left(\frac{1.1318}{2.2635}\right)$$

$$\log(1.166774) = (n-1)\log(0.5009)$$

$$n = 1 + \frac{\log(1.166774)}{\log(0.5009)}$$

$$= 0.777$$

$$\phi = \frac{0.75 + 0.25}{0.777} = 1.072$$

Using equation 6 on either of the two apparent viscosities:

$$\gamma_w = \frac{4(1.1317)}{0.00375}(1.072) = 1294\ \text{s}^{-1}$$

$$\text{Using equation 6: } k = \frac{0.003624}{(1294)^{0.777-1}} = 0.0178\ \text{Pa s}^n$$

(c) At 50 cm^3/s, $\gamma_w = 1294\ \text{s}^{-1}$

$$\text{At 100 cm}^3/\text{s},\ \gamma_w = \frac{4(2.3634)}{0.00375}(1.072) = 2521\ \text{s}^{-1}$$

Evaluation of Wall Effects in Tube Viscometry. The equations derived in the previous sections for evaluating flow properties of fluids by tube or capillary viscometry are based on the assumption that there is zero slip at the wall. This condition may not always be true for all fluids, particularly suspensions.

To determine if slip exists at the wall, it will be necessary to conduct experiments on the same fluid using viscometers of different radius. Kokini and Plutchok (1987) defined a parameter β_c, which can be used to correct for slip at the tube wall. β_c is the slope of a plot of $Q/(\pi R^3 \tau_w)$ against $1/R^2$ obtained on the same fluid, using different viscometers of varying radius at flow rates which would give a constant shear stress at the wall, τ_w. Q = the volumetric rate of flow, and R = the viscometer radius. With β_c known, a corrected volumetric flow rate Q_c is calculated as follows:

$$Q_c = Q - \pi R \tau_w \beta_c \tag{23}$$

The corrected flow rate Q_c is then used instead of the actual flow rate in determining the mean velocity and shear rates in calculating the flow behavior index and the consistency index of fluids.

The experiments involve a determination of pressure drop at different rates of flow on at least four tubes, each with a different radius. The value of β_c may vary at different values of τ_w; therefore, for each tube, data are plotted as log Q against log τ_w. The four linear plots representing the data obtained on the four tubes are drawn on the same graph, and at least four values of τ_w are selected. Points on each line at a constant value of τ_w are then selected. At each value of τ_w, a plot is made of $Q/(\pi R^3 \tau_w)$ against $1/R^2$. The plot will be linear with a slope β_c. The value of β_c may be different at different values of τ_w; therefore, plots at four different values of τ_w are needed. After determining β_c values at each τ_w, equation 23 is used to determine Q_c, which is then used to determine γ_w. An example is given in Kokini and Plutchok's (1987) article.

Glass Capillary Viscometer Used as a Forced Flow Viscometer. Fluids which exhibit non-Newtonian behavior but have low consistency indices are usually difficult to study using tube viscometers because of the low pressure drop. A glass capillary viscometer may be used as a forced flow viscometer by attaching a constant-pressure source to the small leg of the viscometer. A simple setup is shown in Fig. 6.8. The pressure source is a water column of movable height. To operate, the fluid is pipetted into the large leg of the viscometer. The fluid level in the u-tube is made level with the fluid in the large bulb of the viscometer by adjusting the height of the dropping funnel, and the stopcock is shut off. The viscometer is disconnected from the pressure source, and fluid is drawn to the top of the second bulb on the small viscometer leg by suction. After reconnecting the viscometer to the pressure source, the height of the dropping funnel is raised and pressure is applied by opening the stopcock. The time of efflux is then measured as the time for the fluid meniscus to pass through the first and second marks on the viscometer. Pressure is discontinued by closing the stopcock, and the difference in fluid level between the dropping funnel and the large bulb of the viscometer is measured. The flow behavior and consistency indices are calculated from the efflux times at different levels of fluid in the pressure source.

In a capillary viscometer, the volume of fluid which passes through the capillary is the volume which fills the section between the two marks on the viscometer. Two other critical factors are the length of the capillary (L) and the height available for free fall (H). These measurements are made on the vis-

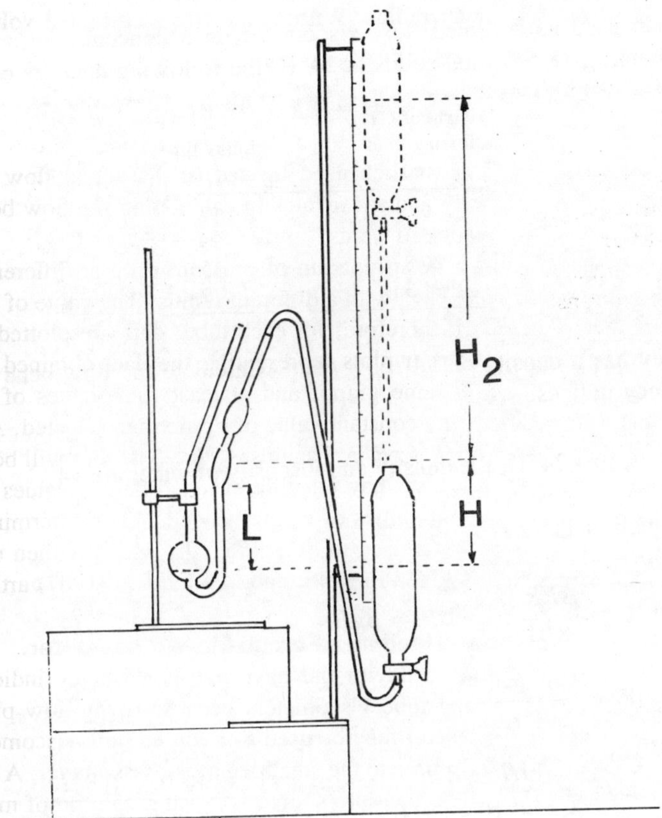

Fig. 6.8. Diagram of a system used to adapt a glass capillary viscometer for forced flow viscometry.

cometer, as indicated in Fig. 6.8. The radius of the capillary may be calculated from the viscometer constant using equation 19. When used as a forced flow viscometer, the pressure drop will be the pressure equivalent to the height of the forcing fluid, as shown in Fig. 6.8, and the pressure equivalent to the height available for free fall of the fluid in the viscometer. The calculations are shown in the following example:

EXAMPLE: A Cannon-Fenske-type glass capillary viscometer has a height available for free fall of 8.81 cm and a capillary length of 7.62 cm. When it is used on distilled water at 24°C, an efflux time of 36 s is measured. The volume of fluid which drains between the two marks on the viscometer is 3.2 cm^3.

(a) Calculate the radius of the capillary.

(b) When used as a forced flow viscometer on a concentrated acid whey containing 18.5% total solids at 24°C, the following data are collected:

Height of (H_2) forcing fluid (cm)	Efflux time (t_e)(s)
0	92
7.1	29
20.8	15
34.3	9

The whey has a density of 1.0763 g/cm³. Calculate the flow behavior and consistency indices.

Solution: (a) At 24°C, the density and viscosity of water are 947 kg/m³ and 0.00092 Pa · s, respectively.
Using equation 11:

$$\mu = \frac{\Delta P R^2}{8L\overline{V}}; \qquad \Delta P = \rho g H$$

$$\overline{V} = \frac{\text{Volume}}{\pi R^2 t_e}$$

$$\mu = \frac{\rho g H R^2 \pi R^2 t_e}{3.2 \times 10^{-6}(8)L}; \qquad R^4 = \frac{(3.2 \times 10^{-6})(8)L\mu}{\rho g H \pi t_e}$$

Substituting known values and solving for R:

$$R = \left[\frac{0.00092(3.2 \times 10^{-6})(8)(0.0762)}{997(9.8)(0.0881)(\pi)(36)}\right]^{0.25}$$

$$= 3.685 \times 10^{-4} \text{ m}$$

(b) The total pressure forcing the fluid to flow is $\Delta P = g(\rho_1 H + \rho_2 H_2)$.

$$\Delta P = 9.8[(1076.3)(0.0881) + (997)(H_2)]$$

$$= 9.8(94.822 + 997 H_2) \text{ Pa}$$

Let q = volumetric rate of flow.

$$q = \frac{3.2}{t_e} \text{ cm}^3/\text{s}$$

The pressure-inducing flow and the volumetric rate of flow calculated using the above equations are:

ΔP (Pa)	q (cm^3/s)
929	0.0564
1623	0.110
2961	0.213
4280	0.355

A log-log plot of ΔP against q shown in Fig. 6.9 gives $n = 0.84$. The consistency index will be calculated from the rates of shear and the shear stress.

$$\overline{V} = \frac{q \times 10^{-6}}{\pi(R^2)} = 2.3466q$$

$$\gamma_w = \frac{4\overline{V}}{R}\left[0.75 + \frac{0.25}{n}\right]$$

$$= \frac{4(2.3466)q}{3.683 \times 10^{-4}}\left[0.75 + \frac{0.25}{0.84}\right]$$

$$= 26699q$$

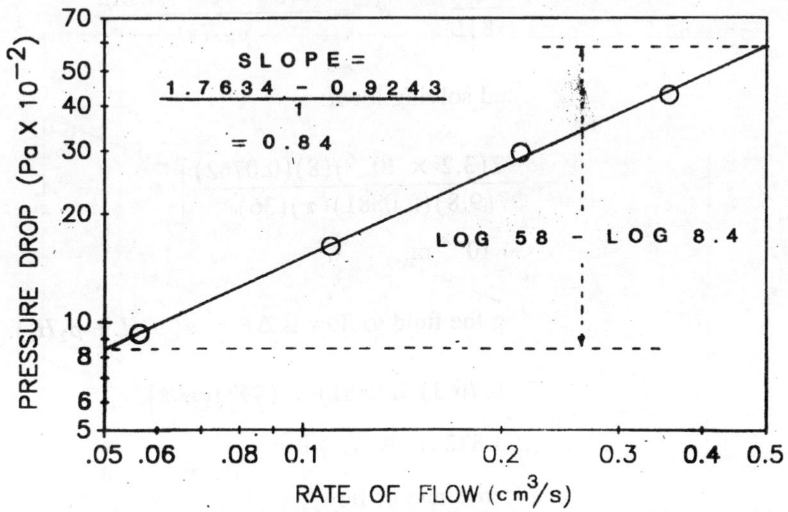

Fig. 6.9. Log-log plot of pressure drop against rate of flow to determine the flow behavior index, n, from the slope.

The shear stress is calculated from ΔP. Substituting known values in equation 22:

$$\tau_w = \frac{\Delta PR}{2L} = \frac{\Delta P(3.683 \times 10^{-4})}{2(0.0762)}$$
$$= 24.1666 \times 10^{-4}\ \Delta P$$

The shear stress and shear rates are:

τ_w (Pa)	γ_w (1/s)
2.245	1505
3.922	2936
7.155	5687
10.343	9478

Figure 6.10 shows a log-log plot of shear stress against shear rate. A point on the line can be used to calculate K. The first data point (2.245,1505) falls directly on the line.

$$K = \frac{\tau_w}{(\gamma_w)^n} = \frac{2.245}{(1505)^{0.84}} = 0.0048\ \text{Pa} \cdot \text{s}^n$$

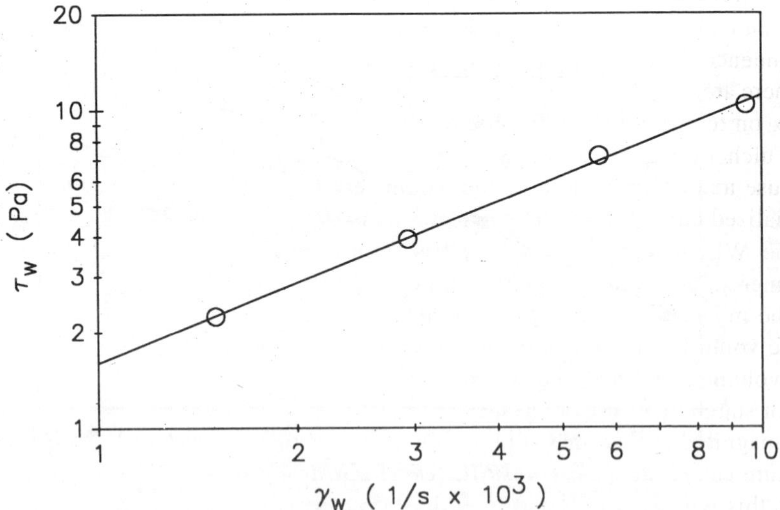

Fig. 6.10. Log-log plot of shear stress against shear rate to determine a point on the line which can be used to calculate the consistency index.

Effect of Temperature on Rheological Properties. Temperature has a strong influence on the resistance to flow of a fluid. It is very important that temperatures be maintained constant when making rheological measurements. The flow behavior index, n, is relatively constant with temperature unless components of the fluid undergo chemical changes at certain temperatures. The viscosity and the consistency index, on the other hand, are highly temperature dependent.

The temperature dependence of the viscosity and the consistency index can be expressed in terms of the Arrhenius equation:

$$\ln\left(\frac{\mu}{\mu_1}\right) = \frac{E_a}{R}\left(\frac{1}{T} - \frac{1}{T_1}\right) \tag{24}$$

where μ = the viscosity at absolute temperature T; θ_1 = viscosity temperature T_1; E_a = the activation energy, J/gmole; R = the gas constant, 8.314 J/(gmole · K); and T is the absolute temperature.

Equation 24 is useful in interpolating between values of μ at two temperatures. When data at different temperatures are available, equation 24 may be expressed as:

$$\ln \mu = A + \frac{B}{T} \tag{25}$$

where the constants B and A are slope and intercept, respectively, of the plot of ln (μ) against $1/T$. The same expressions may be used for the temperature dependence of the consistency index, K.

There are other expressions for the dependence of the viscosity or consistency index on temperature such as the Williams-Landel-Ferry (WLF) equation and the Fulcher equation (Rao et al. 1987), but these also have limitations for general use and determination of the parameters from experimental data requires specialized curve-fitting computer programs.

The WLF equation, however, has implications for the flow behavior of polymer solutions and the viscoelastic properties of gels, therefore, the basis for the model will be briefly discussed. The model is based on the concept of a free volume. Any polymer, by virtue of the size of the molecule, contains a free volume. The free volume may include the space within the helix in a protein or starch molecule or the space formed when sections of random coil mesh. The magnitude of the free volume is temperature dependent. At a specific temperature called the *glass transition temperature*, the free volume is essentially zero; this will be manifested by a drastic change in the thermophysical properties of the material. The free volume is best manifested by the specific volume. A plot of the specific volume with temperature above the glass transition temperature is continuous, and the slope defines the thermal expansion coefficient. The point on the curve where a discontinuity occurs is the glass transition temperature. The WLF model relates the ratio, μ/μ_g, where μ = the viscosity at temperature T and μ_g = the viscosity at the glass transition temperature, T_g, to a function of $T - T_g$.

The Arrhenius equation is the simplest to use and is widely applied in expressing the temperature dependence of a number of factors.

EXAMPLE: The flow behavior index of applesauce containing 11% solids is 11.6 and 9.0 Pa · s^n at 30° and 82°C, respectively. Calculate the flow behavior index at 50°C.

Solution: Using equation 24: T = 303°K; T_1 = 355°K:

$$\frac{E_a}{R} = \frac{\ln (11.6/9.0)}{1/303 - 1/355} = 524.96$$

Using equation 24 with k_1 = 11.6, T_1 = 303°K, and T = 323°K.

$$K = 11.6[e]^{524.96(1/323 - 1/303)}$$

$$= 11.6(.898) = 10.42 \text{ Pa} \cdot \text{s}^n$$

Back Extrusion. Back extrusion is another method for evaluating the flow properties of fluids. It is particularly useful with materials which have the consistency of a paste and with suspensions which have large particles suspended in them, since the suspended solids tend to intensify wall effects when the fluid is flowing in a small tube. When a rotational viscometer is used, pulsating torque readings are usually exhibited when a lumpy fluid is being evaluated.

The key to accurate determination of fluid flow properties using back extrusion is the assurance of annular flow, i.e., the plunger must remain in the center of the larger stationary cylinder throughout the test. Figure 6.11 shows a back extrusion device designed for use on an Instron universal testing machine. A stanchion positioned at the center of the large cylinder which is machined to a close tolerance to fit an opening at the center of the plunger guides the movement of the plunger and ensures concentric positioning of the plunger and the outer cylinder at all times during the test.

The equations that govern flow during back extrusion have been derived by Osorio and Steffe (1987).

Let F_t = the force recorded at the maximum distance of penetration of the plunger, L_m. The height of the column of fluid which flows through the annulus, L, is:

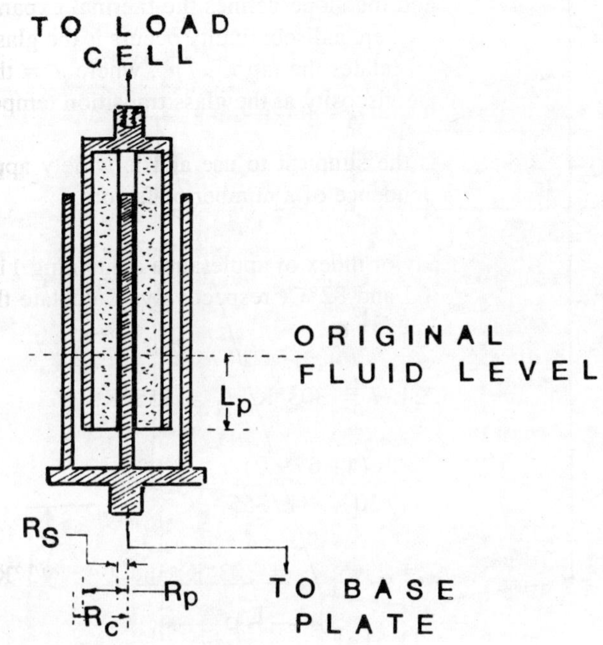

Fig. 6.11. Diagram of a back extrusion cell suitable for measuring flow properties of fluids with lumpy or paste-like consistency.

$$L = \frac{A_p L_p}{A_a}$$

where A_p is the area of the plunger which displaces fluid, and A_a is the cross-sectional area of the annulus.

$$A_p = \pi(R_p^2 - R_s^2); \qquad A_a = \pi(R_c^2 - R_p^2)$$

where R_c = the radius of the cylinder, R_p = the radius of the plunger, and R_s = the radius of the stanchion which positions the plunger in the center of the cylinder. L_p cannot be measured easily on the plunger and cylinder assembly itself, since the fluid level in the cylinder is not visible outside, but it can be read easily from the force tracing on the Instron chart by measuring the distance from the initiation of the rise in force and the point of maximum force on the chart. Figure 6.12 shows a typical back extrusion force-displacement curve.

Part of the total force, F_t, counteracts the force of gravity on the mass of fluid in the annulus.

$$\text{Force of gravity on annular fluid} = \rho g L \pi(R_c^2 - R_p^2)$$

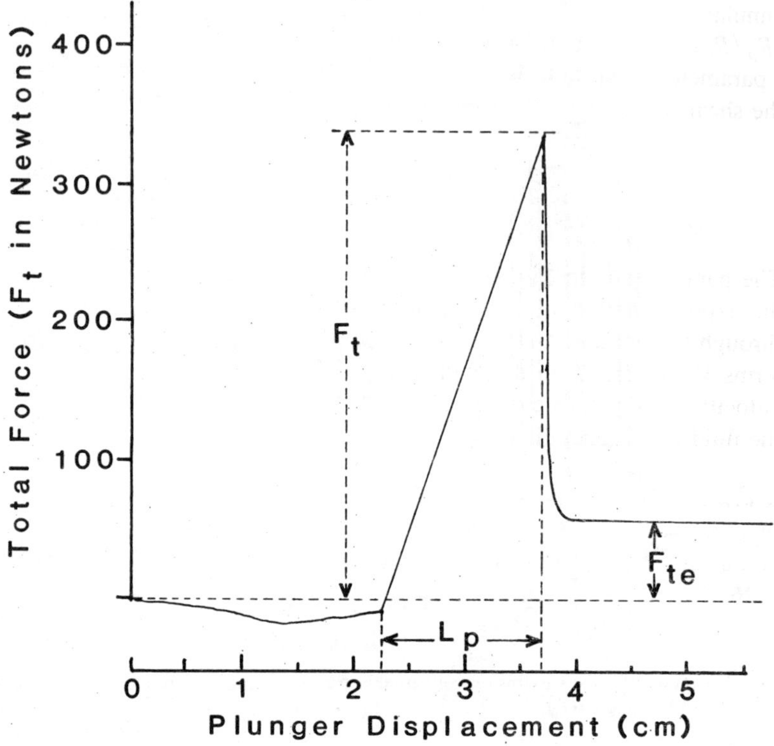

Fig. 6.12. Typical force-displacement diagram obtained for back extrusion showing how total force, plunger displacement, and yield force are obtained.

The net force, F_n, which is exerted by the fluid to resist flow through the annulus is the total force corrected for the force of gravity.

$$F_n = F_t - \rho g L \pi (R_c^2 - R_p^2)$$

where ρ = density of the fluid.

The flow behavior index, n, is the slope of a log-log plot of F_n/L against the velocity of the plunger, V_p.

Determination of the shear rate at the plunger wall and the shear stress at this point involves a rather complex set of equations which were derived by Osorio and Steffe (1987). The derivation is based on a dimensionless radius, λ, which is the ratio r_m/R_c, where r_m = radius of a point in the flow stream where velocity is maximum. The differential momentum and mass balance equations were solved simultaneously to obtain values of λ for different values for the flow behavior index of fluids and different annular gaps. The authors presented results of the calculations as a table of λ for different n and annular gap size expressed as σ, the ratio of plunger to cylinder radius (R_p/R_c). Table 6.1 shows these values. Figure 6.13 relates the value of σ to a parameter Θ for fluids with different n. The parameter Θ is used to calculate the shear rate at the wall from the plunger velocity, V_p.

$$\text{Let: } \sigma = \frac{R_p}{R_c}; \qquad \sigma_s = \frac{R_s}{R_c}; \qquad \lambda = \frac{r_m}{R_c}$$

The parameter Θ in Fig. 6.12 is a dimensionless parameter which satisfies the expressions for the integral of the dimensionless point velocity of fluid through the annulus and the quantity of fluid displaced by the plunger. The terms V_p, P, and K in the expression of Θ in Fig. 6.13 are: V_p = plunger velocity, P = pressure drop per unit distance traveled by the fluid, and K is the fluid consistency index.

Knowing the flow behavior index of the fluid and the dimensionless plunger radius, σ, a value for λ is obtained from Table 6.1. Figure 6.13 is then used to obtain a value of Θ. The shear rate at the plunger wall is calculated using equation 26.

$$\left.\frac{dV}{dr}\right|_{R_p} = \frac{V_p}{R_c\Theta}(\sigma^2 - \sigma_s^2)\left(\frac{\lambda^2}{\sigma} - \sigma\right)^{1/n} \tag{26}$$

The shear stress at the plunger wall is calculated from the net force, F_n, exerted on the base of the column of fluid to force it up the annulus.

The shear stress at the wall, τ_w, is calculated using equation 27. Equations 26 and 27 are derived to account for the fact that part of the plunger area being occupied by the stanchion which positions the plunger in the center of the concentric cylinder does not displace fluid up through the annular gap.

$$\tau_w = \frac{F_n}{2\pi L R_p}\left(\frac{\lambda^2 - \sigma^2}{\lambda^2 - \sigma^2}\right) \tag{27}$$

The consistency index of the fluid can then be determined from the intercept of a log-log plot of shear stress and shear rate.

The yield stress can be calculated from the residual force after the plunger has stopped, F_{ne}, corrected for the hydrostatic pressure of the height of fluid in the annulus.

$$\tau_0 = \frac{(F_{ne})}{2\pi(R_p + R_c)L} \tag{28}$$

Table 6.1. Values of λ for Different Values of σ and n

σ	n									
	0.1	0.2	0.3	0.4	0.5	0.6	0.7	0.8	0.9	1.0
0.1	0.4065	0.4889	0.5539	0.6009	0.6344	0.6586	0.6768	0.6907	0.7017	0.7106
0.2	0.5140	0.5680	0.6092	0.6398	0.6628	0.6803	0.6940	0.7049	0.7138	0.7211
0.3	0.5951	0.6313	0.6587	0.6794	0.6953	0.7078	0.7177	0.7259	0.7326	0.7382
0.4	0.6647	0.6887	0.7068	0.7206	0.7313	0.7399	0.7469	0.7527	0.7575	0.7616
0.5	0.7280	0.7433	0.7547	0.7636	0.7705	0.7761	0.7807	0.7846	0.7878	0.7906
0.6	0.7871	0.7962	0.8030	0.8082	0.8124	0.8158	0.8186	0.8209	0.8229	0.8246
0.7	0.8433	0.8480	0.8516	0.8544	0.8566	0.8584	0.8599	0.8611	0.8622	0.8631
0.8	0.8972	0.8992	0.9007	0.9019	0.9028	0.9035	0.9042	0.9047	0.9052	0.9055
0.9	0.9493	0.9498	0.9502	0.9504	0.9507	0.9508	0.9510	0.9511	0.9512	0.9513

From: Osorio and Steffe (1987); used with permission.

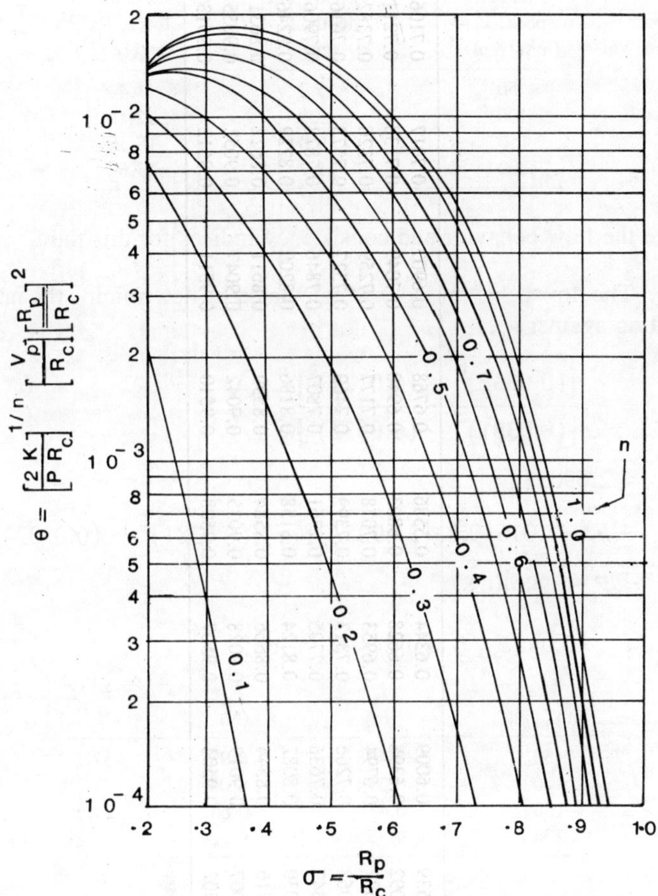

Fig. 6.13. Graph of the parameter θ is used to determine the shear rate from the plunger velocity. (Source: Osorio, F. A. amd Steffe, J. F. 1987. Back extrusion of power law fluids. J. Tex Studies. 18:43. Used with permission.)

where $F_{ne} = F_{te} - \rho g L \pi (R_c^2 - R_p^2)$ and F_{te} = the residual force after the plunger has stopped.

EXAMPLE: The following data were collected in back extrusion of a fluid through a device similar to that shown in Fig. 6.8. The diameters were 2.54 cm for the stanchion, 6.64 cm for the plunger, and 7.62 cm for the outer diameter. The fluid had a density of 1.016 g/cm^3. The total force; the distance penetrated by the plunger, and the plunger speed in each test are as follows:

Plunger speed, V_p (mm/min)	Penetration depth, L_p (cm)	Total force, F_p (N)
50	15.2	11.44
100	13.9	12.16
150	16.0	15.36
200	13.2	13.60

Calculate the flow behavior and consistency indices for this fluid.

Solution: The flow behavior index is calculated by determining the ratio F_n/L and plotting against V_p.

$$L = L_p \frac{[(0.0332)^2 - (0.0127)^2]}{[(0.0381)^2 - (0.0332)^2]}$$

$$= 2.6905 L_p$$

$$F_n = F_t - 1016(9.8)(2.6905L_p)(\pi)[(0.0381)^2 - (0.0332)^2]$$

$$= F_t - 29.4028L_p$$

Values of F_n, L, and F_n/L at different V_p are:

V_p (mm/min)	L (m)	F_n (N)	F_n/L (N/m)
50	0.408	6.98	17.1
100	0.374	8.08	21.59
150	0.430	10.66	24.79
200	0.355	9.72	27.37

A log-log plot of F_n/L against V_p is shown in Fig. 6.14. The slope is 0.33, and the flow behavior index is n. The value of λ is obtained by interpolation from Table 6.1. $\sigma = 0.871$, $n = 0.33$.

$$n = 3, \sigma = 0.871: \lambda = \frac{0.9007 + (0.9504 - 0.9019)(0.071)}{0.1} = 0.9351$$

$$n = 0.4, \sigma = 0.871: \lambda = \frac{0.9019 + (0.9504 - 0.9019)(0.071)}{0.1} = 0.9363$$

$$n = 0.33: \lambda = \frac{0.9351 + (0.9363 - 0.9351)(0.03)}{0.01} = 0.9354$$

The value of Θ is obtained from Fig. 6.13. The curves for $n = 3$ and $n = 4$ do not intercept $\sigma = 0.871$ in Fig. 6.13. However, the curves are linear at low values of Θ and may be represented by the following equations:

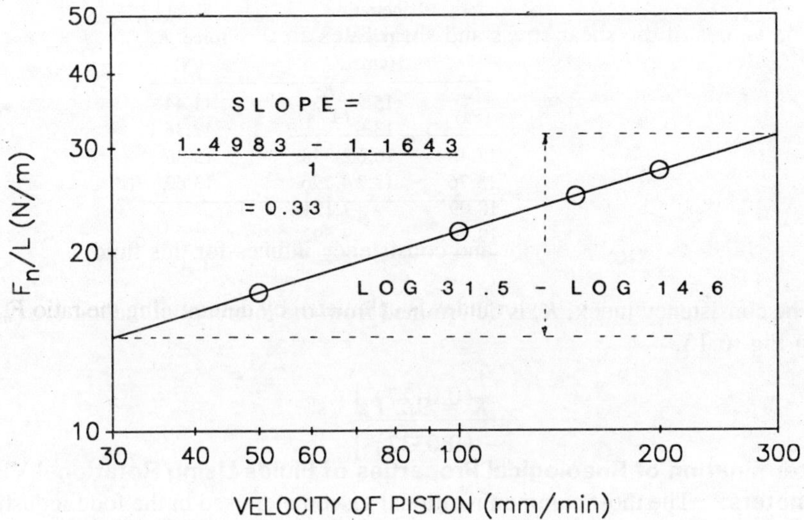

Fig. 6.14. Log-log plot of corrected force to extrude fluid through an annular gap, against piston velocity, to obtain the flow behavior index, n, from the slope.

$$n = 0.3\colon \log \Theta = 1.48353 - 7.5767\sigma$$

$$n = 0.4\colon \log \Theta = 1.9928 - 7.4980\sigma$$

Thus, at $n = 0.3$, $\Theta = 7.665 \times 10^{-6}$; at $n = 0.4$, $\Theta = 2.8976 \times 10^{-5}$. At $n = 0.33$, $\Theta = 1.4058 \times 10^{-5}$ by interpolation. Knowing the value of n, Θ, and λ, τ_w and γ_w can be calculated. Using equation 26: $\sigma = 0.871$, $\lambda = 0.9354$, $\sigma_s = 0.3333$, $1/n = 3$.

$$\gamma_w = \frac{1}{0.381(1.4058 \times 10^{-5})}\left[(0.871)^2 - (0.3333)^2\right]$$

$$\left[\frac{(0.9354)^2}{0.087}\right]^3\left(\frac{1}{60{,}000}\right)V_p$$

$$= 0.04796 V_p; \qquad V_p \text{ is in mm/min}$$

Using equation 27:

$$\tau_w = \frac{(0.9354)^2 - (0.871)^2}{2\pi(0.0332)\left[(0.9354)^2 - (0.3333)^2\right]}\left(\frac{F_n}{L}\right)$$

$$= 0.73\,\frac{F_n}{L}$$

The values of the shear stress and shear rates are:

τ_w (Pa)	λ_w (1/s)
12.48	2.398
15.76	4.796
18.09	7.194
19.98	9.592

The consistency index, K, is determined from the intercept of the log-log plot in Fig. 6.15.

$$K = 9.2 \text{ Pa} \cdot \text{s}^n$$

Determination of Rheological Properties of Fluids Using Rotational Viscometers. The most common rotational viscometer used in the food industry is the concentric cylindrical viscometer. The viscometer shown in Fig. 6.16 consists of concentric cylinders with fluid in the annular space.

The torque needed to rotate one of the cylinders is measured. This torque is proportional to the drag offered by the fluid to the rotation of the cylinder.

If the outer cup is stationary and if the torque measured is A, the force acting on the surface of the inner cylinder to overcome the resistance to rotation will be A/R.

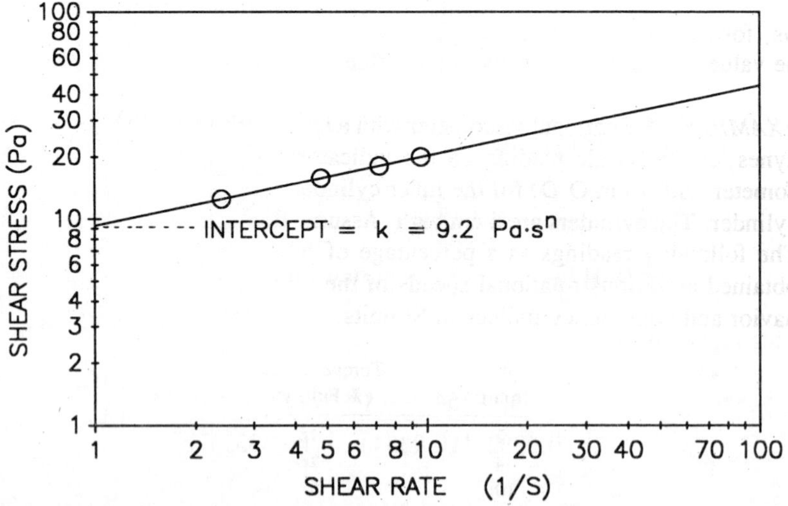

Fig. 6.15. Log-log plot of shear stress against the shear rate to obtain the consistency index, K, from the intercept.

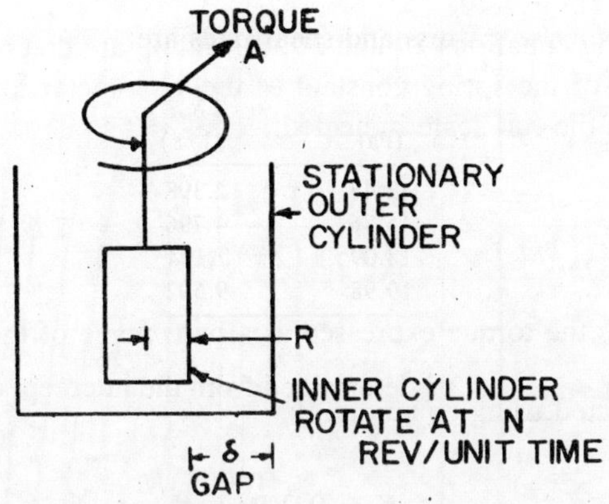

Fig. 6.16. Schematic diagram of a rotational viscometer.

The shear stress τ_w at the wall is:

$$\tau_w = \frac{A}{R}\frac{1}{2\pi RL} = \frac{A}{R^2(2\pi L)} \tag{29}$$

If the gap δ is very small, the shear rate at the wall of the inner cylinder rotating at N revolutions per unit of time is:

$$\gamma_w = \frac{2\pi RN}{\delta} \tag{30}$$

Thus, for a narrow-gap coaxial cylindrical viscometer, it is possible to determine values of τ_w for a corresponding value of γ_w.

EXAMPLE: A rotational viscometer with a spring constant equivalent to 7187 dynes/cm full scale reading on the indicator is used on a narrow-gap viscometer with 1 cm O.D. for the inner cylinder and 1.5 cm I.D. for the outer cylinder. The cylinders are 6 cm high. Assume that end effects are negligible. The following readings as a percentage of full scale on the indicator were obtained at various rotational speeds of the spindle. Determine the flow behavior and consistency indices in SI units.

N (rpm)	Torque Indicated (% Full Scale)
2	15
4	26
10	53
20	9[illegible]

Solution: Using equations 29 and 30, τ_w and γ_w are determined. The torque is the product of the spring constant of the viscometer torque indicator and the fraction of the full scale indicated.

$$\tau_w = \frac{A}{R^2(2\pi L)} = \frac{7187[(\%\ \text{FS})(0.01)]}{(0.5)^2(2\pi)(6)} = (7.6256)(\%\ \text{FS})$$

where % FS is the torque expressed as a percentage of the full scale for the instrument.

γ_w is calculated using equation 30:

$$\gamma_w = \frac{2\pi(0.005)N}{(0.015 - 0.01)(0.5)(60)}$$

N (rev/min)	Torque Indicated (% FS)	γ_w (1/s)	τ_w (dynes/cm^2)
2	15	0.4188	114.38
4	26	0.8377	198.26
10	53	2.0944	404.16
20	93	4.1888	709.18

A log-log plot of τ_w against γ_w, shown in Fig. 6.17, has a slope of 0.784 and τ_w of 225 dynes/cm^2 at $\gamma_w = 1$. Thus, $n = 0.784$ and $K = 225$ in dyne · s/cm^2 units. n is dimensionless; therefore, only K needs to be converted to SI units of Pa · s.

$$\text{Pa} \cdot \text{s} = 225 \frac{\text{dynes} \cdot \text{s}}{\text{cm}^2} \frac{(100)^2\ \text{cm}^2}{\text{m}^2} \frac{1\ \text{Pa} \cdot \text{m}^2}{N} \frac{1 N}{10^5\ \text{dyne}}$$

$$= \frac{225}{10}; \qquad K = 22.5\ \text{Pa} \cdot \text{s}^n$$

Wide-Gap Rotational Viscometer. A wide-gap rotational viscometer consists of a cylinder or a disk-shaped bob on a spindle which rotates in a pool of liquid. A torque transducer attached to the base of the rotating spindle measures the drag offered by the fluid to the rotation of the bob. Different-sized spindles may be used. Based on the spindle size and the rotational speed, the indicated torque can be converted to an apparent viscosity by using an appropriate conversion factor specific for the size of the spindle and the rotational speed. For a shear-thinning fluid, the apparent viscosity increases as the rotational speed decreases. A log-log plot of the apparent viscosity against rotational speed will have a slope equivalent to $n - 1$. If one spindle is used at several rotational speeds, the unconverted torque reading is directly proportional to the shear stress and the rpm is directly proportional to the shear rate. A log-log plot of torque reading against rpm will have a slope equivalent to the flow behavior index n.

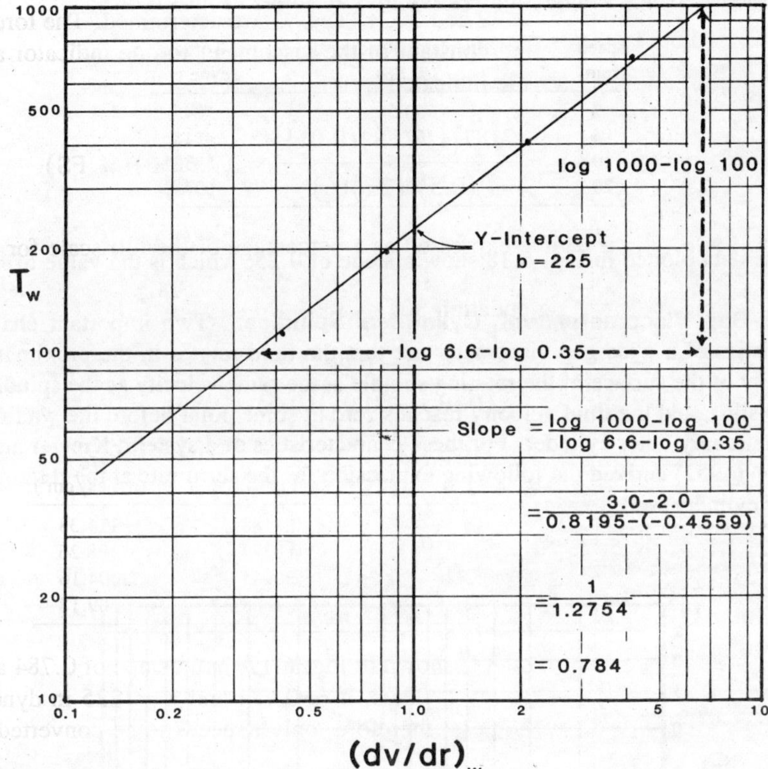

Fig. 6.17. Log-log plot of shear stress against the shear rate to obtain the flow behavior and consistency indices.

EXAMPLE: A Brookfield viscometer model RVF was used to evaluate the apparent viscosity of tomato catsup. One spindle (No. 4) gave readings within the measuring scale of the instrument at four rotational speeds. The viscometer constant was 7187 dynes/cm full scale. The torque reading at various rotational speeds in rpm are as follows:

Rotation Speed (rpm)	Viscometer Indicator Reading (% FS)
2	53.5
4	67
10	80.5
20	97

Evaluate the flow behavior index for this fluid.

Solution: N is the slope of a log-log plot of torque against rpm.

Speed (rpm)	Indicator Reading (% FS)	Torque (dyne-cm)
2	53.5	3845
4	67	4815
10	80.5	5786
20	97	6971

The data plotted in Fig. 6.18 show a slope of 0.25, which is the value of n.

Wide-Gap Viscometer with Cylindrical Spindles. Two important characteristics of a wide-gap viscometer are that the fluid rotates in the proximate vicinity of the surface of the rotating spindle at the same velocity as the spindle (zero slip), and the fluid velocity reaches zero at some point before the wall of the stationary outer cylinder. For these characteristics of a system, Kreiger and Maron (1952) derived the following expression for the shear rate at the rotating inner cylinder wall:

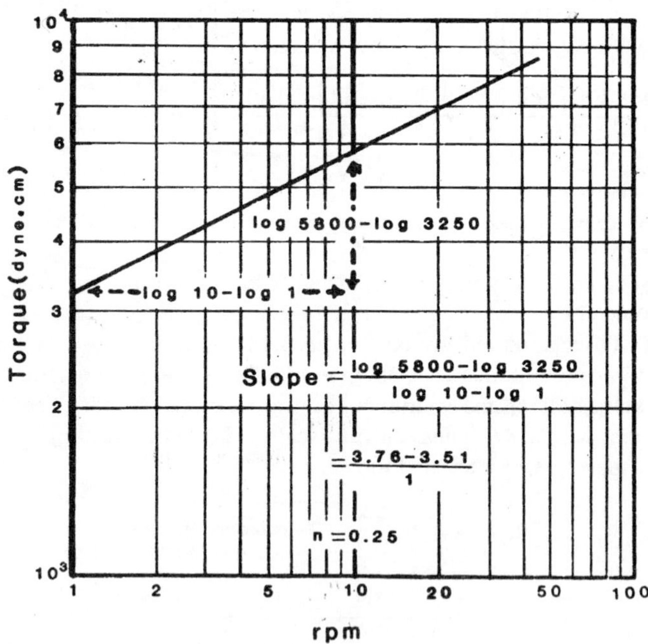

Fig. 6.18. Log-log plot of torque against the rotational speed to obtain the flow behavior index, n, from the slope.

$$\gamma_w = \frac{4\pi N}{n} \tag{31}$$

where N = rotational speed and n is the flow behavior index.

When the torque readings are converted to τ_w using Equation 29, it is possible to determine the flow behavior and consistency indices for the fluid. A log-log plot of the torque indicator reading against rotational speed N will have a slope which is the value of n. Knowing n, equation 31 can then be used to calculate γ_w. A log-log plot of τ_w against γ_w will have K for an intercept.

EXAMPLE: The following data were obtained in the determination of the flow behavior of a fluid using a rotational viscometer. The viscometer spring constant is 7187 dynes/cm full scale. The spindle has a diameter of 0.960 cm and a height of 4.66 cm.

N (rpm)	Torque (% FS)
2	0.155, 0.160, 0.157
4	0.213, 0.195, 0.204
10	0.315, 0.275, 0.204
20	0.375, 0.355, 0.365

Calculate the consistency and flow behavior indices for this fluid.

Solution: Equation 29 will be used to calculate τ_w, and equation 31 for γ_w.

It will be necessary to determine n before equation 31 can be used. The slope of a log-log plot of torque against the rotational speed will be the value of n. This plot in Fig. 6.19 shows a slope of 0.36, which is the value of n. The multiplying factor to convert the rotational speed in rpm to the shear rate at the wall of the rotating cylinder, γ_w, in s^{-1} units is $4(3.1416)/[0.36(60)] = 0.5818$. The multiplying factor to convert the indicated torque as a fraction of viscometer full scale torque to the shear stress at the rotating cylinder wall, τ_w, in Pascals is $7187(10^{-7})/[2(3.1416)(0.0096/2)^2(0.0466)] = 106.5367$.

The values for τ_w and γ_w are:

Shear stress (τ_w) (Pa)	Shear rate (γ_w) (1/s)
16.5, 17.0, 16.7	1.16
22.7, 20.8, 21.7	2.33
33.6, 29.3, 21.7	5.82
39.9, 37.8, 38.9	11.64

The plot of τ_w against γ_w in Fig. 6.20 shows that τ_w = 15.2 Pa at γ_w = 1; thus, K = 15.2 Pa · s^n.

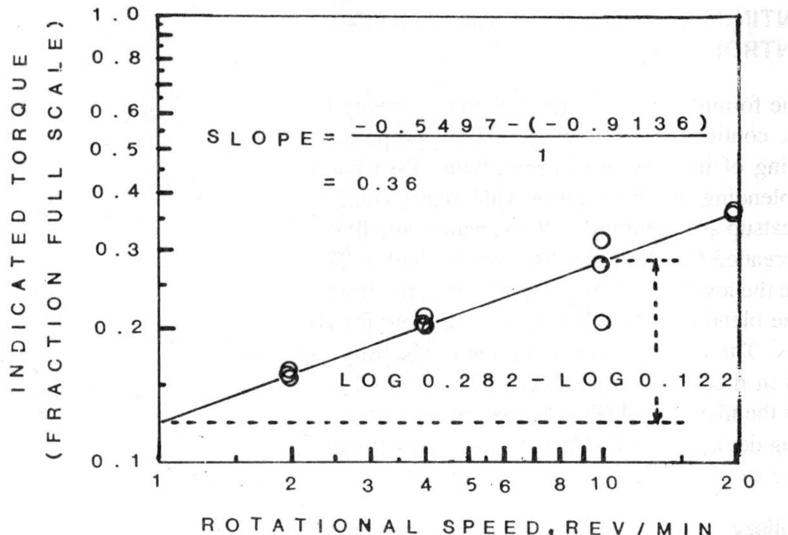

Fig. 6.19. Log-log plot of indicated torque against the rotational speed to obtain the flow behavior index, n, from the slope.

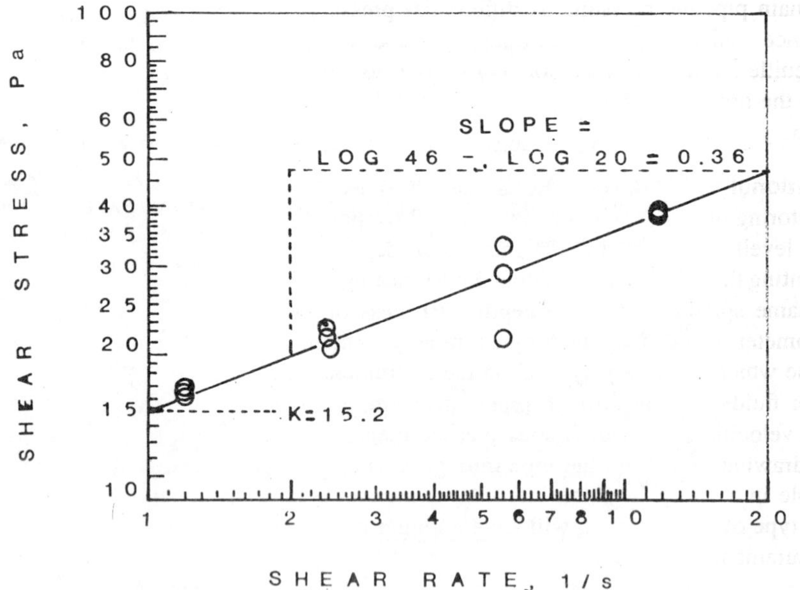

Fig. 6.20. Log-log plot of shear stress against shear rate to obtain the consistency index, K, from the intercept.

CONTINUOUS VISCOSITY MONITORING AND CONTROL

In the formulation of foods in which viscosity or consistency is a quality attribute, continuous monitoring of flow properties and automatic control of the feeding of ingredients are important. Two good examples of these processes are blending of tomato paste with sugar, vinegar, and flavoring in manufacturing catsup and blending of flour, water, and flavoring ingredients to make batter for breaded fried fish, poultry, or vegetables. These processes are rather simple since the level of only one ingredient in the formulation controls the consistency of the blend and therefore is very suitable for automatic control of flow properties. The principles for continuous viscosity monitoring are similar to those used in rheology. Some means must be used to divert a constant flow of fluid from the main pipeline to the measuring device. Several types of viscosity measuring devices are suitable for continuous monitoring in flowing systems. Three of the easiest to adapt on a system are described below.

Capillary Viscometer. A capillary viscometer may be used for continuous monitoring of consistency by tapping into the main piping, using a metering pump to deliver a constant flow through a level capillary, and returning the fluid to the main pipe downstream from the intake point. The metering pump must be a positive displacement pump to prevent flow fluctuations due to changes in the main pipeline pressure. A differential pressure transducer mounted at the entrance and exit of the capillary is used to measure the pressure drop. Poiseuille's equation (equation 11) is used to calculate the apparent viscosity from the pressure difference measured and the flow delivered by the metering pump.

Rotational Viscometer. Rotational viscometers can be used for continuous monitoring of viscosity in open vessels. Provision must be made for changes in fluid levels to ensure that the immersion depth of the spindle is appropriate. Mounting the instrument on a float device resting on the fluid surface will ensure the same spindle immersion depth regardless of fluid level. Some rotational viscometers can display the torque remotely; others give a digital readout of the torque which can be easily read on the instrument.

For fluids flowing through pipes, problems may be encountered with high fluid velocities. Semicontinuous measurements may be made by periodically withdrawing fluid from the pipe through a sampling valve and depositing this sample in a cup, where the viscosity is measured by a rotational viscometer. This type of measurement will have a longer response time than a truly in-line measurement.

Viscosity-Sensitive Rotameter. This device is similar to a rotameter and operates on the principle that the drag offered by an obstruction in a field of flow is proportional to the viscosity and the rate of flow. A small float positioned within a vertical tapered cylinder will assume a particular position when fluid with a fixed flow rate and a particular viscosity is flowing through the tube. This device needs a constant rate of flow to be effective. Usually, a metering pump draws out fluid from a tap on a pipe, delivers the fluid through the viscosity-sensitive rotameter, and returns the fluid to the pipe through a downstream tap.

FLOW OF FALLING FILMS

Fluid films are formed when a fluid flows over a vertical or inclined surface or is withdrawn from a tank at a rapid rate. The differential equation for the shear stress–shear rate relationship for the fluid coupled with a differential force balance will allow a determination of the thickness of the film as a function of fluid flow properties.

Films of Constant Thickness. When the rate of flow of a fluid over a vertical or inclined plane is constant, the film thickness is also constant. Figure 6.21 shows a fluid film flowing down a plane inclined at an angle α with the vertical plane. If the width of the plane is W, a force balance on a macroscopic section of the film of length L and of thickness x, measured from the film surface, is as follows:

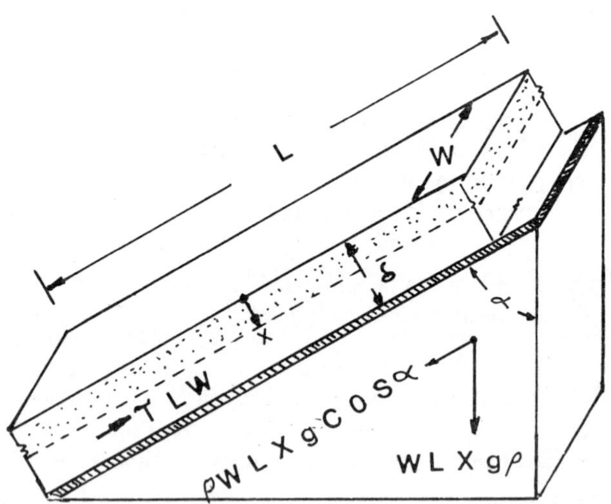

Fig. 6.21. Diagram of a control element used in analysis of fluid film flowing down an inclined surface.

Mass of the differential increment $= LW\rho x$

Force due to gravity on mass $= LW\rho gx$

Component of gravitational force along inclined plane $= LW\rho gx\cos(\alpha)$

Fluid resistance against the flow $= \tau WL$. If the fluid follows the power law equation (equation 2), a force balance results in:

$$WLK\left[-\frac{dV}{dr}\right]^n = LW\rho gx\cos(\alpha)$$

Rearranging:

$$-\frac{dV}{dx} = \left[\frac{\rho gx\cos(\alpha)}{K}\right]^{1/n}$$

Integrating:

$$\int dV = -\left[\frac{\rho g\cos(\alpha)}{K}\right]^{1/n}\int [x]^{1/n}\,dx$$

The integrated expression is:

$$V = -\left[\frac{\rho g\cos(\alpha)}{K}\right]^{1/n}\left[\frac{n}{n+1}\right][x]^{(n+1)/n} + C$$

The boundary condition at $x = \delta$, $V = 0$, when substituted in the above expression, will give the value of the integration constant, C. δ is the film thickness, and since x is measured from the surface on the film, the position $x = \delta$ is the surface of the incline along which the film is flowing. The final expression for the fluid velocity profile is:

$$V = \left[\frac{\rho g\cos(\alpha)}{K}\right]^{1/n}\left[\frac{n}{n+1}\right][\delta]^{(n+1)/n}\left[1 - \left[\frac{x}{\delta}\right]^{(n+1)/n}\right]$$

The thickness of the film δ can be expressed in terms of an average velocity of flow, $\overline{V}$. The expression is derived as follows:

$$W\delta\overline{V} = \int_0^\delta VW\,dx$$

$$= W\left[\frac{\rho g\cos(\alpha)}{K}\right]^{1/n}\left[\frac{n}{n+1}\right][\delta]^{(n+1)/n}\int_0^\delta\left[1 - \left[\frac{x}{\delta}\right]^{(n+1)/n}\right]dx$$

$$\overline{V} = \left[\frac{\rho g\cos(\alpha)}{K}\right]^{1/n}\left[\frac{n}{2n+1}\right][\delta]^{(n+1)/n} \quad (32)$$

The film thickness is:

$$\delta = \left[\frac{\bar{V}(2n+1)K^{1/n}}{n[\rho g\cos(\alpha)]^{1/n}}\right]^{n/(n+1)} \tag{33}$$

EXAMPLE: Applesauce at 80°C (K = 9 Pa · s^n; n = 0.33, ρ = 1030 kg/m^3) is to be deaerated by allowing it to flow as a film down the vertical walls of a tank 1.5 m in diameter. If the film thickness desired is 5 mm, calculate the mass rate of flow of applesauce to be fed into the deaerator. At this film thickness and 80°C, an exposure time of film to a vacuum of 25 in. Hg of 45 s is required to reduce the dissolved oxygen content to a level needed for product shelf stability. Calculate the height of the tank needed for the film to flow over.

Solution: Equation 32 will be used to calculate the average velocity of the film needed to give a film thickness of 5 mm. Since the side is vertical, α = 0 and cos α = 1.

$$\bar{V} = \left[\frac{(1030)(9.8)}{9.0}\right]^{1/0.33}\left[\frac{(0.005)^{4.03}}{6.03}\right]$$

$$= 0.154 \text{ m/s}$$

The mass rate of flow, $\dot{m}$, is:

$$\dot{m} = \frac{0.154 \text{ m}}{\text{s}}(\pi)(0.75)^2 \text{ m}^2\left(\frac{1030 \text{ kg}}{\text{m}^3}\right)$$

$$= 280 \text{ kg/s}$$

The height of the vessel needed to provide the 45 s required exposure for deaeration is:

$$h = \frac{0.154 \text{ m}}{\text{s}}(15 \text{ s}) = 2.31 \text{ m}$$

Time-Dependent Film Thickness. Problems of this type are encountered when fluids cling as a film on the sides of a storage vessel when the vessel is emptied (drainage) or when a solid is passed through a pool of fluid and emerges from the fluid coated with a fluid film (withdrawal). The latter process may be encountered when applying batter over a food product prior to breading and frying. The solution to problems of withdrawal and drainage is the same.

Figure 6.22 shows a fluid film with width W falling down a vertical surface. A control segment of the film having a thickness δ located a distance z from the top, and a length dz along the surface on which it flows, is shown. The solution involves solving partial differential equations, which requires a mass balance across the control volume.

If V is the average velocity of the fluid in the film at any point z distant from the top, the mass of fluid entering the control volume from the top is:

$$M_i = \overline{V}\delta W\rho$$

The mass leaving the control volume from the lower part of the control volume is:

$$M_e = \left[\overline{V}\delta + \frac{\partial}{\partial z}(\overline{V}\delta)\, dz\right] W\rho$$

The accumulation term due to the difference between what entered and left the control volume is:

$$\frac{\partial \delta}{\partial z}\rho W\, dz$$

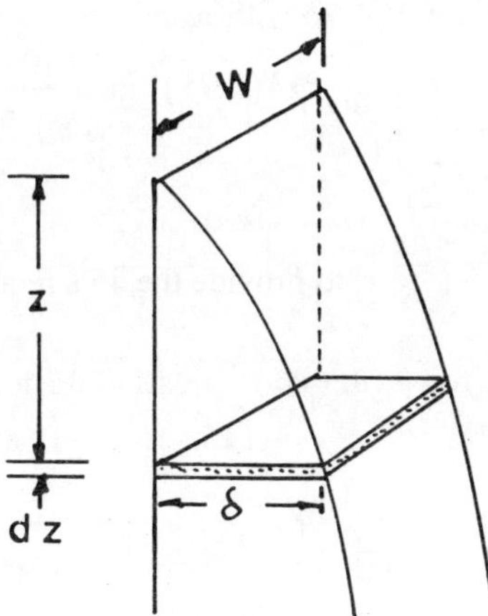

Fig. 6.22. Diagram of a control element used in analysis of time-dependent flow of a fluid film down a surface.

The mass balance across the control volume is:

$$\overline{V}\delta W\rho = \left[\overline{V}\delta + \frac{\partial}{\partial z}(\overline{V}\delta)\, dz\right]\rho W + \frac{\partial \delta}{\partial z}\rho W\, dz$$

Simplification will produce the partial differential equation which describes the system.

$$\frac{\partial}{\partial z}(\overline{V}\delta) = -\frac{\partial \delta}{\partial t} \tag{34}$$

The solution to this partial differential equation will be the product of two functions: $G(t)$, which is a function only of t, and $F(z)$, which is a function only of z.

$$\delta = F(z)G(t) \tag{35}$$

Partial differentiation of equation 35 gives:

$$\frac{\partial \delta}{\partial t} = F(z)\frac{\partial G(t)}{\partial t} \tag{36}$$

$$\frac{\partial \delta}{\partial z} = G(t)\frac{\partial F(z)}{\partial z} \tag{37}$$

Equation 32 represents $\overline{V}$ for a fluid film flowing down a vertical surface ($\cos \alpha = 1$). Partial differentiation of $\overline{V}\delta$ with respect to z gives:

$$\frac{\partial}{\partial z}(\overline{V}\delta) = \frac{\partial}{\partial z}\left[\left[\frac{\rho g}{K}\right]^{1/n}\left[\frac{n}{2n+1}\right]\left[\delta\right]^{(n+1)/n}\delta\right]$$

$$= \left[\frac{\rho g}{K}\right]^{1/n}\left[\delta\right]^{(n+1)/n}\frac{\partial \delta}{\partial z} \tag{38}$$

Substituting equation 34 for $\partial/\partial z(\overline{V}\delta)$, equation 36 for $\partial\delta/\partial t$, and equation 37 for $\partial\delta/\partial z$ in equation 38:

$$-F(z)\frac{\partial G(t)}{\partial t} = \left[\frac{\rho}{K}\right]^{1/n}[F(z)G(t)]^{(n+1)/n}G(t)\frac{FF(z)}{\partial z}$$

Each side of the partial differential equation contains only one variable; therefore, the equation can be separated and each side integrated independently by equating it to a constant. Let the constant $= \lambda$.

The function with the time variable is obtained by integrating the left-hand side of the equation and setting it equal to λ. The function with the z variable is obtained the same way, using the right-side of the equation.

$$G(t) = \left[\lambda\left[\frac{n+1}{n}\right]t\right]^{-n/(n+1)} + C_1$$

$$F(z) = \left[\lambda\left[\frac{K}{\rho g}\right]^{1/n}\left[\frac{n}{n+1}\right]z\right]^{n/(n+1)} + C_2$$

The boundary conditions, $\delta = 0$ at $z = 0$ and $t = \infty$, will be satisfied when the integration constants C_1 and C_2 are zero. When expressions for $F(z)$ and $G(t)$ are combined according to equation 35, the constant λ cancels out. The final expression for the film thickness as a function of time and position is:

$$\delta = \left[\left[\frac{K}{\rho g}\right]^{1/n}\left[\frac{z}{t}\right]\right]^{n/(n+1)} \tag{39}$$

EXAMPLE 1: Calculate the thickness of batter that would be attached to the sides of a brick-shaped food product at the midpoint of its height if the product is immersed in the batter, removed, and allowed to drain for 5 s before breading. The batter has a flow behavior index of 0.75, a density of 1003 kg/m^3, and a consistency index of 15 $Pa \cdot s^n$. The vertical side of the product is 2 cm high.

Solution: $z = 0.001$ m; $t = 5$ s. Using equation 38:

$$\delta = \left[\left[\frac{15}{1003(9.8)}\right]^{1/0.75}\left(\frac{0.001}{5}\right)\right]^{0.75/1.75}$$

$$= 0.000599 \text{ m.}$$

EXAMPLE 2: Applesauce is rapidly drained from a drum 75 cm in diameter and 120 cm high. At the end of 60 s after the fluid is drained from the drum, calculate the mass of applesauce adhering to the walls of the drum. The applesauce has a flow behavior index of 0.34, a consistency index of 11.6 Pa $\cdot$ s^n, and a density of 1008 kg/m^3.

Solution: Equation 39 will be used. However, the thickness of the film will vary with the height; therefore, it will be necessary to integrate the quantity of fluid adhering at the different positions over the height of the vessel. At any given time, t, the increment volume of fluid, dq, over an increment of height dz will be:

$$dq = \pi[r^2 - (r - \delta)^2]\,dz$$

The total volume of film covering the height of the drum will be:

$$q = \int_0^z \pi(2r\delta - \delta^2)\,dz$$

Substituting the expression for δ (equation 39) and integrating over z, holding time constant:

$$q = 2\pi r\left[\frac{2}{\rho g}\right]^{1/(n+1)}\left[\frac{1}{t}\right]^{n/(n+1)}\left[\frac{n+1}{2n+1}\right]\left[z\right]^{(2n+1)/(n+1)}$$

$$-\ \pi\left[\frac{K}{\rho g}\right]^{2/(n+1)}\left[\frac{1}{t}\right]^{2n/(n+1)}\left[\frac{n+1}{3n+1}\right]\left[z\right]^{(3n+1)/(n+1)}$$

Substituting values of known quantities:

$K/\rho g = 11.6/(1008)(9.8) = 0.001174$
$2/(n + 1) = 1.515;\ 2n/(n + 1) = 0.507;\ (n + 1)/(3n + 1) = 0.6633$
$1/(n + 1) = 0.7463;\ n/(n + 1) = 0.254;\ (n + 1)/(2n + 1) = 0.7976$

$$V = 2\pi(0.375)(0.001174)^{0.7463}\left(\frac{1}{60}\right)^{0.254}(0.7976)$$

$$= (1.2)^{1/0.7976} - \pi(0.001174)^{1.515}(1/60)^{0.507}$$

$$= (0.6633)(1.2)^{1/0.6633}$$

$$= 0.0054289 - 0.0000125101 = 0.005416\ \text{m}^3$$

The mass of applesauce left adhering to the walls of the drum is 0.005416 m^3 (1008) kg/m^3 = 5.46 kg.

TRANSPORTATION OF FLUIDS

The basic equations characterizing flow and forces acting on flowing fluids are derived using the laws of conservation of mass, conservation of momentum, and conservation of energy.

Momentum Balance. Momentum is the product of mass and velocity and has units of kg · m/s. When mass is expressed as a mass rate of flow, the product with velocity is the rate of flow of momentum and will have units of kg · m/s^2, the same units as force. When a velocity gradient exists in a flowing fluid, momentum transfer occurs across streamlines, and the rate of momentum transfer per unit area, $d(mV/A)/dt$, the momentum flux, has units of kg/m · s^2. The unit of momentum flux is the same as that of stress or pressure. The pressure drop in fluids flowing through a pipe, which is attributable to fluid resistance, is the result of a momentum flux between streamlines in a direction perpendicular to the direction of flow. The momentum balance may be expressed as:

$$\begin{matrix}\text{Rate of momentum} \\ \text{flow in}\end{matrix} + \sum F = \begin{matrix}\text{rate of momentum} \\ \text{flow out}\end{matrix} + \text{accumulation}$$

ΣF is the sum of external forces acting on the system, (e.g., atmospheric pressure, stress on the confining vessel) or forces exerted by restraints on a nozzle discharging fluid. Since momentum is a function of velocity, which is a vector quantity, i.e., it has both magnitude and direction, all the terms in the above equation are vector quantities. The form of the equation given above implies that a positive sign on ΣF indicates a force applied in a direction entering the system. When making a momentum balance, the component of the force or velocity acting in a single direction (e.g., x, y, or z component) is used in the equation. Thus, in three-dimensional space, a momentum balance may be made for the x, y, and z directions.

The force balance used to derive the Poiseuille equation (equation 9) is an example of a momentum balance over a control volume, a cylindrical shell of fluid flowing within a cylindrical conduit. Equation 9 shows that the momentum flow (momentum flux multiplied by the surface area of the control volume) equals the net force (pressure drop multiplied by the cross-sectional area of the control volume) due to the pressure acting on the control element.

When the control volume is the whole pipe, i.e., the wall of the pipe is the boundary surrounding the system under consideration, it is possible to determine the forces acting on the pipe or its restraints. These forces are significant in instances where a fluid changes direction, as in a bend; when velocity changes, as in a converging pipe section; when a fluid discharges out of a nozzle; or when a fluid impacts a stationary surface.

Momentum flow (rate of change of momentum, $\dot{M}$) has the same units as force and is calculated in terms of the mass rate of flow, $\dot{m}$, kg/s, the velocity, V, or the volumetric rate of flow, q, as follows:

$$\dot{M} = \dot{m}V = AV^4\rho = qV\rho$$

A is the area of the flow stream perpendicular to the direction of flow and ρ is the fluid density.

The momentum flux (rate of change of momentum per unit area) is the quotient, $\dot{M}/A$.

A force balance over a control volume is the same as a balance of momentum flow.

EXAMPLE: Calculate the force acting on the restraints of a nozzle from which fluid is discharging to the atmosphere at the rate of 5 kg/s. The fluid has a density of 998 kg/m³. It enters the nozzle at a pressure of 438.3 kPa above atmospheric pressure. The nozzle has a diameter of 6 cm at the inlet and 4 cm at the discharge.

Solution: Assume that the velocity of the fluid through the nozzle is uniform, i.e., there is no velocity variation across the pipe diameter. The pipe wall may be used as a system boundary for making the momentum balance in order to include the force acting on the restraint. The momentum balance, with subscripts 1 and 4 referring to the inlet and discharge of the nozzle, respectively, is:

$$\dot{m}V_1 + P_1A_1 + F_x = mV_4 + P_4A_4$$

Atmospheric pressure (Pa) acts on the inlet to the nozzle entering the control volume; this is also the pressure at the discharge. Since P_1 is pressure above atmospheric at the inlet, the momentum balance becomes:

$$\dot{m}V_1 + P_aA_1 + P_1A_1 + F_x = mV_4 + P_aA_4$$

Solving for F_x:

$$F_x = m(V_4 - V_1) - P_a(A_1 - A_4) - P_1A_1$$

Substituting known quantities and solving for V_1 and V_4:

$$A_1 = 0.004847 \text{ m}^4; \qquad A_4 = 0.00031415$$

$$V_1 = \frac{5 \text{ kg}}{\text{s}} \left| \frac{\text{m}^3}{998 \text{ kg}} \right| \frac{1}{0.004847 \text{ m}^4} = 1.77 \text{ m/s}$$

$$V_4 = \frac{5 \text{ kg}}{\text{s}} \left| \frac{\text{m}^3}{998 \text{ kg}} \right| \frac{1}{0.00031415 \text{ m}^4} = 15.95 \text{ m/s}$$

$$\begin{aligned} F_x &= 5(15.95 - 1.77) - 438{,}000(0.004847) \\ &= -101{,}300(0.004847 - 0.00031415) \\ &= 70.9 - 947.4 = -856.5N \end{aligned}$$

The negative sign indicates that this force is acting in a direction opposite the momentum flow out of the system; thus, the nozzle is being pushed away from the direction of flow.

The Continuity Principle. The principle of conservation of mass in fluid dynamics is referred to as the *continuity principle*. This principle is applied whenever there is a change in velocity, a change in the diameter of the conduit, split flow, or a change in the density of compressible fluids such as gases. The continuity principle for flow in one direction, expressed in equation form, is as follows:

$$\rho_1 V_1 A_1 = \rho_2 V_2 A_2 + A \frac{\partial}{\partial t} (\rho V) \tag{40}$$

If velocity is changing in all directions, a three-dimensional mass balance should be made. The above equation is a simplified form suitable for use in fluid transport systems where velocity is directed in one direction. At a steady state, the time derivative is zero. If the fluid is incompressible, the density is constant and the continuity equation reduces to:

$$V_1 A_1 = V_2 A_2 \tag{41}$$

EXAMPLE: A fluid having a density of 1005 kg/m^3 is being drawn out of a storage tank 3.5 m in diameter through a tap on the side at the lowest point in the tank. The tap consists of a short length of 1.5 in. nominal sanitary pipe (I.D. = 0.03561) with a gate valve. If fluid flows out of the tap at the rate of 40 L/min, calculate the velocity of the fluid in the pipe and the velocity at which the fluid level recedes inside the tank.

Solution: The mass rate of flow $\dot{m}$, is constant. The mass rate of flow is the product of the volumetric rate of flow, q, and the density:

$$\dot{m} = \frac{40 \text{ L}}{\text{min}} \cdot \frac{0.001 \text{ m}^3}{\text{L}} \cdot \frac{1 \text{ min}}{60 \text{ s}} \cdot \frac{1005 \text{ kg}}{\text{m}^3} = 0.67 \text{ kg/s}$$

$$= V_{\text{pipe}} A_{\text{pipe}} = V_{\text{tank}} A_{\text{tank}}$$

$$V_{\text{pipe}} = \frac{0.67 \text{ kg}}{\text{s}} \cdot \frac{1 \text{ m}^3}{1005 \text{ kg}} \cdot \frac{1}{\pi (0.01781)^2 \text{ m}^2} = 0.669 \text{ m/s}$$

$$V_{\text{tank}} = V_{\text{pipe}} \frac{A_{\text{pipe}}}{A_{\text{tank}}} = 0.669 \frac{(0.03561)^2}{(3.5)^2}$$

$$= 6.92 \times 10^{-5} \text{ m/s}$$

Fluid Flow Regimes. Equation 10 indicates that a Newtonian fluid flowing inside a tube has a parabolic profile. Fluid molecules in the center of the tube flow at the maximum velocity. The equation holds only when each molecule of the fluid remains in the same radial position as the fluid traverses the length of the tube. The path of any molecule follows a well-defined streamline which can be readily shown by injecting a dye into the flow stream. When injected in a very fine stream, the dye will trace a straight line parallel to the direction of flow. This type of flow is called *streamline* or *laminar flow*. Figure 6.23 shows the streamline flow which is observed when a dye is continuously injected into a flow stream (A) and the development of the parabolic velocity profile (D) that is observed when a viscous dye solution is injected into a tube filled with a very viscous fluid at rest (B) and flow is allowed to develop slowly (C).

As the rate of flow increases, molecular collisions occur at a more frequent rate and crossovers of molecules across streamlines occur. Eddy currents develop in the flow stream. This condition of flow is turbulent, and the velocity profile predicted by equation 10 no longer holds. Figure 6.24 shows what can be observed when a dye is continuously injected into a fluid in turbulent flow. Swirling and curling of the dye can be observed. A certain amount of mixing also occurs. Thus, a particle originally introduced at the center of the tube may

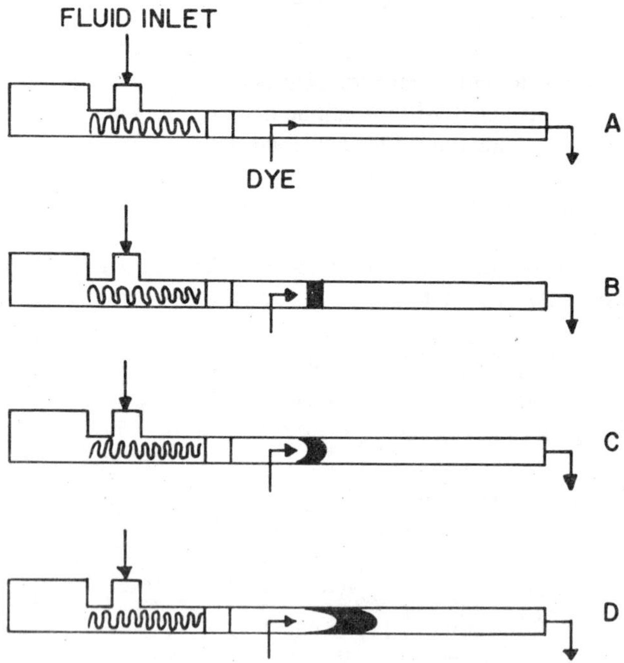

Fig. 6.23. Development of laminar velocity profile during flow of a viscous fluid through a pipe.

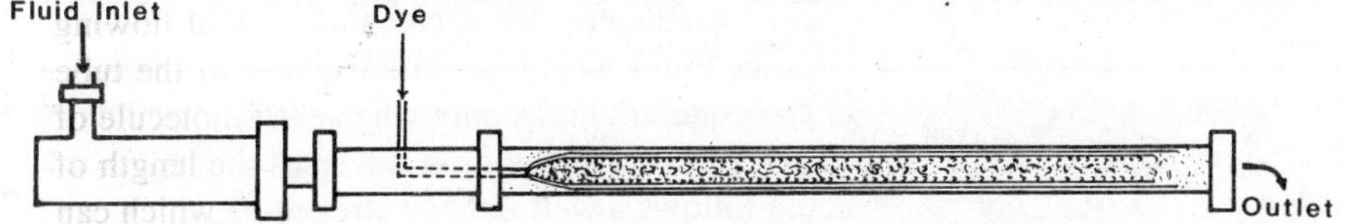

Fig. 6.24. Path traced by a dye injected into a fluid flowing through a tube in turbulent flow.

traverse back and forth between the wall and the center as the fluid travels the length of the tube. The velocity profile in turbulent flow is flat compared to the parabolic profile in laminar flow. The ratio of average to maximum velocity in laminar flow is 0.5 compared to an average ratio of approximately 0.8 in fully developed turbulent flow.

The Reynolds Number. The Reynolds number is a dimensionless quantity that can be used as an index of laminar or turbulent flow. The Reynolds number, represented by *Re*, is a function of the tube diameter D, the average velocity $\bar{V}$, the density of the fluid ρ, and the viscosity μ.

$$Re = \frac{D\bar{V}\rho}{\mu} \tag{42}$$

For values of the Reynolds number below 2100, flow is laminar and the pressure drop per unit length of pipe can be determined using equation 11 for a Newtonian fluid and equation 15 for power law fluids. For Reynolds numbers above 2100, flow is turbulent and pressure drops are obtained using an empirically derived friction factor chart.

Pipes and Tubes. Tubes are thin-walled, cylindrical conduits whose nominal sizes are based on the outside diameter. Pipes are thicker-walled than tubes, and their nominal size is based on the inside diameter. The term *sanitary pipe* used in the food industry refers to stainless steel tubing where the nominal designation is based on the outside diameter. Table 6.2 shows dimensions for steel and sanitary pipes and heat exchanger tubes.

Frictional Resistance to Flow for Newtonian Fluids. Flow through a tubular conduit is accompanied by a drop in pressure. The drop in pressure is equivalent to the stress that must be applied on the fluid to induce flow. This stress can be considered as a frictional resistance to flow. Working against this stress requires the application of power in transporting fluids. An expression for the pressure drop as a function of the fluid properties and the dimension of the conduit would be useful in predicting power requirements for inducing flow

Table 6.2. Pipe and Heat Exchanger Tube Dimensions (Numbers in Parentheses Represent the Dimension in Meters)

	Steel Pipe (Sch. 40)		Sanitary Pipe		Heat Exchanger Tube (18 Ga)	
Nominal size (in.)	ID in./(m)	OD in./(m)	ID in./(m)	OD in./(m)	ID in./(m)	OD in./(m)
0.5	0.622 (0.01579)	0.840 (0.02134)	—	—	0.402 (0.01021)	0.50 (0.0217)
0.75	0.824 (0.02093)	1.050 (0.02667)	—	—	0.652 (0.01656)	0.75 (0.01905)
1	1.049 (0.02664)	1.315 (0.03340)	0.902 (0.02291)	1.00 (0.0254)	0.902 (0.02291)	1.00 (0.0254)
1.5	1.610 (0.04089)	1.900 (0.04826)	1.402 (0.03561)	1.50 (0.0381)	1.402 (0.03561)	1.50 (0.0381)
2.0	2.067 (0.05250)	2.375 (0.06033)	1.870 (0.04749)	2.00 (0.0508)	—	—
2.5	2.469 (0.06271)	2.875 (0.07302)	2.370 (0.06019)	2.5 (0.0635)	—	—
3.0	3.068 (0.07793)	3.500 (0.08890)	2.870 (0.07289)	3.0 (0.0762)	—	—
4.0	4.026 (0.10226)	4.500 (0.11430)	3.834 (0.09739)	4.0 (0.1016)	—	—

ID = inside diameter.
OD = outside diameter.

over a given distance or for determining the maximum distance a fluid will flow with the given pressure differential between two points.

For Newtonian fluids in laminar flow, the Poiseuille equation (equation 11) may be expressed in terms of the inside diameter (D) of a conduit as follows:

$$\frac{\Delta P}{L} = \frac{32\bar{V}\mu}{D^2} \tag{43}$$

Equation 43 can be expressed in terms of the Reynolds number.

$$\frac{\Delta P}{L} = \frac{32\bar{V}\mu}{D^2} \cdot \frac{(D\bar{V}\rho)/\mu}{Re}$$

$$\frac{\Delta P}{L} = \frac{2\left(\frac{16}{Re}\right)(\bar{V})^2(\rho)}{D} \tag{44}$$

Equation 44 is similar to the Fanning equation derived by dimensional analysis, where the friction factor f is $16/Re$ for laminar flow. The Fanning equation, which is applicable in both laminar and turbulent flow, is:

$$\frac{\Delta P}{\rho} = \frac{2f(\overline{V})^2 L}{D} \tag{45}$$

For turbulent flow, the Fanning friction factor f may be determined for a given Reynolds number and a given roughness factor from a friction factor chart, as shown in Fig. 6.25. For smooth tubes, a plot of log f obtained from the lowest curve in Fig. 6.25 against the log of the Reynolds number would give the following relationships between f and Re:

$$f = 0.048(Re)^{-0.20} \quad 10^4 < Re < 10^6 \tag{46}$$

$$f = 0.193(Re)^{-0.35} \quad 3 \times 10^3 < Re < 10^4 \tag{47}$$

EXAMPLE 1: What pressure must be generated at the discharge of a pump that delivers 100 L/min of a fluid having a specific gravity of 1.02 and a viscosity of 100 centipoises? The fluid flows through a 1.50-in. (nominal) sanitary pipe 50 m long. The pipe is straight and level, and the discharge end of the pipe is at atmospheric pressure.

Solution: Given flow rate (q) = 100 L/min; ρ = 1.02 g/cm^3; L = 50 m; D = 1.5 in. (nominal). From Table 6.2, D = 1.402 in. The problem can be solved using either equations 43, 44, or 45. However, it is necessary to determine first if the flow is laminar or turbulent. Expressing all units in the SI system, the dimensional equation that represents the Reynolds number is:

$$Re = \frac{D(m)\overline{V}(\text{m/s})\,\rho(\text{kg/m}^3)}{\mu(\text{Pa} \cdot \text{s})}$$

[Note that the Reynolds number will be dimensionless if the base units for Pa · s, kg/(m · s), is substituted in the dimensional equation.]

$$D = 1.402 \text{ in.} \cdot \frac{0.0254 \text{ m}}{\text{in.}} = 0.0356 \text{ m}$$

$$q = 100 \frac{1}{\text{min}} \cdot \frac{0.001 \text{ m}^3}{L} \cdot \frac{1 \text{ min}}{60 \text{ s}} = 0.00167 \text{ m}^3/\text{s}$$

$$\overline{V} = \frac{q}{\text{area}} = \frac{0.00167 \text{ m}^3}{\text{s}} \cdot \frac{1}{\pi/4(0.0356)^2 \text{ m}^2} = 1.677 \text{ m/s}$$

$$\rho = 1.02 \cdot \frac{1000 \text{ kg}}{\text{m}^3} = 1020 \text{ kg/m}^3$$

$$\mu = (100 \text{ cp}) \cdot \frac{0.001 \text{ Pa} \cdot \text{s}}{\text{cp}} = 0.1 \text{ Pa} \cdot \text{s}$$

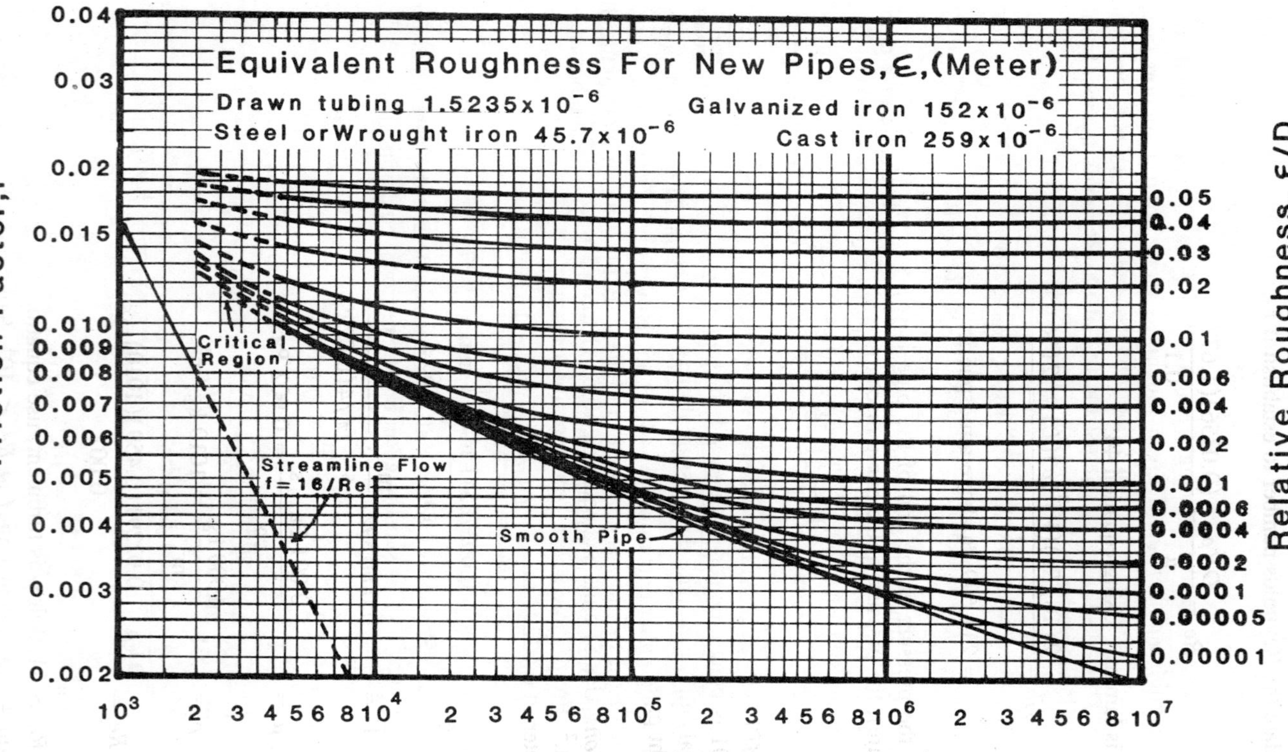

Fig. 6.25. The Moody diagram for the Fanning friction factor. (Based on Moody, *Trans. ASME* 66:671, 1944.)

Calculate the Reynolds number to establish if flow is laminar or turbulent.

$$Re = \frac{D\overline{V}\rho}{\mu} = \frac{(0.0356)(1.677)(1020)}{0.1} = 609$$

Flow is laminar. Using equation 44:

$$\Delta P = \frac{2(16/Re)(\overline{V})^2(\rho)(L)}{D} = \frac{2(16)(1.677)^2(1020)(50)}{609(0.0356)}$$

$$= 211{,}699 \text{ Pa above atmospheric}$$

Since the pressure at the end of the pipe is 0 gage pressure (atmospheric pressure), the pressure at the pump discharge will be

$$P = 211{,}699 \text{ Pa gage}$$

EXAMPLE 2: Milk with a viscosity of 2 centipoises and a specific gravity of 1.01 is being pumped through a 1-in. (nominal) sanitary pipe at the rate of 3 gal/min. Calculate the pressure drop in $lb_f/in.^2$ per foot length of level straight pipe.

Solution: Use SI units and equation 40 or 41 to calculate ΔP and convert to $lb_f/in.^2$. From Table 6.2, a 1-in. (nominal) sanitary pipe will have an inside diameter of 0.02291 m.

$$q = 3\,\frac{\text{gal}}{\text{min}} \cdot \frac{0.00378541\ \text{m}^3}{\text{gal}} \cdot \frac{1\ \text{min}}{60\ \text{s}} = 0.00018927\ \text{m}^3/\text{s}$$

$$\overline{V} = \frac{q}{A} = 0.00018927\,\frac{\text{m}^3}{\text{s}} \cdot \frac{1}{\pi/4(0.02291)^2\ \text{m}^2} = 0.459\ \text{m/s}$$

$$\mu = 2\ \text{cp} \cdot \frac{0.001\ \text{Pa}\cdot\text{s}}{\text{cp}} = 0.002\ \text{Pa}\cdot\text{s}$$

$$\rho = 1.01 \cdot \frac{1000\ \text{kg}}{\text{m}^3} = 1010\ \text{kg/m}^3$$

$$Re = \frac{D\overline{V}\rho}{\mu} = \frac{0.02291(0.459)(1010)}{0.002} = 5310$$

Since $Re > 2100$ flow is turbulent, use equation 45 for calculating ΔP and equation 47 for calculating f at $Re < 10^4$. Sanitary pipe is considered a smooth pipe.

$$f = 0.193(Re)^{-0.35} = 0.193(5310)^{-0.35} = 0.0095$$

$f = 0.0095$ can also be obtained from Fig. 6.24. The pressure drop per unit length of pipe may be calculated as $\Delta P/L$. Since the units used in the calculations are in SI, the result will be Pa/m. This can then be converted to the desired units of $(\text{lb}_f/\text{in}^2)/\text{ft}$. It may also be solved by substituting a value for L in equation 45.

$$L = 1\ \text{ft}(0.3048\ \text{m/ft}) = 0.3048\ \text{m}$$

$$\Delta P = \frac{2f\overline{V}^2 L\rho}{D} = \frac{2(0.0095)(0.459)^2(0.3048)(1010)}{0.02291}$$

$$= 53.79\ \text{Pa}$$

$$\Delta P = 53.79\ \text{Pa} \cdot \frac{1\ \text{lb}_f/\text{in.}^2}{6894.757\ \text{Pa}} = 0.0078\ \frac{\text{lb}_f/\text{in}^2}{\text{ft of pipe}}$$

EXAMPLE 3: Calculate the pressure drop for 100 m of level straight 1-in. (nominal) wrought iron pipe for water flowing at the rate of 10 gal/min. Use a density for water of 62.4 lb/ft^3 and a viscosity of 0.98 centipoise.

Solution: This problem is different from Examples 1 and 2 in that the pipe is not smooth and therefore equation 46 or 47 cannot be used for calculating f. After calculating the Reynolds number, f can be determined from a Moody diagram (Fig. 6.24) if flow is turbulent, and ΔP can be calculated using equation 45.

From Table 6.2: $D = 0.03579$ m.

$$q = \frac{10\ \text{gal}}{\text{min}} \cdot \frac{0.00378541\ \text{m}^3}{\text{gal}} \cdot \frac{1\ \text{min}}{60\ \text{s}} = 0.0006309\ \text{m}^3/\text{s}$$

$$\overline{V} = \frac{q}{A} = \frac{0.0006309\ \text{m}^3/\text{s}}{(\pi/4)(0.03579)^2(\text{m}^2)} = 0.6274\ \text{m/s}$$

$$\rho = 62.4\ \frac{\text{lb}_m}{\text{ft}^3} \cdot \frac{(3.281)^3\ \text{ft}^3}{\text{m}^3} \cdot \frac{0.45359\ \text{kg}}{\text{lb}_m} = 999.7\ \text{kg/m}^3$$

$$\mu = 0.98\ \text{cp} \cdot \frac{0.001\ \text{Pa}\cdot\text{s}}{\text{cp}} = 0.00098\ \text{Pa}\cdot\text{s}$$

$$Re = \frac{D\overline{V}\rho}{\mu} = \frac{0.03579(0.6274)(999.7)}{0.00098} = 22{,}906$$

To use Fig. 6.24, first determine ϵ/D. From Fig. 6.24, $\epsilon = 0.0000457$ m. $\epsilon/D = 0.0000457/0.03579 = 0.00128$. For $Re = 2.29 \times 10^4$ and $\epsilon/D = 0.00128$, $f = 0.0069$.

$$\Delta P = \frac{2(0.0069)(0.6274)^2(100)(999.7)}{0.03579} = 15.17 \text{ kPa}$$

Frictional Resistance to Flow for Non-Newtonian Fluids. For fluids whose flow behavior can be expressed by equation 2 (pseudoplastic or power law fluids), the Rabinowitsch-Mooney equation (equation 18) for the shear rate at the wall was derived in the section "Velocity Profile and Shear Rate for a Power Law Fluid." Substituting the shear stress at the wall (equation 22) for τ_w and the shear rate at the wall, γ_w (equation 18), in equation 2:

$$\frac{\Delta PR}{2L} = K\left[\frac{\bar{V}}{R} \cdot \frac{3n+1}{n}\right]^n$$

$$\Delta P = \frac{2LK(\bar{V})^n}{(R)^{n+1}}\left[\frac{3n+1}{n}\right]^n \quad (48)$$

Equation 44 for the pressure drop of a fluid in laminar flow through a tubular conduit may be written in terms of the radius of the tube and the Reynolds number as follows:

$$\Delta P = \frac{2(16/Re)(\bar{V})^2 \rho L}{2R} \quad (49)$$

Equation 49 is simply another form of equation 44. These equations are very general and can be used for any fluid. In laminar flow, the friction factor, expressed as $16/Re$, may be used as a means of relating the pressure drop of a non-Newtonian fluid to a dimensionless quantity, Re. Thus, Re for a non-Newtonian fluid can be determined from the flow behavior and consistency indices. Equating equations 48 and 49:

$$\frac{2LK(\bar{V})^n}{(R)^{n+1}}\left[\frac{3n+1}{n}\right]^n = \frac{2(16/Re)(\bar{V})^2 L\rho}{2R}$$

$$Re = \frac{8(\bar{V})^{2-n}(R)^n \rho}{K\left[\frac{3n+1}{n}\right]^n} \quad (50)$$

Equation 50 is a general expression for the Reynolds number that applies for both Newtonian ($n = 1$) and non-Newtonian liquids ($n \neq 1$). If $n = 1$, equation 50 is exactly the same as equation 42 when $K = \mu$.

Pressure drops for power law non-Newtonian fluids flowing through tubes can be calculated using equation 49, with the Reynolds number calculated using equation 50, if flow is laminar. If flow is turbulent, the Reynolds number can be calculated using equation 50 and the pressure drop calculated using equation 45 with the friction factor f determined from this Reynolds number using equations 46 or 47. The Moody diagram (Fig. 6.24) can also be used to determine f from the Reynolds number.

EXAMPLE 1: Tube viscometry of a sample of tomato catsup shows that flow behavior follows the power law model (equation 2), with $K = 125$ dyne-sec^n/cm^2 and $n = 0.45$. Calculate the pressure drop per meter length of level pipe if this fluid is pumped through a 1-in. (nominal) sanitary pipe at the rate of 5 gal (US)/min. The catsup has a density of 1.13 g/cm^3.

Solution: First, convert all the given quantities to SI units.

$$n = 0.45 \text{ (dimensionless)}$$

$$K = 125\,\frac{\text{dyne}\cdot \text{s}^n}{\text{cm}^2}\cdot\frac{1(10)^{-5}N}{\text{dyne}}\cdot\frac{(100)^2\ \text{cm}^2}{\text{m}^2}$$

$$k = 12.5 \text{ Pa}\cdot\text{s}^n$$

From Table 6.2, $D = 0.02291$ m; $R = 0.01146$ m.

$$q = \frac{5\ \text{gal}}{\text{min}}\cdot\frac{0.00378541\ \text{m}^3}{\text{gal}}\cdot\frac{1\ \text{min}}{60\ \text{s}} = 0.00031545\ \text{m}^3/\text{s}$$

$$\bar{V} = \frac{q}{A} = \frac{0.00031545\ \text{m}^3/\text{s}}{(\pi/4)(0.02291)^2\ \text{m}^2} = 0.7652\ \text{m/s}$$

$$\rho = 1.13\,\frac{\text{g}}{\text{cm}^3}\cdot\frac{(100)^3\ \text{cm}^3}{\text{m}^3}\cdot\frac{1\ \text{kg}}{1000\ \text{g}} = 1130\ \text{kg/m}^3$$

$$L = 1\ \text{m}$$

Calculate the Reynolds number using equation 50.

$$Re = \frac{8(0.7652)^{2-0.45}(0.01146)^{0.45}(1130)}{12.5\left[\dfrac{3(0.45)+1}{0.45}\right]^{0.45}} = 30.39$$

Flow is laminar, and pressure drop can be calculated using equation 45.

$$\Delta P = \frac{2(16/30.39)(0.7652)^2(1130)(1)}{0.02291} = 30.41 \text{ kPa}$$

EXAMPLE 2: Peach puree having a solids content of 11.9% shows a fluid consistency index K of 72 dyne · s^n/cm^2 and a flow behavior index n = 0.35. If this fluid is pumped through a 1-in. (nominal) sanitary pipe at 50 gal (US)/min, calculate the pressure drop per meter of level straight pipe. Peach puree has a density of 1.07 g/cm^3.

Solution: D = 0.02291 (from Table 6.2); R = 0.01146 m; n = 0.35 (dimensionless).

$$K = 72 \frac{\text{dyne} \cdot \text{s}^n}{\text{cm}^2} \cdot \frac{(10)^{-5} N}{\text{dyne}} \cdot \frac{(100)^2 \text{ cm}^2}{\text{m}^2}$$

$$K = 7.2 \text{ Pa} \cdot \text{s}^n$$

$$q = 50 \frac{\text{gal}}{\text{min}} \cdot \frac{0.00378541 \text{ m}^3}{\text{gal}} \cdot \frac{1 \text{ min}}{60 \text{ s}} = 0.0031545 \text{ m}^3/\text{s}$$

$$\bar{V} = \frac{q}{A} = \frac{0.0031545 \text{ m}^3}{(\pi/4)(0.02291)^2} = 7.6523 \text{ m/s}$$

$$\rho = 1.07 \frac{\text{g}}{\text{cm}^3} \cdot \frac{(100)^3 \text{ cm}^3}{\text{m}^3} \cdot \frac{1 \text{ kg}}{1000 \text{ g}} = 1070 \text{ kg/m}^3$$

Using equation 50:

$$Re = \frac{8(7.6523)^{2-0.35}(0.1146)^{0.35}(1070)}{7.2\left[\frac{3(0.35)+1}{0.35}\right]^{0.35}} = 8618$$

Flow is turbulent. The sanitary pipe is a smooth pipe, so equation 47 will be used to calculate f.

$$f = 0.193(Re)^{-0.35} = 0.193(8618)^{-0.35} = 0.00809$$

P can now be calculated using equation 42. L = 1 m.

$$\Delta P = \frac{2(0.00809)(7.6523)^2(1)(1070)}{0.02291} = 44.251 \text{ kPa/m}$$

Frictional Resistance Offered by Pipe Fittings to Fluid Flow. The resistance of pipe fittings to flow can be evaluated in terms of an equivalent length of straight pipe. Each type of pipe fitting has its specific flow resistance expressed as a ratio of the equivalent length of straight pipe over its diameter (L'/D). Table 6.3 lists the specific resistance of various pipe fittings. The equivalent length of a fitting, which is the product of (L'/D) from Table 6.3 and the pipe diameter, is added to the length of straight pipe within the piping system to determine the total pressure drop across the system.

EXAMPLE: Calculate the pressure drop due to fluid friction across 50 m of a 1-in. (nominal) sanitary pipe which includes five 90° elbows in the piping system. Tomato catsup with the properties described in Example 1 in the section "Frictional Resistance to Flow for Non-Newtonian Fluids" flows through the pipe at the rate of 5 gal (US)/min.

Solution: In Example 1 of the section referred to above, the tomato catsup pumped under these conditions exhibited a pressure drop of 30.41 kPa per meter of straight level pipe.

The equivalent length of five 90° elbows is determined as follows:

$$D \text{ of 1-in. (nominal) sanitary pipe} = 0.02291 \text{ m}$$

$$\frac{L'}{D} \text{ for a 90° elbow in Table 6.3} = 35$$

$$\text{Equivalent length of one fitting} = (0.02291 \text{ m})(35) = 0.80 \text{ m}$$

$$\text{Equivalent length of five 90° elbows} = 5(0.8) = 4 \text{ m}$$

Table 6.3. Specific Resistance of Various Pipe Fittings Expressed as an Equivalent Length of Straight Pipe to Pipe Diameter Ratio

Fitting	L'/D (dimensionless)
90° elbow std.	35
45° elbow std.	15
Tee (used as a coupling), branch plugged	20
Tee (used as an elbow) entering the branch	70
Tee (used as an elbow) entering the tee run	60
Tee branching flow	45
Gate valve, fully open	10
Globe valve, fully open	290
Diaphragm valve, fully open	105
Couplings and unions	Negligible

Total equivalent length of the whole piping system:

= length of straight pipe + equivalent length of fittings

= 50 m + 4 m = 54 m

$$\text{Pressure drop} = \frac{30.41 \text{ kPa}}{\text{m}} (54 \text{ m}) = 1642 \text{ kPa}$$

Mechanical Energy Balance: The Bernoulli Equation. When fluids are transferred from one point to another, a piping system is used. Since fluids exhibit resistance to flow, an energy loss occurs as the fluid travels downstream along the pipe. If the initial energy level is higher than the energy at any point downstream, the fluid will flow spontaneously. On the other hand, if the energy change needed to take the fluid to a desired point downstream exceeds the initial energy level, energy must be applied to propel the fluid through the system. This energy is an input provided by a pump.

An energy balance across a piping system is similar to the energy balance made in Chapter 5. Table 6.4 lists the energy terms involved in fluid flow, their units, and the formulas for calculating them. All energy entering the system, including the energy input, must equal that leaving the system and the energy loss due to fluid friction. The boundaries of the system must be carefully defined to identify the input and exit energies. As in previous problems in material and energy balances, the boundaries may be moved around to isolate the unknown quantities being determined in the problem. The mechanical energies involved are primarily kinetic and potential energy and mechanical energy at the pump.

The potential and kinetic energies may have finite values on both the source and discharge points of a system. Work input should appear on the side of the equation opposite the frictional resistance term. A balance of the intake and exit

Table 6.4. Energy Terms Involved in the Mechanical Energy Balance for Fluid Flow in a Piping System, the Formulas for Calculating Them, and Their Units

Energy Term	Formula	Dimensional Expression	Formula (basis: 1 kg)	Unit
Potential energy				
Pressure	$m\left(\frac{P}{\rho}\right)$	$\frac{\text{kg}(N \cdot \text{m}^{-2})}{\text{kgm}^{-3}}$	$\frac{P}{\rho}$	joule/kg
Elevation	mgh	$\text{kg}(\text{m} \cdot \text{s}^{-2})(\text{m})$	gh	joule/kg
Kinetic energy	$\frac{1}{2}\text{m}V^2$	$\text{kg}(\text{m} \cdot \text{s}^{-1})^2$	$\frac{V^2}{2}$	joule/kg
Work input (from pump)	W		W	joule/kg
Frictional resistance	$\frac{\text{m}\Delta P_f}{\rho}$	$\frac{\text{kg}(N \cdot \text{m}^{-2})}{\text{kg} \cdot \text{m}^{-3}}$	$\frac{\Delta P_f}{\rho}$	joule/kg

energies in any given system, including the work input and resistance terms based on a unit mass of fluid, is:

$$\frac{P_1}{\rho} + gh_1 + \frac{V_1^2}{2} + W_s = \frac{P_2}{\rho} + gh_2 + \frac{V_2^2}{2} + \frac{\Delta P_f}{\rho} \quad (51)$$

Equation 51 is the Bernoulli equation.

EXAMPLE 1: A pump is used to draw tomato catsup from the bottom of a deaerator. The fluid level in the deaerator is 10 m above the level of the pump. The deaerator is being operated at a vacuum of 0.5 kg_f/cm^2 (kilogram force/cm^2). The pipe connecting the pump to the deaerator is a 2.5-in. (nominal) stainless steel sanitary pipe 8 m long with one 90° elbow. The catsup has a density of 1130 kg/m^3, a consistency index k of 10.5 Pa · s^n, and a flow behavior index n of 0.45 (dimensionless). If the rate of flow is 40 L/min, calculate the pressure at the intake side of the pump to induce the required rate of flow.

Solution: Figure 6.26 is a diagram of the system. The problem asks for the pressure at a point in the system before work is applied by the pump to the fluid. Therefore, the term W_s in equation 51 is zero. The following must be done before quantities can be substituted into equation 51:

Assume that atmospheric pressure is 101 kPa. P_1 must be converted to

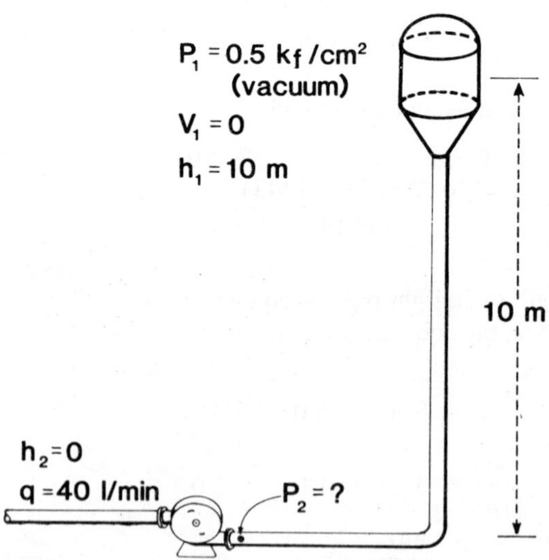

Fig. 6.26. Diagram of a piping system for catsup from a deaerator to a pump.

absolute pressure. The volumetric rate of flow must be converted to velocity ($\bar{V}$). The frictional resistance to flow ($\Delta P_f/\rho$) must be calculated.

$$\text{Vacuum} = 0.5\,\frac{\text{kg}_f}{\text{cm}^2}\cdot\frac{0.8\text{ m}\cdot\text{kg}}{\text{kg}_f\cdot\text{s}^2}\cdot\frac{(100)^2\text{ cm}^2}{\text{m}^2} = 49\text{ kPa}$$

$$P_1 = \text{atmospheric pressure} - \text{vacuum} = 101 - 49 = 52\text{ kPa}$$

$$q = 40\,\frac{\text{L}}{\text{min}}\cdot\frac{0.001\text{ m}^3}{\text{L}}\cdot\frac{1\text{ min}}{60\text{ s}} = 0.0006666\text{ m}^3/\text{s}$$

From Table 6.2, $D = 0.06019$; $R = 0.03009$.

$$\text{Cross-sectional area, } A = \frac{\pi}{4}(0.06019)^2 = 0.002845\text{ m}^2$$

$$\bar{V} = \frac{q}{A} = \frac{0.0006666\text{ m}^3/\text{s}}{0.002845\text{ m}^2} = 0.2343\text{ m/s}$$

The frictional resistance to flow can be calculated using equation 45 after the Reynolds number has been calculated using equation 50.

$$Re = \frac{8(0.2343)^{2-0.45}(0.03009)^{0.45}(1130)}{10.5\left[\dfrac{3(0.45)+1}{0.45}\right]^{0.45}} = 8.921$$

Flow is laminar, and using equation 45:

$$\frac{\Delta P_f}{L\rho} = \frac{2(16/8.920)(0.2343)^2}{0.06019} = 3.272\text{ J/(kg}\cdot\text{m)}$$

where L = length of straight pipe + equivalent length of fitting.

From Table 6.3, the equivalent length of a 90° elbow is 35 pipe diameters.

$$L = 8 + 1(35)(0.06019) = 10.1\text{ m}$$

$$\frac{\Delta P_f}{\rho} = 3.272(10.1) = 33.0\text{ J/kg}$$

Substituting all terms in equation 51:

$$\frac{52{,}000}{1130} + 9.8(10) + 0 + 0 = \frac{P_2}{1130} + 0 + \frac{(0.2343)^2}{2} + 33$$

$$P_2 = \left[\frac{52{,}000}{1130} + 98 - 33.0 - 0.03\right](1130)$$

$$= (46.02 + 98 - 33.0 - 0.03)(1130)$$

$$= 125.4 \text{ kPa absolute}$$

This problem illustrates one of the most important factors, often overlooked, in the design of a pumping system: to provide sufficient suction head to allow fluid to flow into the suction side of the pump. As an exercise, the reader can repeat these calculations using a pipe size of 1.0 in. at the given volumetric flow rate. The calculated absolute pressure at the intake side of the pump will be negative, a physical impossibility.

Another factor that must be considered when evaluating the suction side of pumps is the vapor pressure of the fluid being pumped. If the suction pressure necessary to induce the required rate of flow is lower than the vapor pressure of the fluid at the temperature it is being pumped, boiling will occur, vapor lock will develop, and no fluid will flow into the intake side of the pump. For example, suppose that the pressure at the pump intake is 69.9 kPa. From Appendix Table A.3, the vapor pressure of water at 90°C is 70 kPa. If the height of the fluid level is reduced to 4.8 m, the pressure at the pump intake will be 69 kPa. Under these conditions, if the temperature of the catsup is 90°C or above, it will not be possible to draw the liquid into the intake side of the pump.

EXAMPLE 2: The pump in Example 1 delivers the catsup to a heat exchanger and then to a filler. The discharge side of the pump consists of 1.5-in. stainless steel sanitary pipe 12 m long with two elbows. Joined to this pipe is the heat exchanger, which is 20 m of steam-jacketed, 1-in. stainless steel sanitary pipe with two U-bends. (Assume that the resistance of a U-bend is double that of a 90° elbow.) Joined to the heat exchanger is another 20 m of 1.5-in. pipe with two 90° elbows, and the catsup is discharged into the filler bowl at atmospheric pressure. The pipe elevation at the point of discharge is 3 m from the level of the pump. The piping system before the pump and the rates of flow are the same as in Example 1. Calculate the power requirement for the pump. The diagram of the system is shown in Fig. 6.27.

Solution: From Example 1, $h_1 = 10$ m; $V_1 = 0$; $P_1 = 52$ kPa. $\Delta P_f/\rho$ for the section prior to the pump $= 33$ J/kg. $\Delta P_f/\rho$ for the section at the discharge side of the pump will be calculated in two steps: for the 1.5-in. pipe and for the 1-in. pipe at the heat exchanger. $D = 1.5$-in. pipe $= 0.03561$ m; $R = 0.0178$.

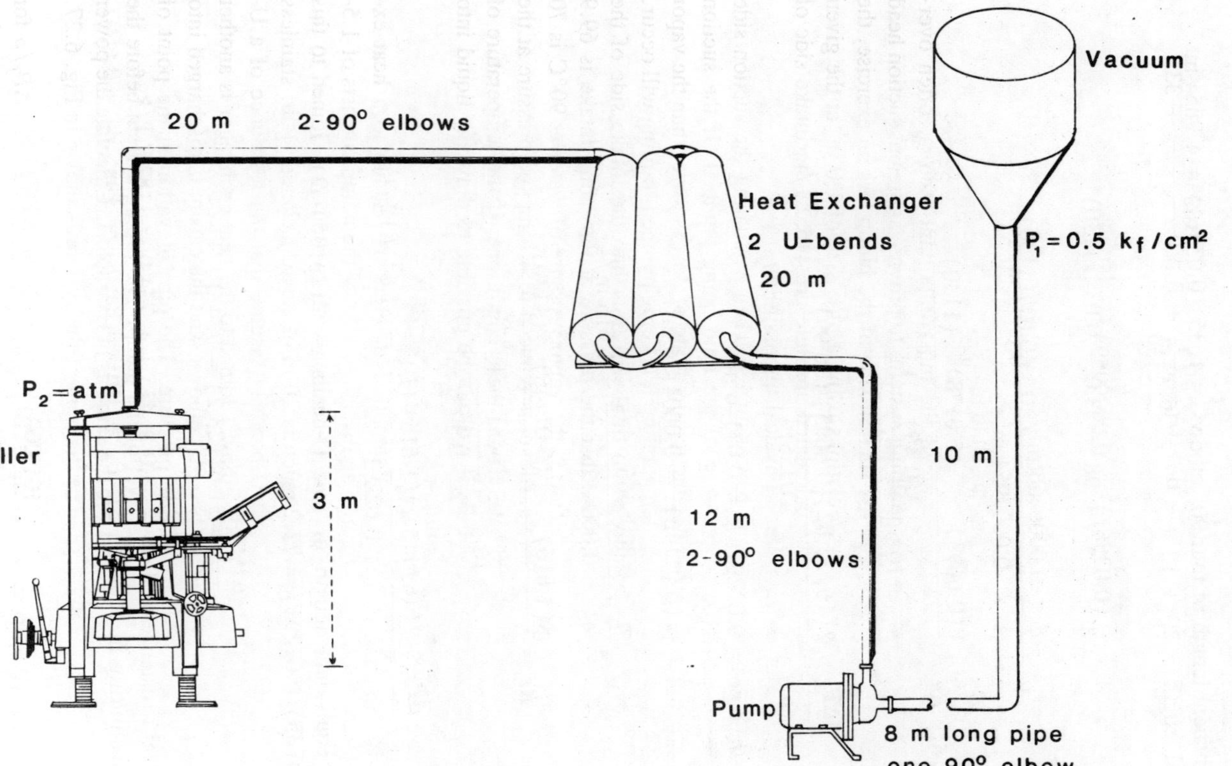

Fig. 6.27. Diagram of a piping system for catsup from a deaerator to a pump and heat exchange and filler.

Equivalent length of four 90° elbows = 4(35)(0.03561) = 4.9854 m.
Total length = 20 + 12 + 4.99 = 36.99 m.

$$A = \frac{\pi}{4}(0.03561)^2 = 0.00099594 \text{ m}^2$$

$$V = \frac{q}{A} = \frac{0.0006666 \text{ m}^3/\text{s}}{0.0009954 \text{ m}^2} = 0.6693 \text{ m/s}$$

$$Re = \frac{8(0.6693)^{2-0.45}(0.0178)^{0.45}(1130)}{10.5\left[\dfrac{3(0.45)+1}{0.45}\right]^{0.45}} = 35.87$$

$$\frac{\Delta P_f}{\rho} = \frac{2(16/35.82)(0.6693)^2(36.99)}{0.03561} = 415.7 \text{ J/kg}$$

D = 1-in. pipe = 0.02291; R = 0.01146 m.

$$A = \pi(0.01146)^2 = 0.0004126 \text{ m}^2$$

$$V = \frac{q}{A} = \frac{0.0006666 \text{ m}^3/\text{s}}{0.0004126 \text{ m}^2} = 1.616 \text{ m/s}$$

$$Re = \frac{8(1.616)^{2-0.45}(0.01146)^{0.45}(1130)}{10.5\left[\dfrac{3(0.45)+1}{0.45}\right]^{0.45}} = 115.3$$

$$\frac{\Delta P_f}{\rho} = \frac{2(16/115.2)(1.616)^2(L)}{0.02291} = 31.64(L) \text{ J/kg}$$

The equivalent length of two U-bends = 2(2)(35)(D); L = 20 + 2(2)(35)(0.02291) = 23.2 m.

$$\frac{\Delta P_f}{\rho} = 31.64(23.2) = 734.0 \text{ J/kg}$$

Total resistance to flow = 33.0 + 415.7 + 734.0 = 1182.7 J/kg.
Substituting in equation 51: P_2 = 101000 Pa; h_2 = 3 m; V_2 = 0.6697 m/s.

$$\frac{52{,}000}{1130} + 9.8(10) + 0 + W_s = \frac{101{,}000}{1130} + 9.8(3) + \frac{(0.6697)^2}{2} + 1183.2$$

$$W_s = \frac{101{,}000}{1130} + 9.8(3) + \frac{(0.6697)^2}{2} + 1183.2 - \frac{52{,}000}{1130} - 9.8(10)$$

$$= 89.4 + 29.4 + 0.2 + 1182.7 - 46 - 98 = 1157.7 \text{ J/kg}$$

$$\text{Mass flow rate} = q(\rho) = 0.0006666(1130) = 0.7532 \text{ kg/s}$$

$$\text{Power} = (1157.7 \text{ J/kg})(0.7532 \text{ kg/s}) = 872 \text{ watts}$$

PROBLEMS

1. The following data were obtained when tomato catsup was passed through a tube having an inside diameter of 1.384 cm and a length of 1.22 m.

Flow Rate (cm^3/sec)	P (dynes/cm^3 × 10^{-4})
107.50	50.99
67.83	42.03
50.89	33.07
40.31	29.62
10.10	15.56
8.80	14.49
33.77	31.00
53.36	35.14
104.41	46.85

Determine the fluid consistency index b and the fluid flow index n of this fluid.

2. Figure 6.28 shows a water tower and the piping system for a small manufacturing plant. If water flows through the system at 40 L/s, what would be the pressure at point B? The pipes are wrought iron pipes. The water has a density of 998 kg/m^3 and a viscosity of 0.8 centipoise. The pipe is 1.5-in. (nominal) wrought iron pipe.

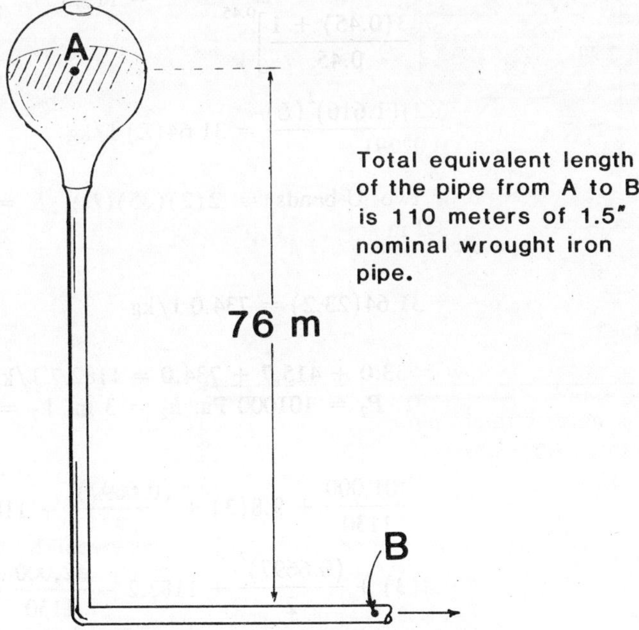

Fig. 6.28. Water tower and piping system for Problem 2.

3. The catsup in Problem 1 ($n = 0.49$, $K = 6.54$ Pa · s^n) is to be heated in a shell and tube heat exchanger. The exchanger has a total of 20 tubes 7 m long arranged parallel inside a shell. Each tube is a 3/4-in. outside diameter, 18-gage heat exchanger tube. (From a table of thickness of sheet metal and tubes, an 18-gage wall is 0.049 in. thick.) It is possible to arrange the fluid flow pattern by the appropriate selection of heads for the shell so that the number of passes and the number of tubes per pass can be varied. Calculate the pressure drop across the heat exchanger (the pressure at the heat exchanger inlet necessary to push the product through the heat exchanger) if a product flow rate of 40 L/min (density = 1013 kg/m^3) is going through the system for (a) a two-pass system (10 tubes/pass) and (b) a five-pass system (four tubes per pass). Consider only the tube resistance, and neglect the resistance at the heat exchanger heads.
4. Figure 6.29 shows a deaerator operated at 381 mm Hg vacuum. (Atmosphere pressure is 762 mm Hg.) It is desired to allow a positive suction flow into the pump. If the fluid has a density of 1040 kg/m^3, the flow rate is 40 L/min, and the pipe is 1.5-in. sanitary pipe, calculate the height h that the bottom of the deaerator must be set above the pump level so that the pressure at the pump intake is at least 5 kPa above atmospheric pressure. The fluid is Newtonian and has a viscosity of 100 centipoises.

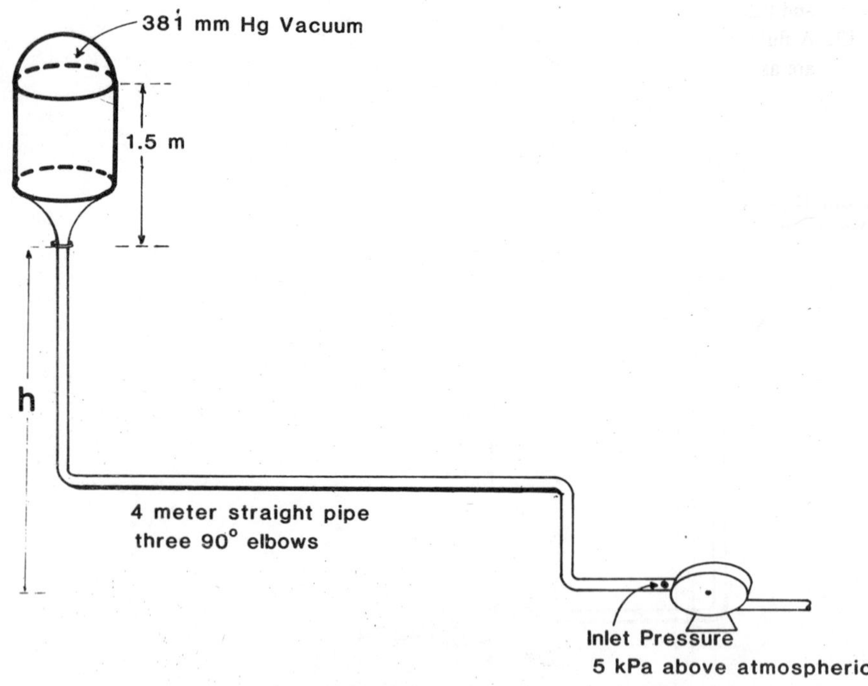

Fig. 6.29. Diagram for Problem 4.

5. Calculate the total equivalent length of 1-in. wrought iron pipe that would produce a pressure drop of 70 kPa due to fluid friction for a fluid flowing at the rate of 50 L/min. The fluid has a density of 988 kg/m^3 and a viscosity of 2 centipoises.
6. Calculate the horsepower required to pump a fluid having a density of 1040 kg/m^3 at the rate of 40 L/min through the system shown in Fig. 6.30. $\Delta P_f/\rho$ calculated for the system is 120 J/kg. Atmospheric pressure is 101 kPa.
7. Calculate the average and maximum velocities of a fluid flowing at the rate of 20 L/min through a 1.5-in. sanitary pipe. The fluid has a density of 2030 kg/m^3 and a viscosity of 50 centipoises. Is the flow laminar or turbulent?
8. Determine the inside diameter of a tube that can be used in a high-temperature, short-time heater sterilizer so that orange juice with a viscosity of 3.75 centipoises and a density of 1005 kg/m^3 will flow at a volumetric flow rate of 4 L/min and have a Reynolds number of 2000 while going through the tube.
9. Calculate the pressure generated at the discharge of a pump that delivers a pudding mix (ρ = 995 kg/m^3; K = 1.0 Pa · s^n; n = 0.6) at the rate of 50 L/min through 50 m of a 1.5-in. straight, level 1.5-in. stainless steel sanitary pipe. What is the equivalent viscosity of a Newtonian fluid that will give the same pressure drop?
10. What pipe diameter will give a rate of flow of 4 ft/min for a fluid delivered at the rate of 2 gal/min?
11. Calculate the viscosity of a fluid that will allow a pressure drop of 35 kPa over a 5-m length of 3/4-in. stainless steel sanitary pipe if the fluid is flowing at 2 L/min and has a density of 1010 kg/m^3. Assume laminar flow.
12. A fluid is evaluated for its viscosity using a Brookfield viscometer. Collected data are as follows:

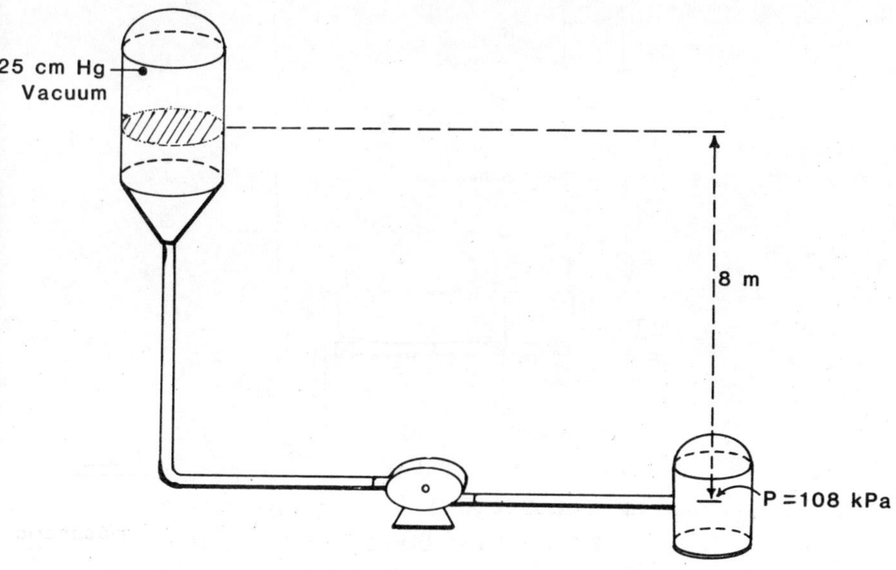

Fig. 6.30. Diagram for Problem 6.

Speed (RPM)	Viscosity (Centipoises)
20	7,230
10	12,060
4	25,200
2	39,500

Is the fluid Newtonian or non-Newtonian? Calculate the fluid flow index (n) for this fluid.

13. A fluid having a viscosity of 0.05 lb_{mass}/ft (sec) requires 30 sec to drain through a capillary viscometer. If this same viscometer is used to determine the viscosity of another fluid and it takes 20 sec to drain, calculate the viscosity of this fluid. Assume that the fluids have the same densities.
14. Figure 6.31 shows a storage tank for sugar syrup which has a viscosity of 15.2 centipoises at 25°C and a density of 1008 kg/m^3. Friction loss includes an entrance loss to the drain pipe, which equals the kinetic energy gain of the fluid, and resistance to flow through the short section of the drain pipe.
    (a) Formulate an energy balance equation for the system.
    (b) Formulate the continuity equation which represents the increment change in fluid level in the tank as a function of the fluid velocity through the drain pipe.
    (c) Solve simultaneously the equations formulated in (a) and (b) to calculate the time required to drain the tank to a residual level 1 m from the bottom of the bank.

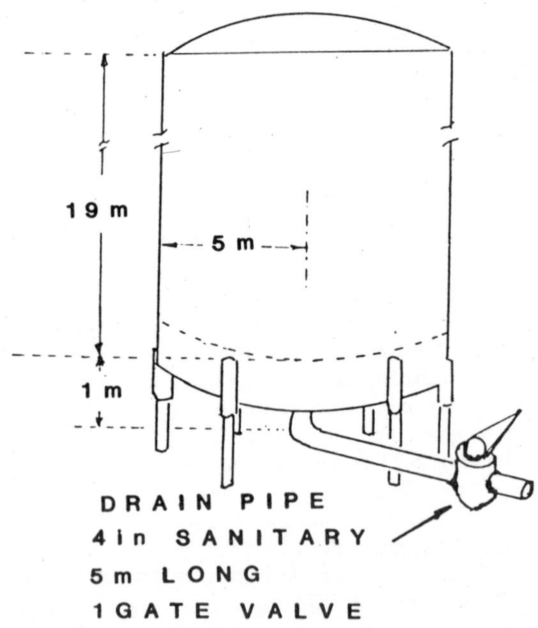

Fig. 6.31. Diagram for Problem 14.

(d) Calculate the amount of residual fluid in the tank, including the film adhering to the side of the tank.

15. A fluid tested on a tube viscometer 0.75 cm in diameter and 30 cm long exhibits a pressure drop of 1200 Pa when the flow rate is 50 cm^3/s.
(a) Calculate the apparent viscosity and the apparent rate of shear under this condition of flow.
(b) If the same fluid flowing at the rate of 100 cm^3/s through a viscometer tube 0.75 cm in diameter and 20 cm long exhibits a pressure drop of 300 Pa, calculate the flow behavior and consistency indices. Assume that wall effects are negligible.

16. Apparent viscosities in centipoises of 4000, 2500, 1205, and 850 were reported on a fluid at rotational speeds of 2, 4, 10, and 20 rpm. This same fluid was reported to require a torque of 900 dynes/cm to rotate a cylindrical spindle 1 cm in diameter and 5 cm high within the fluid at 20 rpm.
(a) Calculate the flow behavior and consistency indices for this fluid.
(b) When exhibiting an apparent viscosity of 4000 centipoises, what would have been the shear rate under which the measurement was made?

17. The flow behavior and consistency indices of whey at a solids content of 24% have been reported to be 0.94 and 4.46×10^{-3} Pa · s^n, respectively. If a rotational viscometer having a full-scale torque of 673.7 dynes/cm is to be used for testing the flow behavior of this fluid, determine the diameter and height of a cylindrical spindle to be used so that at the slowest speed of 2 rev/min. the minimum torque will be 10% of the full-scale reading of the instrument. Assume a length/diameter ratio of 3 for the spindle. Would this same spindle induce a torque within the range of the instrument at 20 rpm?

18. Egg whites having a consistency index of 2.2 Pa · s^n and a flow behavior index of 0.62 must be pumped through a 1.5-in. sanitary pipe at a flow rate which will induce a shear rate at the wall of 150/s. Calculate the rate of flow in L/min and the pressure drop due to fluid flow resistance under these conditions.

19. The system shown in Fig. 6.32 is used to control the viscosity of a batter formulation used on a breading machine. A float indicates the fluid level in the reservoir, and by attaching the float to an appropriate transducer, addition of dry ingredients and water to the mixing tank may be regulated to maintain the batter's consistency. The appropriate fluid level in the reservoir may be maintained when a different consistency of the batter is required by changing the length of the pipe draining the reservoir. If the fluid is Newtonian, with a viscosity of 100 centipoises, and if fluid density is 1004 kg/m^3, calculate the equivalent length of drain pipe needed to maintain the level shown. Assume that entrance loss is negligible compared to fluid resistance through the drain pipe.

20. The system shown in Fig. 6.33 has been reported to be used for disintegrating wood chips in the pulp industry after digestion. It is desired to test the feasibility of using this system on a starchy root crop such as cassava or sweet potatoes in order to disrupt starch granules for easier hydrolysis with enzymes to produce sugars for alcoholic fermentation. In simulating the system on a small scale, two parameters are of importance: the shear rate of the slurry as it passes through the discharge pipe and the impact force of the fluid against the plate. Assume that entrance loss from the tank to the discharge pipe is negligible. Under the conditions shown, calculate

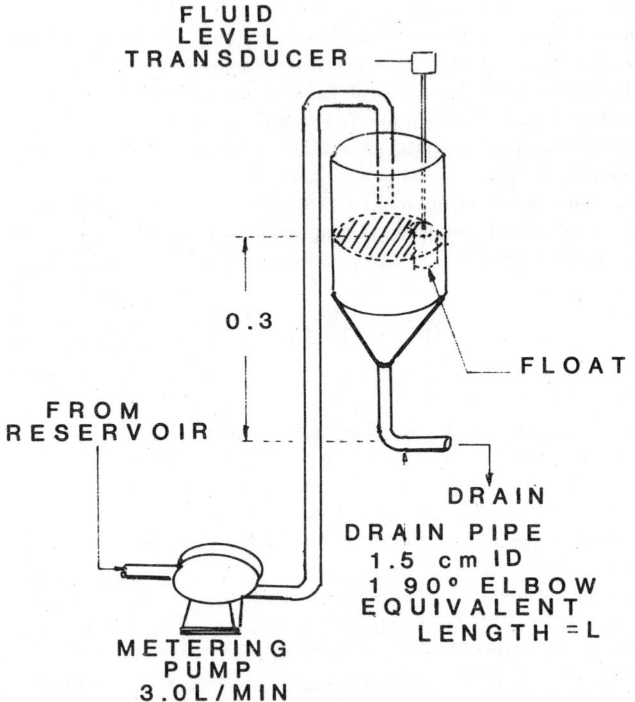

Fig. 6.32. Diagram for Problem 19.

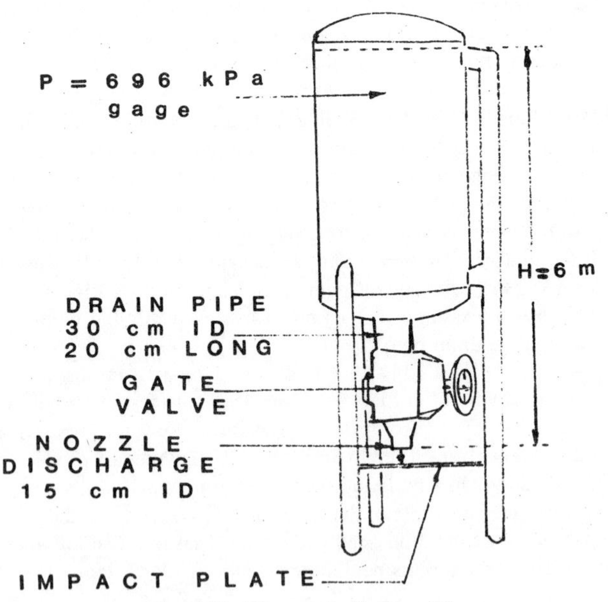

Fig. 6.33. Diagram for Problem 20.

the shear rate through the drain pipe and the impact force against the plate at the time the drain pipe is first opened. The slurry has flow behavior and consistency indices of 0.7 and 0.8 Pa · s^n, respectively, and a density of 1042 kg/m^3

21. In a falling film direct-contact steam heater for sterilization, milk is pumped into a header which distributes the liquid to several vertical pipes, and the liquid flows as a film in laminar flow down the pipe. If the fluid has a density of 998 kg/m^3 and a viscosity of 1.5 centipoises, calculate the flow rate down the outside surface of each of 3.7 cm outside diameter pipes so that the fluid film will flow at a Reynolds number of 500. Calculate the fluid film thickness when flow develops at this Reynolds number.

SUGGESTED READING

Bennet, C. O., and Myers, J. E. 1962. *Momentum, Heat, and Mass Transport*. McGraw-Hill Book Co., New York.

Charm, S. E. 1971. *Fundamentals of Food Engineering*, 2nd ed. AVI Publishing Co., Westport, Conn.

Foust, A. S., Wenzel, L. A., Clump, C. W., Maus, L., and Andersen, L. B. 1960. *Principles of Unit Operations*. John Wiley & Sons, New York.

Heldman, D. R., and Singh, R. P. 1981. *Food Process Engineering*. AVI Publishing Co., Westport, Conn.

Kokini, J. L., and Plutchok, G. J. 1987. Viscoelastic properties of semisolid foods and their biopolymeric components. *Food Technol.* 41(3):89.

Kreiger, I. M., and Maron, S. H. 1952. Direct determination of flow curves of non-Newtonian fluids. *J. Appl. Phys.* 33:147.

Leniger, H. A., and Beverloo, W. A. 1975. *Food Process Engineering*. D. Riedel Publishing Co., Boston.

McCabe, W. L., and Smith, J. C. 1967. *Unit Operations of Chemical Engineering*, 2nd ed. McGraw-Hill Book Co., New York.

Osorio, F. A., and Steffe, J. F. 1987. Back extrusion of power flow fluids. *J. Tex. Studies* 18:43.

Rao, M. A., Shallenberger, R. S., and Cooley, H. J. 1987. Effect of temperature on viscosity of fluid foods with high sugar content. In: *Food Engineering and Process Applications*, M. LeMaguer and P. Jelen (eds.). Elsevier Applied Science Publishing Co., New York, Vol. I, pp. 23–38.

7

Heat Transfer

Heat transfer is the movement of energy from one point to another by virtue of a difference in temperature. Heating and cooling are manifestations of this phenomenon, which is utilized in industrial operations and domestic activities. Increasing energy costs and, in some cases, inadequate availability of energy will require peak efficiency in heating and cooling operations. An understanding of the mechanisms of heat transport is needed to recognize the limitations of heating and cooling systems. This understanding can lead to adoption of practices which circumvent these limitations. In industrial and domestic heating and cooling, energy use audits can be used to determine total energy use and the distribution within the process, to identify areas of high energy use, and to target these areas for energy conservation measures.

MECHANISMS OF HEAT TRANSFER

Heat is transferred from one material to another when there is a difference in their temperatures. The temperature difference is the factor which determines the rate of heat transfer.

Heat Transfer by Conduction. When heat is transferred between adjacent molecules, the process is called *conduction*. This is the mechanism of heat transfer in solids.

Fourier's First Law of Heat Transfer. According to Fourier's first law, the heat flux, in conduction heat transfer, is proportional to the temperature gradient.

$$\frac{q}{A} = -k\frac{dT}{dx} \tag{1}$$

In equation 1, q is the rate of heat flow and A is the area through which heat is transferred. A is the area perpendicular to the direction of heat flow. The expression q/A, the rate of heat transfer per unit area, is called the *heat flux*. The derivative dT/dx is the temperature gradient. The negative sign in equation 1 indicates that positive heat flow will occur in the direction of decreasing temperature. k is the thermal conductivity and is a physical property of a material. Values of the thermal conductivity of common construction and insulating materials and food products are given in Appendix Tables A.9 and A.10, respectively.

Estimation of Thermal Conductivity of Food Products. The thermal conductivity of materials varies with the composition and, in some cases, the physical orientation of components. Foods, being of biological origin, are subject to high variability in composition and structure; therefore, the k value of foods presented in the tables is not always the same for all foods in the category listed. The effect of variations in the composition of a material on thermal conductivity has been reported by Choi and Okos (1987). Their procedure may be used to estimate k from the composition. k is calculated from the thermal conductivity of the pure component, k_i, and the volume fraction of each component, X_{vi}.

$$k = \sum(k_i X_{vi}) \tag{2}$$

The thermal conductivity values in W/(m · K) of pure water (k_w), ice (k_{ic}), protein (k_p), fat (k_f), carbohydrate (k_c), fiber (k_{fi}), and ash (k_a) are calculated at the temperature in °C under consideration, using equations 3 to 9, respectively.

$$k_w = 0.57109 + 0.0017625\,T - 6.7306 \times 10^{-6}\,T^2 \tag{3}$$

$$k_{ic} = 2.2196 - 0.0062489\,T + 1.0154 \times 10^{-4}\,T^2 \tag{4}$$

$$k_p = 0.1788 + 0.0011958\,T - 2.7178 \times 10^{-6}\,T^2 \tag{5}$$

$$k_f = 0.1807 - 0.0027604\,T - 1.7749 \times 10^{-7}\,T^2 \tag{6}$$

$$k_c = 0.2014 + 0.0013874\,T - 4.3312 \times 10^{-6}\,T^2 \tag{7}$$

$$k_{fi} = 0.18331 + 0.0012497\,T - 3.1683 \times 10^{-6}\,T^2 \tag{8}$$

$$k_a = 0.3296 + 0.001401\,T - 2.9069 \times 10^{-6}\,T^2 \tag{9}$$

The volume fraction X_{vi} of each component is determined from the mass fraction, X_i, the individual densities, ρ_i, and the composite density, ρ, as follows:

$$X_{vi} = \frac{X_i \rho}{\rho_i} \tag{10}$$

$$\rho = \frac{1}{[\sum X_i/\rho_i]} \tag{11}$$

The individual densities, in kg/m^3, are obtained using equations 12 to 18, respectively, for water (ρ_w), ice (ρ_{ic}), protein (ρ_p), fat (ρ_f), carbohydrate (ρ_c), fiber (ρ_{fi}), and ash (ρ_a).

$$\rho_w = 997.18 + 0.0031439T - 0.0037574T^2 \tag{12}$$

$$\rho_{ic} = 916.89 - 0.13071T \tag{13}$$

$$\rho_p = 1329.9 - 0.51814T \tag{14}$$

$$\rho_f = 925.59 - 0.41757T \tag{15}$$

$$\rho_c = 1599.1 - 0.31046T \tag{16}$$

$$\rho_{fi} = 1311.5 - 0.36589T \tag{17}$$

$$\rho_a = 2423.8 - 0.28063T \tag{18}$$

A program in BASIC to calculate the thermal conductivity from the composition of a material is given in Appendix Table A.11.

EXAMPLE 1: Calculate the thermal conductivity of lean pork containing 7.8% fat, 1.5% ash, 19% protein, and 71.7% water at 19°C.

Solution: X_{ic}, X_c, and $X_{fi} = 0$. Substituting $T = 19$°C in equations 3 to 9, the individual thermal conductivity of each component, in $W/(m \cdot K)$, is:

$$k_w = 0.602; \quad k_p = 0.200; \quad k_f = 0.128; \quad k_a = 0.355$$

Substituting T in equations 12 to 18, the density of each component, in kg/m^3, is:

$$\rho_w = 996; \quad \rho_p = 1320; \quad \rho_f = 917; \quad \rho_a = 2418$$

Using equation 10:

$$X_{vw} = 0.753; \quad X_{vp} = 0.150; \quad X_{vf} = 0.09; \quad X_{va} = 0.006$$

Using equation 2:

$$k = 0.602(0.753) + 0.200(0.150) + 0.09(0.128) + 0.006(0.355)$$

$$= 0.497 \text{ W/(m} \cdot \text{K)}$$

EXAMPLE 2: Calculate the thermal conductivity of milk which contains 87.5% water, 3.7% protein, 3.7% fat, 4.6% lactose and 0.5% ash at 10°C.

Solution: Using the BASIC program in Appendix Table A.11, the following were obtained: $k_w = 0.588$, $k_p = 0.190$, $k_f = 0.153$, $k_c = 0.215$, and $k_a = 0.343$ W/(m · K). The densities are: $\rho_w = 997$, $\rho_p = 1324$, $\rho_f = 921$, $\rho_c = 1595$, and $\rho_a = 2420$ kg/m^3. The volume fractions are: $X_{vw} = 0.899$, $X_{vp} = 0.029$, $X_{vf} = 0.041$, $X_c^v = 0.029$, $X_{va} = 0.0021$, and $k = 0.547$ W/(m · K).

Estimated values for k using the above procedure appear to be closer to measured values in fluid materials which can be well mixed compared to solids. The reported value of k for pork in Example 1 was 0.453 W/(m · K), and that for milk in Example 2 was 0.530 W/(m · K).

Fourier's Second Law of Heat Transfer. When the rate of heat transfer across a solid is not uniform, i.e., there is a difference in the rate at which energy enters and leaves a control volume, this difference will be manifested as a rate of change of temperature with time within the control volume. This condition is called unsteady-state heat transfer. Figure 7.1 shows a control volume for analyzing heat transfer into a cube with sides dx, dy, and dz.

The following are heat balance equations across the control volume in the x, y, and z directions. Partial differential equations are used since, in each case, only one direction is being considered.

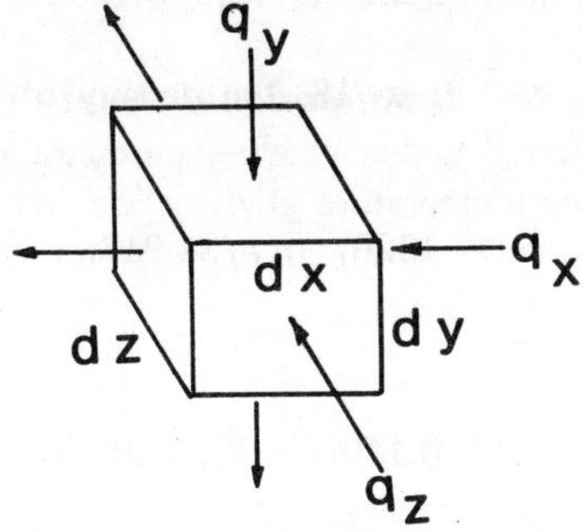

Fig. 7.1. Control volume for analysis of heat transfer into a cube.

$$q = q_x + q_y + q_z = \rho\, dx\, dy\, dz\, C_p \frac{\partial T}{\partial t}$$

$$q_x = k\, dy\, dz \left[\frac{\partial T}{\partial x}\bigg|_1 - \frac{\partial T}{\partial x}\bigg|_2 \right]$$

$$q_y = k\, dx\, dz \left[\frac{\partial T}{\partial y}\bigg|_1 - \frac{\partial T}{\partial y}\bigg|_2 \right]$$

$$q_z = k\, dy\, dx \left[\frac{\partial T}{\partial z}\bigg|_1 - \frac{\partial T}{\partial z}\bigg|_2 \right]$$

Combining:

$$\rho C_p \frac{\partial T}{\partial t} = k \left[\frac{\partial T/\partial x|_1 - \partial T/\partial x|_2}{dx} + \frac{\partial T/\partial y|_1 - \partial T/\partial y|_2}{dy} + \frac{\partial T/\partial z|_1 - \partial T/\partial z|_2}{dz} \right]$$

The difference in the first derivatives divided by dx, dy, or dz is a second derivative; therefore:

$$\frac{\partial T}{\partial t} = \frac{k}{\rho C_p} \left[\frac{\partial^2 T}{\partial x^2} + \frac{\partial^2 T}{\partial y^2} + \frac{\partial^2 T}{\partial z^2} \right] \tag{19}$$

Equation 19 represents Fourier's second law of heat transfer. The rate of change of temperature with time at any position within a solid conducting heat is proportional to the second derivative of the temperature with respect to distance at that particular point. The ratio $k/(\rho \cdot C_p)$ is α, the *thermal diffusivity*. The BASIC program in Appendix Table A.11 can be used to estimate the thermal diffusivity of foods from their composition.

Temperature Profile for Unidirectional Heat Transfer Through a Slab. If heat is transferred under steady-state conditions, and A is constant along the distance, x, the temperature gradient, dT/dx, will be constant, and integration of equation 1 will result in an expression for temperature as a linear function of x.

$$T = -\frac{q/A}{k} \cdot x + C \tag{20}$$

Substituting the boundary condition, $T = T_1$ at $x = x_i$, in equation 20:

$$T = \frac{q/A}{k} \cdot (x_1 - x) + T_1 \tag{21}$$

Equation 21 is the expression for the steady-state temperature profile in a slab where A is constant in the direction of x. The temperature gradient, which is constant in a slab transferring heat at a steady state, is equal to the ratio of the heat flux to the thermal conductivity.

If the temperature at two different points in the solid is known, i.e., $T = T_1$ at $x = x_1$ and $T = T_2$ at $x = x_2$, equation 21 becomes:

$$T_1 - T_2 = -\frac{q/A}{k}(x_1 - x_2)$$

$$\frac{q}{A} = -k \cdot \frac{\Delta T}{\Delta x} \tag{22}$$

Substituting q/A in equation 21:

$$T = -\frac{\Delta T}{\Delta x} \cdot (x_1 - x) + T_1 \tag{23}$$

To keep the convention on the signs and ensure that proper temperatures are calculated at any position within the solid, $\Delta T = T_2 - T_1$ with increasing subscripts in the direction of heat flow, and $\Delta x = x_2 - x_1$ is positively increasing in the direction of heat flow.

Thus, if a temperature drop ΔT across any two points in a solid separated by distance Δx is known, it will be possible to determine the temperature at any other point within that solid using equation 23. Equations 22 and 23 also show that when heat is transferred in a steady state in a slab, the product, $k(\Delta T/\Delta x)$, is a constant, and

$$k\left(\frac{\Delta T}{\Delta x}\right)_1 = k\left(\frac{\Delta T}{\Delta x}\right)_2 = \ldots \tag{24}$$

If the solid consists of two or more layers having different thermal conductivity, the heat flux through each layer is constant; therefore the temperature profile will exhibit a change in slope at each junction between the layers.

A composite slab containing three materials having thermal conductivity k_1, k_2, and k_3 is shown in Fig. 7.2. The temperature profile shows a different temperature gradient within each layer. For a composite slab, equation 24 becomes:

$$k_1\left(\frac{\Delta T_1}{\Delta x_1}\right) = k_2\left(\frac{\Delta T_2}{\Delta x_2}\right) = k_3\left(\frac{\Delta T_3}{\Delta x_3}\right) \tag{25}$$

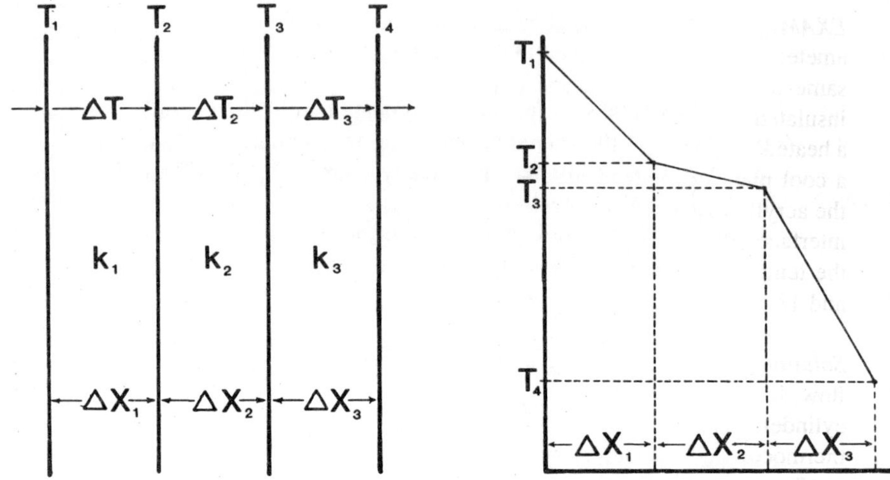

Fig. 7.2. Diagram of composite slab and temperature drop at each layer.

Equation 25 can be used to determine experimentally the thermal conductivity of a solid by placing it between two layers of a second solid and determining the temperature at any two points within each layer of solid after a steady state has been achieved. The existence of a steady state and a unidimensional heat flow can be proven by equality of the first and third terms. Two of the terms involving a known thermal conductivity k_1 and the unknown thermal conductivity k_2 can be used to determine k_2.

EXAMPLE 1: Thermocouples embedded at two points within a steel bar 1 and 2 mm from the surface indicate temperatures of 100° and 98°C, respectively. Assuming no heat transfer occurring from the sides, calculate the surface temperature.

Solution: Since the thermal conductivity is constant, either equation 23 or equation 25 can be used. Following the convention on signs for the temperature and distance differences, $T_2 = 98$, $T_1 = 100$, $x_2 = 2$ mm, and $x_1 = 1$ mm. The temperature gradient $\Delta T/\Delta x = (T_2 - T_1)/(x_2 - x_1) = (98 - 100)/0.001(2 - 1) = -2000$. Equation 23 becomes:

$$T = -(-2000)(x_1 - x) + T_2$$

At the surface, $x = 0$, and at point $x_1 = 0.001$, $T_1 = 100$. Thus:

$$T = 2000(0.001) + 100 = 102°\text{C}$$

EXAMPLE 2: A cylindrical sample of beef 2 cm thick and 3.75 cm in diameter is positioned between two 5-cm-thick acrylic cylinders of exactly the same diameter as the meat sample. The assembly is positioned inside an insulated container such that the bottom of the lower acrylic cylinder contacts a heated surface maintained at 50°C and the top of the upper cylinder contacts a cool plate maintained at 0°C. Two thermocouples each are embedded in the acrylic cylinders, positioned 0.5 cm and 1.5 cm from the sample–acrylic interface. If the acrylic has a thermal conductivity of 1.5 W/(m · K) and the temperatures recorded at steady state are, respectively, 45°, 43°, 15°, and 13°C, calculate the thermal conductivity of the meat sample.

Solution: Equation 25 will be used. Proceeding along the direction of heat flow, let $T_1 = 45$, $T_2 = 43$, T_3 = temperature at the meat and bottom acrylic cylinder interface, $T_5 = 15$, and $T_6 = 10$°C. Let $x_1 = 0$ at the lowermost thermocouple location, $x_2 = 1$, $x_3 = 1.5$, $x_4 = 3.5$, $x_5 = 4$, and $x_6 = 5$ cm. $(T_2 - T_1)/(x_2 - x_1) = (T_6 - T_5)/(x_6 - x_5) = -2/0.01 = -200$. This proves unidimensional heat flow. The heat flux through the system is:

$$\frac{q}{A} = -k\left(\frac{\Delta T}{\Delta x}\right) = -1.5(-200) = 300 \text{ W}$$

Using equation 23, the temperatures at the acrylic and meat sample interfaces are:

$$T_3 = -(-200)(0.01 - 0.015) + 43 = 42°\text{C}$$

$$T_4 = -(-200)(0.04 - 0.035) + 15 = 16°\text{C}$$

Using equation 25, with ΔT as the temperature drop across the meat sample ($T_3 - T_4$) and Δx as the thickness of the meat sample, the thermal conductivity of the meat is:

$$k = \frac{q/A}{\Delta T/\Delta x} = \frac{300}{26/0.05} = 0.5769 \text{ W/(m} \cdot \text{K)}$$

Conduction Heat Transfer Through Walls of a Cylinder. When the direction of heat flow is through the walls of a cylinder, the area perpendicular to the direction of heat flow changes with position. A in equation 1 at any position within the wall is the surface area of a cylinder of radius r, and equation 1 may be expressed as:

$$q = -k(2\pi rL)\frac{dT}{dr}$$

When heat transfer occurs in a steady state, q is constant and:

$$-\frac{dr}{r} = \frac{2\pi Lk}{q}\, dT$$

Integrating and substituting the boundary conditions: at $r = r_1$, $T = T_2$:

$$\ln\left(\frac{r_2}{r_1}\right) = \frac{2\pi Lk}{q}(T_1 - T_2) \tag{26}$$

$$q = \frac{T_1 - T_2}{[\ln(r_2/r_1)/2\pi Lk]} \tag{27}$$

Equation 27 follows the convention established in equation 24: that q is positive in the direction of decreasing temperature. Increasing subscripts represent positions proceeding farther away from the center of the cylinder.

The Temperature Profile in the Walls of a Cylinder in Steady-State Heat Transfer. At any point r, the temperature T may be obtained from equation 27 by substituting r for r_2 and T for T_2.

$$T = \frac{\ln(r/r_1)q}{2\pi Lk} + T_1$$

Substituting equation 27 for q:

$$T = \frac{\ln(r/r_1)}{\ln(r_2/r_1)}(T_1 - T_2) + T_1 \tag{28}$$

or

$$\frac{(T - T_1)}{\ln(r/r_1)} = \frac{(T_1 - T_2)}{\ln(r_2/r_1)} \tag{29}$$

If the wall of the cylinder consists of layers having different thermal conductivities, the temperature profile may be obtained using the same procedure used in deriving equations 28 and 29, except that the thermal conductivities of the different layers are used.

$$\frac{(T - T')}{[\ln(r/r')/k_2]} = \frac{(T_1 - T_2)}{[\ln(r_2/r_1)/k_1]} = \frac{q}{2\pi L} \tag{30}$$

In equation 30, the temperatures T_1 and T_2 must transect a layer bounded by r_1 and r_2, which has a uniform thermal conductivity of k_1. Similarly, the layer bounded by r and r' where the temperatures are T and T' must also have a uniform thermal conductivity of k_2. Each of the terms in equation 30 is proportional to the rate of heat transfer. Equations 30 and 25 represent a fundamental relationship in steady-state heat transfer by conduction, i.e., the rate of heat flow through each layer in a multilayered solid is equal.

EXAMPLE: A 3/4-in. steel pipe is insulated with 2-cm-thick fiberglass insulation. If the inside wall of the pipe is at 150°C and the temperature at the outside surface of the insulation is 35°C, (a) calculate the temperature at the interface between the pipe and the insulation and (b) calculate the rate of heat loss for 1-m length of pipe.

Solution: From Table 6.2 on pipe dimensions, the inside diameter of a 3/4-in. steel pipe is 0.02093 m and the outside diameter is 0.02667 m. From Appendix Table A.9, the thermal conductivity of steel is 45 W/(m · K) and that of fiberglass is 0.035 W/(m · K). From the pipe dimensions, r_1 = 0.01047 m, r_2 = 0.01334 m, and r_3 = (0.01334 + 0.020) = 0.03334 m. T_1 = 150. T_2 = the temperature at the pipe–insulation interface. T_3 = 35°C. T_1 and T_2 transect the metal wall of the pipe, with k = 45 W/(mK). From equation 30: $T = T_2$; $T' = T_s$; $r = r_3$ and $r' = r_2$; k_1 = 0.035 W/(m · K) and k_2 = 45 W/(m · K). Substituting in equation 30:

$$\frac{(T_2 - T_s)k_1}{\ln(r_3/r_2)} = \frac{(T_1 - T_2)}{\ln(r_2/r_1)}$$

$$\frac{(T_2 - 35)(0.035)}{\ln(0.03334/0.01334)} = \frac{(150 - T_2)(45)}{\ln(0.01334/0.01047)}$$

$$0.038210T_2 - 1.33735 = 185.75(150) - 185.75T_2$$

$$T_2(185.75 + 0.03821) = 27{,}864$$

$$T_2 = 149.97°\text{C}$$

Heat Transfer by Convection. This mechanism transfers heat when molecules move from one point to another and exchanges energy with another molecule in the other location. Bulk molecular motion is involved in convection heat transfer. Bulk molecular motion is induced by density changes associated with differences in fluid temperature at different points in the fluid, condensation, or vaporization (natural convection), or when a fluid is forced to flow past a surface by mechanical means (forced convection). Heat transfer by convection is evaluated as the rate of heat exchange at the interface between a fluid and a solid. The rate of heat transfer by convection is proportional to the temperature difference and is expressed as:

$$q = hA(T_m - T_s) = hA\Delta T \tag{31}$$

h = the heat transfer coefficient, A = the area of the fluid–solid interface where heat is being transferred, and ΔT, the driving force for heat transfer, the difference in fluid temperature, T_m, and the solid surface temperature, T_s. Convection heat transfer is often represented as heat transfer through a thin layer of fluid which possesses a temperature gradient at the fluid–surface interface. The temperature, which is assumed to be uniform at T_m in the fluid bulk, gradually changes through the fluid film until it assumes the solid surface temperature past the film. Thus, the fluid film may be considered as an insulating layer which resists heat flow between the fluid and the solid. The fluid film is actually a boundary layer which has different properties and different velocity from the bulk of the fluid. The magnitude of the heat transfer coefficient varies inversely with the thickness of the boundary layer. Conditions which result in a reduction of the thickness of this boundary layer will promote heat transfer by increasing the value of the heat transfer coefficient.

Natural Convection. Natural convection depends on gravity and density and on viscosity changes associated with temperature differences in the fluid to induce convective currents. The degree of agitation produced by the convective currents depends on the temperature gradient between the fluid and the solid surface. When ΔT is small, convective currents are not too vigorous, and the process of heat transfer is referred to as *free convection*. The magnitude of the heat transfer coefficient in free convection is very low, on the order of 60 $W/(m^2 \cdot K)$ for air and 60 to 3000 $W/(m^2 \cdot K)$ for water. When the surface is in contact with a liquid and the surface temperature exceeds the boiling point of the liquid, bubbles of superheated vapor are produced at the solid–liquid interface. As these bubbles leave the surface, the boundary layer is agitated, resulting in very high heat transfer coefficients. This process of heat transfer is called *nucleate boiling*, and the magnitude of the heat transfer coefficient for water is on the order of 5000 to 50,000 $W/(m^2 \cdot K)$. When ΔT is very high, excessive generation of vapor at the interface produces an insulating layer of vapor which hinders heat transfer. This process of heat transfer is called *film boiling*, and the heat transfer coefficient is much lower than that in nucleate boiling.

Another form of natural convection is the transfer of heat from condensing vapors. Condensing vapors release a large amount of energy on condensation; therefore, heat transfer coefficients are very high. When the vapors condense as droplets which eventually coalesce and slide down the surface, the vapor is always in direct contact with a clean surface; and therefore, heat transfer coefficients are very high. This type of heat transfer is called *dropwise condensa-*

tion. Heat transfer coefficients on the order of 10000 W/($m^2 \cdot$ K) are common in dropwise condensation. When vapors condense as a film of liquid on a surface, the liquid film forms a barrier to heat transfer and heat transfer coefficients are lower. This process of heat transfer is called *filmwise condensation*, and the magnitude of the heat transfer coefficient may be on the order of 5000 W/($m^2 \cdot$ K).

Forced Convection. In forced convection heat transfer, heat transfer coefficients depend upon the velocity of the fluid, its thermophysical properties, and the geometry of the surface. In general, heat transfer coefficients for noncondensing gases are about two orders of magnitude lower than those for liquids. Techniques for calculating heat transfer coefficients are discussed in the section "Local Heat Transfer Coefficients."

Heat Transfer by Radiation. Electromagnetic waves traveling through space may be intercepted by a suitable surface and absorbed, raising the energy level of the intercepting surface. When the electromagnetic waves have the frequency of light, the phenomenon is referred to as *radiation*. All bodies at temperatures higher than absolute zero emit energy in proportion to the fourth power of their temperature. In a closed system, bodies exchange energy by radiation until their temperatures equalize. Radiation heat transfer, like convection, is a surface phenomenon; therefore, the conditions at the surface determine the rate of heat transfer.

Types of Surfaces. Surfaces may be classified as black or gray bodies according to their ability to absorb radiation. A black body is one which absorbs all radiation which impinges on its surface. An example of a black body surface is the interior of a hollow sphere which has a small opening to admit radiation. All the energy which enters the small opening is reflected back and forth within the sphere and eventually is totally absorbed. *Emissivity* (ϵ) is a property which is a measure of the ability of a surface to absorb incident radiation. Black bodies have emissivity of 1. Most surfaces do not have the capability of absorbing all incident radiation. Some of the energy impinging on the surface may be reflected or transmitted. If a surface absorbs incident radiation equally at all wavelengths it is considered a *gray body*. Gray bodies may be assigned emissivity values of less than 1. Table 7.1 shows emissivity of some surfaces.

Kirchhoff's Law. A body at constant temperature is in equilibrium with its surroundings, and the amount of energy absorbed by radiation will be exactly the same as that emitted. Thus, the absorptivity of a surface (α) is exactly the same as the emissivity (ϵ), and these two properties may be used interchangeably.

Table 7.1. Emissivity of Various Materials

Material	Temp. (°C)	Emissivity
Aluminum, bright	170	0.04
Aluminum paint	100	0.3
Chrome, polished	150	0.06
Iron, hot rolled	20	0.77
Brick, mortar, plaster	20	0.93
Glass	90	0.94
Oil paints, all colors	212	0.92–0.96
Paper	95	0.92
Porcelain	20	0.93
Stainless steel, polished	20	0.24
Wood	45	0.82–0.93
Water	32–212	0.96

Source: Bolz, R. E., and Tuve, G. L. (eds.). 1970 *Handbook of Tables for Applied Engineering Science*. CRC Press, Cleveland, Ohio.

Stephan-Boltzman Law. The energy flux emitted by a black body is directly proportional to the fourth power of the absolute temperature.

$$\frac{q}{A} = \sigma T^4 \tag{32}$$

σ is the Stephan-Boltzman constant, which has a value of 5.6732×10^{-8} $W/(m^2 \cdot K^4)$. For gray bodies:

$$q = A\sigma\epsilon T^4 \tag{33}$$

Planck's Distribution Law. The intensity of energy flux radiating from a black surface is a function of the wavelength, and a wavelength exists where the intensity of radiation is maximum. The energy flux from a black surface at an absolute temperature T as a function of the wavelength λ is:

$$\frac{q}{A} = \frac{C_1}{\lambda^5} \cdot \frac{1}{[e^{C_2/(\lambda T)} - 1]} \tag{34}$$

C_1 and C_2 are constants. The total flux over the entire spectrum will be:

$$\frac{q}{A} = \int_{-\infty}^{\infty} \left[\frac{C_1 \lambda^{-5}}{e^{C_2/(\lambda T)}} - 1 \right] d\lambda \tag{35}$$

Equation 34 can be used to show that there will be a wavelength where the energy flux is maximum. The total energy flux over the whole spectrum is the energy radiated by a body at temperature T; thus, equation 35 is equivalent to equation 32.

Wein's Displacement Law. The wavelength for maximum energy flux from a body shifts with a change in temperature. The product of the wavelength for maximum flux intensity and absolute temperature is a constant. $\lambda_{max} \cdot T = 2.884 \times 10^{-3}$ m · K.

Lambert's Law. This is also called the *cosine law*. The energy flux over a solid angle ϕ in a direction Θ from a normal drawn toward the surface is a function of cos Θ.

$$\frac{dq}{dA\, d\phi} = \frac{\epsilon \sigma T^4}{\pi} \cos \Theta \tag{36}$$

Equation 36 is the basis for the derivation of angle factors used in calculating the rate of radiation heat transfer between two bodies having the same emissivities. Figure 7.3 shows two area increments transferring heat by radiation and the representation of the cosine law according to equation 36.

The solid angle $d\phi_1$ with which the area element dA_1 is viewed from dA_2 is:

$$d\phi_1 = \frac{\cos \Theta_2\, dA_2}{s^2}$$

Integrating equation 36 after substituting the expression for the solid angle will give the rate of heat transfer from area 1 to area 2.

$$q_{A1-A2} = \frac{\epsilon_1 \sigma (T_1^4 - T_2^4)}{\pi} \int_{A_1} \int_{A_2} \frac{\cos \Theta_2\, dA_2}{s^2} \cdot dA_1 \cos \Theta_1$$

The integral term divided by πA_1 is the shape factor (F_{1-2}) for area 2. The equations for the rate of heat transfer become:

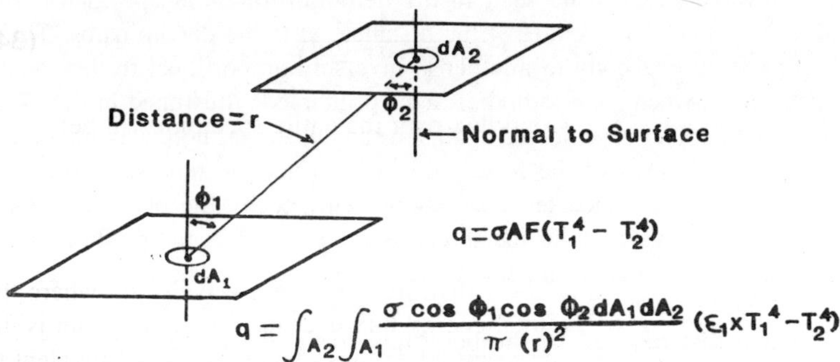

Fig. 7.3. Two area elements and representation of the cosine law for heat transfer by radiation.

$$q_{A1-A2} = (F_{1-2})A_1\epsilon\sigma(T_1^4 - T_2^4) \tag{37}$$

$$q_{A2-A1} = (F_{2-1})A_2\epsilon\sigma(T_1^4 - T_2^4) \tag{38}$$

Since in a steady state the rate of heat transfer from A_1 to A_2 is the same as that from A_2 to A_1, the product of A_1 and the shape factor for A_1 is equal to the product of A_2 and the shape factor for A_2. If the two surfaces have different emissivities, an effective shape factor ($\overline{F}$) derived by Jacob and Hawkins (1957) is:

$$\frac{1}{\overline{F}_{1-2}A_1} = \frac{1}{A_1F_{1-2}} + \frac{1}{A_1}\left(\frac{1}{\epsilon_1} - 1\right) + \frac{1}{A_2}\left(\frac{1}{\epsilon_2} - 1\right) \tag{39}$$

$$\frac{1}{\overline{F}_{2-1}A_2} = \frac{1}{A_2F_{2-1}} + \frac{1}{A_1}\left(\frac{1}{\epsilon_1} - 1\right) + \frac{1}{A_2}\left(\frac{1}{\epsilon_2} - 1\right) \tag{40}$$

The rate of heat transfer based on area A_1 or A_2, using the effective shape factor, is:

$$q = A_1\overline{F}_{1-2}\sigma(T_1^4 - T_2^4) = A_2\overline{F}_{2-1}\sigma(T_1^4 - T_2^4) \tag{41}$$

Shape factors are not easily derived; their derivation is beyond the scope of this text book. However, equations for calculating the shape factor and procedures for applying shape factor algebra for different geometries are available in the literature (e.g., Rohsenow and Hartnett, 1973).

Effect of Distance Between Objects on Heat Transfer. The total energy intercepted by a small body from a large body is dependent only on the angle of sight. However, as can be seen in the definition of the shape factor, the integral term involves the square of the distance, s, in the denominator. Thus, the heat flux from one body to another is inversely proportional to the square of the distance between the two bodies. This principle is illustrated in Fig. 7.4. If the area of an object is projected against a screen, the farther away the object is, the larger is the area of the projection. The projection represents the area intercepting radiation. Since the total energy emitted by the object does not change when the distance from the screen is increased, the energy flux on the screen surface will decrease inversely with the distance. The product of the energy flux and the projected area on the screen, however, will remain constant and will equal that radiated by the body toward the screen. The distance between objects is included in the integral expression for the shape factor.

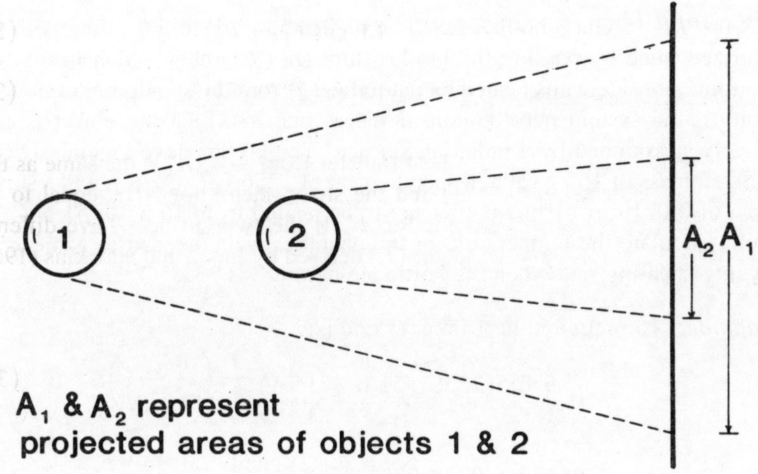

Fig. 7.4. Effect of distance on intensity of energy transfer by radiation.

Shape Factors for Simple Geometries.

1. Small object completely surrounded by a large object:

$$\bar{F}_{1-2} = \epsilon_1; \qquad F_{1-2} = 1$$

2. Large parallel planes with equal areas:

$$\frac{1}{\bar{F}_{1-2}} = \frac{1}{\epsilon_1} + \frac{1}{\epsilon_2} - 1 \tag{42}$$

3. Two parallel disks with centers directly in line (from Rohsenow and Hartnett, 1973):
a = diameter of one disk and b = diameter of the other; c = distance between the disks. A_1 = the area of the larger disk with diameter b.

$$\bar{F}_{1-2} = 0.5[Z - (Z^2 - 4X^2Y_2) \cdot 5] \tag{43}$$

where $X = a/c$, $Y = c/b$, and $Z = 1 + (1 + X^2)Y^2$.

4. Two parallel long cylinders with equal diameters, b, separated by distance 2a (from Rohsenow and Hartnett, 1973):

$$F_{1-2} = \frac{2}{\pi}\left[\sqrt{(X^2 - 1)} - X + \frac{\pi}{2} - \arccos\left(\frac{1}{x}\right)\right] \tag{44}$$

where $X = 1 + a/b$.

EXAMPLE 1: Glass bottles are to be preheated prior to hot filling with pasteurized juice by passing the bottles through a chamber which has its top, bottom, and side walls heated by natural gas. From the standpoint of the glass bottles, the system may be considered as one in which the object is completely surrounded by a radiating surface. The bottles have an emissivity of 0.94, a mass of 155 g each, a specific heat of 1256 J/(kg · K), and a surface area of 0.0219 m^2. If the bottles are to be heated from 15.5° to 51.6°C in 1 min, calculate the temperature of the chamber walls needed to achieve this average heating rate when the bottle temperature is 33.6°C.

Solution: The required heat transfer rate is:

$$q = \frac{0.155(1256)(51.6 - 15.5)}{60} = 117.1 \text{ W}$$

The heat transfer by radiation using equation 37:

$$F_{1-2} = 1; \qquad T_2 = 33.6 + 273 = 303.6 \text{ K}$$

$$q = Ae\sigma(T_1^4 - T_2^4)$$

$$q = 0.0219(0.94)(5.6732e - 08)(T^4 - (306.6)^4)$$

$$T^4 = \frac{117.1}{0.0219(0.94)(5.6732e - 08)} + (306.6)^4$$

$$T = (1.091034 \times 10^{11})^{0.25} = 574.7 \text{ K}$$

EXAMPLE 2: Cookies traveling on a conveyor inside a continuous baking oven occupy most of the area on the surface of the conveyor. The top wall of the oven directly above the conveyor has an emissivity of 0.92, and the cookies have an emissivity of 0.8. If the top wall of the oven has a temperature of 175°C, calculate the average rate of heat transfer by radiation between the cookies per unit area on the side which faces the top wall of the oven when the cookie surface temperature is 70°C.

Solution: Assume that the layer of cookies on the conveyor and the top oven wall constitute a set of long parallel plates. The shape factor $\bar{F}_{1-2}$ is calculated using equation 42:

$$\frac{1}{\bar{F}_{1-2}} = \frac{1}{0.92} + \frac{1}{0.80} - 1 = 1.3369$$

$$\bar{F}_{1-2} = \frac{1}{1.3369} = 0.748$$

$$T_1 = 70 + 273 = 343 \text{ K}; \qquad T_2 = 175 + 273 = 448 \text{ K}$$

Using equation 41:

$$\frac{q}{A} = 0.748(5.6732 \times 10^{-8})(448^4 - 343^4)$$

$$= 1122\ \text{W/m}^2$$

Microwave and Dielectric Heating. Microwaves, like light, are a form of electromagnetic vibration. Heat transfer is dependent on the degree of excitability of the molecules in the absorbing medium and the frequency of the field to which the medium is exposed. *Dielectric heating* is the term employed when relatively low frequencies are used and the material is placed between two electrodes to which an electric current is passed. Frequencies from 60 Hz to 100 MHz may be used for dielectric heating. *Microwave heating* refers to the use of electromagnetic waves of very high frequency, making it possible to transmit the energy through space. The most common frequencies used for microwave heating are 2450 and 915 MHz. Domestic microwave ovens operate at 2450 MHz. The equations which govern heat transfer by microwave and dielectric systems are the same.

Energy Absorption by Foods in a Microwave Field. The energy absorbed by a body is:

$$\frac{q}{V} = 0.556(10^{-12})\, fE^2 e \tan(\delta) \tag{45}$$

where q/V = energy absorbed, W/cm^3; f = frequency, Hz; e = dielectric constant, an index of the rate at which energy penetrates a solid, dimensionless; $\tan(\delta)$ = dielectric loss factor, an index of the extent to which energy entering the solid is converted to heat, dimensionless; and E = field strength in volts/cm^2. e and $\tan(\delta)$ are properties of the material and are functions of composition and temperature. f and E are dependent upon the frequency of the generator used. Table 7.2 shows the dielectric constant and the dielectric loss factor for foods, food components, and some packaging materials. Large pieces of metals are opaque to microwaves, i.e., microwaves bounce off the surface and very little heat energy is absorbed. However, small strips of metal such as wires will absorb microwaves and act as resistance wires, thereby heating very rapidly in a microwave field. A continuous metal sheet will have very low electrical resistance and will not heat up in a microwave field. However, a discontinuous metal sheet such as metallized plastic contains many small areas of metal, which present a large resistance to electrical current flow; therefore, intense heating occurs in these materials. Glass and plastic are practically transparent to microwaves, i.e., they transmit microwaves, and very little energy is absorbed.

Table 7.2. Dielectric Properties of Food and Other Materials

Material	Temperature °C	e''	tan (δ)
Beef (raw)	−15	5.0	0.15
Beef (raw)	25	40	0.30
Beef (roast)	23	28	0.20
Peas (boiled)	−15	2.5	0.20
Peas (boiled)	23	9.0	0.50
Pork (raw)	−15	6.8	1.20
Pork (roast)	35	23.0	2.40
Potatoes (boiled)	−15	4.5	0.20
Potatoes (boiled)	23	38.0	0.30
Spinach (boiled)	−15	13.0	0.50
Spinach (boiled)	23	34.0	0.80
Suet	25	2.50	0.07
Porridge	−15	5.0	0.30
Porridge	23	47.0	0.41
Pyrex	25	4.80	0.0054
Water	1.5	80.5	0.31
Water	25	76.7	0.15
0.1 M NaCl	25	75.5	0.24

Sources: Copson, 1971. *Microwave Heating*. AVI Publishing Co. Westport, Conn.; Schmidt, W. 1960. *Phillips Tech. Rev*. 3:89.

The frequency of the microwave power generated by a microwave generator is declared on the name plate of the unit. The power output is also supplied by the manufacturer for each unit. The coupling efficiency of a microwave unit is expressed as the ratio of power actually supplied to the unit and the actual power absorbed by the material heated. When the quantity of material being heated is large, the power generated by the system limits the power absorbed, rather than the value predicted by equation 45. The time it takes to heat a large sample of material can be used to determine the microwave power output of a microwave unit.

If a maximum q/v is determined by varying the quantity of food heated until further reduction results in no further increase in q/v, this value is the limiting power absorption and is dependent upon the dielectric loss properties of the material according to equation 45. If the dielectric constant and loss tangent of the material are known, it is possible to determine the electromagnetic field strength, and differences in the heating rates of different components in the food mixture can be predicted using equation 45.

Relative Heating Rates of Food Components. When the power output of a microwave unit limits the rate of energy absorption by the food, components with different dielectric properties will have different heating rates. Using subscripts 1 and 2 to represent component 1 and 2, e'' to represent the product of e and tan (δ), and C to represent the constant, equation 45 becomes:

$$q_1 = \frac{m_1}{\rho_1} CfE^2 e_1''$$

$$q_2 = \frac{m_2}{\rho_2} CfE^2 e_2''$$

Since $P = q_1 + q_2$ and $q = mC_p\, dT/dt$:

$$CfE^2 = \frac{P}{(m_1/\rho_1)e_1'' + (m_2/\rho_2)e_2''}$$

$$\frac{dT_1}{dt} = \frac{\rho_2 e_1'' P}{C_{p1}(\rho_2 e_1'' m_1 + \rho_1 e_2'' m_2)} \tag{46}$$

$$\frac{dT_2}{dt} = \frac{\rho_1 e_2'' P}{C_{p2}(\rho_2 e_1'' m_1 + \rho_1 e_2'' m_2)} \tag{47}$$

The relative rate of heating is:

$$\frac{dT_1}{dT_2} = \frac{\rho_2 e_1'' C_{p2}}{\rho_1 e_2'' C_{p1}} \tag{48}$$

Similar expressions may be derived for more than two components.

EXAMPLE: The dielectric constant of beef at 23°C and 2450 MHz is 28, and the loss tangent is 0.2. The density is 1004 kg/m^3, and the specific heat is 3250 $J/(kg \cdot K)$. Potato at 23°C and 2450 MHz has a dielectric constant of 38 and a loss tangent of 0.3. The density is 1010 kg/m^3, and the specific heat is 3720 $J/(kg \cdot K)$.

(a) A microwave oven has a rated output of 600 W. When 0.25 kg of potatoes was placed in the oven, the temperature rise after 1 min of heating was 38.5°C. When 60 g of potato was heated in the oven, a temperature rise of 40°C was observed after 20 s. Calculate the average power output of the oven and the mass of potatoes which must be present so that the output of the oven limits the rate of power absorption rather than the capacity of the material to absorb the microwave energy.

(b) If potatoes and beef are heated simultaneously, what are the relative rates of heating?

Solution: (a) Assume that a 0.25-kg mass of product is sufficient to make microwave power availability the rate-limiting factor for microwave absorption.

$$P = 0.25 \text{ kg}\left[\frac{3720 \text{ J}}{\text{kg} \cdot \text{K}}\right](38.5 \text{ K})(1 \text{ min})\frac{1 \text{ min}}{60 \text{ s}}$$

$$= 596.75 \text{ W}$$

When a small amount of material is heated:

$$P = 0.06 \text{ kg}\left[\frac{3720 \text{ J}}{\text{kg} \cdot \text{K}}\right](40 \text{ K})\left[\frac{1}{20 \text{ s}}\right] = 444.6 \text{ W}$$

The amount absorbed with the small mass in the oven is much smaller than when a larger mass is present; therefore, it may be assumed that power absorption by the material limits the rate of heating.

$$\frac{q}{V} = \left[\frac{446.6 \text{ W}}{0.06 \text{ kg } 1 \text{ m}^3/1010 \text{ kg}}\right]\frac{1 \text{ m}^3}{10^{-6} \text{ cm}^3} = 7.5177 \text{ W/cm}^3$$

$$= \frac{P(\rho)10^{-6}}{\text{m}}$$

Since with a small load in the oven the power absorption is only 7.5177 W/cm^3, this may be assumed to be $(q/V)_{\text{lim}}$, the maximum rate of power absorption by the material. If the power output of the oven as calculated above is 596.75 W, the mass present where the rate of power absorption equals the power output of the oven is:

$$m = \frac{P(\rho)10^{-6}}{(q/V)_{\text{lim}}} = \frac{596.75(1010)}{7.5177} = 0.08 \text{ kg}$$

Thus, any mass greater than 0.08 kg in this microwave oven will heat at the rate determined by the power output of the oven.

(b) Using equation 48: Let subscript 1 refer to beef and subscript 2 refer to potato.

$$\frac{dT_1}{dT_2} = \frac{1010(28)(0.2)(3720)}{1004(38)(0.3)(3260)} = 0.563$$

Thus, the beef will be heating more slowly than the potatoes.

TEMPERATURE-MEASURING DEVICES

Temperature is defined as the degree of thermal agitation of molecules. Changes in the molecular motion of a gas or liquid change its volume or pressure, and a solid undergoes a dimensional change such as expansion or contraction. Certain metals lose electrons when molecules are thermally excited, and when they are paired with another metal whose molecules could receive these displaced electrons, an electromotive force will be generated. These responses of materials to changes in temperature are utilized in the design of thermometers, electronic and mechanical temperature-measuring devices. The calibration of all temperature-measuring devices is based on the conditions of use. Some electronic instruments have ambient temperature compensation integrated in the circuitry, and others do not. Similarly, fluid-filled thermometers are calibrated for certain immersion depths; therefore, partial immersion of a thermometer calibrated for total immersion leads to errors in measurement. Thus, conditions of use which correspond to the calibration of the instrument must be known for accurate measurements. For other conditions, the instruments will need recalibration. Thermometers filled with fluids other than mercury may have to be recalibrated after a certain period of use to ensure that breakdown of the fluid, which might change its thermal expansion characteristics, has not occurred. In the food industry, a common method of calibration involves connecting the thermometers to a manifold which contains saturated steam. The pressure of the saturated steam can then be used to determine the temperature from steam tables and compared to the thermometer readings. Recalibration of fluid-filled thermometers is normally not done. Thermometers used at critical control points in processes must be replaced when they lose accuracy. Electronic temperature-measuring devices must be recalibrated frequently against a fluid-filled thermometer.

The temperature indicated by a measuring device represents the temperature of the measuring element itself rather than the temperature of the medium in contact with the element. The accuracy of the measurement depends on how heat is transferred to the measuring element. The temperature registered by the instrument is that of the measuring element after heat exchange approaches equilibrium.

Thermometers will not detect oscillating temperatures if the mass of the temperature-measuring element is such that the lag time for heat transfer equilibration is large in relation to the period of the oscillation. The response time of temperature-measuring devices is directly proportional to the mass of the measuring element; the smaller the mass, the more accurate the reading.

Measurement of the surface temperature of a solid in contact with either a liquid or a gas is not very accurate if the measuring element is simply laid on the solid surface. Depending upon the thickness of the measuring element, the temperature indication will be intermediate between that of the surface and the fluid temperature at the interface. Accurate solid surface temperatures can only be determined by embedding two thermocouples a known distance from the surface and extrapolating the temperature readings toward the surface to obtain the true surface temperature.

Measurements of the temperatures of gases can be influenced by radiation. An unshielded thermometer or thermocouple in a gas stream exposed to surfaces having temperatures higher than that of the gas stream will read a higher temperature than the true gas stream temperature because of radiation. Shielding of measuring elements is necessary for accurate measurements of gas temperatures.

Liquid-in-Glass Thermometers. Liquid-in-glass thermometers commonly employ mercury for general use and mineral spirits, ethanol, or toluene for low-temperature use. When properly calibrated, any temperature-stable liquid may be used. Non-mercury-filled thermometers are now available for use as temperature indicators in food processing facilities.

Fluid-Filled Thermometers. These temperature-measuring devices consist of a metal bulb with a long capillary tube attached to it. The end of the tube is attached to a device which causes a definite movement with pressure transmitted to it from the capillary. The capillary and bulb are filled with fluid which changes in pressure with changes in temperature. Capillary length affects the calibration; so does the ambient temperature surrounding the capillary. Thus, temperature-measuring devices of this type must be calibrated in the field. Figure 7.5b shows a typical fluid-filled thermometer with a spiral at the end of the capillary. The movement of the spiral is transmitted mechanically through a rack-and-pinion arrangement or a lever to a needle, which moves to indicate the temperature on

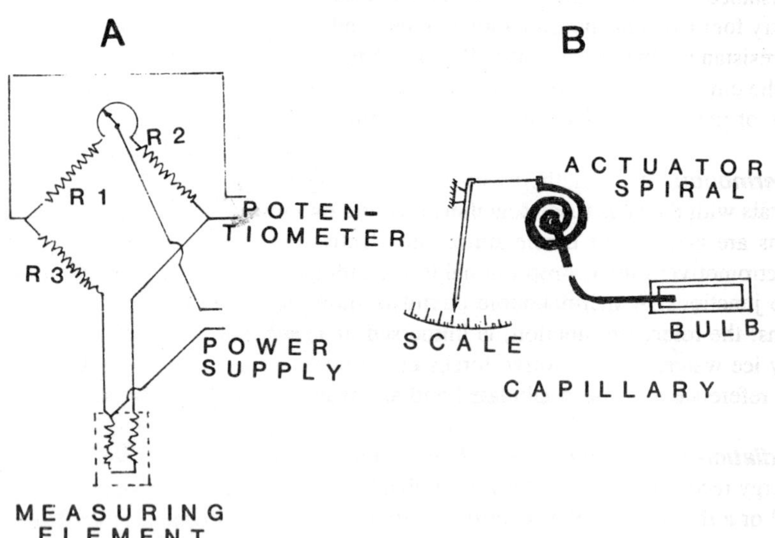

Fig. 7.5. Schematic diagram of temperature-measuring devices. Fluid-filled thermometer (A) and RTD circuit (B).

a dial. Some stainless steel–cased, dial-type thermometers are of this type. Proper performance of these instruments requires total immersion of the bulb.

Bimetallic Strip Thermometers. When two metals having dissimilar thermal expansion characteristics are joined, a change in temperature causes a change in shape. For example, thermostats for domestic space cooling and heating utilize a thin strip of metal fixed at both ends to a thicker strip of another metal. An increase in temperature causes the thinner metal, which has a higher thermal coefficient of expansion, to bend. The change in position may open or close properly positioned electrical contacts, which then activates heating or cooling systems. In some configurations, the bimetallic strip may be wound in a helix or spiral. The strip is fixed at one end, and the free end moves with changes in temperature to position an indicator needle on a temperature scale. This type of thermometer requires total immersion of the bimetallic strip for proper performance.

Resistance Temperature Devices (RTDs). The principle of RTDs is based on the change in resistance of a material with changes in temperature. Figure 7.5a shows an RTD circuit. The resistance strip may be an insulated coil of resistance wire encased in a metal tube (resistance bulb) or a strip of metal such as platinum. Other materials, such as semiconductors, exhibit a large drop in resistance with increasing temperatures. The whole instrument consists of the resistance or semiconductor element for sensing the temperature, electronic circuitry for providing the excitation voltage, and a measuring circuit for the change in resistance. Response is usually measured as a change in voltage or current in the circuit. RTDs vary in size and shape, depending upon use. Total immersion of the sensing element is necessary for accurate results.

Thermocouples. A thermocouple is a system of two separate, dissimilar metals with the ends fused together, forming two junctions. When the two junctions are at different temperatures, an electromotive force is generated. This electromotive force is proportional to the difference in temperature between the two junctions. A thermocouple circuit is shown in Fig. 7.6. One of the junctions, the reference junction, is immersed in a constant temperature bath, usually ice water. Electromotive forces at various temperatures, with ice water as the reference have been tabulated and are available in handbooks.

Radiation Pyrometers. This type of temperature-sensing device focuses the energy received from a surface through a lens on a device such as a photovoltaic cell or a thermopile. This energy stimulates the production of an electromotive force, which can then be measured. Since the amount of energy received per unit area by a receiving surface from a source varies inversely with the distance, these types of measuring devices have distance compensation features. Distance

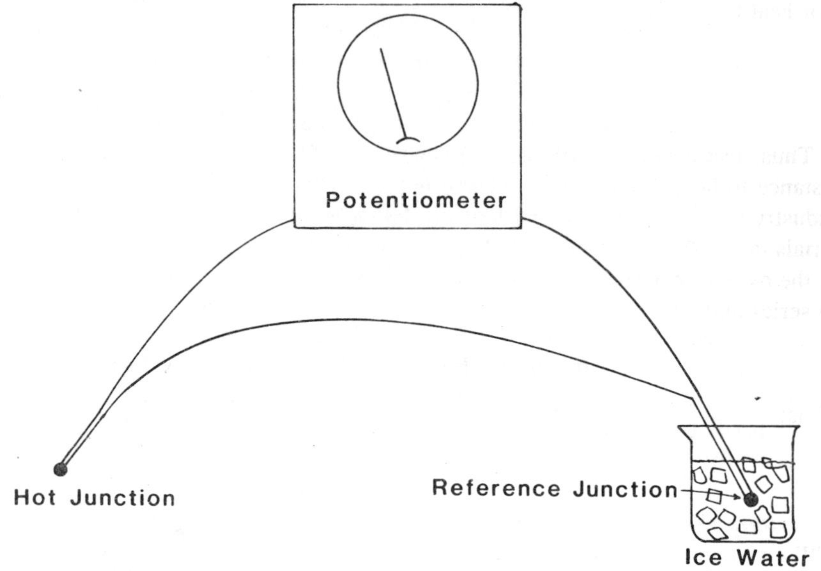

Fig. 7.6. Thermocouple circuit showing ice water as the reference.

compensation is achieved by focusing the energy received on the lens of the instrument on a smaller area with an energy-sensitive surface. Radiation pyrometers are useful for sensing very high temperatures such as those in furnaces, flames, and red hot or molten metals. Some, which have been calibrated for operation at near ambient temperatures, are very useful for noncontact monitoring of the surface temperatures of processing equipment.

STEADY-STATE HEAT TRANSFER

The Concept of Resistance to Heat Transfer. Equation 25, derived from Fourier's first law, shows that in a steady-state system, the quantity of heat passing through any part of the system must equal that passing through any other part and must equal the total quantity of heat passing through the system.

The problem of heat transfer through multiple layers can be analyzed as a problem involving a series of resistances to heat transfer. The transfer of heat can be considered as analogous to the transfer of electrical energy through a conductor. ΔT is the driving force equivalent to the voltage E in electrical circuits. The heat flux q is equivalent to the current I.

Ohm's law for electrical circuits is:

$$I = \frac{E}{R} \tag{49}$$

For heat transfer through a slab:

$$\frac{q}{A} = \frac{\Delta T}{[\Delta X/k]} = \frac{\Delta T}{R} \tag{50}$$

Thus, in comparing equations 40 and 50, R is equivalent to $\Delta X/k$. The resistance to heat transfer is $\Delta X/k$ and is the "R" rating used in the insulation industry to rate the effectiveness of insulating materials. For multilayered materials in the geometry of a slab where A is constant in the direction of increasing x, the overall resistance to heat transfer is the sum of the individual resistances in series and:

$$R = R_1 + R_2 + R_3 + \cdots R_n$$

Thus:

$$\frac{1}{R} = \frac{\Delta x_1}{k_1} + \frac{\Delta x_2}{k_2} + \cdots \frac{\Delta x_n}{k_n}$$

and:

$$\frac{q}{A} = \frac{\Delta T}{[\Delta x_1/k_1 + \Delta x_2/k_2 \cdots \Delta x_n/k_n]} \tag{51}$$

Equation 50 can also be written as:

$$\frac{q}{A} = \frac{\Delta T_1}{R_1} = \frac{\Delta T_2}{R_2} = \frac{\Delta T_3}{R_3} = \cdots \frac{\Delta T_n}{R_n} = \frac{\Delta T}{R} \tag{52}$$

Equation 52 is similar to equation 25 previously derived using Fourier's first law.

For heat transfer through a cylinder, the heat transfer rate in equation 27 results in an expression for the heat transfer resistance as:

$$R = \frac{\ln (r_2/r_1)}{2\pi LK} \tag{53}$$

For resistances in series:

$$q = \frac{\Delta T}{R}$$

$$= \frac{\Delta T}{[\ln (r_2/r_1)/2\pi Lk_1 + \ln (r_3/r_2)/2\pi Lk_2 + \cdots \ln (r_{n+1}/r_n)/2\pi Lk_n]} \tag{54}$$

and:

$$\frac{\Delta T_1}{[\ln (r_2/r_1)/2\pi Lk_1]} = \frac{\Delta T_2}{[\ln (r_3/r_2)/2\pi Lk_2]} = \cdots \frac{\Delta T_n}{[\ln (r_{n+1}/r_n)/2\pi Lk_n]} \tag{55}$$

For convection heat transfer:

$$q = hA\Delta T = \frac{\Delta T}{(1/hA)}$$

$$R = \frac{1}{hA} \tag{56}$$

Combined Convection and Conduction, the Overall Heat Transfer Coefficient. Most problems encountered in practice involve heat transfer by combined convection and conduction. Usually, the temperatures of fluids on both sides of a solid are known and the rate of heat transfer across the solid is to be determined. Heat transfer involves convective heat transfer between a fluid on one surface, conductive heat transfer through the solid, and convective heat transfer again at the opposite surface to the other fluid. The rate of heat transfer may be expressed in terms of U, the overall heat transfer coefficient, or in terms of R, an overall resistance.

Consider a series of resistances involving n layers of solids and n fluid–surface interfaces. The thermal conductivity of the solids are $k_1 \cdots k_2 \cdots k_3 \cdots k_n$, and the heat transfer coefficients are $h_1 \cdots h_2 \cdots h_n$, with subscript n increasing along the direction of heat flow.

$$q = UA\Delta T = \frac{\Delta T}{R} \tag{57}$$

$$R = \frac{1}{UA} \tag{58}$$

For a slab:

$$R = \Sigma\left[\frac{1}{h_n A}\right] + \Sigma\left[\frac{X_n}{k_n A}\right] = \frac{1}{UA}$$

Since A is the same across a slab:

$$\frac{1}{U} = \frac{1}{h_1} + \frac{\Delta x_1}{k_2} + \frac{\Delta x_2}{k_3} + \cdots \frac{\Delta x_n}{k_n} + \cdots \frac{1}{h_n} \tag{59}$$

or:

$$U = \frac{1}{[\Sigma(1/h_n) + \Sigma(X_n/k_n)]} \tag{60}$$

For a cylinder:

$$R = \Sigma\left[\frac{\ln (r_{n+1}/r_n)}{2\pi L k_n}\right] + \Sigma\left[\frac{1}{h_n A_n}\right] = \frac{1}{UA}$$

If the A used as a multiplier for U is the outside area, then $U = U_0$, the overall heat transfer coefficient based on the outside area. U_i = overall heat transfer coefficient based on the inside area. Using h_i = inside heat transfer coefficient and h_0 = outside heat transfer coefficient, r_i and r_0 are the inside and outside radius of the cylinder, respectively.

$$(2\pi r_0 L)\, U_0 = \frac{1}{1/2\pi L \Sigma[\ln (r_{n+1}/r_n)/k_n] + 1/2\pi L[1/h_0 r_0 + 1/h_i r_i]}$$

$$U_0 = \frac{1}{r_0 \Sigma[\ln (r_n/r_{n-1})/k_n] + [r_0/r_i h_i] + [1/h_0]}$$

or:

$$\frac{1}{U_0} = \frac{r_0}{r_i h_i} + \frac{r_0 \ln (r_2/r_1)}{k_1} + \frac{r_0 \ln (r_3/r_2)}{k_2} + \cdots \frac{r_0 \ln (r_n/r_{n-1})}{k_n} + \frac{1}{h_0}$$

$$U_i = \frac{1}{r_i \Sigma[\ln (r_n/r_{n-1})/k_n] + [r_i/r_0 h_0] + [1/h_i]} \tag{61}$$

or:

$$\frac{1}{U_i} = \frac{1}{h_i} + \frac{r_i \ln (r_2/r_1)}{k_1} + \frac{r_i \ln (r_3/r_2)}{k_2} + \cdots \frac{r_i \ln (r_n/r_{n-1})}{k_n} + \frac{r_i}{r_0 h_0} \tag{62}$$

EXAMPLE 1: Calculate the rate of heat transfer across a glass pane which consists of two 1.6-mm-thick glass layers separated by an 0.8-mm layer of air. The heat transfer coefficient on one side, which is at 21°C, is 2.84 W/(m^2 · K), and that on the opposite side, which is at −15°C, is 11.4 W/(m^2 · K). The thermal conductivity of glass is 0.52 W/(m · K), and that of air is 0.031 W/(m · K).

Solution: When stagnant air is trapped between two layers of glass, convective heat transfer is minimal and the stagnant air layer will transfer heat by conduction. There are five resistances to heat transfer. R_1 is the convective resistance at one surface exposed to air, R_2 is the conductive resistance of one 1.6-mm-thick layer of glass, R_3 is the conductive resistance of the air layer between the glass, R_4 is the conductive resistance of the second 1.6-mm-thick layer of glass, and R_5 is the convective resistance of the opposite surface exposed to air. Using equation 9:

$$\frac{1}{U} = \frac{1}{h_1} + \frac{x_i}{k_1} + \frac{x_2}{k_2} + \frac{x_3}{k_3} + \frac{1}{h_2}$$

$$= \frac{1}{2.84} + \frac{1.6 \times 10^{-3}}{0.52} + \frac{0.8 \times 10^{-3}}{0.031} + \frac{1.6 \times 10^{-3}}{0.52} + \frac{1}{11.4}$$

$$= 0.352 + 0.0031 + 0.0258 + 0.0031 + 0.0877$$

$$= 0.4718$$

$$U = 2.12 \frac{\mathrm{W}}{\mathrm{m}^2 \cdot \mathrm{K}}$$

$$\frac{q}{A} = U\Delta T = 2.12(21 - (-15))$$

$$= 76.32 \frac{\mathrm{W}}{\mathrm{m}^2}$$

EXAMPLE 2: (a) Calculate the overall heat transfer coefficient for a 1-in. (nominal), 16-gage heat exchanger tube when the heat transfer coefficient is 568 W/(m^2 · K) inside and 5678 W/(m^2 · K) outside. The tube wall has a thermal conductivity of 55.6 W/(m · K). The tube has an inside diameter of 2.21 cm and a 1.65-mm wall thickness. (b) If the temperature of the fluid inside the tube is 80°C and 120°C on the outside, what is the inside wall temperature?

Solution:

$$r_i = 2.21/2 = 1.105 \text{ cm}$$

$$r_0 = 1.105 + 0.1651 = 1.2701 \text{ cm}$$

(a) Using equation 61:

$$\frac{1}{U_0} = \frac{1}{5678} + \frac{1.2701 \times 10^{-2} \ln(1.2701/1.105)}{55.6}$$

$$+ \frac{1.2701 \times 10^{-2}}{(1.105 \times 10^{-2})(568)}$$

$$= 1.76 \times 10^{-4} + 0.318 \times 10^{-4} + 20.236 \times 10^{-4}$$

$$= 22.315 \times 10^{-4}; \qquad U_0 = 448 \text{ W/m}^2 \cdot \text{K}$$

(b) Temperature on the inside wall:

q(overall) = q(across inside convective heat transfer coefficient).

Let T_f = temperature of fluid inside tube and T_w = temperature of the wall.

$$U_0 A_0 \Delta T = h_i A_i (T_w - T_f)$$

$$448(2\pi r_0 L)(120 - 80) = 568(2\pi r_i L)(T_w - 80)$$

$$(T_w - 80) = \frac{448(r_0)(40)}{568(r_i)}$$

$$= 80 + \frac{448(1.2701)(40)}{568(1.105)}$$

$$= 80 + 36.3 = 116.3°\text{C}$$

Heat Exchange Equipment. Heat exchangers are equipment for transferring heat from one fluid to another. In its simplest form, a heat exchanger is a copper tube exposed to air, where fluid flowing inside the tube is cooled by transferring heat to ambient air. When used to transfer large quantities of heat per unit time, heat exchangers take several forms to provide for efficient utilization of the heat contents of both fluids exchanging heat and to allow for compactness of the equipment. The simplest heat exchanger to make is one of tubular design. Heat exchangers commonly used in the food industry include the following:

Swept Surface Heat Exchangers. The food product passes through an inner cylinder, and the heating or cooling medium passes through the annular space between the inner cylinder and an outer cylinder or jacket.

A rotating blade running the whole length of the cylinder continuously agitates the food as it passes through the heat exchanger and, at the same time, continuously scrapes the walls through which heat is being transferred (the heat

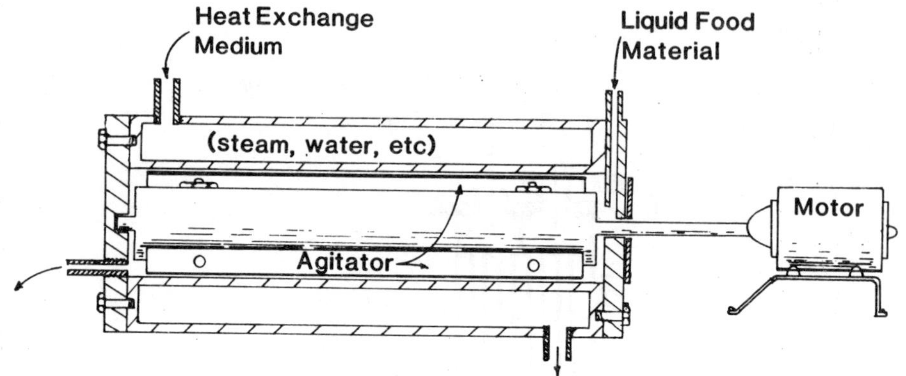

Fig. 7.7. Diagram of a swept surface heat exchanger.

transfer surface). This type of heat exchanger can be used to heat, cool, or provide heat to concentrate viscous food products. Figure 7.7 shows a swept surface heat exchanger.

Double-Pipe Heat Exchanger. This heat exchanger consists of one pipe inside another. The walls of the inner pipe form the heat transfer surface. This type of heat exchanger is usually built and installed in the field. A major disadvantage is the relatively large space it occupies for the quantity of heat exchanged compared to other types of heat exchangers. Figure 7.8 shows a double-pipe heat exchanger.

Shell-and-Tube Heat Exchanger. This heat exchanger consists of a bundle of tubes enclosed by a shell. The head arrangement allows for one-tube pass or multiple passes for the product. In a one-pass arrangement, the product enters at one end and exists at the opposite end. In a multipass arrangement, the product may travel back and forth through different tubes with each pass before finally leaving the heat exchanger. The heat exchange medium on the outside of the tubes is usually distributed using a system of baffles. Figure 7.9 shows two types of shell-and-tube heat exchanges.

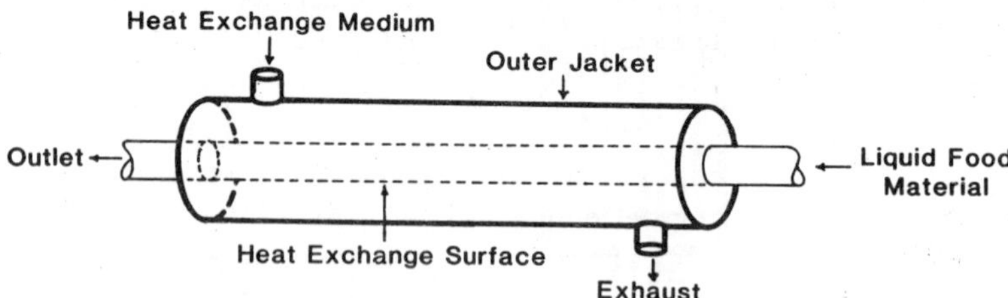

Fig. 7.8. Diagram of a double-pipe heat exchanger.

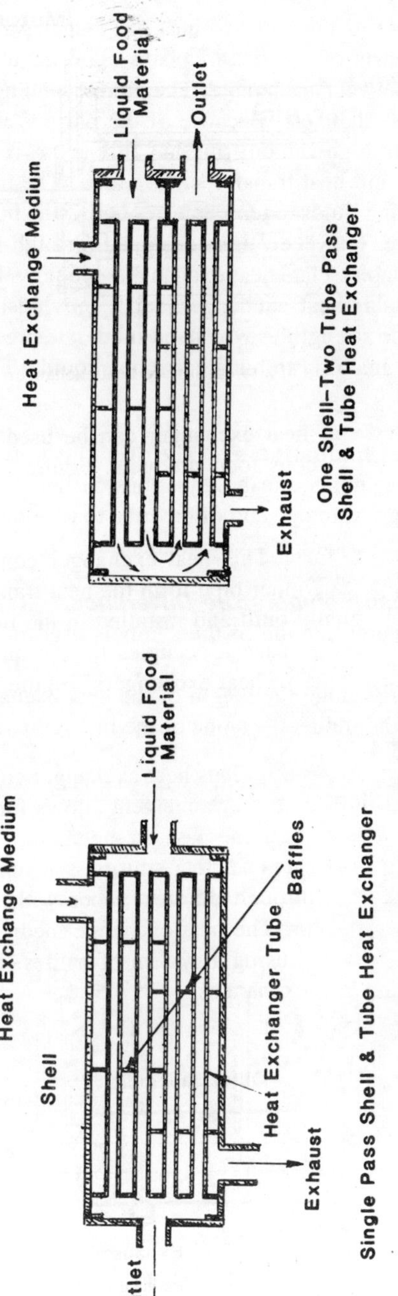

Fig. 7.9. Diagram of a single-pass and a multiple-pass shell-and-tube heat exchanger.

Plate Heat Exchanger. This type of heat exchanger was developed for the dairy industry. It consists of a series of plates clamped together on a frame. Channels are formed between each plate. The product and heat transfer medium flows through alternate channels. Because of the narrow channel between the plates, the fluid flows at high velocity and in a thin layer, resulting in very high heat transfer rates per unit heat transfer surface area. The plate heat exchanger is mostly used for heating fluids to temperatures below the boiling point of water at atmospheric pressure. However, units designed for high-temperature service are commercially available. Plate heat exchangers are now used in virtually any application where tubular heat exchangers were previously commonly used. Newer designs have the strength to withstand moderate pressure or vacuum. A major limitation is the inability to handle viscous liquids. Figure 7.10 shows a plate heat exchanger.

Heat Transfer in Heat Exchangers. The overall heat transfer coefficient in tubular exchangers is calculated using equations 61 and 62. Since in heat exchangers ΔT can change from one end of the tube to the other, a mean ΔT must be determined for use in equation 57 to calculate q.

The Logarithmic Mean Temperature Difference. Refer to Fig. 7.11 for the diagram and the representation of the symbols used in the following derivation.

If ΔT changes linearly along the length of the heat exchanger relative to the temperature of one of the fluids, the slope of the line representing ΔT vs. T_B is:

$$\text{Slope} = \frac{d(\Delta T)}{dT_B} = \frac{\Delta T_2 - \Delta T_1}{T_{B2} - T_{B1}}$$

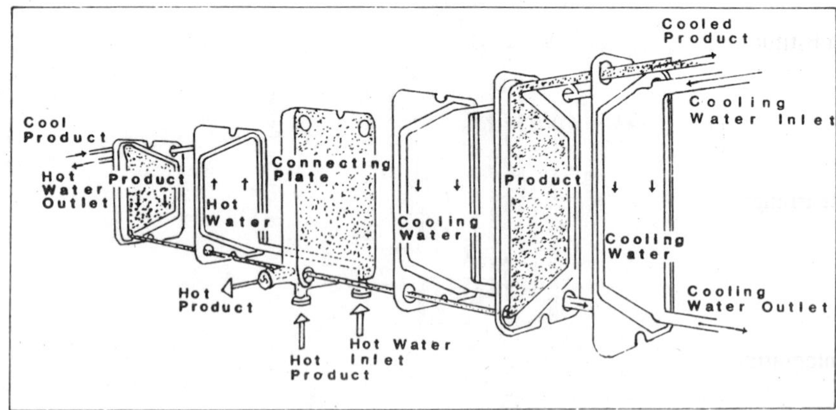

Fig. 7.10. Diagram of a plat heat exchanger showing alternating paths of processed fluid and heat exchange medium.

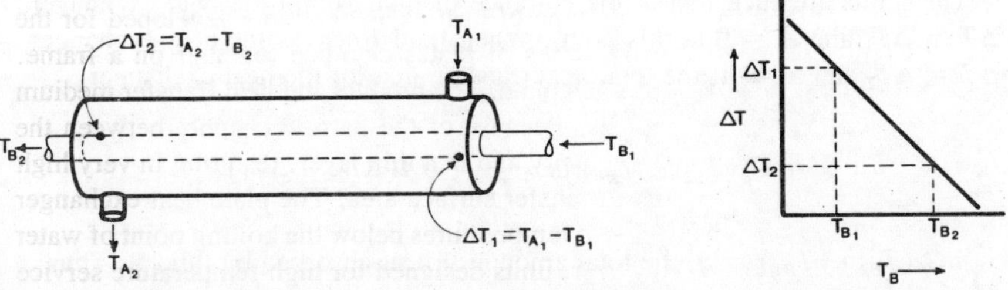

Fig. 7.11. Diagram of the temperature profiles of fluid along the length of a heat exchanger tube.

Rearranging:

$$dT_B = d\Delta T \frac{T_{B2} - T_{B1}}{\Delta T_2 - \Delta T_1}$$

The amount of heat transferred across any point in the exchanger is:

$$q = U\, dA\Delta T$$

This transferred heat will cause a rise in product temperature, dT_B, and can be also expressed as:

$$q = mC_p\, dT_B$$

Equating:

$$U\, dA\Delta T = mC_p\, dT_B$$

Substituting the expression for dT_B:

$$U\, dA\Delta T = mC_p\left[\frac{d\Delta T(T_{B2} - T_{B1})}{(\Delta T_2 - \Delta T_1)}\right]$$

Rearranging:

$$mC_p(T_{B2} - T_{B1})\frac{d\Delta T}{\Delta T} = U(\Delta T_2 - \Delta T_1)\,(dA)$$

Integrating:

$$mC_p(T_{B2} - T_{B1})\int_{\Delta T_1}^{\Delta T_2}\frac{d\Delta T}{\Delta T} = U(\Delta T_2 - \Delta T_1)\int_0^A dA$$

The limits are such that at the entrance of fluid B into the heat exchanger, $\Delta T = \Delta T_1$ and $A = 0$ at this point. When fluid B leaves the heat exchanger, $\Delta T = \Delta T_2$ and $A = A$, the total heat transfer area for the heat exchanger.

$$mC_p(T_{B2} - T_{B1}) \ln \frac{\Delta T_2}{\Delta T_1} = UA(\Delta T_2 - \Delta T_1)$$

$mC_p(T_{B2} - T_{B1}) = q$, the total amount of heat absorbed by fluid B. Thus:

$$q = UA \frac{\Delta T_2 - \Delta T_1}{\ln \Delta T_2/\Delta T_1} = UA\overline{\Delta T_L}$$

where ΔT_L is the logarithmic mean ΔT expressed as:

$$\overline{\Delta T_L} = \frac{\Delta T_2 - \Delta T_1}{\ln \Delta T_2/\Delta T_1} \qquad (63)$$

When heat is exchanged between two liquids or between a liquid and a gas, the temperatures of both fluids change as they travel through the heat exchanger. If both fluids enter on the same end of the unit, flow will be in the same direction and this type of flow is cocurrent. Countercurrent flow exists when the fluids flow in opposite directions through the unit. True countercurrent or cocurrent flow occurs in double-pipe, plate, and multiple-tube, single-pass shell, and tube heat exchangers. When true cocurrent or countercurrent flow exists, the rate of heat transfer can be calculated by:

$$q = U_iA_i\Delta T_L \qquad \text{or} \qquad q = U_0A_0\Delta T_L \qquad (64)$$

The log mean temperature difference is evaluated between the inlet and exit ΔT values.

In multipass heat exchangers, the fluid inside the tubes travels back and forth across zones of alternating high and low temperatures. In these heat exchangers, a correction factor is used on the $\overline{\Delta T_L}$. The correction factor depends upon the manner in which shell fluid and tube fluid pass through the heat exchanger.

$$q = U_iA_iF\overline{\Delta T_L} \qquad \text{or} \qquad q = U_0A_0F\overline{\Delta T_L} \qquad (65)$$

where F is the correction factor. The reader is referred to a heat transfer textbook or handbook for graphs showing correction factors to $\overline{\Delta T_L}$ for multipass shell and tube heat exchangers.

EXAMPLE 1: Applesauce is being cooled from 80° to 20°C in a swept surface heat exchanger. The overall coefficient of heat transfer based on the inside surface area is 568 $W/m^2 \cdot K$. The applesauce has a specific heat of 3817 $J/kg \cdot K$ and is being cooled at the rate of 50 kg/h. Cooling water enters in countercurrent flow at 10°C and leaves the heat exchanger at 17°C. Calculate (a) the quantity of cooling water required and (b) the required heat transfer surface area for the heat exchanger.

Solution: The diagram of the heat exchanger in countercurrent flow is shown in Fig. 7.12.

(a) The rate of heat transfers =

$$50 \frac{\text{kg}}{\text{h}} \cdot 3187 \frac{\text{J}}{\text{kg} \cdot \text{K}} \cdot (80 - 20)\,°\text{K} = 9{,}561{,}000 \frac{\text{J}}{\text{h}} = 2.656 \text{ kW}$$

Let x = quantity of water used. The specific heat of water is 4186 J/kg K.

$$x(4186)(17 - 10) = 9{,}561{,}000$$

$$x = \frac{9{,}561{,}000}{4186(7)} = 326 \text{ kg/h}$$

(b) $\Delta T_2 = 20 - 10 = 10°\text{C}$
$\Delta T_1 = 80 - 17 = 63°\text{C}$

Using equation 63:

$$\overline{\Delta T_L} = \frac{63 - 10}{\ln 63/10} = 53/\ln 6.3 = 28.8$$

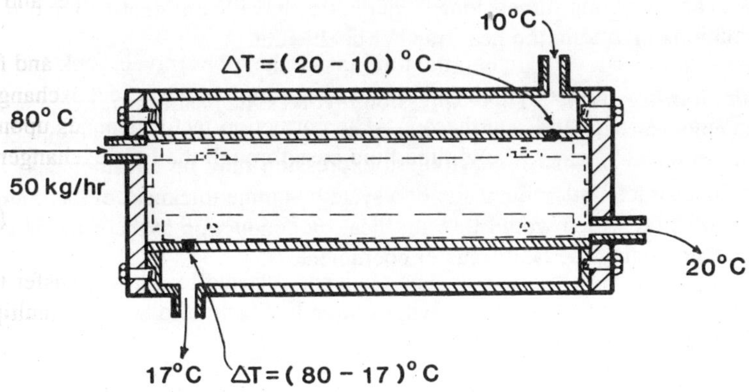

Fig. 7.12. Diagram of a heat exchanger in a counterflow configuration for the fluids exchanging heat.

The rate of heat transfer is calculated using equation 64.

$$q = U_i A_i \overline{\Delta T_L} = 568\ A_i(28.8)$$

$$(568)(A_i)(28.8) = 2656$$

$$A_i = \frac{2656}{568(28.8)} = 0.162\ \text{m}^2$$

LOCAL HEAT TRANSFER COEFFICIENTS

A major problem in heat transfer is the estimation of heat transfer coefficients to be used for design purposes. These values of h must be determined from the properties of the fluid and the geometry of the system.

Dimensionless Quantities. The use of dimensionless quantities arises from the principle of similarity. Equations that describe different systems having similar characteristics can be superimposed on each other to form a single expression suitable for all systems. Thus, if the physical characteristics of a fluid and the conditions that exist in an experiment are expressed in terms of dimensionless quantities, it is possible to extrapolate the results of an experiment to other fluids and other conditions. The principle of similarity makes it unnecessary to establish equations experimentally for heat transfer to each fluid. A general correlation equation is suitable for all fluids. Determination of the different dimensionless quantities involved in the relationships between the variables describing a system is done by a method called *dimensional analysis*. In dimensional analysis, an equation relating the various variables is first assumed, and by performing an analysis of the base dimensions of the variables to make the equation dimensionally consistent, specific groupings of the variables are formed. The following dimensionless quantities have been identified and used in correlations involving the heat transfer coefficient.

Nusselt Number (Nu). This expression involves the heat transfer coefficient (h), the characteristic dimension of the system (d), and the thermal conductivity of the fluid (k). This dimensionless expression may be considered as the ratio of the characteristic dimension of a system and the thickness of the boundary layer of fluid which would transmit heat by conduction at the same rate as that calculated using the heat transfer coefficient.

$$Nu = h\,\frac{d}{k} \tag{66}$$

Reynolds Number (Re). This expression involves the characteristic dimension of the system (d), the velocity of the fluid (V), the density (ρ), and the viscosity (μ). It may be considered as the ratio of inertial forces to the frictional force. The Reynolds number was discussed in Chapter 6.

Prandtl Number (Pr). This expression involves the specific heat (C_p), the viscosity (μ), and the thermal conductivity (k). It may be considered as the ratio of the rate of momentum exchange between molecules and the rate of energy exchange between molecules that lead to the transfer of heat.

$$Pr = \frac{C_p \mu}{k} \tag{67}$$

Grashof Number (Gr). This quantity involves the characteristic dimension of a system (d), the acceleration due to gravity (g), the thermal expansion coefficient (β), the density of the fluid (ρ), the viscosity (μ), and the temperature difference ΔT between a surface and the fluid temperature beyond the boundary layer. This number may be considered as a ratio of the force of gravity to the buoyant forces which arise as a result of a change in the temperature of a fluid.

$$Gr = \frac{d^3 g \beta \rho^2 \Delta T}{\mu^2} \tag{68}$$

Peclet Number (Pe). This dimensionless number is the product of the Reynolds number and the Prandtl number.

$$Pe = RePr = \frac{\rho V C_p d}{k} \tag{69}$$

Rayleigh Number (Ra). This dimensionless number is the product of the Grashof number and the Prandtl number.

$$Ra = GrPr = \frac{d^3 g \beta C_p \rho^2 \Delta T}{\mu k} \tag{70}$$

Graetz Number (Gz). This is similar to the Peclet number. It was derived from an analytical solution to the equations of heat transfer from a surface to a fluid flowing along that surface in laminar flow. The Graetz number is:

$$Gz = \frac{\pi}{4}\left[Re \cdot Pr \cdot \left(\frac{d}{L}\right)\right] = \frac{\dot{m} C_p}{kL} \tag{71}$$

where $\dot{m}$ is the mass rate of flow, kg/s.

Equations for Calculating Heat Transfer Coefficients. These equations generally take the form:

$$Nu = f\left(Re,\ Pr,\ \left(\frac{L}{d}\right),\ Gr,\ \left(\frac{\mu}{\mu_w}\right)\right) \tag{72}$$

The Grashof number is associated with free convection, and the length/diameter ratio (L/d) appears when flow is laminar. When calculating these dimensionless quantities, the thermophysical properties of fluids at the arithmetic mean temperature at the inlet and exit are used. The viscosity of fluid at the wall (μ_w) affects heat transfer and is included in the correlation equation to account for the difference in the cooling and heating processes. Most of the English literature on correlation equations for heat transfer is based on the general expression:

$$Nu = \alpha(Re)^{\beta}(Pr)^{\gamma}\left[\frac{L}{d}\right]^{\delta} \tag{73}$$

where α, β, γ, and δ are constants obtained from correlation analysis of experimental data.

A number of correlation equations suitable for use under various conditions have been published. A summary of the equations suitable for use under commonly encountered conditions in food processing is given in Appendix Table A.12. In order to illustrate the use of these equations in the design of a heating or cooling system, some equations are also given in this chapter.

Simplified Equations for Natural Convection to Air or Water. For natural convection to air or water, Table 7.3 lists simplified equations based on those given by McAdams (1954).

Fluids in Laminar and Turbulent-Flow Inside Tubes. Equation 74, used for laminar flow, was originally derived by Leveque and is discussed in a number of heat transfer textbooks. Equation 75 is another form of equation 74. Equation 76 , commonly referred to as the *Dittus-Boelter equation*, is used for turbulent flow.

$$Nu = 1.615\left[RePr\left(\frac{d}{L}\right)\right]^{0.33}\left(\frac{\mu}{\mu_w}\right)^{0.14} \tag{74}$$

$$Nu = 1.75(Gz)^{0.33}\left(\frac{\mu}{\mu_w}\right)^{0.14} \tag{75}$$

$$Nu = 0.023(Re)^{0.8}(Pr)^{0.33}\left(\frac{\mu}{\mu_w}\right)^{0.14} \tag{76}$$

Table 7.3. Equations for Calculating Heat Transfer Coefficients in Free Convection from Water or Air

Surface Conditions	Equation	Value of C for: Air	Water
Horizontal cylinder heated or cooled	$h = C\left(\frac{\Delta T}{D}\right)^{0.25}$	1.3196	291.1
Fluid above heated horizontal plate	$h = C(\Delta T)^{0.25}$	2.4493	
Fluid below heated horizontal plate	$h = C(\Delta T)^{0.25}$	1.3154	
Fluid above cooled horizontal plate	$h = C(\Delta T)^{0.25}$	1.3154	
Fluid below cooled horizontal plate	$h = C(\Delta T)^{0.25}$	2.4493	
Vertical cylinder heated or cooled	$h = C\left(\frac{\Delta T}{D}\right)^{0.25}$	1.3683	127.1
Vertical plate heated or cooled	$h = C\left(\frac{\Delta T}{L}\right)^{0.25}$	1.3683	127.1

Source: Calculated from values reported in McAdams, W. H. 1954. *Heat Transmission Co.*, New York. 3rd ed. McGraw-Hill Book Co., New York.

Sieder and Tate (1936) introduced the use of the ratio of viscosity of fluid bulk and that at the wall as a multiplying factor to account for the difference in the heat transfer coefficient between a fluid and a heated or a cooled surface.

Heat Transfer to Non-Newtonian Fluids in Laminar Flow. Metzner et al. (1957) used Leveque's solution to heat transfer to a moving fluid in laminar flow as a basis for non-Newtonian flow heat transfer. They defined the delta function, a factor $\Delta^{0.33}$, as the ratio Nu_n/Nu (non-Newtonian Nusselt number/Newtonian Nusselt number).

Using the delta function as a multiplying factor for the Newtonian Nusselt number, equation 74 can be used for non-Newtonian fluids as follows:

$$Nu = 1.75\,\Delta^{0.33}\,Gz^{0.33}\left(\frac{\mu}{\mu_w}\right)^{0.33} \tag{77}$$

When the flow behavior index is greater than 0.4 for all values of Gz, $\Delta^{0.33} = [(3n + 1)/4n]^{0.33} = \delta^{0.33}$ When $n < 0.4$, $\Delta^{0.33}$ is determined as follows:

$$\Delta^{0.33} = -0.24n + 1.18 \qquad Gz = 5,\ 0 < n < 0.4 \tag{78}$$

$$\Delta^{0.33} = -0.60n + 1.30 \qquad Gz = 10,\ 0 < n < 0.4 \tag{79}$$

$$\Delta^{0.33} = -0.72n + 1.40 \qquad Gz = 15,\ 0 < n < 0.4 \tag{80}$$

$$\Delta^{0.33} = -0.35n + 1.57 \qquad Gz = 25,\ 0 < n < 0.4\ldots \tag{81}$$

Equation 81 can also be used when $Gz > 25$ $0.1 < n < 0.4$

For $Gz > 25$ and $n < 0.1$, the relationship between $\Delta^{0.33}$ and n is nonlinear and the reader is referred to Metzner et al.'s (1957) paper for determination of the delta function. Most food fluids have $n > 0.1$; therefore, the equations for the delta function given above should be adequate to cover most problems encountered in food processing.

Adapting Equations for Heat Transfer Coefficients to Non-Newtonian Fluids. Metzner et al.'s adaptation of correlation equations for heat transfer coefficients derived for Newtonian fluids to non-Newtonian fluids allows the use of practically any of the extensive correlations derived for Newtonian fluids for systems involving non-Newtonian fluids. For non-Newtonian fluids, the viscosity is not constant in the radial direction of the pipe. It is possible, however, to utilize the equations derived for Newtonian fluids by using an equivalent viscosity for the non-Newtonian fluid that causes the same pressure drop at the flow rate under consideration. A similar approach was used by Metzner et al. (1957) for heat transfer to non-Newtonian fluids in turbulent flow.

For a non-Newtonian fluid that follows the power law equation:

$$\tau = K(\gamma)^n$$

where τ = the shear stress, γ = the shear rate, K = the consistency index, and n = flow behavior index.

The Reynolds number at an average rate of flow V is:

$$Re = \frac{8(V)^{2-n}R^n\rho}{K(3 + 1/n)^n}$$

The viscosity to be used in the Prandtl number and the Sieder-Tate viscosity correction term for the fluid bulk are determined as follows:

For a Newtonian fluid:

$$Re = \frac{DV\rho}{\mu}$$

$$\mu = \frac{DV\rho}{Re}$$

Substituting the Reynolds number of the non-Newtonian fluid:

$$\begin{aligned}\mu &= \frac{DV\rho}{8(V)^{2-n}R^n\rho/[K(3 + 1/n)^n]} \\ &= K\left(\frac{3n + 1}{n}\right)^n\left[\frac{2RV\rho}{8V^{2-n}R^n\rho}\right] \qquad (82) \\ &= \frac{K}{4}\left(\frac{3n + 1}{n}\right)^n (R)^{1-n}(V)^{n-1}\end{aligned}$$

At the wall, the viscosity μ_w used in the Sieder-Tate viscosity correction term is determined from the apparent viscosity (equation 6, Chapter 6) as follows:

$$\mu_w = K(\gamma_w)^{n-1}$$

The shear rate at the wall, γ_w, is calculated using the Rabinowitsch-Mooney equation (Chapter 6, equation 18):

$$\gamma_w = \frac{8V}{D}\left[\frac{3n+1}{4n}\right] = \frac{2V}{D}\left[\frac{3n+1}{n}\right]$$

$$\mu_w = K\left(\frac{2V}{D}\right)^{n-1}\left[\frac{3n+1}{n}\right]^{n-1} \qquad (83)$$

Equations 82 and 83 for the viscosity terms can be used in correlation equations for heat transfer coefficients to Newtonian fluids, adapting these equations for non-Newtonian fluids.

As an example, equation 74 will be used to determine an equivalent equation for non-Newtonian fluids. Since Leveque's derivation of equation 74 was based on the fluid velocity profile at the boundary layer near the surface, the viscosity term in the Reynolds number should be the apparent viscosity at the shear rate existing at the surface. The viscosity term in the Prandtl number is based on viscous dissipation and conduction in the bulk fluid and should be the apparent viscosity based on the average velocity. Substituting equation 83 for the Reynolds number μ and equation 82 for the Prandtl number μ in equation 74:

$$Nu = 1.615\left[\frac{DV\rho}{K[2V/D]^{n-1}[3n+1/n]^{n-1}} \frac{C_p[K/4[3n+1/n]^n[D/2]^{1-n}V^{n-1}]}{k}\frac{D}{L}\right]^{0.33}\left[\frac{\mu}{\mu_w}\right]^{0.14}$$

Simplifying:

$$Nu = 1.615\left[\frac{3n+1}{4n}\right]^{0.33}\left[\frac{D^2V\rho C_p}{kL}\right]^{0.33}\left[\frac{\mu}{\mu_w}\right]^{0.14}$$

Substituting $\dot{m} = \left[\pi D^2 \frac{V}{4}\right][\rho]; \quad [D^2V\rho] = \left(\frac{4}{\pi}\right)\dot{m}$

$$Nu = 1.615\left[\frac{4}{\pi}\right]^{0.33}\left[\frac{\dot{m}C_p}{kL}\right]^{0.33}\left[\frac{3n+1}{4n}\right]^{0.33}\left[\frac{\mu}{\mu_w}\right]^{0.14}$$

This expression is the same as equation 77.

EXAMPLE 1: Calculate the rate of heat loss from a 1.524-m inside diameter horizontal retort 9.144 m long. Steam at 121°C is inside the retort. Ambient air is at 25°C. The retort is made out of steel ($k = 42$ W/m · K) and has a wall thickness of 0.635 cm.

Solution: The heat transfer coefficient from the outside surface to ambient air controls the rate of heat transfer; therefore, variations in the steam side heat transfer coefficient have little effect on the answer. Assume that the steam side heat transfer coefficient is 6000 W/(m^2 · K).

The system is a horizontal cylinder with a vertical plate at each end. The outside heat transfer coefficient can be calculated using the appropriate equations from Table 7.3.

For a horizontal cylinder to air:

$$h = 1.3196\left[\frac{\Delta T}{D_0}\right]^{0.25}$$

For a vertical plate to air:

$$h = 1.3683\left[\frac{\Delta T}{L}\right]^{0.25}$$

ΔT is evaluated from the outside wall temperature. The problem can be solved, using a trial-and-error procedure, by first assuming a heat transfer coefficient, calculating a wall temperature, and recalculating the heat transfer coefficient until the assumed values and calculated values converge.

The magnitude of the heat transfer coefficient in free convection to air is on the order of 5 W/m^2 · K. Equation 62 can be used to calculate an overall heat transfer coefficient for the cylindrical surface.

$$\frac{1}{U_i} = \frac{r_i}{r_0 h_0} + \frac{r_i \ln r_0/r_i}{k} + \frac{1}{h_i}$$

$$= \frac{0.762}{0.76835(5)} + \frac{0.762 \ln (0.76835/0.762)}{42} + \frac{1}{6000}$$

$$= 0.1984 + 0.00015 + 0.000167$$

$$U_i = 5.03 \text{ W/(m}^2 \cdot \text{K)}$$

$$U_i A_i \Delta T = h_0 A_0 \Delta T_w$$

$$U_i(2\pi r_i L)\Delta T = h_0(2\pi r_0 L)\Delta T_w$$

$$U_i r_i \Delta T = h_O r_O \Delta T_w$$

$$\Delta T_w = \frac{U_i r_i \Delta T}{h_O r_O} = \frac{5.03(0.762)(121 - 25)}{5(0.76835)}$$

$$= 95.84°C$$

$$h_O = 1.3196\left(\frac{\Delta T_w}{D_O}\right)^{0.25}$$

$$= 1.3196\left(\frac{95.82}{1.5367}\right)^{0.25}$$

$$= 3.71 \text{ W}/(\text{m}^2 \cdot \text{K})$$

The calculated value is less than the assumed h_O. Use this calculated value to recalculate U_i. All the terms in the expression for $1/U_i$ previously used are the same except for the first term. Assume that $h_O = 3.71 \text{ W}/(\text{m}^2 \cdot 2)$.

$$\frac{1}{U_i} = \frac{0.762}{0.76835(3.71)} + 0.00015 + 0.000167$$

$$= 3.7365 \text{ W}/(\text{m}^2 \cdot \text{K})$$

$$\Delta T_w = \frac{3.73(0.762)(121 - 25)}{3.71(0.76835)} = 95.89$$

$$h_O = 1.3196\left(\frac{95.89}{1.5367}\right)^{0.25}$$

$$= 3.71 \text{ W}/(\text{m}^2 \cdot \text{K})$$

Since the assumed and calculated values are the same, the correct h_O must be $h_O = 3.71 \text{ W}/(\text{m}^2 \cdot \text{K})$ and:

$$q = U_i A_i \Delta T = 3.7365(\pi)(1.524)(9.144)(121 - 25)$$

$$= 15{,}704 \text{ W}$$

For the ends, examination of the equation for the heat transfer coefficient reveal that h with vertical plates is approximately 4% higher than h with horizontal cylinders for the same ΔT if L is approximately the same as D. Assume that $h = 3.77(1.04) = 3.86 \text{ W}/(\text{m}^2 \cdot \text{K})$. Since the end is a vertical flat plate, U can be calculated using equation 59.

$$\frac{1}{U} = \frac{1}{3.86} + \frac{0.00635}{42} + \frac{1}{6000}$$

$$U = 3.855 \text{ W}/(\text{m}^2 \cdot \text{K})$$

$$UA\Delta T = h_0 A \Delta T_w$$

$$\Delta T_w = \frac{U\Delta T}{h_0} = \frac{3.855(121 - 25)}{3.86} = 95.88$$

$$h_0 = 1.3683\left[\frac{95.87}{1.5367}\right]^{0.25}$$

$$= 3.846 \text{ W}/(\text{m}^2 \cdot \text{K})$$

The assumed and calculated values are almost the same. Use $h_0 \doteq 3.846$ W/(m^2 · K):

$$\frac{1}{U} = \frac{1}{3.846} + 0.000156 + 0.000167$$

$$U = 3.835 \text{ W}/(\text{m}^2 \cdot \text{K})$$

Since there are two sides, the area will be $A = 2\pi R^2$.

$$q = UA\Delta T = (3.835)(\pi)(1.5367)^2(2)(121 - 25)$$

$$= 5463 \text{ W}$$

Total heat loss = 15,704 + 5463 = 21,167 W.

EXAMPLE 2: Calculate the overall heat transfer coefficient for applesauce heated from 20° to 80°C in a stainless steel tube 5 m long with an inside diameter of 1.034 cm and a wall thickness of 2.77 mm. Steam at 120°C is outside the tube. Assume a steam side heat transfer coefficient of 6000 W/(m^2 · K). The rate of flow is 0.1 m/s. Applesauce has a density of 995 kg/m^3, and this density is assumed to be constant with temperature. The value for n is 0.34, and the values for K are 11.6 at 30°C and 9.0 at 82°C in Pa · s units. Assume that K changes with temperature according to an Arrhenius-type relationship as follows:

$$\log K = A + \frac{B}{T}$$

where A and B are constants and T is the absolute temperature. The thermal conductivity may be assumed to be constant with temperature at 0.606 W/(m

· K). The specific heat is 3817 J/(kg · K). The thermal conductivity of the tube wall is 17.3 W/(m · K).

First, determine the temperature dependence of K. B is the slope of a plot of log K against $1/T$.

$$B = \frac{\log 11.6 - \log 9.0}{(1/303 - 1/355)} = \frac{1.064458 - 0.954243}{0.0033 - 0.00282}$$

$$= 227.99$$

$$= \log 11.6 - \frac{227.99}{303} = 1.064458 - 0.752433$$

$$= 0.3120$$

$$\log K = 0.3143 + \frac{227.99}{T}$$

The arithmetic mean temperature for the fluid is $(20 + 80)/2 = 50°C$. At this temperature, $K = \log^{-1}(0.3143 + 227.3/323) = 10.42$.

$$G = \frac{0.1 \text{ m}}{\text{s}} \frac{1}{\pi(0.00517)^2 \text{ m}^2} = 1190.88 \text{ kg}/(\text{s} \cdot \text{m}^2)$$

The equivalent Newtonian viscosity is calculated using equation 82.

$$\mu = \frac{10.42}{4}\left(\frac{3(0.34) + 1}{0.34}\right)(0.00517)^{1-0.34}(0.1)^{0.34-1}$$

$$= \frac{10.42}{4}(0.8328)(0.030967)(4.5709)$$

$$= 0.676 \text{ Pa} \cdot \text{s}$$

The Reynolds number is:

$$Re = \frac{0.01034(0.1)(955)}{0.676} = 1.521; \qquad \text{flow is laminar}$$

Equation 77 must be used, since $n < 0.4$ and the delta function deviates from $\delta = (3n + 1)/4n$.

Solving for the Graetz number:

$$m = \pi\left(\frac{D^2}{4}\right)(V)(\rho) = \pi(0.00517)^2(0.1)(995) = 0.008355$$

$$Gz = \frac{0.008355(3817)}{[(0.606)(5)]} = 10.52$$

Using equation 79 for $n < 0.4$ and $Gz = 10$:

$$\Delta^{0.33} = -0.6(0.34) + 1.096 = 1.096$$

The Nusselt number is:

$$Nu = 1.75(1.096)(10.525)^{0.33}\left(\frac{\mu}{\mu_w}\right)^{0.14}$$

$$= 4.170\left(\frac{\mu}{\mu_w}\right)^{0.14} = \frac{hD}{k}$$

Solving for h:

$$h = \frac{4.17[\mu/\mu_w]^{0.14}(0.606)}{0.01034} = 244.39\left[\frac{\mu}{\mu_w}\right]^{0.14}$$

Assume that most of the temperature drop occurs across the fluid side resistance. Assume that $T_w = 116°C$.

K at 116°C = $\log^{-1}(0.3126 + 227.99/389) = 7.91$ Pa · s

Using equation 83:

$$\mu_w = 7.91\left[\frac{(2)(0.1)}{0.01034}\right]^{0.34-1}\left[\frac{3(0.34)+1}{0.34}\right]^{0.34-1}$$

$$= 7.91(0.141548)(0.30849) = 0.3454$$

$$h_i = 244.39\left[\frac{0.676}{0.3454}\right]^{0.14} = 268.48 \text{ W/(m}^2 \cdot \text{K)}$$

Using equation 62 for the overall heat transfer coefficient:

$$\frac{1}{U_i} = \frac{r_i}{r_0 h_0} + \frac{r_i \ln r_0/r_i}{k} + \frac{1}{h_i}$$

$$r_i = 0.00517; \qquad r_0 = 0.00517 + 0.00277 = 0.00794$$

$$\frac{1}{U_i} = \frac{0.00517}{0.00794(6000)} + \frac{0.00517 \ln(0.00794/0.00517)}{17.3} + \frac{1}{275.3}$$

$$= 0.000109 + 0.000128 + 0.003725$$

$$= 0.003962; \qquad U_i = 252.42$$

Check if wall temperature is close to assumed temperature.

$$U_i A_i \Delta T = h A_i \Delta T_w$$

$$\Delta T_w = \frac{U_i A_i \Delta T}{h_i A_i} = \frac{252.42}{268.48}(120 - 50) = 65.8$$

$$T_w = 50 + 65.8 = 115.8$$

This is close to the assumed value of 116°C; therefore:

$$h_i = 268.48\ \text{W}/(\text{m}^2 \cdot \text{K}) \qquad \text{and} \qquad U_i = 252.42\ \text{W}/(\text{m}^2 \cdot \text{K})$$

EXAMPLE 3: The applesauce in Example 2 is being cooled at the same rate from 80° to 120°C in a double-pipe heat exchanger, with water flowing in the annular space outside the tube at a velocity of 0.5 m/s. The inside diameter of the outer jacket is 0.04588 m. Water enters at 8°C and leaves at 16°C. Calculate the food side and the water side heat transfer coefficients and the overall heat transfer coefficient.

Solution: First, calculate the water side heat transfer coefficient. At a mean temperature of 12°C water has the following properties: $k = 0.58$ W/(m · K); $\mu = 1.256 \times 10^{-3}$ Pa · s; $\rho = 1000$ kg/m^3; $C_p = 4186$ J/(kg · K); and $D_0 = 0.01588$ m. For an annulus, the characteristic length for the Reynolds number is 4(cross-section area/wetted perimeter).

$$D = \frac{(\pi)(d_2^2 - d_1^2)}{(\pi)(d_1 + d_2)}$$

$$= d_2 - d_1 = 0.03$$

$$Re = \frac{D_0 \bar{V} \rho}{\mu} = \frac{(0.030)(0.5)(1000)}{1.256 \times 10^{-3}} = 11{,}942$$

$$Pr = \frac{C_p \mu}{k} = \frac{4186(1.256 \times 10^{-3})}{0.58} = 9.06$$

For water in turbulent flow in an annulus, Monod and Allen's equation from Appendix Table A.12 with the viscosity factor is:

$$\frac{hD_0}{k} = 0.02 Re^{0.8} Pr^{0.33} \left[\frac{d_2}{d_1}\right]^{0.53} \left[\frac{\mu}{\mu_w}\right]^{0.14}$$

$$= 0.02(11942)^{0.8}(9.06)^{0.33}\left[\frac{0.04588}{0.01588}\right]^{0.53}\left[\frac{\mu}{\mu_w}\right]^{0.14}$$

$$h_0 = \frac{0.58}{0.03}(0.02)(1827)(2.069)(1.754)\left[\frac{\mu}{\mu_w}\right]^{0.14}$$

$$= 2565\left[\frac{\mu}{\mu_w}\right]^{0.14}$$

Assume that T_w = 2°C higher than the arithmetic mean fluid temperature. T_w = 15°C; $\mu_w = 1.215 \times 10^{-3}$ Pa · s.

$$h_0 = 2565\left[\frac{1.256}{1.215}\right]^{0.14} = 2577 \text{ W}/(\text{m}^2 \cdot \text{K})$$

Now calculate the food fluid side heat transfer coefficient. Since the fluid arithmetic mean temperature is the same as in Example 2:

$$h_i = 256.7\left[\frac{\mu}{\mu_w}\right]^{0.14}$$

The overall ΔT between the water and the applesauce, based on the mean temperature, is 50 − 12 = 38°C. T_w = 15°C.

$$K_w = \log^{-1}\left(\frac{0.3126 + 227.99}{288}\right) = 12.69 \text{ Pa} \cdot \text{s}$$

For Example 2: μ = 0.676 Pa · s. Using equation 83:

$$\mu_w = 12.69\left[\frac{2(0.1)}{0.01034}\right]^{0.34-1}\left[\frac{3(0.34)+1}{0.34}\right]^{0.34-1}$$

$$= 0.554$$

$$h_i = 256.7\left[\frac{0.676}{0.554}\right]^{0.14} = 263.95 \text{ W}/(\text{m}^2 \cdot \text{K})$$

Calculating for U_i using equation 62:

$$\frac{1}{U_i} = \frac{0.00517}{0.00794(2577)} + \frac{0.00517 \ln(0.00794/0.00517)}{17.3} + \frac{1}{263.95}$$

$$= 0.000253 + 0.000128 + 0.003789 = 0.004170$$

$$U_i = 239.83 \text{ W}/(\text{m}^2 \cdot \text{K})$$

Solving for the wall temperature:

$$\Delta T_{wi} = \frac{U_i \Delta T}{h_i} = \frac{(239.83)(38)}{263.95} = 34.5°\text{C}$$

$$T_w = 50 - 34.5 = 15.5°\text{C}$$

The small difference in the calculated and assumed wall temperatures will not alter the calculated h_i and h_0 values significantly; therefore, a second iteration is not necessary. The values for the heat transfer coefficients are:

Water side: $h_0 = 2577 \text{ W}/(\text{m}^2 \cdot \text{K})$

Food fluid side: $h_i = 263.95 \text{ W}/(\text{m}^2 \cdot \text{K})$

Overall heat transfer coefficient: $U_i = 239.83 \text{ W}/(\text{m}^2 \cdot \text{K})$

UNSTEADY-STATE HEAT TRANSFER

Unsteady-state heat transfer occurs when food is heated or cooled under conditions in which the temperature at any point within the food or the temperature of the heat transfer medium changes with time. In this section, procedures for calculating the temperature distribution within a solid during a heating or cooling process will be discussed.

Heating of Solids Having Infinite Thermal Conductivity. Solids with very high thermal conductivity have a uniform temperature within the solid. A heat balance between the rate of heat transfer and the increase in the sensible heat content of the solid gives:

$$mC_p \frac{dT}{dt} = hA(T_m - T)$$

Solving for T and using the initial condition, $T = T_0$ at $t = 0$:

$$\ln\left(\frac{T_m - T}{T_m - T_0}\right) = -\frac{hA}{mC_p}t \tag{84}$$

Equation 84 shows that a semilogarithmic plot of $\theta = (T_m - T)/(T_m - T_0)$ against time will be linear, with a slope $-hA/mC_p$. If a solid is heated or cooled, the dimensionless temperature ratio, θ, plotted semilogarithmically against time, will show an initial lag before becoming linear. If the extent of the lag time is defined such that the intersection point of the linear portion of the heating curve with the ordinate is defined, it is possible to calculate point temperature changes in a solid for long heating times from an empirical value of the slope of the heating curve. This principle is utilized in thermal process calculations for foods in Chapter 8.

Fluids in a well-stirred vessel exchanging heat with another fluid which contact the vessel walls change in temperature as in equation 84. A well-mixed steam-jacketed kettle used for cooking foods is an example of a system where temperature change follows equation 84. For a well-mixed steam-jacketed kettle, U, the overall heat transfer coefficient between the steam and the fluid inside the kettle is used instead of the local heat transfer coefficient, h, used in deriving equation 84.

EXAMPLE 1: A steam-jacketed kettle consists of a hemispherical bottom having a diameter of 69 cm and a cylindrical side 30 cm high. The steam jacket of the kettle is over the hemispherical bottom only. The kettle is filled with a food product which has a density of 1008 kg/m^3 to a point 10 cm from the rim of the kettle. If the overall heat transfer coefficient between the steam and the food in the jacketed part of the kettle is 1000 W/(m^2 · K), and steam at 120°C is used for heating in the jacket, calculate the time for the food product to heat from 20° to 98°C. The specific heat of the food is 3100 J/(kg · K).

Solution: The surface area of a hemisphere is $(1/2)\pi d^2$; the volume is $0.5(1/6)\pi d^3$. Use equation 84 to solve for t. A = area of the hemisphere with $d = 0.69$ m. $A = 0.5(\pi)(d^2) = 0.7479\ \text{m}^2$.

$$\text{Volume} = 0.5(0.1666)(\pi)(0.69^3) + \pi[(0.69)(0.5)]^2(0.30 - 0.10)$$

$$= 0.08597 + 0.07478 = 0.16075\ \text{m}^3$$

$$= 0.16075\ \text{m}^3(1008\ \text{kg/m}^3) = 162.04\ \text{kg}$$

$$T_m = 120;\ T_0 = 20;\ U = 1000;\ C_p = 3100$$

$$\ln\left(\frac{120 - 98}{120 - 20}\right) = -\frac{1000(0.7479)}{162.04(3100)}t = -0.001489t$$

$$t = \frac{-\ln(0.22)}{0.001489} = 1016.9\ \text{s} = 16.95\ \text{min}$$

Solids with Finite Thermal Conductivity. When k is finite, a temperature distribution exists within the solid. Heat transfer follows Fourier's second law, which was derived in the section "Heat Transfer by Conduction" (equation 19) for a rectangular parallelopiped. For other geometries, the following differential equations represent the heat balance (Carslaw and Jaeger, 1959).

For cylinders with temperature symmetry in the circumferential direction:

$$\frac{\partial T}{\partial t} = \alpha\left(\frac{\partial^2 T}{\partial r^2} + \frac{\partial T}{r\partial r} + \frac{\partial^2 T}{\partial Z^2}\right) \quad (85)$$

For spheres with symmetrical temperature distribution:

$$\frac{\partial T}{\partial t} = \alpha\left[\frac{\partial^2 (rT)}{r\partial r^2} + \frac{1}{r^2 \sin\phi} + \frac{\partial}{\partial\phi}\left(\sin\phi\,\frac{\partial T}{\partial\phi}\right)\right] \quad (86)$$

These equations may be solved analytically by considering one-dimensional heat flow and obtaining a composite solution by a multiplicative superposition technique. Thus, the solution for a brick-shaped solid is the product of the solutions for three infinite slabs, and that for a finite cylinder is the product of the solution for an infinite cylinder and an infinite slab. Analytical solutions are not easily obtained for certain conditions in food processing, such as an interrupted process where a change in boundary conditions occurred at a midpoint in the process or when the boundary temperature is an undetermined function of time. Equations 19, 85, and 86 may be solved using a finite difference technique if analytical solutions are not suitable for the conditions specified. Techniques for solving partial differential equations are discussed and analytical solutions are given in Carslaw and Jaeger (1959).

The Semi-Infinite Slab with Constant Surface Temperature. A semi-infinite slab is defined as one with infinite width, length, and depth. Thus, heat is transferred in only one direction, from the surface toward the interior. This system is also referred to as a *thick solid*. Under certain conditions, such as very short times of heating, the surface of a solid and a point very close to the surface would have the temperature response of a semi-infinite solid to a sudden change in the surface conditions. Equation 19 may be solved using the boundary conditions; at time 0, the slab is initially at T_0 and the surface is suddenly raised to temperature T_s. The temperature T at any point x, measured from the surface, expressed as a dimensionless temperature ratio, θ, is:

$$\theta = erf\left[\frac{x}{(4\alpha t)^{0.5}}\right] \quad (87)$$

$\theta = (T_s - T)/(T_s - T_0)$ and *erf* is the error function. The error function is a well-studied mathematical function, and the values are found in standard mathematical tables. Values of the error function are given in Table 7.4.

The error function approaches 1.0 when the value of the argument is 3.6. In a solid, the point where *erf* = 1 is undisturbed by the heat supplied at the surface. Thus, a penetration depth may be calculated beyond which the thermal conditions are undisturbed. If this penetration depth is less than the half-thick-

Table 7.4. Values of the Error Function

x	erf(x)	x	erf(x)	x	erf(x)
0	0	0.70	0.677801	1.7	0.983790
0.05	0.056372	0.75	0.711156	1.8	0.989091
0.10	0.112463	0.80	0.742101	1.9	0.992790
0.15	0.157996	0.85	0.770668	2.0	0.995322
0.20	0.222703	0.90	0.796908	2.2	0.998137
0.25	0.276326	0.95	0.820891	2.4	0.999311
0.30	0.328627	1.00	0.842701	2.6	0.999764
0.35	0.37938	1.1	0.880205	2.8	0.999925
0.40	0.428392	1.2	0.910314	3.0	0.999978
0.45	0.47548	1.3	0.934003	3.2	0.999994
0.50	0.520500	1.4	0.952790	3.4	0.999998
0.55	0.563323	1.5	0.966105	3.6	1.00000
0.60	0.603856	1.6	0.976348	3.8	1.00000
0.65	0.64202			4.0	1.00000

Source: Kreyzig, E. 1963. Advanced Engineering Mathematics, John Wiley & Sons, NY.

ness of a finite slab, the temperature distribution near the surface may be approximated by equation 87. A finite body may exhibit a thick body response if the penetration depth is much less than the half-thickness. The thick body response to a change in surface temperature is the temperature distribution expressed by equation 87 from the surface ($x = 0$) to a point in the interior δ distant from the surface when $\delta = 3.8(4\alpha t)^{0.5}$ is much less than the half-thickness of a finite solid.

If the surface heat transfer coefficient is finite, the solution to equation 19, with the boundary condition $T = T_0$ at time zero, heat is transferred at the surface from a fluid at temperature T_m, with a heat transfer coefficient h:

$$\Theta = erf\left[\frac{x}{\sqrt{(4\alpha t)}}\right] + [e]^{hx/k + (h/k)^2\alpha t}\left[erfc\left[\frac{x}{\sqrt{(4\alpha t)}} + \frac{h}{k}\sqrt{(\alpha t)}\right]\right] \quad (88)$$

where $erfc[F(x)] = 1 - erf[F(x)]$.

Equation 88 becomes equation 87 when h is infinite, since the complementary error function has a value of zero when the argument of the function is very large. Surface conductance plays a role along with thermal conductivity, in determining if a body will exhibit a thick body response. Schneider (1973) defined a critical Fourier number when a body ceases to exhibit a thick body response:

$$Fo_{\text{critical}} = 0.00756\ Bi^{-0.3} + 0.02 \qquad \text{for } 0.001 \leq Bi \leq 1000$$

The Fourier number is $Fo = \alpha t/(L)^2$; L = thickness/2; Bi = Biot number = hL/k.

EXAMPLE: A beef carcass at 38°C is introduced into a cold room at 5°C. The thickness of the carcass is 20 cm. Calculate the temperature at a point 2 cm from the surface after 20 min. The density is 1042 kg/m^3, the thermal conductivity is 0.44 W/(m · K), the specific heat is 3558 J/(kg · K), and the surface heat transfer coefficient is 20 W/(m^2 · K).

Solution: For a short exposure time, the material may exhibit a thick body response. This condition is affirmed by calculating the Biot number and the Fourier number.

$$L = 20(0.5) = 10 \text{ cm} = 0.10 \text{ m}$$

$$Bi = \frac{hL}{k} = \frac{20(0.10)}{0.44} = 4.55$$

$$Fo_{\text{critical}} = 0.00756(4.55)^{-0.3} + 0.02 = 0.024$$

$$Fo = \frac{\alpha t}{L^2} = \frac{0.44}{1042(3558)} \frac{20(60)}{(0.10)^2} = 0.0142$$

The actual Fourier number is less than the critical value for the solid to cease exhibiting a thick body response; therefore, the error function solution can be used to calculate the temperature at the designated point. Using equation 88, $x = 0.02$ m from the surface.

$$\alpha = \frac{0.44}{(1042)(3558)} = 1.187 \times 10^{-7}$$

$$\frac{x}{(2)\sqrt{(\alpha t)}} = \frac{0.02}{(2)[(1.187 \times 10^{-7})(20)(60)]^{0.5}}$$

$$= 0.838$$

$$\frac{hx}{k} = \frac{20(0.02)}{0.44} = 0.909$$

$$\left(\frac{h}{k}\right)^2 \alpha t = \left(\frac{20}{0.44}\right)(1.187 \times 10^{-7})(20)(60) = 0.294$$

$$\left(\frac{h}{k}\right)\sqrt{(\alpha t)} = \left(\frac{20}{0.44}\right)[(1.187 \times 10^{-7})(20)(60)]^{0.5}$$

$$= 0.542$$

$$\Theta = erf(0.838) + [e]^{0.909 + 0.294}[erfc(0.838) + 0.5421]$$

$$= 0.7663812 + (3.3301)(0.059674) = 0.9335$$

$$\text{For cooling: } \Theta = \frac{T - T_m}{T_0 - T_m}$$

$$T = T_m + \Theta(T_0 - T_m) = 5 + (0.9335)(38 - 5) = 35.8°\text{C}$$

The Infinite Slab. This is a slab with thickness $2L$ extending to infinity at both ends.

When h is infinite:

$$\Theta = 2 \sum_{n=0}^{\infty} \left[\frac{(-1)^n}{(n + 0.5)\pi} [e]^{-(n+0.5)^2(\pi^2 \alpha t/L^2)} \right] \left[\cos \left[\frac{(n + 0.5)\pi x}{L} \right] \right] \quad (89)$$

When h is finite:

$$\Theta = 2 \sum_{n=1}^{\infty} [e]^{-\delta_n^2 \alpha t/L^2} \left[\frac{\sin(\delta_n)\cos(\delta_n x/L)}{\delta_n + \sin(\delta_n)\cos(\delta_n)} \right] \quad (90)$$

δ_n are the positive roots of the transcendental equation:

$$\delta_n \tan(\delta_n) = \frac{hL}{k}$$

The center (Θ_c) and surface (Θ_s) temperatures for an infinite solid with surface heat transfer are obtained by setting $x = 0$ and $x = L$ in equation 90:

$$\Theta_c = 2 \sum_{n=1}^{\infty} \frac{\sin(\delta_n)[e]^{-(\delta_n/L)^2 \alpha t}}{\delta_n + \sin(\delta_n)\cos(\delta_n)} \quad (91)$$

$$\Theta_s = 2 \sum_{n=1}^{\infty} \frac{\sin(\delta_n)\cos(\delta_n)[e]^{-(\delta_n)^2 \alpha t}}{\delta_n + \sin(\delta_n)\cos(\delta_n)} \quad (92)$$

Temperature Distribution for a Brick-Shaped Solid. The solution to the differential equation for a brick-shaped solid is the product of the solution for infinite slabs of dimensions L_1, L_2, and L_3. The roots of the transcendental equation are different for each dimension of the brick and are designated δ_{n1}, δ_{n2}, and δ_{n3}, respectively, for sides with half-thickness L_1, L_2, and L_3.

Let F = Fourier number, $\alpha t/L_2$.

$$F(x\delta_{ni}) = \frac{\sin(\delta_{ni})\cos(\delta_{ni} x/L)}{\delta_{ni} + \sin(\delta_{ni})\cos(\delta_{ni})}$$

$$\Theta = 8 \sum_{n=1}^{\infty} F(x\delta_{n1})F(x\delta_{n2})F(x\delta_{n2})[e]^{-\Sigma(\delta_{ni})^2 Fo_i} \quad (93)$$

Of interest to food scientists and engineers are center temperature and surface temperature. Let:

$$F(c\delta_{ni}) = \frac{\sin(\delta_{ni})}{\delta_{ni} + \sin(\delta_{ni})\cos(\delta_{ni})}$$

$$F(s\delta_{ni}) = \frac{\sin(\delta_{ni})\cos(\delta_{ni})}{\delta_{ni} + \sin(\delta_{ni})\cos(\delta_{ni})}$$

The center or surface temperature may be calculated using $F(c\delta_{ni})$ or $F(s\delta_{ni})$ in place of $F(x\delta_{ni})$ in equation 93. Appendix Table A.13 lists a computer program in BASIC for calculating the surface and center temperature of a brick-shaped solid. An example of the output of the program is shown in Fig. 7.13. The surface temperature is affected by the thickness of the solid and does not assume the heating medium temperature immediately after the start of the heating process.

Use of Heisler and Gurney-Lurie Charts. Before the age of personal computers, calculations involving the transient temperature response of solids were very laborious. Solutions to the partial differential equations were plotted and arranged in a form which made it easy to obtain solutions. Two of the transient

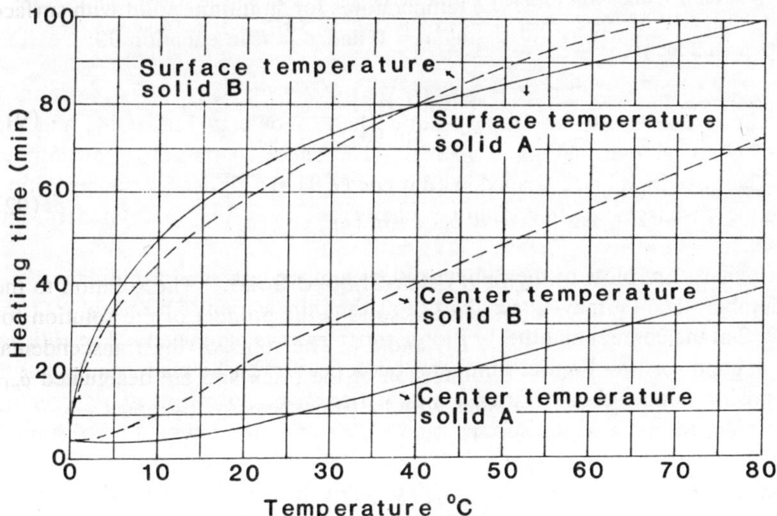

Fig. 7.13. Temperature at the surface and at the geometric center of two brick-shaped solids heated in an oven at 177°C from an initial temperature of 4°C. Solid A dimensions: 20.32 × 10.16 × 5.08 cm. Solid B dimensions: 30.48 cm × 15.24 × 5.08 cm. Parameters: σ = 1085 kg/m^3; C_p = 4100 J/(kg · K); k = 0.455 W/(m · K); h = 6.5 W/(m^2 · K).

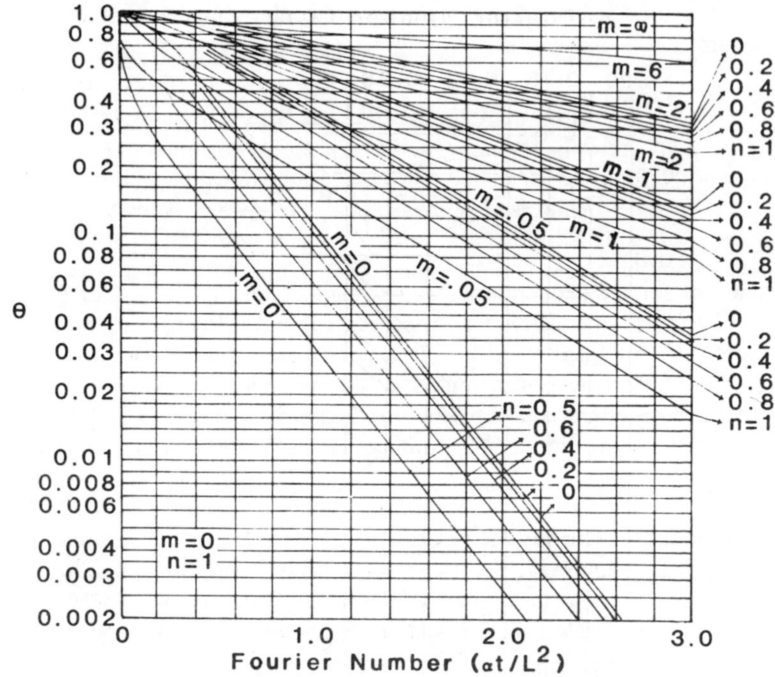

Fig. 7.14. Gurnie-Lurie chart for the temperature response of an infinite slab. (Source: Adapted from McAdams, W. H. 1954. *Heat Transmission*, 3rd. ed. McGraw-Hill. Used with permission of McGraw-Hill Inc.)

temperature charts are the Gurney-Lurie chart, shown in Fig. 7.14, and the Heisler chart, shown in Figs. 7.15 and 7.16. These charts are for an infinite slab. Their use for a brick-shaped solid involves the multiplicative superposition technique discussed earlier. Similar charts are available for cylinders and spheres.

Both charts are plots of the dimensionless temperature ratio against the Fourier number. The Gurney-Lurie chart plots the dimensionless position ($x/L = n$; $n = 0$ at the center) and the 1/Biot number $= m$ as parameters. The Heisler chart is good for low Fourier numbers, but a chart must be made available for each position under consideration. The Heisler chart shown in Fig. 7.15 is the temperature response at the surface, and Fig. 7.16 is the temperature response at the center of an infinite slab.

Use of these charts involves calculating the Biot number and the Fourier number. The appropriate curve which corresponds to the Biot number and the position under consideration is then selected. The charts are used to obtain a value for the dimensionless temperature ratio, Θ. If the solid is brick-shaped, Fourier and Biot numbers are obtained for each direction, and the values of Θ

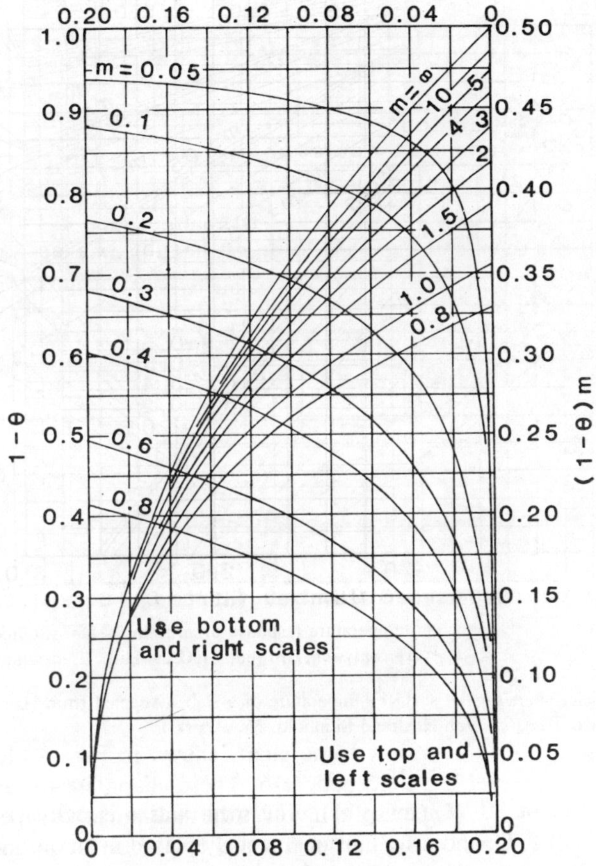

Fig. 7.15. Heisler chart for the center temperature of a slab. (Adapted from Hsu, S. T. 1963. *Engineering Heat Transfer*. Van Nostrand Reinhold, New York.)

obtained for each direction are multiplied to obtain the net temperature response from all three directions.

EXAMPLE 1: Calculate the temperature at the center of a piece of beef 3 cm thick, 12 cm wide and 20 cm long after 30 min of heating in an oven. The beef has a thermal conductivity of 0.45 W/(m · K), a density of 1008 kg/m^3, and a specific heat of 3225 J/(kg · K). Assume a surface heat trans-

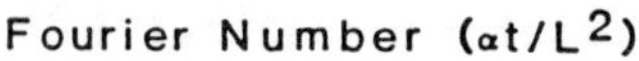

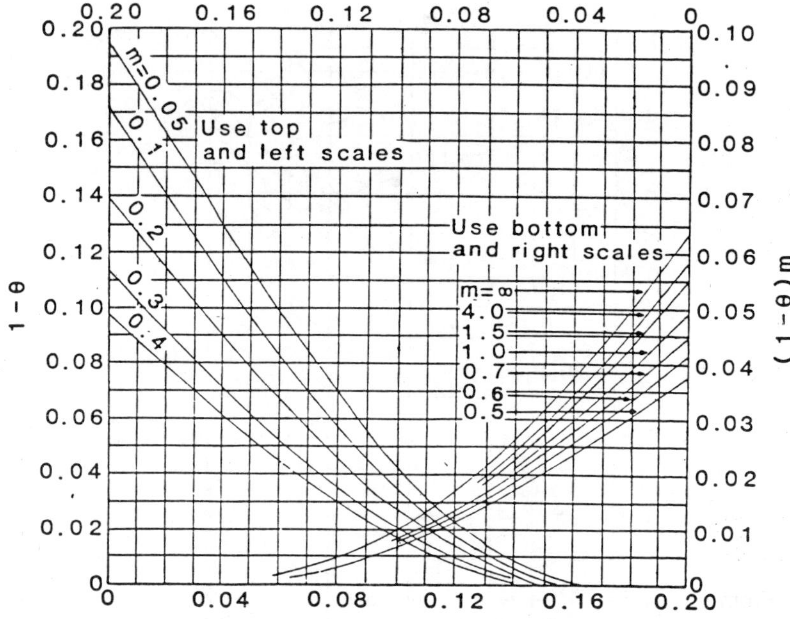

Fig. 7.16. Heisler chart for the surface temperature of a slab. (Adapted from Hsu, S. T. 1963. *Engineering Heat Transfer*. Van Nostrand Reinhold, New York.)

fer coefficient of 20 W/($m^2 \cdot$ K). The meat was originally at a uniform temperature of 5°C, and it was instantaneously placed in an oven maintained at 135°C. Neglect radiant heat transfer and assume no surface evaporation.

Solution: Consider each side separately. Consider the x direction as the thickness ($L = 0.5(0.03) = 0.015$ m), the y direction as the width ($L = 0.5(0.12) = 0.06$ m), and the z direction as the length ($L = 0.5(0.2) = 0.1$ m).

$$\alpha = \frac{k}{\rho C_p} = \frac{0.45}{(1008)(3225)} = 1.3843 \times 10^{-7}\ \mathrm{m^2/s}$$

$$Fo = \frac{\alpha t}{L^2}$$

$$Fo_x = \frac{1.3843 \times 10^{-7}(30)(60)}{(0.015)^2} = 1.107$$

$$Fo_y = \frac{1.3843 \times 10^{-7}(30)(60)}{(0.06)^2} = 0.0692$$

$$Fo_z = \frac{1.3843 \times 10^{-7}(30)(60)}{(0.1)^2} = 0.013$$

$$Bi = \frac{hL}{k}$$

$$Bi_x = \frac{20(0.015)}{0.45} = 0.667; \quad m = 1.5$$

$$Bi_y = \frac{20(0.06)}{0.45} = 2.667; \quad m = 0.374$$

$$Bi_z = \frac{20(0.1)}{0.45} = 4.444; \quad m = 0.225$$

At the center, $n = 0$

From Fig. 7.14, Θ for $m = 1.5$ is obtained by interpolation between the values for $m = 1$ and $m = 2$. From Fig. 7.14, $Fo_x = 1.107$, $n = 0$, $m = 1$, $\Theta = 0.52$.

$$m = 0; \quad m = 2; \quad \Theta = 0.7$$

$$\text{For } m = 1.5, \Theta = \frac{0.52 + (0.70 + 0.52)(1.5 - 1)}{1} = 0.61$$

The values of Θ for $Fo_y = 0.0692$ and Fo_z appear to be almost 1.0 from Fig. 7.14. To verify, use Fig. 7.16 to see that at $m = 0.4$ and $Fo = 0.07$, $1 - \Theta \equiv 0$, and at $m = 0.2$ and $Fo = 0.01$, $1 - \Theta = 0$. Thus, heating for this material occurs primarily from one dimension.

$$T = T_m - \Theta(T_m - T_0)$$

$$= 135 - 0.61(135 - 5) = 55.7°\text{C}$$

Calculating Surface Heat Transfer Coefficients from Experimental Heating Curves. If heating proceeds for a long time, the series represented by equation 93 converges rapidly and the first term in the series is adequate. Equation 93 then becomes:

$$\Theta_c = 2F(c\delta_1)F(c\delta_2)F(c\delta_3)[e]^{-Fo_1\delta_1^2 - Fo_2\delta_2^2 - Fo_3\delta_3^2}$$

Let $F(c\delta) = 2F(c\delta_1)F(d\delta_2)F(c\delta_3)$. Taking the logarithms of both sides of the equation:

$$(\Theta_c) = \log\left[F(c\delta)\right] - \left[\alpha t \log(e)\right]\left[\frac{\delta_1^2}{L_1^2} + \frac{\delta_2^2}{L_2^2} + \frac{\delta_3^2}{L_3^2}\right] \quad (94)$$

Equation 94 shows that a plot of log (Θ_c) against t is linear, and the surface heat transfer coefficient can be determined from the slope if α is known.

$$\text{Slope} = -\left[\alpha \log(e)\right]\left[\frac{\delta_1^2}{L_1^2} + \frac{\delta_2^2}{L_2^2} + \frac{\delta_3^2}{L_3^2}\right] \quad (95)$$

A computer program in BASIC shown in Appendix Table A.14 can be used to determine the average heat transfer coefficient from the center temperature heating curve of a brick-shaped solid.

Similar approaches may be used for cylindrical solids.

Freezing Rates. The temperature distribution in solids exposed to a heat exchange medium at temperatures below the solid's freezing point is complicated by the change in phase and the unique properties of the frozen and unfrozen zones in the solid. In addition, the ice front advances toward the interior from the surface, and at the interface between the two zones a tremendous heat sink exists in the form of the heat of fusion of water. A number of approaches have been used to model the freezing process mathematically but the most successful one in terms of simplicity and accuracy is the refinement by Cleland and Earle (1984) of the empirical equation originally developed by Plank (1913).

Plank's original equation was:

$$t_f = \frac{\lambda}{T_f - T_a}\left[\frac{PD}{h} + \frac{RD^2}{k}\right] \quad (96)$$

where t_f = freezing time for a solid with a freezing point T_f and a thermal thickness D which is the full thickness in the case of a slab or the diameter in the case of a sphere. P and R are shape constants ($P = 1/6$ for a sphere, $1/4$ for a cylinder, and $1/2$ for a slab; $R = 1/24$ for a sphere, $1/16$ for a cylinder, and $1/8$ for a slab). λ = latent heat of fusion per unit volume, h = heat transfer coefficient, and k = thermal conductivity of the frozen solid. Shape factors P and R have been developed for a brick-shaped solid and presented as a graph based on the dimensions of the brick. This graph can be seen in Charm (1971).

Plank's equation has been refined by Cleland and Earle (1984) to account for the facts that not all water freezes at the freezing point and that the freezing process may proceed to a specific final product temperature. The effect of an initial temperature different from the freezing point has also been included in the analysis. Cleland and Earle's (1984) equation is:

$$t_f = \frac{\Delta H_{10}}{(T_f - T_a)(EHTD)}\left[P\frac{D}{h} + R\frac{D^2}{k_s}\right]\left[1 - \frac{1.65\,\mathrm{STE}}{k_s}\ln\left(\frac{T_{\mathrm{fin}} - T_a}{-10 - T_a}\right)\right] \tag{97}$$

ΔH_{10} = enthalpy change to go from T_f to $-10°\mathrm{C}$ in $\mathrm{J/m^3}$

T_a = freezing medium temperature

T_{fin} = final temperature

EHTD = equivalent heat transfer dimensionality, defined as the ratio of the time to freeze a slab of half-thickness *D* to the time required to freeze the solid having the same *D*. *EHTD* = 1 for a slab with large width and length such that heat transfer is effectively only in one direction and 3 for a sphere. Bricks and cylinders have an *EHTD* between 1 and 3

h = surface heat transfer coefficient

k_s = thermal conductivity of the frozen solid

P, *R* = parameters which are functions of the Stephan number (*STE*) and the Plank number (*PK*).

$$STE = C_s\frac{(T_f - T_a)}{\Delta H_{10}}; \qquad PK = C_1\frac{(T_i - T_f)}{\Delta H_{10}}$$

C_s = volumetric heat capacity of frozen material, $\mathrm{J/(m^3 \cdot K)}$

C_1 = volumetric heat capacity of unfrozen material, $\mathrm{J/(m^3 \cdot K)}$

T_i = initial temperature of the solid

$$P = 0.5[1.026 + 0.5808PK + STE(0.2296PK + 0.105)]$$

$$R = 0.125[1.202 + STE(3.41PK + 0.7336)]$$

The major problem in using equation 97 is the determination of *EHTD*. *EHTD* can be determined experimentally by freezing a slab and the solid of interest and taking the ratio of the freezing times. *EHTD* considers the total dimensionality of the material instead of just one dimension.

EXAMPLE: Calculate the freezing time for blueberries in a belt freezer where the cooling air is at $-35°\mathrm{C}$. The blueberries have a diameter of 0.8 cm and will be frozen from an initial temperature of 15°C to a final temperature of −20°C. They contain 10% soluble solids, 1% insoluble solids, and 89% water. Use Choi and Okos' (1987) correlation for determining the thermal conductivity, Chang and Tao's (1981) correlation for determining the freezing point and enthalpy change below the freezing point, and Siebel's equation for determining the specific heat above and below freezing. The blueberries have a density of 1070 kg/m^3 unfrozen and 1050 kg/m^3 frozen. The heat transfer coefficient is 120 W/(m^3 · K)

Solution: For the specific heat above freezing, use equation 11, Chapter 5:

$$C_1' = 837.36(0.11) + 4186.8(0.89) = 3818.3 \text{ J/(kg} \cdot \text{K)}$$

Converting to J/(m^3 · K):

$$C_1 = \frac{3818 \text{ J}}{\text{kg K}} \frac{1070 \text{ kg}}{\text{m}^3} = 4.0856 \times 10^6 \text{ J/kg}$$

For the specific heat below freezing, use equation 13, Chapter 5:

$$C_s' = 837.36(0.11) + 2093.4(0.89) = 1955.63 \text{ J/(kg} \cdot \text{K)}$$

Converting to J/(m^3 · K):

$$C_s = \frac{1955.63 \text{ J}}{\text{kg} \cdot \text{K}} \frac{1050 \text{ kg}}{\text{m}^3} = 2.053 \times 10^6 \text{ J/(m}^3 \cdot \text{K)}$$

Use Chang and Tao's correlation for the enthalpy change between freezing point and −10°C (Chapter 5, the section "Enthalpy Change with a Change in Phase.") From equation 20, Chapter 5:

$$T_f = 287.56 - 49.19(0.89) + 37.07(0.89)^2 = 273 \text{ K}$$

Solving for the enthalpy at the freezing point: Equation 22, Chapter 5: $H_f = 9292.46 + 405{,}096(0.89)$

$$H_f = 370327.9 \text{ J/kg}$$

$$T = -10°\text{C} = 263 \text{ K}$$

$$T_r = \frac{(263 - 227.6)}{(273 - 227.6)} = 0.776$$

$$H = H_f\left[aT_r + (1 - a)(T_r)^b\right]$$

Solving for a: equation 17, Chapter 5.

$$a = 0.262 + 0.0498(0.89 - 0.73) - 3.465(0.89 - 0.73)^2$$
$$= 0.362 + 0.007968 - 0.088704 = 0.281$$

Solving for b: equation 18, Chapter 5.

$$b = 27.2 - 129.04(0.89 - 0.23) - 481.46(0.89 - 0.23)^2$$
$$= 27.2 - 6.6151 - 1.265276 = 19.3196$$

Solving for H at $-10°C$, $T_r = 0.776$: equation 23, Chapter 5.

$$H = H_f\left[(0.281)(0.776) + (1 - 0.281)(0.776)^{19.3196}\right]$$
$$= 0.2234 H_f; \qquad \Delta H_{10} = H_f(1 - 0.2234) = 0.7766(370327.9)$$
$$\Delta H_{10} = 2.876 \times 10^5 \text{ J/kg}$$

Choi and Okos' correlation, using the BASIC program in Appendix Table A.11, is used to determine k_s. The amount of water unfrozen at the freezing point needs to be determined. This is done by determining the enthalpy at the freezing point, using the base temperature for Chang and Tao's enthalpy correlations of 227.3 K as the reference and assuming that all the water is in the form of ice. The heat of fusion of ice is 334,944 J/kg.

The specific heat below freezing calculated above was 1955.23 J/(kg · K).

$$H_f' = 0.89(334{,}944) + 1955.23(273 - 227.6)$$
$$= 386{,}867$$

Mass fraction unfrozen water at the freezing point

$$\frac{(386{,}867 - 370{,}329.9)}{334{,}944} = 0.049$$

Using the BASIC program in Appendix Table A.11, the following are entered with the prompts: $X_{water} = 0.049$; $X_{ice} = 0.89 - 0.049 = 0.8406$; $X_{carb} = 0.10$; $X_{fiber} = 0.01$; all other components are zero. $T = -10°C$. The output is: $k_s = 2.067$ W/(m · K). Solving for freezing time using equation 97:

$$\Delta H_{10} = 2.876 \times 10^5 \text{ J/kg } (1070 \text{ kg/m}^3) = 3.077 \times 10^8 \text{ J/m}^3$$

$$T_a = -35°C; \qquad T_{fin} = -20°C$$

$$h = 120\ \text{W}/(\text{m}^2 \cdot \text{K}); \qquad EHDT = 3 \text{ for a sphere}; \qquad D = 0.008 \text{ m}$$

$$k_s = 2.067\ \text{W}/(\text{m} \cdot \text{K}); \qquad T_i = 15°\text{C};\ C_s = 2.953 \times 10^6\ \text{J}/\text{m}^3$$

$$C_1 = 4.0856\ \text{J}/\text{m}^3$$

$$STE = 2.053 \times 10^6 \left[\frac{0 - (-35)}{6.685 \times 10^7}\right] = 1.0749$$

$$PK = 4.056 \times 10^6 \left[\frac{15 - 0}{6.685 \times 10^7}\right] = 0.9167$$

$$\begin{aligned} P &= 0.5\big[1.026 + 0.5808(0.9167) \\ &\quad + 1.0749[0.22696(0.9167) + 0.105]\big] \\ &= 0.9488 \end{aligned}$$

$$\begin{aligned} R &= 0.125\big[1.202 + 1.0749[(3.41)(0.91674) + 0.7336]\big] \\ &= 0.6688 \end{aligned}$$

Substituting values in equation 97:

$$\begin{aligned} t_f &= \frac{3.077 \times 10^8}{[0 - (-35)](3)} \left[0.9488 \frac{0.008}{120} + 0.6688 \frac{(0.008)^2}{2.067}\right] \\ &\quad \left[1 - \frac{1.65(1.0749)}{2.067} \ln\left[\frac{-20 - (-35)}{-10 - (-35)}\right]\right] \\ &= 0.029395 \times 10^8 [0.00006325 + 0.00002071](1.4383) \\ &= 353.9 \text{ s} \end{aligned}$$

PROBLEMS

1. How many inches of insulation are required to insulate a ceiling such that the surface temperature of the ceiling facing the living area is within 2°C of the room air temperature? Assume a heat transfer coefficient on both sides of the ceiling of 2.84 $\text{W}/(\text{m}^2 \cdot \text{K})$ and a thermal conductivity of 0.0346 $\text{W}/(\text{m} \cdot \text{K})$ for the insulation. The ceiling is 1.27 cm thick plasterboard with a thermal conductivity of 0.433 $\text{W}/(\text{m} \cdot \text{K})$. Room temperature is 20°C and attic temperature is 49°C.
2. If a heat transfer coefficient of 2.84 $\text{W}/(\text{m}^2 \cdot \text{K})$ exists on each of the two inside faces of two sheets of 6.35-mm-thick glass separated by an air gap, calculate the gap that can be used such that the rate of heat transfer by conduction through the air gap equals the rate of heat transfer by convection. What is the rate of heat transfer if this gap is exceeded.

The thermal conductivity of air of 0.0242 W/(m · K). Solve for temperature of 20°C and −12°C on the outside surfaces of the glass. Would the calculated gap change in value for different values of the surface temperatures? Would there be any advantage to increasing the gap beyond this calculated value?

3. A walk-in freezer 4 × 6 m and 3 m high is to be built. The walls and ceiling consist of 1.7-mm-thick stainless steel (k = 14.2 W/(m · K)), 10-cm-thick foam insulation (k = 0.34 W/(m · K)), a thickness of corkboard (k = 0.043 W/(m · K)), and 1.27-cm-thick wood siding (k = 0.43 W/(m · K)). The inside of the freezer is maintained at −40°C. Ambient air outside the freezer is at 32°C. The heat transfer coefficient is 5 W/(m^2 · K) on the wood side of the wall and 2 W/(m^2 · K) on the stainless steel side.
   (a) If the outside air has a dew point of 29°C, calculate the thickness of the corkboard insulation that will prevent condensation of moisture on the outside wall of this freezer.
   (b) Calculate the rate of heat transfer through the walls and ceiling of this freezer.

4. (a) Calculate the rate of heat loss to the surroundings and the quantity of steam that would condense per hour per meter of a 1.5-in. (nominal) schedule 40 steel pipe containing steam at 130°C. The heat transfer coefficient on the steam side is 11,400 W/(m^2 · K), and that on the outside of the pipe to air is 5.7 W/m^2 · K. Ambient air averages 15°C in temperature for the year. The thermal conductivity of the steel pipe wall is 45 W/(m · K).
   (b) How much energy would be saved in 1 year (365 days 24 h/day) if the pipe is insulated with 5-cm-thick insulation having a thermal conductivity of 0.07 W/(m · K). The heat transfer coefficients on the steam and air sides are the same as in (a).

5. The rate of insolation (solar energy impinging on a surface) on a solar collector is 475 W/(m^2 · K). Assuming that 80% of the impinging radiation is absorbed (the rest is reflected), calculate the rate at which water can be heated from 15° to 50°C in a solar hot water heater consisting of a spiral wound, horizontal, 1.9-cm-inside-diameter polyethylene pipe 30 m long. The pipe has a wall thickness of 1.59 mm. The projected area of a horizontal cylinder receiving radiation is the diameter multiplied by the length. Assume a heat transfer coefficient of 570 W/(m^2 · K) on the water side and 5 W/(m^2 · K) on the air side when calculating heat loss to the surroundings after the radiant energy from the sun is absorbed. The thermal conductivity of the pipe wall is 0.3 W/m · K. Ambient temperature is 7°C.

6. A swept surface heat exchanger cools 3700 kg of tomato paste per hour from 93° to 32°C. If the overall heat transfer coefficient based on the inside surface area is 855 W/m^2 · K, calculate the heating surface area required for concurrent flow and countercurrent flow. Cooling water enters at 21°C and leaves at 27°C. The specific heat of tomato paste is 3560 J/(kg · K).

7. Design the heating and cooling section of an aseptic canning system that process 190 L per minute of an ice cream mix. The material has a density of 1040 kg/m^3 and a specific heat of 3684 J/(kg · K).
   (a) Calculate the number of units of swept surface heat exchangers required to heat the material from 39° to 132°C. Each unit has an inside heat transfer surface area of 0.97 m^2. The heating medium is steam at 143°C. Previous experience

with a similar unit on this material has shown that an overall heat transfer coefficient of 1700 W/(m^2 · K) based on the inside surface area may be expected.

(b) Calculate the number of units (0.9 m^2 inside surface area per unit) required to cool the sterilized ice cream mix from 132° to 32°C. The cooling jacket of the swept surface heat exchangers is cooled by freon refrigerant from a refrigeration system at −7°C. Under these conditions, a heat transfer coefficient of 855 W/(m^2 · K) based on the inside surface area may be expected.

8. A small swept surface heat exchanger having an inside heat transfer surface area of 0.11 m^2 is used to test the feasibility of cooking a slurry in a continuous system. When the slurry was passed through the heat exchanger at a rate of 168 kg/h, it was heated from 25° to 72°C. Steam at 110°C was used. The slurry has a specific heat of 3700 J/(kg · K).

(a) What is the overall heat transfer coefficient in this system?

(b)If this same heat transfer coefficient is expected in a larger system, calculate the rate at which the slurry can be passed through a similar swept surface heat exchanger having an inside heat transfer surface area of 0.75 m^2 if the inlet and exit temperatures are 25° and 72°C, respectively, and steam at 120°C is used for heating.

9. A steam-jacketed kettle has an inside heat transfer surface area of 0.43 m^2 that is completely covered by the product. The product needs to be heated from 10° to 99°C. The product contains 80% water and 20% nonfat solids. Previous experience has established that an overall heat transfer coefficient based on the inside surface are of 900 W/(m^2 · K) may be expected. The kettle holds 50 kg of product. Condensing steam at 120°C is in the heating jacket. The contents are well stirred continuously during the process.

(a) Calculate the time required for the heating process to be completed.

(b) Determine the nearest nominal size steel pipe that can be used to supply steam to this kettle if the rate of steam flow through the pipe is to average a velocity of 12 m/s.

10. A processing line for a food product is beinig designed. It is necessary to estimate the number of kettles required to provide a production capacity of 500 kg/h. The cooking process involving the kettles requires heating the batch from 27° to 99°C, simmering at 99°C for 30 min, and filling the cans with hot product. Filling and emptying the kettles requires approximately 15 min. The specific heat of the product is 3350 J/(kg · K). The density is 992 kg/m^3. Available for heating are cylindrical vessels with hemispherical bottoms with the hemisphere completely jacketed. The height of the cylindrical section is 25 cm. The diameter of the vessel is 0.656 m. Assume that the vessels are filled to 85% of capacity each time. The overall heat transfer coefficient based on the inside surface area averages 600 K/(m^2 · K). How many kettles are required to provide the desired production capacity? Steam condensing at 120°C is used in the jacket for heating.

11. A process for producing frozen egg granules is proposed in which a refrigerated rotating drum which has a surface temperature maintained at −40°C contacts a pool of liquid eggs at 5°C. The eggs freeze on the drum surface, and the frozen material is scraped off the surface at a point before the surface reenters the liquid eggs. The frozen material, if thin enough, will be collected as frozen flakes. In this process, the thickness of frozen egg which forms on the drum surface is determined by the

dwell time of the drum within the pool of liquid. An analogy of the process, which may be solved using the principles discussed in the section on freezing water, 7.5h, is the freezing of a slab directly in contact with a cold surface, i.e., h is infinite. It is desired that the frozen material be 2 mm thick on the drum surface.

(a) Calculate the dwell time of the drum surface within the liquid egg pool. Assume that on emerging from the liquid egg pool, the frozen material temperature will be 2°C below the freezing point.

(b) If the drum has a diameter of 50 cm and travels 120° after emerging from the liquid egg pool before the frozen material is scraped off, calculate the rotational speed of the drum needed to satisfy the criterion stipulated in (a), and calculate the average temperature of the frozen material at the time it is scraped off the drum. The density of liquid eggs is 1012 kg/m^3, and that of frozen eggs is 1009 kg/m^3. Calculate the thermophysical properties based on the following compositional data: 75% water, 12% protein, 12% fat, 1% carbohydrates.

12. In an experiment for pasteurization of orange juice, a 0.25-in. outside diameter tube with a 1/32-in.-thick wall was made into a coil and immersed in a water bath maintained at 95°C. The coil was 2 m long, and when the juice was pumped at the rate of 0.2 L/min, its juice temperature changed from 25° to 85°C. The juice contained 12% total solids. Calculate:

(a) The overall heat transfer coefficient.

(b) The inside local heat transfer coefficient if the ratio h_0/h_i is 0.8.

(c) The inside local heat transfer coefficient, h_i is directly proportional to the 0.8th power of the average velocity. If the rate at which the juice is pumped through the system is increased to 0.6 L/min, calculate the tube length needed to raise the temperature from 25° to 90°C.

13. Calculate the surface area of a heat exchanger needed to pasteurize 100 kg/h of catsup by heating in a one-pass shell and tube heat exchanger from 40° to 95°C. The catsup has a density of 1090 kg/m^3, a flow behavior index of 0.5, and a consistency index of 0.5 and 0.35 Pa · s^n at 25°C and 50°C, respectively. Estimate the thermal conductivity and specific heat using the correlations discussed in Chapters 5 and 7. The catsup contains 0.4% fiber, 33.8% carbohydrate, and 2.8% ash; the balance is water. The catsup is to travel within the heat exchanger at a velocity of 0.3 m/s. Steam condensing at 135°C is used for heating, and a heat transfer coefficient of 15,000 W/(m^2 · K) may be assumed for the steam side. The heat exchanger tubes are type 304 stainless steel, with an inside diameter of 0.02291 m and an outside diameter of 0.0254 m.

14. A box of beef which was tightly packed was originally at 0°C. It was inadvertently left on a loading dock during the summer, when ambient conditions were 30°C. Assuming that the heat transfer coefficient around the box averages 20 W/(m^2 · K), calculate the surface temperature and the temperature at a point 2 cm deep from the surface after 2 h. The box consisted of 0.5-cm-thick fiberboard with a thermal conductivity of 0.2 W/(m · K) and had dimensions of 47 × 60 × 30 cm. Since the box material has very little heat capacity, it may be assumed to act as a surface resistance, and an equivalent heat transfer coefficient may be calculated such that the resistance to heat transfer will be the same as the combined conductive resistance of the cardboard and the convective resistance of the surface heat transfer coefficient. The meat has a density of 1042 kg/m^3, a thermal conductivity of 0.44 W/(m · K), and a specific heat of 3558 J/(kg · K).

15. In operating a microwave oven, reduced power application to the food is achieved by alternatively cutting on and off the power applied. To minimize excessive heating in some parts of the food being heated, the power application must be cycled such that the average temperature rise with each application of power does not exceed 5°C followed by a 10-s pause between power applications. If 0.5 kg of food is being heated in a microwave oven having a power output of 600 W, calculate the fraction of full power output which must be set on the oven controls such that the above temperature rise and pause cycles are achieved. The food has a specific heat of 3500 J/(kg · K).

SUGGESTED READING

ASHRAE. 1965. *Ashrae Guide and Data Book. Fundamentals and Equipment for 1965 and 1966.* American Society of Heating, Refrigerating and Air Conditioning Engineers.

Ball, C. O., and Olson, F. C. W. 1957. *Sterilization in Food Technology.* McGraw-Hill Book Co., New York.

Bennet, C. O., and Myers, J. E. 1962. *Momentum, Heat, and Mass Transport.* McGraw-Hill Book Co., New York.

Carslaw, H. S., and Jaeger, J. C. 1959. *Conduction of Heat in Solids*, 2nd ed. Oxford University Press, London.

Charm, S. E. 1971. *Fundamentals of Food Engineering*, 2nd ed. AVI Publishing Co., Westport, Conn.

Chang, H. D. and Tao, L. C. 1981. Correlation of enthalpy of food systems *J. Food Sci.* 46:1493.

Choi, Y. and Okos, M. R. 1987. Effects of temperature and composition on thermal properties of foods. In: Food Engineering and Process Applications, M. leMaguer and P. Jelen eds. Vol. 1. Elsevier Appl Science Publishers, New York.

Cleland, A. C., and Earle, R. L. 1984. Freezing time prediction for different final product temperature. *J. Food Sci.* 49:1230.

Heldman, D. R. 1973. *Food Process Engineering.* AVI Publishing Co., Westport, Conn.

Holman, J. P. 1963. Heat Transfer. McGraw-Hill Book Co., New York.

Jacob, M., and Hawkins, G. A. 1957. *Heat Transfer*, Vol. II. John Wiley & Sons, New York.

Leniger, H. A., and Beverloo, W. A. 1975. *Food Process Engineering.* D. Riedel Publishing Co., Boston.

McAdams, W. H. 1954. *Heat Transmission*, 3rd ed. McGraw-Hill Book Co., New York.

McCabe, W. L., and Smith, J. C. 1967. *Unit Operations of Chemical Engineering*, 2nd ed. McGraw-Hill Book Co., New York.

McCabe, W. L., Smith, J. C., and Harriott, P. 1985. *Unit Operations in Chemical Engineering*, 4th ed. McGraw-Hill Book Co., New York.

Metzner, A. B., Vaughn, R. D., and Houghton, G. L. 1957. Heat transfer to non-Newtonian fluids. *AIChE J.* 3:92.

Peters, M. S. 1954. *Elementary Chemical Engineering.* McGraw-Hill Book Co., New York.

Plank, R. 1913. De gefrierdauer von eisblocken. *Z. fur die gesamte Kalte-Ind.* 20(6):109.

Rohsenow, W. M., and Hartnett, J. P. 1973. *Handbook of Heat Transfer.* McGraw-Hill Book Co., New York.

Schneider, P. J. 1973. Conduction. In: *Handbook of Heat Transfer*, W. M. Rohsenow and J. P. Hartnett (eds). McGraw-Hill Book Co., New York.

Siebel, J. E. 1918. *Compend of Mechanical Refrigeration and Engineering*, 9th ed. Nickerson and Collins, Chicago. Cited in: ASHRAE 1966. *ASHRAE Guide and Data Book Applications for 1966-67*, American Society for Heating, Refrigerating and Air Conditioning Engineers, Atlanta, GA.

Sieder, E. N., and Tate, G. E. 1936. *Ind. Eng. Chem.* 28:1429. Heat transfer and pressure drop of liquids in tubes.

Watson, E. L., and Harper, J. C. 1989. *Elements of Food Engineering*. AVI Publishing Co., Westport, Conn.

8

Kinetics of Chemical Reactions in Foods

Chemical reactions occur in foods during processing and storage. Some reactions result in loss of quality and must be minimized, while others produce a desired flavor or color and must be optimized to obtain the best quality. The science of *kinetics* involves the study of chemical reaction rates and mechanisms. An understanding of reaction mechanisms coupled with quantification of rate constants will facilitate the selection of the best conditions of a process or storage so that the desired characteristics will be present in the product.

THEORY OF REACTION RATES

Two theories have been advanced as a basis for reaction rates. The *collision theory* attributes chemical reactions to the collision between molecules which have high enough energy levels to overcome the natural repulsive forces among the molecules. In gases, chemical reaction rates between two reactants have been successfully predicted using the equations derived for the kinetic energy of molecules and the statistical probability for collisions between certain molecules which possess an adequate energy level for the reaction to occur at a given temperature. The *activation theory* assumes that a molecule possesses a labile group within its structure. Normally, this labile group may be stabilized by oscillating within the molecule or by steric hindrance by another group within the molecule. The energy of the labile group may be raised by an increase in temperature to a level which makes the group metastable. Finally, a chemical reaction results which releases the excess energy and reduces the energy level of the molecule to another stable state. The energy level which a molecule must achieve to initiate a chemical reaction is called the *activation energy*. Both theories for reaction rates give a reaction rate constant which is a function of the number of reacting molecules and the temperature.

Reactions may be reversible. Reversible reactions are characterized by an equilibrium constant which establishes the steady-state concentration of the product and the reactants.

TYPES OF REACTIONS

Unimolecular Reactions: One type of chemical reaction which occurs during degradation of food components involves a single compound undergoing change. Part of the molecule may split off, or molecules may interact with each other to form a complex molecule, or internal rearrangement may occur to produce a new compound. These types of reactions are unimolecular and may be represented as:

$$A \xrightarrow{k_1} \text{products} \tag{1}$$

The reaction may occur in more than one step, and in some cases the intermediate product may also react with the original compound.

$$A \xrightarrow{k_1} B \xrightarrow{k_2} \text{products} \tag{2}$$

$$A \xrightarrow{k_1} B + A \xrightarrow{k_3} \text{products} \tag{3}$$

The reaction rate, r, may be considered the rate of disappearance of the reactant, A, or the rate of appearance of a reaction product. In reactions 1 and 2, the rate of disappearance of A, dA/dt, is proportional to a function of the concentration of A, while in reaction 3, dA/dt is dependent on a function of the concentration of A and B. In reaction 1, the rate of formation of products equals the rate of disappearance of the reactant, but in reactions 2 and 3, accumulation of intermediate products results in a lag between product formation and disappearance of the original reactant. When intermediate reactions are involved, the rate of appearance of the product depends on the rate constant k_2.

The rate constant, k, is the proportionality constant between the reaction rate and the function of the reactant concentration, $F(A)$ or $F(B)$. Thus, for reactions 1 and 2:

$$\frac{dA}{dt} = kF(A) \tag{4}$$

For reaction 3:

$$\frac{dA}{dt} = k_1F(A) + k_3[F(A) + F(B)] \tag{5}$$

In equations 4 and 5, A and B represent the concentrations of reactants A and B, and it is obvious that the reaction rate will increase with increasing reactant concentration. The concentration function, which is proportional to the reaction rate, depends upon the reactant and can change with the conditions under which the reaction is carried out. When studying reaction rates, either the rate of appearance of a product or the rate of disappearance of reactants is of interest. On the other hand, if intermediate reactions are involved in the process of transforming compound A into a final reaction product, and the rate constants k_2 and k_1 are affected differently by conditions used in the process, it is necessary to postulate rate mechanisms and evaluate an overall rate constant based on existing conditions in order to optimize the process effectively.

Most of the reactions involving degradation of food nutrients are of the type shown in reaction 1.

Bimolecular Reactions. Another type of reaction involves more than one molecule. The second step of reaction 3 above is one example of a bimolecular reaction. In general, a bimolecular reaction is as follows:

$$aA + bB \xrightarrow{k_3} cC + dD \tag{6}$$

In this type of reaction, the rate may be based on one of the compounds, either the reactant or the product, and the change in concentration of other compounds may be determined using the stoichiometric relationships in the reaction.

$$r_{3C} = \frac{dC}{dt} = -\frac{c\,dB}{b\,dt} = -\frac{c\,dA}{a\,dt} = \frac{c\,dD}{d\,dt} = kF(A)F(B) \tag{7}$$

An example of the use of the relationship shown in equation 7 is the expression of productivity of fermentation systems as a reduction of substrate concentration, an increase in product concentration, or an increase in the mass of the microorganism involved in the fermentation.

Reversible Reactions. Some reactions are reversible.

$$nA \underset{k_2}{\overset{k_1}{\rightleftarrows}} bB + cC \tag{8}$$

$$r_1 = -\frac{dA}{dt} = k_1F(A); \qquad r_2 = \frac{dA}{dt}k_2F(B)F(C)$$

Again, $F(A)$, $F(B)$, and $F(C)$ are functions of the concentrations of A, B, and C, respectively.

The net reaction rate expressed as a net disappearance of A is:

$$r = -\frac{dA}{dt} = k_1F(A) - k_2F(B) \tag{9}$$

Expressing B and C in terms of A: $B = (n/b)(A_0 - A)$ and $C = (n/c)(A_0 - A)$, where A_0 = initial concentration of A. Let $F(A) = A$, $F(B) = B$, and $F(C) = C$, i.e., the reaction rate is directly proportional to the concentrations of the reactants.

$$r = -\frac{dA}{dt} = k_1A - k_2\left\{\left[\left(\frac{n}{b}\right)(A_0 - A)\right]\left[\left(\frac{n}{c}\right)(A_0 - A)\right]\right\} \tag{10}$$

At equilibrium, $r_1 = r_2$ and:

$$k_1A = k_2\left\{\left[\left(\frac{n}{b}\right)(A_0 - A)\right]\left[\left(\frac{n}{c}\right)(A_0 - A)\right]\right\}$$

$$k_{eq} = \frac{k_1}{k_2} = \frac{[(n/b)(A_0 - A)][(n/c)(A_0 - A)]}{A} \tag{11}$$

Once a constant of equilibrium of known, it is possible to determine the concentrations of A, B, and C from the stoichiometric relationships of the reaction. An example of this type of reaction is the dissociation of organic acids and their salts.

ENZYMATIC REACTIONS

Enzymatic reactions encountered in food processing occur at a rate which is limited by the concentration of the enzymes present. There is usually an abundance of the substrate (reactant), so that changes in substrate concentration do not affect the reaction rate. The activity of an enzyme is defined as the rate at which a specified quantity of the enzyme will convert a substrate to product. The reaction is followed by either measurement of the loss of reactant or the appearance of product. The specific activity of an enzyme is expressed as activity per mass of protein. When using an enzyme in a process, the activity of added enzyme must be known in order that the desired rate of substrate conversion will be achieved.

If there is no product inhibition:

$$-\frac{dS}{dt} = a \tag{12}$$

where a is the enzyme activity. Thus, an enzymatic reaction without product inhibition suggests a linear change in substrate concentration in the early stages of the reaction, when the substrate concentration is so high that the enzyme concentration is rate limiting. Most enzymatic reactions, however, proceed in a curvilinear pattern which approaches a maximum value of infinite time. Consider an enzymatic reaction in which product inhibition exists. The reaction rate will be:

$$-\frac{dS}{dt} = a_0 - B; \qquad B = \text{enzyme bound to the product}$$

Assuming that B is proportional to the amount of product formed, and using k_i as the constant representing the product inhibitory capacity

$$-\frac{dS}{dt} = a_0 - k_i P; \qquad P = \text{product concentration}$$

Consider a reaction in which $S \rightarrow P$. If f = the fraction of substrate converted to product, $S = S_0(1 - f)$ and $P = S_0 f$.

Substituting into the rate equation and letting S_0 = initial substrate level and a_0 = initial enzyme activity:

$$S_0 = \frac{df}{dt} = a_0 - k_i S_0 f \tag{13}$$

$$\frac{df}{dt} = \frac{a_0}{S_0} - k_i f \tag{14}$$

The differential equation can be easily integrated by separating the variables if the transformation $f' = a_0/S_0 - k_i f$ is used.

$$\frac{df}{dt} = -\frac{1}{k_i}\frac{df'}{dt}; \qquad \frac{df'}{dt} = -k_i f'$$

Integrating, substituting f' for f, and using the boundary condition $f = 0$ at $t = 0$:

$$\ln\left(1 - \frac{k_i f S_0}{a_0}\right) = -k_i t; \qquad f = \frac{a_0}{k_i S_0}(1 - e^{-k_i t}) \tag{15}$$

A plot of the fraction of substrate converted against time shows a curvilinear plot which levels off at a certain value of f at infinite time. The maximum conversion will be f_{max}.

$$f_{max} = \frac{a_0}{k_i S_0}; \qquad \ln\left(1 - \frac{f}{f_{max}}\right) = k_i t$$

Depending upon the enzyme and the substrate concentration, the conversion may not immediately follow the exponential expression. For example, if the initial enzyme activity is quite high relative to the substrate concentration, the fraction converted may initially be a linear function of time. However, once product accumulates and the substrate concentration drops, the influence of product inhibition becomes significant and substrate conversion occurs exponentially, as derived above. With product inhibition, the product competes for active sites on the enzyme, making these sites unavailable for forming a complex with the substrate. Eventually, all enzyme active sites are occupied by the product and the reaction stops.

Enzyme activities are determined as the slope of the substrate conversion vs. time curves at time zero (initial reaction velocity) to avoid the effect of product inhibition.

REACTION ORDER

Reaction order is the sum of the exponents of the reactant concentration terms in the rate equation. Table 8.1 lists various deteriorative reactions in foods and the order of the reaction.

Zero-Order Reactions.

$$r = \frac{dA}{dt} = k$$
$$A = A_0 + kt \tag{16}$$

A characteristic of a zero order reaction is a linear relationship between the concentration of the reactant or product and the time of the reaction, t.

Table 8.1. Kinetic Constants of Reactions Occurring in Foods

Reaction	Substrate	Order	Temp. (°C)	D (min)	z (°C)
Thiamin degradation	Meats	1	121.1	158	31
Carotene degradation	Liver	1	121.1	43.6	25.5
Ascorbic acid degradation	Peas	1	121.1	246	50.5
Nonenzymatic browning	Milk	1	121.1	12.5	26
Chlorophyll degradation	Peas	1	121.1	13.2	38.3
Sensory quality loss	Various foods	1	121.1	5–500	26

Sources: Burton, J., *Dairy Res.* 21:104, 1954; Holdsworth, J., *Food Eng.* 4:89, 1985; Lund, D. B., *Food Technol.* 31(2):71, 1977; Rao et. al., *J. Food Sci.* 46:636, 1981; Wilkinson et al., *J. Food Sci.* 46:32, 1981.

First-Order Reactions.

$$r = -\frac{dA}{dt} = kA$$
$$\ln\left(\frac{A}{A_0}\right) = kt \tag{17}$$

A_0 is the concentration of A at time $= 0$. A first-order reaction is characterized by a logarithmic change in the concentration of a reactant with time.

Most of the reactions involved in the processing of foods, as shown in Table 8.1, are first-order reactions.

Second-Order Reactions.

$$r = -\frac{dA}{dt} = kA^2$$
$$\frac{1}{A} - \frac{1}{A_0} = kt \tag{18}$$

Second-order unimolecular reactions are characterized by a hyperbolic relationship between the concentration of the reactant or product and time. A linear plot is obtained if $1/A$ is plotted against time. Second-order bimolecular reactions may also follow the following rate equation:

$$r = -\frac{dA}{dt} = kAB$$

where A and B are the reactants. The differential equation may be integrated by holding B constant to give:

$$\ln\left(\frac{A}{A_0}\right) = -k't \tag{19}$$

k' is a pseudo-first order rate constant. $k' = kB$.

A second-order bimolecular reaction yields a plot of the concentration of the reactant against time similar to that of a first-order unimolecular reaction, but the reaction rate constant varies with different concentrations of the second reactant. An example of a second-order bimolecular reaction is the aerobic degradation of ascorbic acid. Oxygen is a reactant and a family, of pseudo-first-order plots is obtained when ascorbic acid degradation is studied at different levels of oxygen availability.

***n*th-Order Reactions.**

$$r = -\frac{dA}{dt} = kA^n; \qquad n > 1$$
$$A^{1-n} - A_0^{1-n} = -(1-n)kt \qquad (20)$$

Evaluation of reaction order is a trial-and-error process which involves assuming various values for n and determining which value would result in the best fit with the nth order equation above.

Reactions Where Product Concentration Is Rate Limiting. These types of reactions are usually followed not in terms of the concentration of the reactant, but by some manifestation of the completion of the reaction in terms of a physical property change. Examples are protein gelation, measured as an increase in the strength of the gel; the nonenzymatic browning reaction in solid foods; textural changes during cooking; and sensory flavor scores during storage. The magnitude of the attribute measured usually levels off not because of depletion of the reactants but because the measuring technique can no longer detect any further increase in intensity. Reactions of this type can be fitted to the following equation:

$$\ln\left(1 - \frac{C}{C*}\right) = \pm kt \qquad (21)$$

where $C*$ is the value of the measured attribute when it remains constant over long reaction times, and C is the value at any time during the transient stage of the process.

EXAMPLE: The following data show the firmness of a protein gel as a function of time of heating. Derive an appropriate equation to fit these data, and determine the rate constant for the reaction.

Time (min)	0	1	2	3	4	5	10	20
Firmness (g)	0	6.01	8.41	9.36	9.75	9.90	10.2	10.2

Solution: The data show that firmness reaches a constant value at 10 min of heating. The data are fitted to the equation:

$$\ln\left(1 - \frac{F}{F*}\right) = -kt$$

F = firmness value and F^* = final firmness value. The following are the transformed data, which will be analyzed by linear regression to obtain the slope which will be the value of k.

x (time)	$y[\ln(1 - F/F^*)]$
0	0
1	−0.889
2	−1.74
3	−2.497
4	−3.121
5	−3.526

Linear regression of x and y gives a slope of −1.3743.

$$k = 1.3743 \text{ min}^{-1}$$

A plot of the data shows a linear fit except for the last point, which deviates slightly from linearity.

THE REACTION RATE CONSTANT

The reaction rate constant defines the reaction rate. There are several ways in which the speed of a chemical reaction can be reported for first-order reactions which predominate in food systems.

Rate Constant, *k*, for an Exponential Model of Concentration Change. This rate constant has units of reciprocal time and is the slope of a plot of ln (c) against time. This rate constant is defined for the various types of reactions in the section "Reaction Order."

The *D* Value. This method of representing the rate constant for a reaction had its origins in thermobacteriology, where the inactivation rate of microorganisms during heating is expressed as a decimal reduction time. This approach was later applied to chemical reactions so that the same computational scheme could be used to determine microbial inactivation and nutrient degradation during a thermal process for sterilization of foods. The D value is defined as:

$$\log \frac{C}{C_0} = -\frac{t}{D} \tag{22}$$

Thus, the D value is the negative reciprocal of the slope of a plot of log (C) against t. C in the above equation is the concentration of a reactant. D is based on common logarithms, in contrast to k, which is based on natural logarithms. D and k are related as follows:

$$\ln\left(\frac{C}{C_0}\right) = \ln(10)\log\left(\frac{C}{C_0}\right) = -kt$$

Thus:

$$\frac{1}{D} = \frac{k}{\ln(10)}; \quad D = \frac{\ln(10)}{k} \tag{23}$$

The Half-Life. This method of expressing the rate of a reaction is commonly used in radioisotope decay. It is easier to visualize the rate of the reaction when expressed as a half-life rather than as a rate constant based on natural logarithms. The half-life is the time required for the reactant to lose half of its original concentration. The half-life is related to k and D as follows:

$$\ln(10.5) = -k(t_{0.5}); \quad t_{0.5} = -\frac{\ln(0.5)}{k}$$

$$\log(0.5) = -\frac{t_{0.5}}{D}; \quad t_{0.5} = -D\log(0.5) \tag{24}$$

TEMPERATURE DEPENDENCE OF REACTION RATES

The Arrhenius Equation. The activated complex theory for chemical reaction rates is the basis for the Arrhenius equation, which relates reaction rate constants to the absolute temperature. The Arrhenius equation is:

$$k = A_0[e]^{\frac{-E_a}{RT}} \tag{25}$$

E_a is the activation energy, and A_0 is the rate constant as T approaches infinity. Another form of the Arrhenius equation involves the reaction rate constant at a reference temperature. Let T_0 = the reference temperature at which $k = k_0$.

$$k_0 = A_0[e]^{-E_a/RT_0}; \quad k = A_0[e]^{-E_a/RT}$$

Taking the ratio of the two equations:

$$\frac{k}{k_0} = [e]^{-E_a/R\left[\frac{1}{T} - \frac{1}{T_0}\right]} \tag{26}$$

The negative sign is placed on the exponent of the Arrhenius equation so that a positive activation energy will indicate an increasing reaction rate constant with increasing temperature. Using equation 26, the rate constant at any temperature can be determined from the activation energy and the rate constant k_0 at a reference temperature, T_0.

The Q_{10} Value. The Q_{10} value of a reaction is often used for reporting the temperature dependence of biological reactions. It is defined as the number of times a reaction rate changes with a 10°C change in temperature. If a reaction rate doubles with a 10°C change in temperature, $Q_{10} = 2$. For reactions such as enzymatically induced color or flavor change in foods, degradation of natural pigments, nonenzymatic browning, and microbial growth rate, Q_{10} is usually around 2. Thus the rule of thumb in food storage is that a 10°C reduction in storage temperature will increase the shelf life by a factor of 2. The relationship between the Q_{10} value and the activation energy is derived as follows:

Let k_1 = rate constant at T_1 and k_2 = rate constant at T_2. From the definition of Q_{10}:

$$k_2 = k_1[Q_{10}]^{(T_2 - T_1)/10} \tag{27}$$

Taking the logarithm of equation 27:

$$\ln\left(\frac{k_2}{k_1}\right) = \frac{T_2 - T_1}{10} \ln Q_{10} \tag{28}$$

Substituting k_2 for k, k_1 for k_0, T_2 for T, and T_1 for T_0 in equation 26:

$$\frac{k}{k_1} = [e]^{-E_a/R[1/T_2 - 1/T_1]} \tag{29}$$

Taking the logarithm of equation 29:

$$\ln\left[\frac{k_2}{k_1}\right] = \frac{-E_a}{R}\left[\frac{1}{T_2} - \frac{1}{T_1}\right] \tag{30}$$

$$\ln\left[\frac{k_2}{k_1}\right] = \frac{-E_a}{R}\left[\frac{T_1 - T_2}{T_2 T_1}\right] \tag{31}$$

The negative sign on E_a in equation 31 drops out when the signs on T_1 and T_2 in the numerator are reversed. Equating equations 28 and 31 and solving for E_a/R:

$$\frac{E_a}{R} = \frac{\ln (Q_{10})}{10} T_2 T_1 \tag{32}$$

$$Q_{10} = [e]^{E_a/R[10/T_2 T_1]} \tag{33}$$

The Q_{10} value is temperature dependent and should not be used over a very wide range of temperatures.

The *z* Value. The *z* value had its origins in thermobacteriology and was used to represent the temperature dependence of the microbial inactivation rate. *z* was defined as the temperature change needed to change the microbial inactivation rate by a factor of 10. The *z* value has also been used to express the temperature dependence of degradative reactions occurring in foods during processing and storage. The *z* value expressed in terms of the reaction rate constant is as follows:

$$k_2 = k_1 [10]^{T_2 - T_1/z} \tag{34}$$

Taking the logarithm:

$$\ln \left[\frac{k_2}{k_1}\right] = \frac{T_2 - T_1}{z} \ln (10) \tag{35}$$

Equating the right-hand side of equations 31 and 35:

$$\frac{\ln (10)}{z} = \left[\frac{E_a}{R}\right]\left[\frac{1}{T_2 T_1}\right]$$

$$z = \frac{\ln (10)}{(E_a/R)} T_1 T_2 \tag{36}$$

Solving for E_a/R in equations 32 and 36 and equating:

$$z = \frac{10 \ln (10)}{\ln (Q_{10})} \tag{37}$$

EXAMPLE: McCord and Kilara (*J. Food Sci.* 48:1479, 1983) reported the kinetics of inactivation of polyphenol oxidase in mushrooms to be first order, and the rate constants at 50°, 55°, and 60°C are 0.019, 0.054, and 0.134 min^{-1}, respectively. Calculate the activation energy, the *z* value, and the Q_{10} value for the inactivation of polyphenol oxidase in mushrooms.

Solution: The absolute temperatures corresponding to 50°, 55°, and 60°C are 323, 328, and 333 K, respectively. A regression of ln (k) against $1/T$ gives a slope of −21009.6; thus, $-E_a/R = -21009.6$ $E_a/R = 21009$ K^{-1}; $R = 1.987$ Cal/(gmole · K); and $E_a = 41.746$ kcal/gmole.

$$\ln(Q_{10}) = 10\left(\frac{E_a}{R}\right)\left(\frac{1}{T_1T_2}\right) = \frac{10(21{,}009)}{323(333)} = 1.95$$

$$Q_{10} = 7.028$$

$$z = \frac{\ln(10)}{E_a/R}T_1T_2 = \frac{323(333)\ln(10)}{21{,}009} = 11.8°\text{C}$$

PROBLEMS

1. Nagy and Smoot (*J. Agr. Food Chem.* 25:135, 1977) reported the degradation of ascorbic acid in canned orange juice to be first order. The following first order rate constants can be calculated from their data: at T = 29.4°, 37.8°, and 46.1°C, k on day^{-1} is 0.00112, 0.0026, and 0.0087, respectively. Calculate the activation energy, Q_{10}, the D value, and the half-life at 30°C.
2. The following data were collected for the sensory change in beef stored while exposed directly to air at −23°C (from Gokalp et al., *J. Food Sci.* 44:146, 1979): Sensory scores were 8.4, 6.2, 5.5, and 5.1 at 0, 3, 6, and 9 months in storage, respectively. Plot the data and determine an appropriate form of an equation to which the data can be fitted to obtain the reaction rate constant.
3. Accelerated shelf life testing is often done to predict how food products will behave in the retail network. If a food product is expected to maintain acceptable quality in the retail network for 6 months at 30°C, how long should this product be stored at 40°C prior to testing so that the results will be equivalent to those obtained at 6 months at 30°C. Assume that the temperature dependence of the sensory changes in the product is similar to that for the nonenzymatic browning reaction in Table 8.1.
4. Ascorbic acid degradation in sweet potatoes at a water activity of 0.11 is first order, with a rate constant of 0.001500 h^{-1} at 25°C. If Q_{10} for this reaction is 1.8, calculate the amount of ascorbic acid remaining in dried sweet potato stored at 30°C after 3 months in storage if the initial ascorbic acid content was 33 mg/100 g.

SUGGESTED READING

Holdsworth, S. D. 1985. Optimization of thermal processing, a review. *J. Food Eng.* 4:89.
Lund, D. B. 1977. Maximizing nutrient retention. *Food Technol.* 31(2):71.
Perry, R. H., Chilton, C. H., and Kirkpatrick, S. D. 1963. *Chemical Engineers Handbook*, 4th ed. McGraw-Hill Book Co., New York.
Skinner, G. B. 1974. *Introduction to Chemical Kinetics*. Academic Press, New York.

9

Thermal Process Calculations

Inactivation of microorganisms by heat is a fundamental operation in food preservation. The concepts presented in this chapter are applicable not only in canning but in any process where heat is used to inactivate microorganisms and induce chemical changes which affect quality. The term *sterilization*, as used in this chapter, refers to the achievement of commercial sterility, defined as a condition in which microorganisms which cause illness, and those capable of growing in food under normal nonrefrigerated storage and distribution, are eliminated.

PROCESSES AND SYSTEMS FOR STABILIZATION OF FOODS FOR SHELF STABLE STORAGE: SYSTEMS REQUIREMENTS

Different systems are available for treating foods to make them shelf stable. The suitability of a system depends upon the type of food processed, production rates, availability of capital, and labor costs. Product quality and economics are the major factors to be considered in selecting a system. Since the major production costs are overhead and labor, plants with high production capacity are inclined to use systems which have high capitalization and low labor requirements. Products with superior quality result from systems capable of high-temperature, short-time treatments.

In-Can Processing. The simplest and oldest method of modern food preservation involves placing a product in a container, sealing it, and heating the sealed container under pressure. Different types of pressure vessels or retorts are used.

Stationary Retorts. These retorts are cylindrical vessels oriented vertically or horizontally. Crates are used to facilitate loading and unloading of cans. The cans are stacked vertically in the crates, and perforated metal dividers separate the layers of cans. In vertical retorts, the crates are lowered or raised using electric hoists. In horizontal retorts, the crates have a rectangular profile, with dimensions to fit the cylindrical retort. The crates are mounted on a carrier with wheels, and tracks within the retort guide the wheels of the crate carrier during introduction and retrieval.

Stationary retorts for processing of canned foods must be equipped with an accurate temperature controller and a recording device. In addition, steam must be uniformly distributed inside the retort. A steam bleeder continuously vents small amounts of steam and promotes steam flow within the retort. A fluid-in-glass thermometer is required to provide visual monitoring of the retort temperature by the operator. At the start of the process, the retort is vented to remove air and ensure that all cans are in contact with saturated steam. Figure 9.1 shows a horizontal stationary retort and crate.

Pressure-resistant hatches for the retorts are of various design. Some are secured with hinged, bolt-like locks, but more recent designs facilitate opening and closing of the retort. A wheel-type lock advances or retracts locking bars which slip into a retaining slot to secure the cover. Another cover design consists of a locking ring which can be engaged or disengaged with a turn of a lever. A locking ring-type cover assembly is shown in Fig. 9.1.

Typically, stationary retorts are operated by loading the cans, venting the retort, and processing for a specified time at a specified temperature. The time from introduction of the steam to attainment of the processing temperature is

Fig. 9.1. Horizontal stationary retort and crate.

called the *retort come-up time*. The process is "timed" when the retort reaches the specified processing temperature. A timed record of the retort temperature for each batch processed is required to be maintained in a file. Cooling may be done inside the retort. However, slowly cooling cans may be removed as soon as the internal pressure, has dropped to slightly above atmospheric pressure, and cooling is completed in canals where circulating cold chlorinated water contacts the crates, which are suspended and moved through the water by overhead conveyors.

Hydrostatic Cooker. This type of retort is shown in Fig. 9.2. It consists of two water legs which seal steam pressure in the main processing section. When processing at 121.1°C the absolute pressure of steam is 205,740 Pa; therefore, if the atmospheric pressure is 101.325 Pa, a column of water 10.7 m high must be used to counteract the steam pressure. Thus, hydrostatic cookers are large structures which are often in the open. With nonagitating hydrostatic cooker, heat penetration parameters for thermal process calculations are obtained using a stationary retort, and specified processes are similar to those for a stationary retort. The specified process is set by adjusting the speed of the conveyors which carry cans in to and out of the retort such that the residence time in the steam chamber equals the specified process time.

Continuous Agitating Retorts. One type of continuous agitating retort consists of a cylindrical pressure vessel equipped with a rotating reel which carries cans on its periphery. When the reel rotates, cans alternately ride on the reel and roll along the cylinder wall. Figure 9.3 shows a three-shell continuous retort viewed from the end which shows the drive system for the reel. Also shown at the top of the retort is the rotary valve which receives the cans and introduces them continuously into the retort without losing steam from the retort, and the transfer valves which transfer cans from one retort to the other. Figure 9.4 is a cutaway view of the reel and the automatic can transfer valve. Agitation is induced by shifting of the headspace as the cans roll. Agitation is maximum in fluid products with small suspended particles, and no agitation exists with semi-solid products such as canned pumpkin. Agitation minimizes heat-induced changes in a product during thermal processing when products are of low viscosity. The speed of rotation of the reel determines the rate of heating and the residence time of the cans in the retort. Therefore, the heat penetration parameters must be obtained at several reel speeds to match the residence time at a given reel speed to the procesing time calculated using heat penetration parameters from the same reel speed. The processing time (t) in a continuous retort is determined as follows:

$$t = \frac{N_t}{N_p \Omega} \tag{1}$$

Fig. 9.2. Hydrostatic cooker. (Courtesy of Food Machinery Corporation, Canning Machinery Division.)

where N_t = total number of cans in the retort if completely full, N_p = number of pockets around the periphery of the reel, and Ω = rotational speed of the reel.

A simulator called the *steritort* is used to determine heat penetration parameters at different rotational speeds of the reel.

Cans enter a continuous retort and are instantaneously at the processing temperature; therefore, the process time is equal to the residence time of the cans within the retort.

Fig. 9.3. Multiple-shell, continuous rotary retort. (Courtesy of Food Machinery Corporation, Canning Machinery Division.)

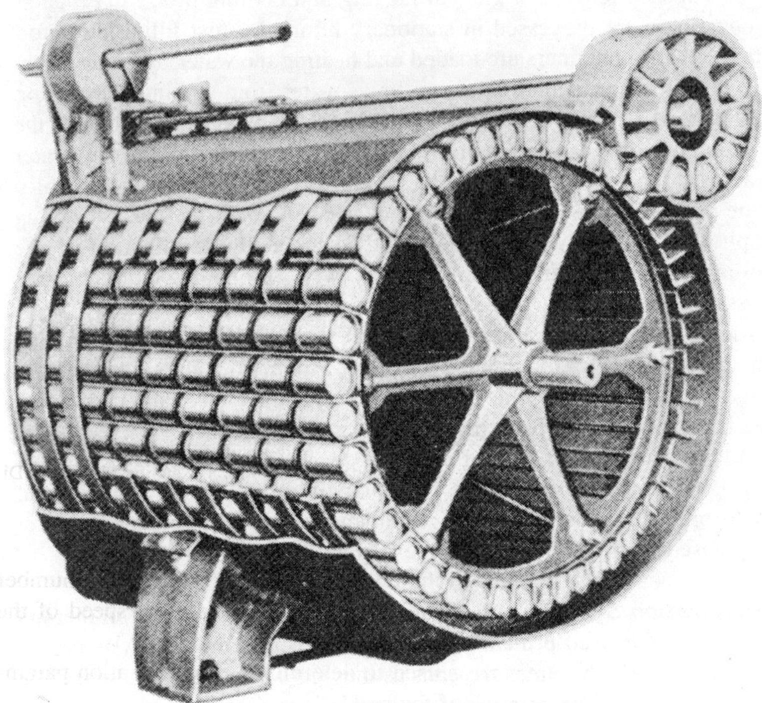

Fig. 9.4. Cross section of a continuous rotary retort showing can positioning on the reel and the can transfer valve, which continuously introduces the cans into the retort without releasing the pressure. (Courtesy of Food Machinery Corporation, Canning Machinery Division.)

Crateless Retorts. The crateless retort has a labor-saving feature over conventional stationary retorts, and it appeals to processors whose level of production cannot economically justify the high initial cost of a hydrostatic or continuous rotational agitating retort. Figure 9.5 shows a diagram of a crateless retort. A system of pumps and hydraulic-operated locks alternately opens the retort, fills it with water, receives the cans which drop into the retort at random, seals the retort for pressure processing with steam, introduces cold water for cooling, and drops the cans and cooling water into a pool of cold chlorinated water for final cooling and retrieval by a conveyor. Energy is saved if the hot water used initially to fill the retorts to receive the cans is stored and reused. Steam waste by venting is eliminated, since steam displaces water at the initial phase of the process. The largest saving from this system is in labor and elimination of the maintenance cost of the retort crates. Thermal process parameters are determined in the same manner as for stationary retorts.

Processing in Glass Containers. The inability of glass to withstand sudden temperature changes requires a gradual heating and cooling process. Products in glass containers are processed in stationary retorts by first filling the retort with water after the containers are loaded and heating the water by direct injection of live steam. A recirculating pump draws water from a point near the top of the water level in the retort and forces it back into the retort through the bottom. This procedure ensures uniform water temperature and uniform water velocity across all containers in the retort. Water is heated slowly, thereby prolonging the come-up time and eliminating thermal shock to the glass. Cooling is accomplished by slowly introducing cold water at the termination of the scheduled process. Water temperature drops slowly, eliminating thermal shock to the glass.

Evaluation of thermal processes in hot water systems is best done using the general method for integrating process lethality. Minimum come-up time to the processing temperature, hold time at the specified temperature, and cool-down time must be part of the process specifications.

Retortable pouches are processed similarly to glass in a hot water system. However, products packed in flexible containers must be secured in individual restraints to prevent changes in package dimensions and minimize stresses at the seams caused by product expansion during heating.

Flame Sterilization System. This relatively recent development in thermal processing systems is used primarily for canned mushrooms. The system consists of a conveyor which rotates the cans as they pass over an open flame. The cans themselves act as the pressure vessel which sterilizes the contents. The fluid inside the cans must be of low viscosity, such as brine, water, or low-sugar syrups, since rapid heat exchange between the can walls and contents is needed to prevent scorching of product on the inner can surface. Internal vac-

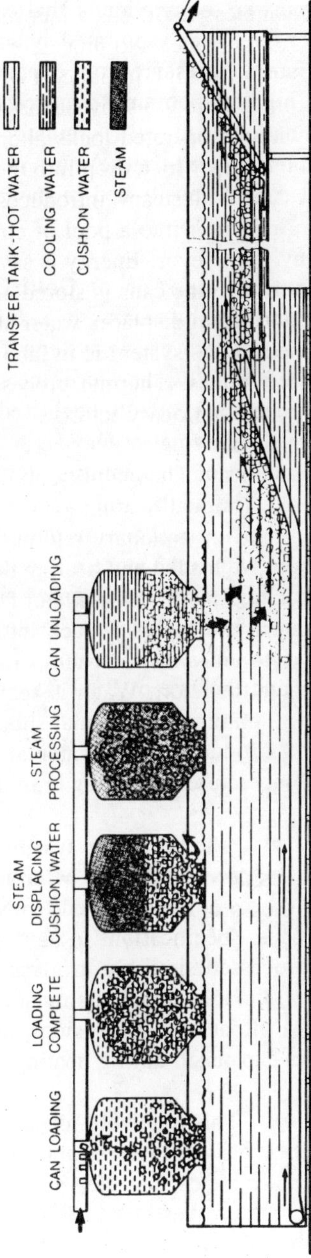

Fig. 9.5. Schematic diagram of the operation of a crateless retort. (Courtesy of Food Machinery Corporation, Canning Machinery Division.)

uum at the time of filling must be the maximum that can be achieved without paneling of the cans. This ensures that a saturated steam atmosphere will exist inside the can and that the internal pressure from expanding air and steam will not be excessive during the high temperatures required for sterilization.

Thermal process determination requires a simulator which rotates the can over an open flame. The internal temperature must be monitored in the geometric center of the largest particle positioned in the geometric center of the can.

Continuous Flow Sterilization: Aseptic or Cold Fill. Fluids and small particle suspensions can be sterilized by heating in heat exchangers. Figure 9.6 is a schematic diagram of an aseptic processing system. The liquid phase reaches the processing temperature very rapidly; therefore, the small sterilization value of the heating phase of the process is generally neglected. The specified process for sterilization in continuously flowing systems is a time of residence in a holding tube, an unheated section of the piping system which leads the fluid from the heat exchangers for heating to the heat exchangers for cooling. A back pressure valve or a positive displacement timing pump is positioned after the cooler to maintain the pressure within the system at a level needed to keep the products boiling temperature higher than the processing temperature. After cooling, the sterile product must be handled in a sterile atmosphere; therefore, the process is called *aseptic processing*. The time of residence is set by the volume of the holding tube and the rate of fluid flow delivered by a positive displacement pump.

$$t_{avg} = \frac{A_c L}{Q} \quad (2)$$

where t_{avg} = average fluid residence time, A_c = cross-sectional area of the holding tube, L = length of the holding tube, and Q = volumetric rate of flow.

The average velocity ($V_{avg} = Q/A_c$) may also be used to calculate the time in the holding tube.

$$t_{avg} = \frac{L}{V_{avg}} \quad (3)$$

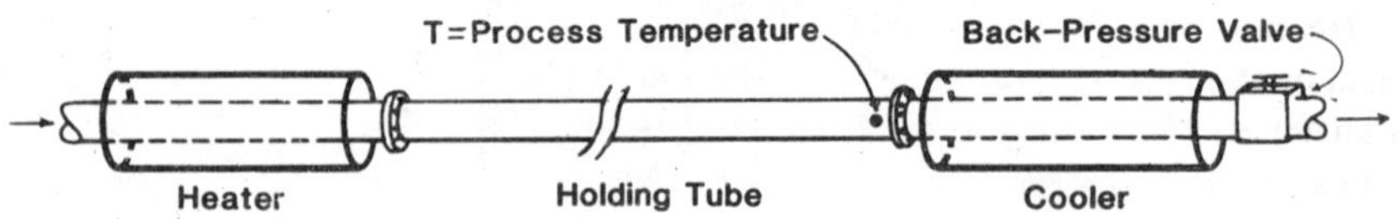

Fig. 9.6. Schematic diagram of an aseptic processing system for product sterilization.

In most cases, however, the residence time of the fastest-flowing portion of the fluid is used as the required hold time in the thermal process calculations. This is because the highest probability of survivors from the thermal process is contributed by the section of fluid flowing close to the geometric center of the tube. The minimum time is:

$$t_{min} = \frac{L}{V_{max}} \tag{4}$$

where the maximum velocity is:

For Newtonian fluids in laminar flow:

$$V_{max} = 2V_{avg} \tag{5}$$

For power law fluids in laminar flow:

$$V_{max} = \frac{(3n + 1)}{(n + 1)} V_{avg} \tag{6}$$

For Newtonian fluids in turbulent flow, the following equation was derived by Edgerton and Jones (1970) for V_{max} as a function of the Reynolds number based on the average velocity:

$$V_{max} = \frac{V_{avg}}{0.0336 \log (Re) + 0.662} \tag{7}$$

An equation similar to equation 7 can be derived by performing a regression analysis on data by Rothfus et al. (*AIChE J.* 3:208, 1957) for a Reynolds number (based on an average velocity) greater than 10^4.

Steam-Air Mixtures for Thermal Processing. A recent development in thermal processing is the use of a mixture of steam and air instead of water or saturated steam for heating. This system has been touted as ideal in the processing of products in retortable pouches and glass. The advantages are elimination of the need for exhausting and the absence of sudden pressure changes on heating or cooling, preventing breakage of the fragile container.

Heating rates on which the scheduled process depends, are strongly dependent on the heat transfer coefficient when steam-air is used for heating. The heat transfer coefficient is a function of the velocity and the mass fraction of steam. Thus, a retort designed for steam-air heating must be equipped with a blower system to generate adequate flow within the retort to maintain uniform velocity and temperature. Accurate separate controllers must be used for pressure and temperature. The mass fraction steam (X_s) in a steam-air mixture operated at a total pressure P is given by:

$$X_s = \frac{P_s}{P}\left(\frac{18}{29}\right) \tag{8}$$

P_s is the saturation pressure of steam at the temperature used in the process.

MICROBIOLOGICAL INACTIVATION RATES AT CONSTANT TEMPERATURE

Rate of Microbial Inactivation. When a suspension of microorganisms is heated at constant temperature, the decrease in the number of viable organisms follows a first order reaction. Let N = number of viable organisms.

$$-\frac{dN}{dt} = kN \tag{9}$$

k is the first-order rate constant for microbial inactivation. Integrating equation 9 and using the initial condition, $N = N_0$ at $t = 0$:

$$\ln\left(\frac{N}{N_0}\right) = -kt \tag{10}$$

Equation 10 suggests a linear semilogarithmic plot of N against t. Equation 10 expressed in common logarithms is:

$$2.303 \log\left(\frac{N}{N_0}\right) = -kt; \quad \log\left(\frac{N}{N_0}\right) = \frac{-kt}{2.303}$$

or:

$$\log\left(\frac{N}{N_0}\right) = \frac{-t}{D} \tag{11}$$

Equation 11 defines D, the decimal reduction time, the time required to reduce the vaiable population by a factor of 10. $D = 2.303/k$. Thus, the decimal reduction time and the first-order kinetic rate constant can be easily converted for use in equations requiring the appropriate form of the kinetic parameter.

N, the number of survivors, is considered to be the probability of spoilage if the value is less than 1. Any value of $N \geq 1$ means certain spoilage (probability of spoilage = 1).

Shape of Microbial Inactivation Curves. Microbial inactivation proceeds as a logarithmic function with time according to equation 11. However, although the most common inactivation curve is the linear semilogarithmic plot shown in Fig. 9.7A, several other shapes are encountered in practice. Fig. 9.7B shows an initial rise in numbers followed by first order inactivation. This has been observed with very heat-resistant spores, and may be attributed to heat activation of some spores which otherwise would not germinate and form colonies before the heat treatment reached the severity needed to kill the organism. Figure 9.7C shows an inactivation curve which exhibits an initial lag or induc-

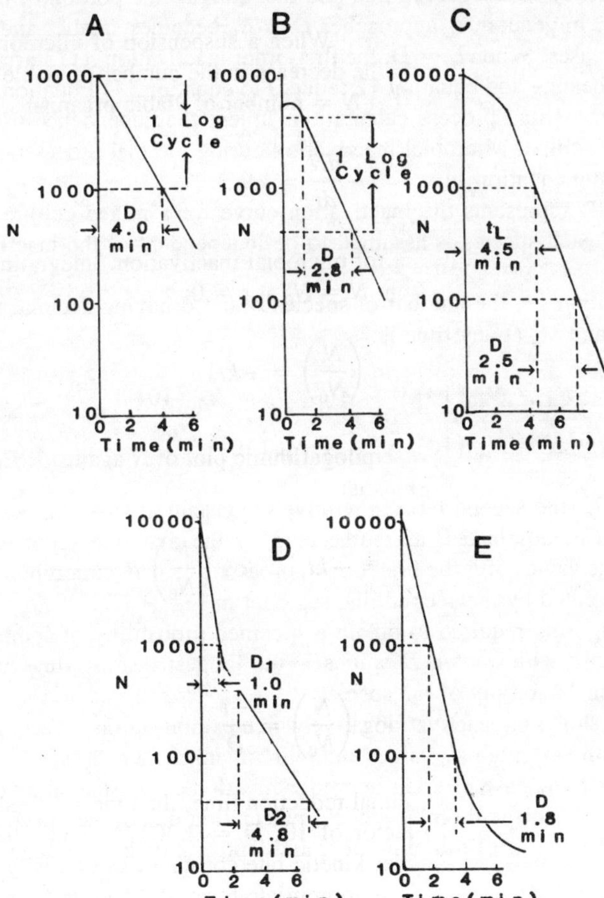

Fig. 9.7. Microbial inactivation curves. (A) First-order inactivation rate. (B) Initial rise in numbers followed by first-order inactivation. (C) Initial lag in the inactivation curve. (D) Inactivation curve exhibited by a mixed culture. (E) Tailing of an inactivation curve.

tion period. Very little change in numbers occurs during the lag phase. The curve represented by Fig. 9.7C can be expressed as:

$$\log \frac{N_0}{N} = 1 + \left(\frac{t - t_L}{D}\right); \qquad t > t_L \tag{12}$$

where t_L = the lag time, defined as the time required to inactivate the first 90% of the population. In most cases, the curved section of the inactivation curve does not extend beyond the first log cycle of inactivation. Therefore, defining t_L as in equation 12 eliminates the arbitrary selection of the lag time from the point of tangency of the curved-line and the straight-line portion of the inactivation curve. In general, t_L approaches D as N_0 becomes smaller and the temperature increases. When $t_L = D$, the first-order inactivation rate starts from the initiation of heating and equation 12 reduces to equation 11. Equation 12 is not often used in thermal process calculations unless the dependence of t_L on N_0 and T are quantified. Microbial inactivation during thermal processing is often evaluated using equation 11.

Figure 9.7D represents the inactivation curve for a mixed culture. The inactivation of each species is assumed to be independent of the inactivation of others.

From equation 11, the number of species A and B having decimal reduction times of D_A and D_B at any time is:

$$N_A = N_{A0}(10)^{-(t/D_A)}; \qquad N_B = N_{B0}(10)^{-(t/D_B)}$$

$$N = N_{A0}(10)^{-(t/D_A)} + N_{B0}(10)^{-(t/D_B)} \tag{13}$$

If $D_A < D_B$, the second term is relatively constant at small values of t and the first term predominates, as represented by the first line segment in Fig. 9.7D. At large values of t, the first term approaches zero and microbial numbers will be represented by the second line segment in Fig. 9.7D.

The heating time required to obtain a specified probability of spoilage from a mixed species with known D values is the longest heating time calculated using equation 11 for any of the species.

Figure 9.7 shows an inactivation curve which exhibits tailing. Tailing is often associated with very high N_0 values and with organisms which have a tendency to clump. As in the case of a lag in the inactivation curve, the effect of tailing is not considered in the thermal process calculation unless the curve is reproducible and the effects of initial number and temperature can be quantified.

EXAMPLE 1. Figure 9.8 shows data on inactivation of spores of F.S. 1518 reported by Berry et al. (*J. Food Sci.* 50:815, 1985). When 6×10^6 spores were inoculated into a can containing 400 g of product and processed at

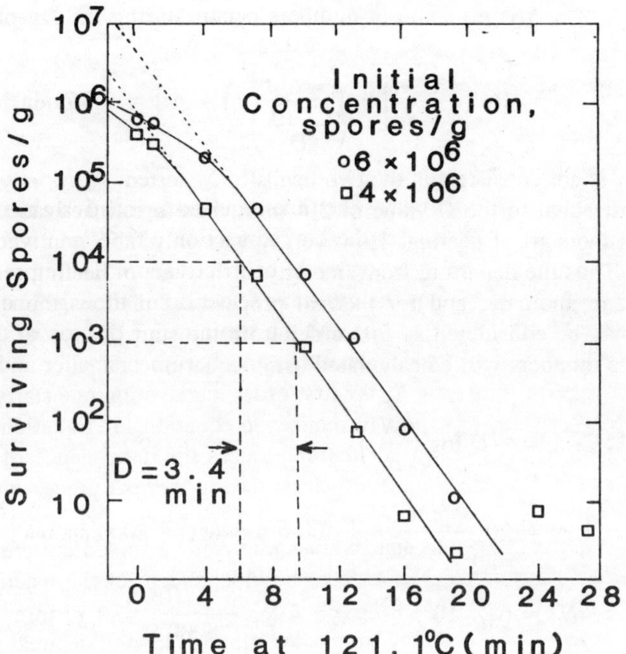

Fig. 9.8. Inactivation curves for spores of FS 1518. (From Berry, M. R., et al., *J. Food Sci.* 50:815, 1985.)

121.1°C, the processed product contained 20 spores per gram. Calculate the equivalent heating time at 121.1°C to which the product was subjected.

Solution: Both lines in Fig. 9.8 are parallel, and the D value of 3.4 min is independent of the initial number, as would be expected from either equations 11 or 12. However, both lines show a departure from linearity at the initial stage of heating. To determine if a lag time should be considered when establishing the number of survivors from a heating process, data from both thermal inactivation curves will be fitted to equation 12 to determine if t_L is consistent with different initial numbers. A point on each plot is arbitrarily picked to obtain a value for N and t. Choosing a value of $N = 10/g$, the time required to reduce the population from N_0 to N is 20 and 16.2 min, respectively, for $N_0 = 6 \times 10^6$ and 4×10^6. Using equation 12:
For $N_0 = 6 \times 10^6$:

$$t_L = 20 - 3.4\left[\ln\left(\frac{6 \times 10^6}{10}\right) - 1\right] = 3.75 \text{ min}$$

For $N_0 = 4 \times 10^6$:

$$t_L = 16.2 - 3.4\left[\ln\left(\frac{4 \times 10^6}{10}\right) - 1\right] = 0.56 \text{ min}$$

Thus, t_L is not consistent at the two levels of N_0 tested. At $N_0 = 6 \times 10^6$, t_L is almost equal to the D value of 3.4 min; therefore, inactivation was first order at the start of heating. However, t_L was only 0.56 min when $N_0 = 4 \times 10^6$. Thus the departure from linearity at the start of heating may be considered an anomaly, and for thermal process calculations the inactivation curve may be considered as first order from the start of heating. Reduction of viable numbers will be calculated using equation 11.

$$t = -D \log\left(\frac{N}{N_0}\right)$$

$$N = 20 \frac{\text{spores}}{\text{g product}} \times 400 \text{ g product} = 8000 \text{ spores}$$

$$N_0 = 6 \times 10^6; \qquad t = 3.4 \log \frac{6 \times 10^6}{8000} = 9.77 \text{ min}$$

Thus, the equivalent lethality of the process is 9.77 min at 121.1°C.

EXAMPLE 2: A suspension containing 3×10^5 spores of organism A having a D value of 1.5 min at 121.1°C and 8×10^6 spores of organism B having a D value of 0.8 min at 121.1°C is heated at a uniform constant temperature of 121.1°C. Calculate the heating time for this suspension at 121.1°C needed to obtain a probability of spoilage of 1/1000.

Solution: Using equation 11:

$$\text{For organism A: } t = 1.5 \log\left(\frac{3 \times 10^5}{0.001}\right) = 12.72 \text{ min}$$

$$\text{For organism B: } t = 0.8 \log\left(\frac{8 \times 10^6}{0.001}\right) = 7.92 \text{ min}$$

Thus, the required time is 12.72 min.

Sterilizing Value or Lethality of a Process. The basis for process lethality is equation 11, the destruction of biological entities in a heated material. The following may be used as a means of expressing the sterilizing value of a process:

$$S = \text{number of decimal reduction} = \frac{N_0}{N}$$

F_T = Process time at constant temperature T which has the equivalent lethality of the given process. Usually F values are expressed at a reference temperature (121.1°C for a sterilization process or 82.2°C for a pasteurization process). The F and D value at 121.1°C are F_0 and D_0, respectively.

In a constant temperature process, S and F values can be easily interconverted using equation 11, with F values substituted for t.

$$S = \frac{F_T}{D_T} \tag{14}$$

However, if the suspension is heated under changing temperature conditions such as the interior of a can during thermal processing, equation 14 can be used only if the F value is calculated using the same z value as the biological entity represented by D_T. The z value is the parameter for temperature dependence of the inactivation rate and will be discussed in more detail in the section "Effect of Temperature on Thermal Inactivation of Microorganisms."

The use of S values to express process lethality is absolute, i.e., S is the expected effect of the thermal process. However, an S value represents a specific biological entity, and when several entities must be inactivated, the S value for each one may be calculated from the F value using equation 14. The use of F values to express process lethality and its calculation under conditions of changing temperature during a process will be discussed in more detail in the section "Sterilizing Value of Processes Expressed as F_0."

Acceptable Sterilizing Value for Processes. A canned food is processed to achieve commercial sterility. Commercial sterility implies the inactivation of all microorganisms that endanger public health to a very low probability of survival. For canned foods, the critical organism is *Clostridium botulinum*. The 12 decimal (12D) concept as a minimum process for inactivation of *C. botulinum* in canned foods is accepted in principle by regulatory agencies and the food industry. However, its interpretation has changed from a literal 12 decimal reduction to what is now generally accepted as a probability of survival of 10^{-12}.

The latter interpretation signifies a dependence of minimum processes, according to the 12 D concept, on initial spore loads. Thus, packaging materials which have very low spore loads will not require as severe a process as products such as mushrooms, which may have very high spore levels. Table 9.1 shows N_0 and N values which may be used as a guide in selecting target N_0/N values for thermal processing.

Spoilage from microorganisms which pose no danger to public health is called *economic spoilage*. Spoilage microorganisms often have higher heat resistance

Table 9.1. N and N_0 Values Used to Obtain Target ln (N_0/N) Values for Commercial Sterility of Canned Foods

Factor	N	N_0	D_{m0}
Public health	10^{-9}	General 10 Meats 10^2 Mushrooms 10^4 Packaging 10^{-5}	0.2
Mesophilic spoilage	10^{-6}	General 10 Meats 10^3	0.5
Thermophilic spoilage	10^{-2}	General 10^2	1.5

Source: From Pflug, I. V., *J. Food Protect.* 50:342, 50:347, 50:528, 1987. Reprinted from Toledo, R. T., *Food Technol.* 44(2):72, 1990.

than *C. botulinum*, and their inactivation is the basis for the thermal process design. A high level of spoilage is expected from the minimum process based on the 12 D concept for *C. botulinum* inactivation.

EXAMPLE 1: The F value at 121.1°C equivalent to 99.999% inactivation of a strain of *C. botulinum* is 1.2 min. Calculate the D_0 value of this organism.

Solution: A 99.999% inactivation is 5 D reductions (one survivor from 100,000). $S = 5$. Using equation 14:

$$D_0 = \frac{F_0}{S} = \frac{1.2}{5} = 0.24 \text{ min}$$

EXAMPLE 2: Calculate F_0 based on the 12 D concept using the D_0 value of *C. botulinum* in Example 1 and a most likely spore load in the product of 100.

Solution: $S = \log 100 - \log(10^{-12}) = 14.$

$$F_0 = 14\,(0.24) = 3.30 \text{ min}$$

EXAMPLE 3: The sterilizing value of a process has been calculated to be an F_0 of 2.88. If each can contains 10 spores of an organism having a D_0 of 1.5 min, calculate the probability of spoilage from this organism. Assume that the F_0 value was calculated using the same z value as the organism.

Solution: Using equation 11 for a process time of 2.88 min:

$$\log \frac{N_0}{N} = \frac{2.88}{1.5}; \qquad N = N_0[10]^{-(2.88/1.5)}$$

$$N = 10(10^{-1.92}) = 0.12; \qquad P_{\text{spoilage}} = 12 \text{ in } 100 \text{ cans}$$

EXAMPLE 4: The most probable spore load in a canned food is 100, and the D_0 of the spore is 1.5 min. Calculate a target F_0 for a thermal process such that the probability of spoilage is 1 in 100,000. If under the same conditions *C. botulinum* type B has a D_0 of 0.2 min, would the target F_0 value satisfy the minimum 12 D process for *C. botulinum*? Assume an initial spore load of 1 per can for *C. botulinum*.

Solution: For the organism, $S = \log(100/10^{-5}) = 7$. Using equation 14: $F_0 = 7(1.5) = 1.5$ min. for *C. botulinum*: $S = \log(1/10^{-12}) = 12$. Using equation 14: F_0, min $= 12(0.2) = 2.4$ min. The F_0 for the spoilage organism satisfies the minimum process for 12 D of *C. botulinum*.

Selection of Inoculation Levels in Inoculated Packs. In order to be reasonably sure of the safety of a process, products may be inoculated with an organism having known heat resistance, processed, and the extent of spoilage compared to the probability of spoilage designed into a process. The organism used must have a higher heat resistance than the background microflora in the product. The level of spoilage and the inoculation level are set such that spoiled cans can be easily evaluated. Use of an organism which produces gas facilitates the detection of spoiled cans, since spoilage will be manifested by swelled cans. If a flat sour organism is used, it will be necessary to open the cans after incubation to determine the number of spoiled cans. If the whole batch of inoculated cans is incubated, the fraction spoiled will be equivalent to the decimal equivalent of the number of surviving organisms. Inoculation levels can be calculated based on equation 11.

The two examples below represent procedures used for validation of thermal processes using inoculation test. The first example represents an incubation test, and the second example represents a spore count reduction test.

EXAMPLE 1: A process was calculated such that the probability of spoilage from an organism with a D_0 value of 1 min is 1 in 100,000 from an initial spore load of 100. To verify this process, an inoculated pack is made. Calculate the level of inoculum of an organism having a D_0 value of 1.5 min which must be used on 100 cans such that a spoilage rate of 5 cans will be equivalent in lethality to the calculated process.

Solution: The F_0 of the calculated process is determined using equation 11 with F_0 substituted for *t*.

$$F_0 = 1\left[\log\left(\frac{100}{1 \times 10^{-5}}\right)\right] = 7 \text{ min}$$

For the inoculum:

$$\log N_0 - \log\left(\frac{5}{100}\right) = \frac{F_0}{D} = 7/1.5 = 4.667$$

$$N_0 = 0.05(10)^{4.667} = 2323 \text{ spores}$$

EXAMPLE 2: In an incidence of spoilage, the isolated spoilage organism was found to have a D_0 value of 1.35 min. It is desired that the probability of spoilage from this organism be 1 in 100,000. Initial spore loads were generally on the order of 10 per can. Calculate the required F_0 for this process to achieve the desired probability of spoilage. If an inoculated pack of *FS* 1518 is to be made and an initial inoculation level of 4×10^5 spores is made into cans which contained 200 g of product, what will be the spore count in the processed product such that the lethality received by the can contents will be equivalent to that achieved by the desired process for eliminating spoilage from the isolated organism? The D_0 value of FS 1518 = 2.7 min.

Solution: The F_0 of the process can be calculated from the heat resistance and spore reduction needed for the spoilage organism.

$$F_0 = D(\log N_0 - \log N) = 1.35\left[\log 10 - \log\left(\frac{1}{100000}\right)\right]$$

$$= 8.1 \text{ min}$$

For FS 1518:

$$\log N - \log(5 \times 10^5) = \frac{-8.1}{2.7} = -3.00$$

$$N = 10\,(-3.00 + 5.698) = 500$$

$$\text{Spore count after processing} = 500 \text{ spores}/200 \text{ g} = 2.5/\text{g}$$

Determination of *D* Values Using the Partial Sterilization Technique. This technique, developed by Stumbo et al. (*Food Technol.* 4:321, 1950) and by Schmidt (*J. Bacteriol.* 59:433, 1950) allows the determination of D values using survivor data at two heating times. Appropriate selection of heating times to exceed the lag time needed for the heated suspension to reach the specified test temperature eliminates the need to correct for temperature changes occurring during the transient period of heating. When this procedure is used, the lag time needed to bring the suspension to the desired temperature must be established, and the two heating times used for D value determination must exceed the lag time. If t_1 and t_2 are the heating times and N_1 and N_2 are the respective number of survivors, the D value is determined by:

$$D = \frac{t_2 - t_1}{\log (N_1) - \log (N_2)} \tag{15}$$

EXAMPLE: Sealed tubes containing equal numbers of spores of an isolate from a spoiled canned food were heated for 10 and 15 min at 115.5°C. The survivors were, respectively, 4600 and 160. Calculate the D value. The lag time for heating the tubes to 115.5°C was established in prior experiments to be 0.5 min.

Solution: Since the heating times are greater than the lag time for the tubes to attain the desired heating temperature, equation 15 can be used.

$$D = \frac{15 - 10}{\log (4600) - \log (160)} = \frac{5}{1.458} = 3.42 \text{ min}$$

The Heat Resistance of Spoilage Microorganisms. The heat resistance of microorganisms is expressed in terms of a D value at a reference temperature and the z value, the temperature dependence of the thermal inactivation rate. The use of the z value in determining D values at different temperatures from D at the reference temperature is discussed in the section "Effect of Temperature on Thermal Inactivation of Microorganisms." Reference temperatures are 121.1°C (250°F) for heat-resistant spores to be inactivated in commercial sterilization processes and 82.2°C (180°F) for vegetative cells and organisms of low resistance which are inactivated in pasteurization processes. D at 121.1°C (250°F) is D_0.

Tables 9.2 and 9.3 lists the resistance of microorganisms involved in food spoilage.

The type of substrate surrounding the organisms during the heating process affects heat resistance. For a more detailed discussion of techniques for determination of microbial heat resistance and the effect of various factors on thermal inactivation rates, the reader is referred to Stumbo's (1973) book and the NFPA (1968) work listed in the Suggested Reading list at the end of this chapter.

F_0 Values Used in Commercial Sterilization of Canned Foods. D_0, N_0 and N values which can be used as a guide for determining F_0 values for food sterilization are listed in Table 9.1. F_0 must be based on the microorganisms involved in economic spoilage, since these have a higher heat resistance

Table 9.2. Heat Resistance of Spoilage Microorganisms in Low-Acid Canned Foods

Organism	Product	D_0 (min)	z (°F)	z (°C)
Clostridium botulinum 213-B	Phosphate buffer (pH 7)	0.16	18	10
	Green beans	0.22	22	12
	Peas	0.22	14	8
Clostridium botulinum 62A	Phosphate buffer (pH 7)	0.31	21	12
	Green beans	0.22	20	11
	Corn	0.3	18	10
	Spinach	0.25	19	11
Clostridium spp. PA 3679	Phosphate buffer (pH 7)	1.45	21	12
	Asparagus	1.83	24	13
	Green beans	0.70	17	9
	Corn	1.20	18	10
	Peas	2.55	19	10
	Shrimp	1.68	21	12
	Spinach	2.33	23	13
Bacillus stearothermophilus FS 1518	Phosphate buffer (pH 7)	3.28	17	9
	Asparagus	4.20	20	11
	Green beans	3.96	18	10
	Corn	4.32	21	12
	Peas	6.16	20	11
	Pumpkin	3.50	23	13
	Shrimp	3.90	16	9
	Spinach	4.94	21	12

Source: Reed, J. M., Bohrer, C. W. and Cameron, E. J., *Food Res.* 16:338–408.
Reprinted from: Toledo R. T. 1980. Fundamentals of Food Engineering. AVI Pub. Co., Westport, CT.

Table 9.3. Heat Resistance of Spoilage Microorganisms in Acid and in Pasteurized Foods

Organism	Temperature °F	Temperature °C	D (min)	z °F	z °C
Bacillus coagulans	250	121.1	0.07	18	10
Bacillus polymyxa	212	100	0.50	16	9
Clostridium pasteurianum	212	100	0.50	16	9
Mycobacterium tuberculosis	180	82.2	0.0003	10	6
Salmonella spp.	180	82.2	0.0032	12	7
Staphylococcus spp.	180	82.2	0.0063	12	7
Lactobacillus spp.	180	82.2	0.0095	12	7
Yeasts and molds	180	82.2	0.0095	12	7
Clostridium botulinum Type E	180	82.2	2.50	16	9

Source: (1) Anderson, E. E., Esselen Jr., W. B. and Fellers, C. R. *Food Res.* 14:499–510, 1949. (2) Crissley, F. D., Peeler, J. T., Angelotti, R. and Hall, H. E., *J. Food Sci.* 33:133–137, 1968. (3) Stumbo, C. R. *Thermobacteriology in Food Processing*, Academic Press, New York, 1973. (4) Townsend, C. T. *Food Res.* 4:231–237, 1939. (5) Townsend, C. T., and Collier, C. P. *Proc. Technical Session of the 48th Annual Convention of the National Canners Association (NCA). NCA information Newsl.* No. 1526, February 28, 1955. (6) Winter, A. R., Stewart, G. F., McFarlane, V. H. and Soloway, M. *Am. J. Pub. Health* 36:451–460, 1946. (7) Zuccharo. J. B., Powers, J. J., Morse, R. E. and Mills W. C. *Food. Res.* 16:30–38, 1951.
Reprinted from: Toledo, 1980. Fundamentals of Food Process Engineering, 1st. ed. AVI Pub. Co. Westport, Conn.

Table 9.4. Values of F_0 for Some Commercial Canning Processes

Product	Can sizes	F_0 (min)
Asparagus	All	2–4
Green beans, brine packed	No. 2	3 · 5
Green beans, brine packed	No. 10	6
Chicken, boned	All	6–8
Corn, whole kernel, brine packed	No. 2	9
Corn, whole kernel, brine packed	No. 10	15
Cream style corn	No. 2	5–6
Cream style corn	No. 10	2 · 3
Dog Food	No. 2	12
Dog Food	No. 10	6
Mackerel in brine	301 × 411	2 · 9–3 · 6
Meat loaf	No. 2	6
Peas, brine packed	No. 2	7
Peas, brine packed	No. 10	11
Sausage, Vienna, in brine	Various	5
Chili con carne	Various	6

Source: Alstrand, D. V., and Ecklund, O. F., *Food Technol.* 6(5):185, 1952.

than *C. botulinum*. Table 9.4 lists F_0 values previously used commercially for different types of foods in various size containers. The data in Tables 9.1 and 9.4 may be used as a base for selection of the F_0 values needed to calculate thermal process schedules for sterilization.

Surface Sterilization. Packaging materials used in aseptic packaging systems and surfaces of equipment may be sterilized using moist heat, dry heat, hydrogen peroxide, high-intensity ultraviolet light, and ionizing radiation from either gamma rays or high-energy electron beams. The last three methods have not been adopted in commercial food packaging, but various forms of heat and hydrogen peroxide combined with heat are commercially utilized.

Dry heat includes superheated steam and hot air. Resistances of microorganisms in these heating media are shown in Table 9.5. Inactivation occurs at a slower rate in dry heat compared to moist heat at the same temperature. The z value in dry heat is also higher than in moist heat. A similar principle is utilized in evaluating microbial inactivation in dry heat and in moist heat. The D values shown in Table 9.5, and the recommended N and N_0 values for commercial sterilization in Table 9.1 may be used to determine exposure times to the sterilant. In dry heat sterilization of surfaces, surface temperatures must be used as the basis for the process rather than the temperature of the medium. The surface heat transfer coefficient and the temperature on the opposite side of the surface sterilized determine the actual surface temperature.

Hydrogen peroxide is the only chemical sterilant allowed for use on food contact surfaces. This compound at 35% (w/w) concentration is applied to the

Table 9.5. Resistance of Microorganisms to Microbicidal Agents

Organism	Heating Medium	$D_{176.6°C}$ (min)	$D_{121.1°C}$ (min)	z (°C)
Bacillus subtilis	Superheated steam	0.57	137	23.3
Bacillus stearothermophilus (FS 1518)	Superheated steam	0.14	982	14.4
Bacillus polymyxa	Superheated steam	0.13	484	15.6
Clostridium sporogenes (P.A. 3679)	Air	0.30	109	21.7
Clostridium botulinum	Air	0.21	9	33.9
Bacillus subtilis	N_2 + He	0.17	285	17.2
Clostridium sporogenes (P.A. 3679)	He	0.45	161	21.7
Bacillus subtilis	A + CO_2 + O_2	0.13	218	17.2

Source: Adapted from Miller, B. M., and Litskey, W. eds, 1976. *Industrial Microbiology*, McGraw-Hill. Used with permission of McGraw-Hill.

surface by atomizing, by spraying, or by dipping, in the case of packaging materials in sheet form, followed by heating to vaporize the hydrogen peroxide and eliminate the residue from the surface. A maximum tolerance of residual hydrogen peroxide in the package of 0.1 part per million is required by federal regulations in the United States. Resistance of various microorganisms in hydrogen peroxide is summarized in Table 9.6. The inactivation rate is also temperature dependent.

EFFECT OF TEMPERATURE ON THERMAL INACTIVATION OF MICROORGANISMS

Microbial inactivation is a first-order chemical reaction, and the temperature dependence of the rate constant can be expressed in terms of an activation energy or a z value (see the section "The Reaction Rate Constant"). Equation 21 in Chapter 8 expresses the rate constant for inactivation in terms of the activa-

Table 9.6. Resistance of Food Spoilage Microorganisms to Inactivation in Hot Hydrogen Peroxide

Organism	$D_{80°C}$ (min)	z (°C)
Clostridium botulinum 169B	0.05	29
Bacillus subtilis ATCC 9372	0.063	41
Bacillus subtilis A	0.037	25.5
Bacillus subtilis	0.027	27
Bacillus stearothermophilus	0.07	22

Source: Toledo, R. T., *AIChE Symp. Ser.* 78(218):81, 1982. Reproduced by permission of the American Institute of Chemical Engineers, © 1982, AIChE.

tion energy. The activation energy is negative for reactions which increases in rate with increasing temperature.

$$\frac{k}{k_0} = [e]^{-E_a/R[1/T - 1/T_0]} \tag{16}$$

Since $k = 2.303/D$ from the section "Rate of Microbial Inactivation," the temperature dependence of the D value in terms of the activation energy is

$$\frac{D}{D_0} = [e]^{E_a/R[1/T - 1/T_0]} \tag{17}$$

Since E_a is negative when reaction rates increase with temperature, equation 17 represents a decrease in D value with increasing temperature.

Thermobacteriologists prefer to use the z value to express the temperature dependence of chemical reactions. The z value may be used on the target F value for microbial inactivation or on the D value to determine required heating times for inactivation at different temperatures. It may also be used on heating times at one temperature to determine equivalence in lethality at a reference temperature. A semilogarithmic plot of heating time for inactivation against temperature is called the *thermal death time* plot. Therefore, equations based on this linear semilogarithmic plot are called *thermal death time* model equations for microbial inactivation at different temperatures.

$$\log \frac{F}{F_0} = \frac{T_0 - T}{z} \tag{18}$$

$$\log \frac{D}{D_0} = \frac{T_0 - T}{z} \tag{19}$$

$$\log \frac{t_0}{t_T} = -\frac{T_0 - T}{z} \tag{20}$$

t_0 is the equivalent heating time at $T_0 = 250°F$ of heating for time t_T at temperature T. When $t_T = 1$, t_0 is the lethality factor, L, the equivalent heating time at 250°F for 1 min at T.

$$L = [10]^{T - T_0/z} \tag{21}$$

The inverse of L is the heating time at T equivalent to 1 min at 250°F and is the parameter F_i used in thermal process calculations.

$$F_i = [10]^{T_0 - T/z} \tag{22}$$

Use of the z value is not recommended when extrapolating D values over a very large temperature range. Equation 17 shows that D will deviate from a linear semilogarithmic plot against temperature when the temperature range between T and T_0 is very large.

EXAMPLE 1: The F_0 for 99.999% inactivation of *C. botulinum* type B is 1.1 min. Calculate F_0 for 12 D inactivation and the F value at 275°F (135°C) when $z = 18°F$.

Solution: Here 99.999% inactivation is equivalent to $S = 5$. For $S = 12$, $F_0 = SD_0 = 12(0.22) = 2.64$ min. Thus the D_0 value is $1.1/5 = 0.22$ min. F_{275} can be calculated using equation 18 or using equation 19 on the D value to obtain D at 275°F.
Using equation 18: $(T_0 - T)/z = -25/18 = -1.389$.

$$F_{275} = F_0(10^{-1.389}) = 2.64(0.0408) = 0.1078 \text{ min}$$

Using equation 19: $D_{275} = 0.22(10^{-1.389}) = 0.00898$ min.

$$F_{275} = SD = 12(0.00898) = 0.1077 \text{ min}$$

EXAMPLE 2: The D_0 for PA 3679 is 1.2 min, and the z value is 10°C. Calculate the process time for 8 D inactivation of PA 3679 at 140.5°C using the thermal death time model (equation 18) and the Arrhenius equation (equation 17). The z value was determined using data on D at 115.5° to 121.1°C.

Solution: Using equation 18:
$(T - T_0)/z = (140.5 - 121.1)/10 = -1.94$.
$F_0 = 8(1.2) = 9.6$ min. $F_{140.5} = 9.6(10^{-1.94})$
$= 9.6(0.01148) = 0.11$ min.
From equation 25, Chapter 8:

$$z = \frac{\ln(10)R}{E_a} T_1 T_2; \qquad E_a = \frac{\ln(10)RT_2T_1}{z}$$

$$T_2 = 121.1 + 273 = 394.1 \text{ K}; \qquad T_1 = 115.5 + 273 = 388.5°\text{K};$$

$$R = \frac{1.987 \text{ cal}}{(\text{gmole} \cdot \text{K})}$$

$$E_a = \frac{[\ln(10)](1.987)(388.5)(394.1)}{10}$$

$$= 70{,}050 \text{ cal/gmole}; \quad \frac{E_a}{R} = 35{,}254$$

$$T = 140.5 + 273 = 413.5\text{ K}; \qquad T_0 = 394.1\text{ K}$$

$$\left(\frac{1}{T} - \frac{1}{T_0}\right) = -0.000119$$

$$F_{140.5} = F_0[e]^{35354(-0.000119)}$$

$$= 9.6(0.015067) = 0.1446\text{ min}$$

These calculations show that when extrapolating D values to high sterilization temperatures using the z value based on data obtained below 250°F (121.1°C), use of the Arrhenius equation will result in a safer process compared to the theımal death time model.

STERILIZATION OF CONTINUOUSLY FLOWING FLUIDS

The system for sterilization or pasteurization of continuously flowing fluids is discussed in the section "Continuous Flow Sterilization: Aseptic or Cold Fill." The process is essentially a constant temperature heating process, with the residence time in the holding tube considered as the processing time and the temperature at the point of exit from the holding tube as the processing temperature at the point of exit from the holding tube as the processing temperature.

Sterilization of Fluids. The velocity distribution in a fluid flowing through a pipe has been derived for a power law fluid in laminar flow. Equation 16 of Chapter 6 is:

$$V = \overline{V}\left(\frac{3n + 1}{n + 1}\right)\left[1 - \left[\frac{r}{R}\right]^{(n+1)/n}\right] \tag{23}$$

Consider an area element of thickness dr within a tube of radius R: If N_0 = the total number of organisms entering the tube and N = the number leaving, n_0 = the organisms per unit volume entering and n = organisms per unit volume leaving:

$$N_0 = n_0\left[2\pi \int_0^R Vr\, dr\right] \tag{24}$$

The expression in brackets in equation 24 is the volumetric rate of flow, $\pi R^2 \overline{V}$. Thus equation 24 becomes:

$$N_0 = \pi R^2 n_0 \overline{V} \tag{25}$$

The residence time of fluid within the area element is L/V, and the number of survivors $N = N_0[10^{-L/(VD)}]$. Thus:

$$N = 2\pi n_0 \int_0^R Vr[10]^{-L/(VD)}\, dr \tag{26}$$

In equation 23 let: $A = (3n + 1)/(n + 1)$; $B = (n + 1)/n$; and $y = r/R$. Equation 23 becomes:

$$V = A(1 - y^B)\bar{V} \tag{27}$$

The ratio $L/(\bar{V} \cdot D)$ is the lethality, S_v, based on the average velocity. Substituting equation 27 into equation 26:

$$S_v = \frac{L}{(\bar{V} \cdot D)}; \qquad r = yR; \qquad dr = R\, dy$$

$$N = 2\pi n_0 \int_0^1 \bar{V}(A)(1 - y^B)[10]^{-S_v/[A(1 - y^B)]} R^2 y\, dy \tag{28}$$

Dividing equation 25 by equation 28:

$$\frac{N_0}{N} = \frac{\pi R^2 n_0 \bar{V}}{2\pi n_0 \int_0^1 \bar{V}(A)(1 - y^B)[10]^{-S_v/[A(1 - y^B)]} R^2 y\, dy}$$

Cancelling $\pi R^2 n_0 \bar{V}$, taking the logarithm of both sides, and making the logarithm of the denominator of the right-hand side negative to bring it to the numerator:

$$\log\left(\frac{N_0}{N}\right) = -\log\left[\int_0^1 [2A(1 - y^B)[10]^{-S_v/[A(1 - y^B)]} y\, dy]\right] \tag{29}$$

$\log(N_0/N)$ in equation 29 is the integrated lethality, S_i.

Equation 29 is evaluated by graphical integration. The integrated sterility in the holding tube is independent of the size of the tube and the average velocity if the length and average velocity are expressed as S_v, the sterilization value based on the average velocity. Thus equation 29 can be solved and the solution will yield a generalized table for the integrated lethality of heat in the holding tube when fluid is flowing through that tube in laminar flow.

Values of S_i at different values of n and S_v obtained by graphical integration

Table 9.7. Integrated Lethality in the Holding Tube of a Continuous Sterilization System for Fluids in Laminar Flow

	Integrated Lethality, S_i						
	Fluid flow behavior index, n						
S_v	0.4	0.5	0.6	0.7	0.8	0.9	1.0
0.1	0.093	0.093	0.092	0.091	0.091	0.091	0.091
0.5	0.426	0.419	0.414	0.409	0.406	0.401	0.401
1	0.809	0.792	0.779	0.768	0.759	0.747	0.747
2	1.53	1.49	1.46	1.44	1.42	1.40	1.38
4	2.92	2.82	2.75	2.69	2.64	2.59	2.56
6	4.27	4.11	3.99	3.89	3.81	3.74	3.68
8	5.60	5.38	5.20	5.06	4.95	4.86	4.78
10	6.92	6.63	6.41	6.23	6.08	5.96	5.86
12	8.23	7.88	7.60	7.38	7.20	7.05	6.92
14	9.54	9.12	8.79	8.52	8.31	8.13	7.98
16	10.84	10.35	9.97	9.66	9.41	9.20	9.03
18	12.14	11.58	11.14	10.79	10.51	10.27	10.07
20	13.44	12.81	12.32	11.93	11.60	11.34	11.11
22	14.73	14.03	13.49	13.05	12.69	12.40	12.15
24	16.03	15.26	14.66	14.18	13.79	13.46	13.18
26	17.32	16.48	15.83	15.30	14.87	14.52	14.22

S_v = sterilization value as number of decimal reductions based on the average velocity $(L/D \cdot \overline{V})$.

of equation 29 are shown in Table 9.7. Examination of the data in Table 9.7 shows that the ratio S_i/S_v is approximately equal to $\overline{V}/V_{max}$. Thus, the use of the maximum velocity for calculating the lethality of heat in a holding tube results in much closer values to the integrated lethality value compared to the use of the average velocity.

EXAMPLE 1: A fluid food product with a viscosity of 5 cP and a density of 1009 kg/m^3 is to be pasteurized in a continuous system which heats the food to 85°C followed by holding in a 1.5-in. sanitary pipe, from which it leaves at 82.2°C. The process should give 12 D reductions of *Staphylococcus aureus*, which has a $D_{82.2°C}$ of 0.0063 min. Calculate the length of the holding tube if the flow rate is 19 L/min.

Solution: From Chapter 6, Table 6.2, the inside diameter of a 1.5-in. sanitary pipe is 0.03561 m.

$$\overline{V} = \frac{19\ \text{L}}{\text{min}}\left|\frac{1\ \text{m}^3}{1000\ \text{L}}\right|\frac{1\ \text{min}}{60\ \text{s}}\left|\frac{1}{\pi[0.5(0.03561)]^2\ \text{m}^2}\right.$$

$\bar{V} = 0.318$ m/s; $Re = (0.03561)(0.318)(1009)/5(0.001) = 2285$. Flow is in the transition zone from laminar to turbulent flow. Assume that flow is laminar and $V_{max} = 2\bar{V} = 2(0.318) = 0.636$ m/s.
Using equation 14: The F value is the required process time which must equal the residence time in the tube for the fastest-flowing particle. $F = t_{min} = S \cdot D = 12(0.0063) = 0.0756$ min. Using equation 4:

$$L = 0.0756 \text{ min } (60 \text{ s/min})(0.636 \text{ m/s}) = 2.88 \text{ m}$$

Another method for determining the length of the holding tube is to assume that the integrated lethality received by the total volume of fluid going through the tube must equal the required lethality. The required lethality is $S_i = 12$ D reductions of *S. aureus*. From Table 9.7, to obtain $S_i = 12$ when $n = 1$ requires interpolation between $S_v = 22$ and $S_v = 20$. $S_v = 22 - 0.15(12.15 - 11.11)/2 = 21.92$. Since $S_v = L/(\bar{V} \cdot D)$, $L = S_v \bar{V} D$. Substituting $\bar{V}$ in m/s, S_v, and D: $L = 21.92(0.318)(0.0063)(60) = 2.63$ m.

If the average velocity is used to calculate the length of the holding tube, $S_v = 12$ and $L = 12(0.318)(0.0063)(60) = 1.44$.

It is obvious that basing the lethality of heat in the folding tube on the average velocity greatly underestimates the extent of microbial inactivation. The integrated lethality gives a length much closer to that based on the maximum velocity.

At the Reynolds number in the range 3000 to 5000, the maximum velocity cannot be obtained using equation 7. Thus, the safest approach to determining the holding tube length in this range of Reynolds numbers is to assume laminar flow.

EXAMPLE 2: An ice cream mix having a viscosity of 70 cP and a density of 1015 kg/m^3 is being canned aseptically in a system which uses a 100-ft-long, 1.0-in. sanitary pipe for a holding tube. The flow rate through the system is 5 gal/min. The fluid temperature at the exit from the holding tube is 285°F (140.6°C).
Calculate (a) the sterilizing value of this process for PA 3679 based on the maximum velocity and (b) and integrated sterilizing value expressed as a decimal reduction of PA 3679 ($D_0 = 1.83$ min; $z = 24$°F).

Solution: From Table 6.2: $D = 0.02291$ m; $R = 0.01146$ m.
Solving for the average velocity:

$$\bar{V} = \frac{5 \text{ gal}}{\text{min}} \left| \frac{3.78541 \times 10^{-3} \text{ m}^3}{\text{gal}} \right| \frac{1 \text{ min}}{60 \text{ s}} \left| \frac{1}{\pi(0.01146)^2 \text{ m}^2} \right.$$

$\overline{V} = 0.765$ m/s. $Re = 0.02291(0.765)(1015)/70(0.001) = 254$. Flow is laminar. L = 100 ft(1 m/3.281 ft) = 30.48 m.

(a) Solving for V_{max}:

$$V_{max} = 2\overline{V} = 2(0.785) = 1.57 \text{ m/s}$$

The residence time of the fastest-flowing particle is:

$$t = \frac{L}{V_{max}} = 19.41 \text{ s}$$

The process occurs at 285°F; therefore, the number of survivors from the process is calculated using the D value at 285°F for D in equation 11. The D value at 285°F is determined using equation 19:

$$\log D = \frac{\log D_0 + (T_0 - T)}{z}$$

$$= \frac{\log[(1.83)(60)] + (250 - 285)}{24} = 2.0406 - 1.458$$

$$D = 10^{0.5826} = 3.8247 \text{ s at } 285°\text{F}$$

Using equation 11: $S = 19.41/3.8247 = 5.075$.

Since t determined using V_{max} is the minimum residence time for any particle flowing through the holding tube, the calculated lethality using this minimum time will be the least possible. At least 5.075 D reduction of PA 3679 will result from the process.

(b) The number of decimal reductions based on the integrated lethality will now be calculated:

$$S_v = \frac{\text{L}}{(\overline{V} \cdot D)} = \frac{30.48}{[(0.765)(3.8247)]} = 10.42$$

From Table 9.6, for $S_v = 10.42$ and $n = 1$, the value of S_i will be obtained by interpolation between $S_v = 10$ and $S_v = 12$. $S_i = 5.86 + 0.42(6.92 - 5.86)/2 = 6.08$. The integrated lethality is 6.08 D reductions of PA 3679.

Nutrient Degradation. Nutrient degradation can be calculated in the same way as for microbial inactivation. Equations 24 to 29 are also applicable for loss of nutrients if the appropriate kinetic parameters of the D and z values are used.

Nutrient retention with increasing processing temperature can be derived by simultaneously solving the rate equations for microbial inactivation and nutrient degradation. Using the subscripts c and m to signify parameters for nutrient degradation and microbial inactivation, respectively, equations 11 and 19 may be combined to determine the heating time at various temperatures for microbial inactivation.

$$t = D_{m0}\left[\log\left(\frac{N_0}{N}\right)\right][10]^{T_0 - T/z_m} \tag{30}$$

A similar expression can be formulated for nutrient degradation.

$$\log\left(\frac{C}{C_0}\right) = -\frac{t}{D_{c0}[10]^{T_0 - T/z_c}} \tag{31}$$

Substituting S_v for log (N_0/N) and combining equations 30 and 31:

$$\log\left(\frac{C}{C_0}\right) = -\left[\frac{D_{m0}}{D_{c0}}S_v\right][10]^{(T_0 - T)(1/z_m - 1/z_c)} \tag{32}$$

Equation 32 is based on constant temperature processes and is applicable for estimating nutrient retention in holding tubes of aseptic processing systems. The derivation is based on plug flow of fluid at the average velocity through the tube, and adjustments must be made to obtain the integrated lethality and nutrient degradation. Table 9.7 can be used to determine the S_v needed to obtain the microbial lethality required, expressed as S_i. $-$Log (C/C_0) will be the S_v for nutrient degradation which can be converted to S_i using Table 9.7.

EXAMPLE 1: The data for thiamin inactivation at 95° to 110°C (from Morgan et al., *J. Food Sci.* 51:348, 1986) in milk show a $D_{100°C}$ value of 3×10^4 s and a z value of 28.4°C. Chocolate milk with a flow behavior index of 0.85 and a consistency index of 0.06 Pa · s^n is to be sterilized at 145°C. The density is 1006 kg/m^3. If the rate of flow is 40 L/min and the holding tube is 1.5-in. sanitary pipe, calculate the holding tube length necessary to give an integrated lethality of 7 D reductions of an organism having a D_0 value of 0.5 min and a z value of 10°C. Assume that the z value of the organism was determined in the temperature range which included 145°C. Calculate the retention of thiamin after this process.

Solution: Since the z value for thiamin was determined at low temperatures, equation 17 will be used to extrapolate the D value to 145°C.
From equation 25, Chapter 8:

$$E_a = \frac{\ln(10)R}{z} T_1 T_2 = \frac{[\ln(10)](1.987)(368)(383)}{28.4}$$

$E_a = 22706$; $E_a/R = 11427$; $T = 145 + 273 = 418$ K; $T_r = 100 + 273 = 373$ K.

Using equation 17 and the subscript c to represent chemical degradation:

$$\left(\frac{1}{T} - \frac{1}{T_r}\right) = \left(\frac{1}{418} - \frac{1}{373}\right) = -0.000289$$

$$D_c = 3 \times 10^4 (e)^{11427(-0.000289)} = 1109 \text{ s}$$

Equation 19 is used to determine D_m at 145°C.

$$\frac{T_0 - T}{z} = \frac{121.1 - 145}{10} = -2.39$$

$$D_m = 0.5(10)^{-2.39} = 0.002037 \text{ min} = 0.1222 \text{ s}$$

Equation 30 will be used, but since D_m and D_c are already calculated at 145°C, $T = T_0$ and equation 30 becomes:

$$\log\left(\frac{C}{C_0}\right) = -\frac{D_m}{D_c} S_{vm} \qquad (30a)$$

S_{vm} is the sterilizing value for the microorganisms based on the average velocity.

The Reynolds number is:

$$Re = \frac{8V^{2-n}R^n \rho}{K[3 + 1/n]^n}$$

From Table 6.2, $R = 0.5(0.03561) = 0.017805$ m.

$$\bar{V} = \frac{40 \text{ L}}{\text{min}} \left| \frac{1 \text{ min}}{60 \text{ s}} \right| \frac{\text{m}^3}{1000 \text{ L}} \left| \frac{1}{\pi(0.017805)^2 \text{ m}^2} \right. = 0.669 \text{ m/s}$$

$$\frac{3n + 1}{n} = 4.176$$

$$Re = \frac{8(0.669)^{1.15}(0.017805)^{0.85}(1006)}{0.06(4.176)^{0.85}} = 816.7$$

Flow is laminar, and the integrated lethality can be evaluated using Table 9.7.

From Table 9.7, obtain S_v for S_i and 7 and $n = 0.85$ by interpolation between $S_v = 10$ and 12 and $n = 0.8$ and 0.9:

$$n = 0.8; \qquad S_v = 10 + \left(\frac{2}{1.12}\right)(0.92) = 11.64$$

$$n = 0.9; \qquad S_v = 10 + \left(\frac{2}{1.09}\right)(1.04) = 11.90$$

$$n = 0.85; \qquad S_v = 11.64 + \left(\frac{0.26}{0.1}\right)(0.05) = 11.77$$

Using equation 30a: $D_c = 1109$ s, $D_m = 0.1222$ s, and $S_{vm} = 11.77$.

Solving equation 30a, (C/C_0) is the nutrient retention based on the average velocity of fluid flow.

$$\log\frac{C}{C_0} = -\frac{0.1222}{1109}11.77 = -0.001317$$

It is necessary to determine the nutrient retention based on the integrated lethality. Table 9.7 is used to convert $\log(C/C_0)_v$ based on the average fluid velocity to an integrated $\log(C/C_0)_i$ based on the total volume of fluid flowing through the tube. However, an examination of Table 9.7 shows that when S_v is less than 0.1, S_i approaches the value of S_v. Thus, for this example problem, $S_v = S_i$ for the nutrient.

Thus $C/C_0 = 10^{-0.001317} = 0.997$, or 99.7% retention.

EXAMPLE 2: Tomato paste ($n = 0.5$, $K = 7.9$ Pa · s^n, $\rho = 1085$ kg/m^3) is sterilized at 95°C, using a holding tube with an inside diameter of 0.03561 m. The system operates at the rate of 50 L/min. Each package to be filled with the sterilized product contains 200 L, and the probability of spoilage to be expected from the process is 1 in 10,000 from spores of *Bacillus polymyxa*, which has $D_{80°C} = 0.5$ min and $z = 9$°C.

The unprocessed paste contains four spores per milliliter. Calculate the length of the holding tube necessary to achieve an integrated sterility which is the desired spoilage probability. Calculate the extent of nonenzymatic browning which occurs during this process, expressed as a percentage increase over the brown color before processing. Assume that the D_0 value of nonenzymatic browning is 125 min at 80°C and that the z value is 16°C.

Solution: Determine if flow is laminar or turbulent.

$$\bar{V} = \frac{50\ \text{L}}{\text{min}}\left|\frac{1\ \text{min}}{60\ \text{s}}\right|\frac{\text{m}^3}{1000\ \text{L}}\left|\frac{1}{\pi[(0.5)(0.03561)]^2\ \text{m}^2}\right.$$

$$= 0.837\ \text{m/s}$$

$$Re = \frac{8(\bar{V})^{2-n}R^n\rho}{K[3 + 1/n]^n}$$

$$\left[\frac{(3n+1)}{n}\right]^n = 2.24$$

$$Re = \frac{8(0.837)^{1.5}(0.017805)^{0.5}(1085)}{7.9(2.24)} = 50$$

Flow is laminar. For each 200-L package, $N_0 = 80{,}000$, $N = 1/10{,}000$, and $\log(N_0/N) = 9.9$. This is based on the desired spoilage probability to be evaluated as an integrated sterility. Thus, $S_i = 9.9$. From Table 9.7, to obtain $S_i = 9.9$ for $n = 0.5$, S_v is obtained by interpolation between $S_v = 14$ and $S_v = 16$. $S_v = 14 + (2/1.23)(9.9 - 9.12) = 15.27$. Since $S_v = L/(\bar{V} \cdot K) = 15.27$, $L = S_v\bar{V}D$.

Solving for D at 95°C from D_{80} and the z value:

$$D_{95} = D_{80}[10]^{(80-95)/9} = 0.0215(0.5) = 0.01077\ \text{min}$$

Substituting and solving for L:

$$\text{L} = S_v\bar{V}D = 15.27(0.837\ \text{m/s})(0.01077\ \text{min})(60\ \text{s/min})$$

$$= 8.25\ \text{m}$$

Equation 32 is used for determining the increase in brown color which results from the process. Since the reaction involves the appearance of a brown color, the negative sign in equation 32 is changed to positive. The reference temperature, T_0, = 80°C; T = 95°C, $z_m = 9$, and $z_c = 16$.

$$\log\left(\frac{C}{C_0}\right) = \frac{0.5}{125}(15.26)[10]^{(80-95)(1/9-1/16)}$$

$$= 0.011$$

The derivation of equation 32 is based on the average velocity; therefore, the value of $\log(C/C_0)$ calculated above is the equivalent of S_v. Table 9.7 can be used to convert this to the integrated browning change, S_i; however, at very small values of S, $S_v \equiv S_i$. Therefore, $S_i = S_v = 0.011$. $\text{Log}(C/C_0) = 0.011$; $C/C_0 = 10^{0.011} = 1.026$. An increase of 2.6% in the intensity of browning will be expected in the process.

Calculation of integrated lethality for fluids in turbulent flow is not possible without an expression for velocity distribution within the tube. Currently available expressions for velocity distributions in turbulent flow are too unwieldy for a generalized treatment of integrated sterility, as was done with equation 27. A possible approach to the determination of an integrated lethality is to determine experimentally the fluid residence time distribution and express this as a distribution function for velocity within the tube relative to the average velocity. In the absence of velocity distribution functions, lethality in the holding tube of continuous sterilization systems in turbulent flow must be calculated using the maximum velocity (equation 7, $Re > 5000$). The previous examples of nutrient degradation during high-temperature, short-time sterilization demonstrate very low values for log (C/C_0) such that the integrated value approaches the value based on the average velocity. Thus, log (C/C_0) can be based on the average velocity.

Sterilization of Fluids Containing Discrete Particulates. Discrete particulates within a flowing fluid are heated by heat transfer from the suspending fluid. Thus, heat transfer coefficients between the particles and the fluid play a significant role in determining the rate of heating. Simplified equations for heat transfer are not applicable, since the fluid's temperature is not constant as the mixture passes through the heater, and the temperature of a fluid in an unheated holding tube may not be constant because of heat exchange between the fluid and the suspended particles. Taking the conservative approach of ignoring the heat absorbed by the particles in the heater can result in significant overprocessing, particularly if the suspended particles have less than 0.5 cm as the thickness of the dimension with the largest area for heat transfer. Finite element or finite difference methods for solving the heat transfer equations with appropriate substitutions for changes in the boundary conditions, when they occur, is the only correct method to determine the lethal effect of heat in the holding tube. The residence time distribution of particles must also be considered, and as in the case of fluids in turbulent flow, the use of a probability distribution function for the residence time in the finite difference or finite element method allows calculation of integrated sterility.

A discussion of the finite difference methods for evaluating heat transfer is beyond the scope of this textbook.

STERILIZING VALUE OF PROCESSES EXPRESSED AS F_0

The sterilizing value of a process, expressed as the number of decimal reductions of a specific biological entity, was discussed in the section ''Selection of Inoculation Levels in Inoculated Packs.'' When comparing various processes for their lethal effect, it is sometimes more convenient to express the lethality as an equivalent time of processing at a reference temperature. The term F_0 is a reference process lethality expressed as an equivalent processing time at 121.1°C based on a z value of 10°C (18°F). If the z value used in the determination of F is other than 10°C, the z value is indicated as a superscript, F_0^z.

For constant temperature processes, F_0 is obtained by calculating L in equation 21 and multiplying L by the heating time at T. $F_0 = Lt$.

For processes which involve changing temperature, such as the heating or cooling stage in a canning process, an integrated lethality is calculated. Small time increments, Δt, are taken at different times in the process, the average temperature at each time increment is used to calculate L in equation 21, and the F_0 is $\Sigma L_T \Delta t$. A z value of 10°C is used in equation 21 to calculate L if F_0 is to be determined. If the F_0 value is to be used later to express lethality as the number of decimal reductions of a particular biological entity, the appropriate z value for that entity has to be used to calculate F_0^z.

THERMAL PROCESS CALCULATIONS FOR CANNED FOODS

When sterilizing foods contained in sealed containers, the internal temperature changes with the time of heating. Lethality of the heating process may be calculated using the *general method*, which is a graphical integration of the lethality-time curve, or by *formula methods*, which utilize previously calculated tabular values of parameters in an equation for the required process time or process lethality. Problems in thermal process calculation can involve either (I) the determination of the process time and temperature needed to achieve a designed lethality or (II) the evaluation of a process's time and temperature. These are referred to by some authors as *Type I* and *Type II* problems. Fundamental in the evaluation of thermal process schedules is heat transfer data, an equation of experimental data for temperature in the container as a function of time. Lethality may be expressed (1) as the value achieved at a single point, i.e., the slowest heating point in the container, or (2) as integrated lethality. For microbial inactivation, where microbial numbers are nil at regions within the container nearest the wall, lethality at a single point is adequate and results in the safest process schedule or in the most conservative estimate of the probability of spoilage. However, when evaluating quality factor degradation, integrated lethality must be determined since a level of the factor in question exists at all points in the container.

The General Method. Process lethality is calculated by graphical integration of the lethality value (equation 21) using time-temperature data for the process.

$$F_0^z = \int_0^t L_t \, dt$$

Equation 14 may also be used to determine process lethality. However, since D is not constant, the sterilizing value expressed as the number of decimal reductions of microorganisms will be integrated over the process time, using the value of D at various temperatures in the process.

$$S = \int_0^t \frac{dt}{D_t}$$

If a process schedule is to be determined, the heating and cooling curves are used to determine the lethality curve, which is then graphically integrated to obtain either F_0 or S. The process lethality must equal the specified value for F_0 or S for the process to be adequate. If the lethality value exceeds that specified, the heating time is scaled back, a cooling curve parallel to the original is drawn from the scaled-back heating time, and the area is recalculated. The process is repeated until the specified and calculated values match. Evaluation of process lethality is done directly on the time-temperature data. Simpson's rule may be used for integration. From the section "Graphical Integration," Simpson's rule is applied to thermal process determination by the general method as follows: Select time increments δt such that at the end of process time t, $t/\delta t$ will be an even number. Using i as the increment index, with $i = 0$ at $t = 0$, $i = 1$ at $t = \delta t$, $i = 2$ at $t = 2\ \delta t$, $i = 3$ at $t = 3\ \delta t$, and so on:

$$A = \left(\frac{\delta t}{3}\right)[L_0 + 4L_1 + 2L_2 + 4L_3 + 2L_4 + \ldots 2L_{i-2} + 4L_{i-1} + L_t]$$

The area under the cooling curve may be evaluated separately from that under the heating curve.

EXAMPLE 1: The following data represent the temperature at the slowest heating point in a canned food processed at a retort temperature of 250°F. Calculate the F_0 value for this process. What process time is required for a lethality equal to an F_0 of 9 min?

Time (min)	Temp. (°F)	Time (min)	Temp. (°F)
0	140	55	238
5	140	60	241
10	140	65	235
15	140	70	245
20	163	75	246.3
25	185	80	247.3 (cool)
30	201	85	247.0
35	213	90	245.2
40	224	95	223.5
45	229.4	100	175
50	234.5	105	153

Solution: From $t = 0$ to $t = 80$ min, $\delta t = 5$ min will give 16 increments. L is calculated using equation 21. $T_0 = 250°F$. $L = 10^{(1/18)(T-250)} = 10^{0.0555(T-250)}$. The lethality values are as follows:

$$L_0 = 10^{-0.05555(110)} = 0 = L_1 = L_2 = L_3$$

$$L_4 = 10^{-0.05555(87)} = 1.5 \times 10^{-5}$$

$$L_5 = 10^{-0.05555(65)} = 2.45 \times 10^{-4}$$

$$L_6 = 10^{-0.05555(54)} = 0.001001$$

$$L_7 = 10^{-0.05555(32)} = 0.016688$$

$$L_8 = 10^{-0.05555(26)} = 0.035938$$

$$L_9 = 10^{-0.05555(20.6)} = 0.071725$$

$$L_{10} = 10^{-0.05555(15.5)} = 0.1377$$

$$L_{11} = 10^{-0.05555(12)} = 0.2155$$

$$L_{12} = 10^{-0.05555(9)} = 0.3163$$

$$L_{13} = 10^{-0.05555(6.5)} = 0.4354$$

$$L_{14} = 10^{-0.05555(5)} = 0.5275$$

$$L_{15} = 10^{-0.05555(3.7)} = 0.6229$$

$$L_{16} = 10^{-0.05555(2.7)} = 0.7079$$

The area under the heating curve will be:

$$4(L_1 + L_3 + L_5 + L_7 + L_9 + L_{11} + L_{13} + L_{15}) = 5.4498$$

$$2(L_2 + L_4 + L_6 + L_8 + L_{10} + L_{12} + L_{14}) = 2.0363$$

$$A = \left(\frac{5}{3}\right)(0 + 5.4498 + 2.0363 + 0.7079) = 13.57$$

The area under the cooling curve will be:

$$L_0 = 10^{-0.05555(2.7)} = 0.70797$$

$$L_1 = 10^{-0.05555(3)} = 0.61829$$

$$L_2 = 10^{-0.05555(4.8)} = 0.54187$$

$$L_3 = 10^{-0.05555(26.5)} = 0.03371$$

$$L_4 = 10^{-0.05555(75)} = 0$$

$$L_5 = 0$$

$$4(L_1 + L_3 + L_5) = 2.8600$$

$$2(L_2 + L_4 + L_6) = 1.08374$$

$$A = \left(\frac{5}{3}\right)(0.70797 + 2.8600 + 1.0837 + 0) = 7.753$$

$$\text{Total area} = 13.57 + 7.753 = 21.32$$

The cooling curve contributed about one-third of the total lethality in this example.

The calculated total lethality is much higher than the specified F_0 of 9 min. Thus, it will be necessary to reduce the heating time. Reduction of the heating time will result in a reduction of the can temperature prior to cooling. Let the heating time be equal to 60 min. There are now only 12 area increments. The can temperature at 60 min of heating is 241°F. The cooling curve will start at 241°F. The cooling temperature will be parallel to the cooling curve of the original process (Fig. 9.9). Using Simpson's rule on the new heating and cooling curves, Table 9.8 may be constructed:

The area under the heating curve is $(1.17479 + 0.35019 + 0.316228)(5/3) = 1.8420(5/3) = 3.07$.

The area under the cooling curve is $(1.114881 + 0.430887 + 0.316228)(5/3) = 1.96199(5/3) = 3.27$.

The total area $= F_0 = 6.3$.

Ten more minutes of heating will add approximately 3 min to the F_0, since 1 min of heating at 241°F is equivalent to 0.31 min at 250°F. Thus the heating time will be 70 min to give an F_0 of approximately 9 min.

This example underscores the importance of the contribution of the cooling part of the process to the total lethal value. The relative contribution of the cooling curve to total lethality increases when product characteristics or processing conditions result in a slow rate of cooling.

Heat Transfer Equations and Time-Temperature Curves for Canned Foods. In the section "Heating of Solids Having Infinite Thermal conductivity" in Chapter 7, the transient temperature of a solid having an infinite thermal conductivity was derived. Equation 84, Chapter 7, may be used to represent the temperature at a single point in a container. If the point considered is at the interior of the container, a time lag will exist from the start of heating to the time the temperature at the point considered actually changes. The following symbols are used for thermal process heat penetration parameters:

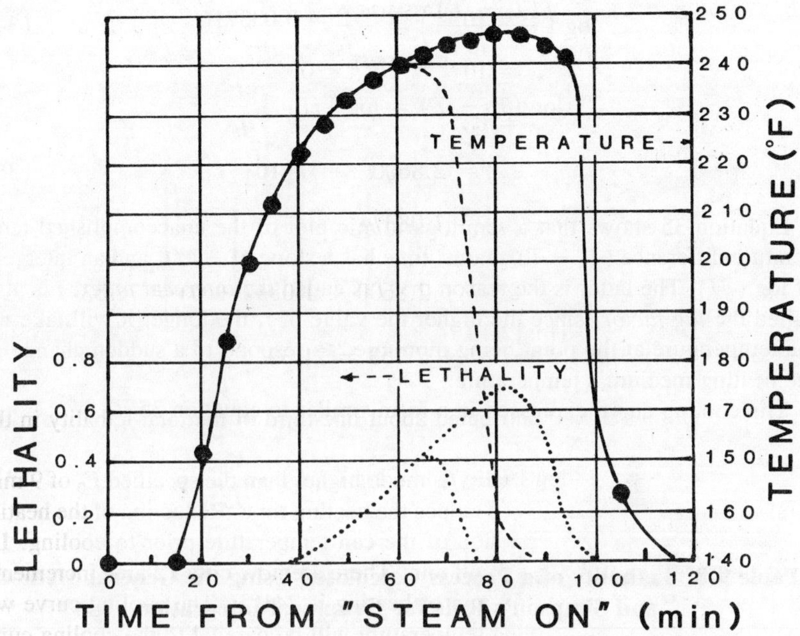

Fig. 9.9. Graph showing retort temperature, lethality of the heat treatment, and the procedure for adjusting the processing time to obtain the specified process lethality.

I = initial temperature difference = $(T_r - T_0)$; T_r = heating medium temperature = retort temperature; T_0 = can temperature at the start of the heating process.

g = unaccomplished temperature difference at the end of a specified heating time = $(T_r - T)$; T = temperature at the point considered at any time, t, during the heating process.

j = lag factor, also known as the intercept index for the linear semilogarithmic temperature–time plot of the heating curve. j_h refers to the heating curve and j_c refers to the cooling curve.

f = the slope index of the linear semilogarithmic temperature–time plot of the heating curve. If the heating curve consists of n line segments, f_i (i = 1 to n) is used to represent the slope index of each line segment, with 1 representing the first line segment from the start of the heating process. f_c refers to the cooling curve.

Expressing equation 85, Chapter 7, in terms of the above parameters:

$$\log\left(\frac{g}{jI}\right) = -\frac{t}{f_h} \tag{33}$$

$$\log\left(\frac{(T_r - T)}{jI}\right) = -\frac{t}{f_h} \tag{34}$$

$$\log(T_r - T) = \log(jI) - \frac{t}{f_h} \tag{35}$$

$$T = T_r - jI[10]^{-t/f_h} \tag{36}$$

Equation 35 shows that a semilogarithmic plot of the unaccomplished temperature difference $(T_r - T)$ against time has a slope of $-1/f_h$ and an intercept of $\log(jI)$. The latter is the reason that j is called the *intercept index*. j is also called the *lag factor*, since the higher the value of j, the longer it will take for the temperature at the point being monitored to respond to a sudden change in the heating medium's temperature.

The cooling curve expressed in a form similar to equation 35 is:

$$\log(T - T_c) = \log(j_c I_c) - \frac{t_c}{f_c} \tag{37}$$

Table 9.8. Lethality of a Process Calculated Using the General Method and Simpson's Rule for Graphical Integration

Time (min)	Temp (°F)	L	$4L$	$2L$	L
0	140	0	—	—	0
5	140	0	0	—	—
10	140	0	—	0	—
15	140	0	0	—	—
20	163	0	—	0	—
25	185	0.000245	0.000980	—	—
30	201	0.001896	—	0.003791	—
35	213	0.008799	0.035214	—	—
40	244	0.035938	—	0.071858	—
45	229.4	0.071706	0.286824	—	—
50	234.5	0.137686	—	0.275371	—
55	238	0.215443	0.861774	—	—
60	241	0.316228	—	—	0.316228
Sum			1.17479	0.351019	0.316228
60	241	0.316228	—	—	0.316228
65	240	0.278256	1.113024	—	—
70	238	0.215443	—	0.430887	—
75	190	0.000464	0.001857	—	—
80	149	0.052079	—	0	—
85	142	0	0	—	—
90	140	0	—	—	0
Sum			1.114881	0.430887	0.316228

where $I_c = (T_g - T_c)$ and T_g = temperature at the end of the heating process $= (T_r - g)$.

Equation 37 shows that a semilogarithmic plot of log $(T - T_c)$ vs. time (t_c with $t_c = 0$ at the start of cooling) will be linear and the slope will be $-1/f_c$. The temperature at any time during the cooling process is:

$$T = T_c + j_c I_c [10]^{-t_c/f_c} \tag{38}$$

Equation 38 represents only part of the cooling curve and is not the critical part which contributes significantly to the total lethality. The initial segment just after the introduction of cooling water is nonlinear and accounts for most of the lethality contributed by the cooling curve. Thus, a mathematical expression which correctly fits the curved segment of the temperature change on cooling is essential to accurate prediction of the total process lethality.

The initial curved segment of cooling curves has been represented as hyperbolic, circular, and trigonometric functions. A key parameter in any case is the intersection of the curved and linear segments of the cooling curve. The linear segment represented by equation 38 can be easily constructed from f_c and j_c, and the temperature at any time within this segment can be calculated easily using equation 38 from the point of intersection of the curved and linear segments. Hayakawa (*Food Technol.* 24:1407, 1970) discussed the construction of the curved segment of the cooling curve using a trigonometric function. The equations for $1 \leq j_c \leq 3$ are:

$$T = T_c + [T_g - T_c]^{\cos(Bt_c)} \tag{39}$$

$$B = \frac{1}{t_L}\left[\arccos\left[\frac{\log(j_c I_c) - t_L/f_c}{\log(I_c)}\right]\right] \tag{40}$$

The cosine function in equation 39 should use the value of the angle in radians. The arccos function in equation 40 should return the value of the angle in radians.

t_L = the time when the curved and linear segments of the cooling curve intersect. t_L may be derived from the intersection of a horizontal line drawn from the temperature at the initiation of cooling and the linear segment of the cooling curve represented by equation 38. At the intersection, $(T - T_c) = I_c$, and $t_c = t_L$. Substituting in equation 38, solving for t_L, and introducing a factor k to compensate for the curvature in the cooling curve:

$$t_L = f_c \log\left(\frac{j_c}{k}\right) \tag{41}$$

Factor k in equation 41 may be determined from the actual cooling curves, when plotting heat penetration data. $k = 0.95$ has been observed to be common in experimental cooling curves for canned foods. Equation 41 represents cooling data for canned foods better than the equation for t_L originally given by Hayakawa (1970).

Plotting Heat Penetration Data. Raw time-temperature data may be plotted directly on semilogarithmic graphing paper to produce the linear plot needed to determine f_h and j by rotating the paper 180°. The numbers on the graph which mark the logarithmic scale are marked as $(T_r - T)$, and the can temperature is marked on the opposite side of the graphing paper. Figures 9.10 and 9.11 show how the can temperature is marked on three-cycle semilogarithmic graphing paper for retort temperatures of 250° and 240°F, respectively.

Determination of f_h and j. Can temperature is plotted on the modified graphing paper, and a straight line is drawn connecting as many of the experimental data points as possible. There is an initial curvature in the curve, but the straight line is drawn all the way to $t = 0$.

In any simulator used for heat penetration data collection, the retort, when heated, does not immediately reach the designated processing temperature. The time from introduction of steam to the time when the processing temperature (T_r) is reached is the retort come-up time, $t_{come\text{-}up}$. Sixty percent of the retort come-up time is assumed to have no heating value; therefore, heating starts from a pseudo-initial time t_{pi}, which is 0.6 $t_{come\text{-}up}$. The pseudo-initial temper-

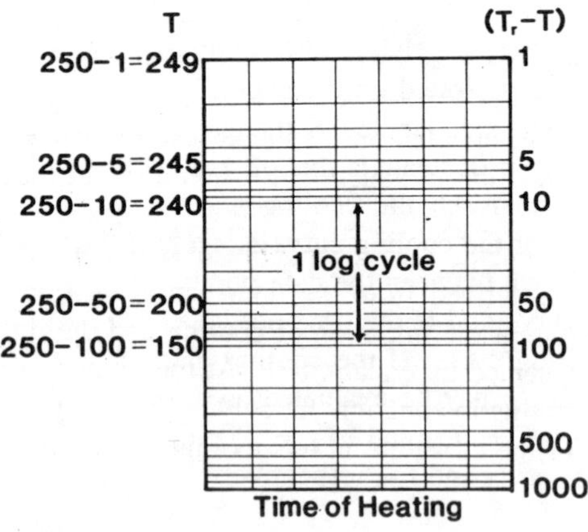

Fig. 9.10. Diagram showing how the axis of semilogarithmic graphing paper is marked for plotting heat penetration data. Retort temperature = 250°F.

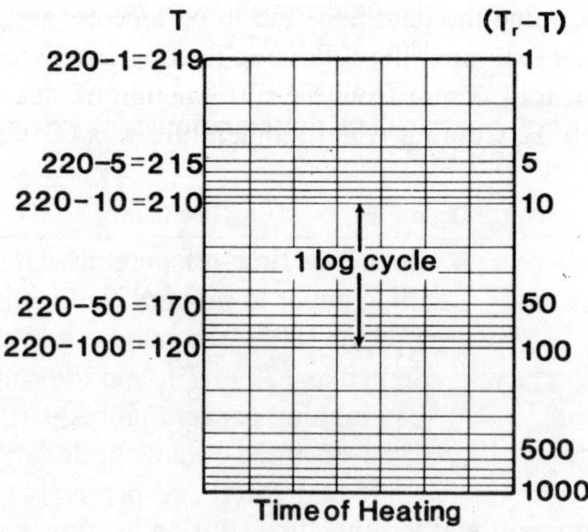

Fig. 9.11. Diagram showing how the axis of semilogarithmic graphing paper is marked for plotting heat penetration data. Retort temperature = 220°F.

ature, T_{pi}, is the intersection of the line drawn through the points and the line representing $t = t_{pi}$. $(T_r - T_{pi}) = jI$. The intercept index, j, is:

$$j = \frac{T_r - T_{pi}}{T_r - T_0} = \frac{jI}{I} \tag{42}$$

The slope index, f_h, is the time needed for the linear section of the heating curve to traverse one log cycle on the graph.

Determination of f_c and j_c. Using 180° rotated semilogarithmic graphing paper, the cooling curve is plotted on the paper, with the marked side labeled $(T - T_c)$ and the opposite side labeled T, the can temperature. The abscissa is the cooling time, t_c. At $t_c = 0$, steam is shut off and cooling water is introduced. The retort is assumed to reach the cooling water temperature immediately; therefore, the intercept of the cooling curve is evaluated at $t_c = 0$. $j_c I_c$ = the intercept of the line drawn between the data points and $t_c = 0$. f_c = the slope index of the cooling curve and is the time needed for the linear section of the curve to traverse one log cycle. If the cooling data do not complete one log cycle within the graph, f_c may be evaluated as the negative reciprocal of the slope of the line. Let $(T_1 - T_c)$ and $(T_2 - T_c)$ represent the unaccomplished temperature difference at t_{c1} and t_{c2}, respectively:

$$f_c = -\frac{t_{c1} - t_{c2}}{\log (T_1 - T_c) - \log (T_2 - T_c)} \tag{43}$$

EXAMPLE: Determine the heat penetration parameters j, f_h, j_c, and f_c for a canned food which exhibited the following heating data when processed in a retort at 250°F. It took 3 min from the introduction of steam to the time the retort reached 250°F. Cooling water temperature was 60°C.

Time (min)	Temp (°F)	Time (min)	Temp (°F)
0	180	30	245
5	190	30 cool	245
10	210	35	235
15	225	40	175
20	235	45	130
25	241	50	101

Calculate the temperature at various times during heating and cooling of this product processed at $T_r = 251$°F if $T_0 = 160$°F and $t = 35$ min from steam introduction. $T_c = 70$°F. Retort come-up time = 3 min. Calculate the F_0 of this process using the general method and Simpson's rule for graphical integration of the lethality.

Solution: The heating and cooling curves are plotted in Figs. 9.12 and 9.13, respectively. The value for $f_h = 22$ min, and the way it is determined is shown in Fig. 9.12. $T_{pi} = 152$ min is read from the intersection of the line through the data points, and $t_{pi} = 0.6t_{\text{come-up}} = 1.8$ min. jI can be read by projecting the intersection to the axis labeled $(T_r - T)$. $jI = 98$°F or is obtained by subtracting T_{pi} from T_r. $I = (T_r - T_0) = 250 - 180 = 70$°F. $j = 98/70 = 1.40$.

The cooling curve is shown in Fig. 9.13. The intercept of the linear portion of the curve with $t_c = 0$ is projected to the side marked $(T - T_c)$, and j_cI_c is read to be 333°F. The initial temperature difference for cooling, I_c, is 245 − 60 = 185°F. Thus $j_c = j_cI_c/I_c = 333/185 = 1.8$. The slope index for the cooling curve is calculated from the points ($t_c = 0$; $(T - T_c) = 333$) and ($t_c = 20$; $(T - T_c) = 41$).

$$f_c = -\frac{0 - 20}{\log[333] - \log(41)} = 22 \text{ min}$$

The curved section of the cooling curve intersects the linear section at $t_c = 6$ min. Thus, $t_L = 6$ min, and for this cooling curve, k in equation 41 is:

$$k = \frac{j_c}{[10]^{t_L/f_c}} = \frac{1.8}{(10)^{6/22}} = 0.96$$

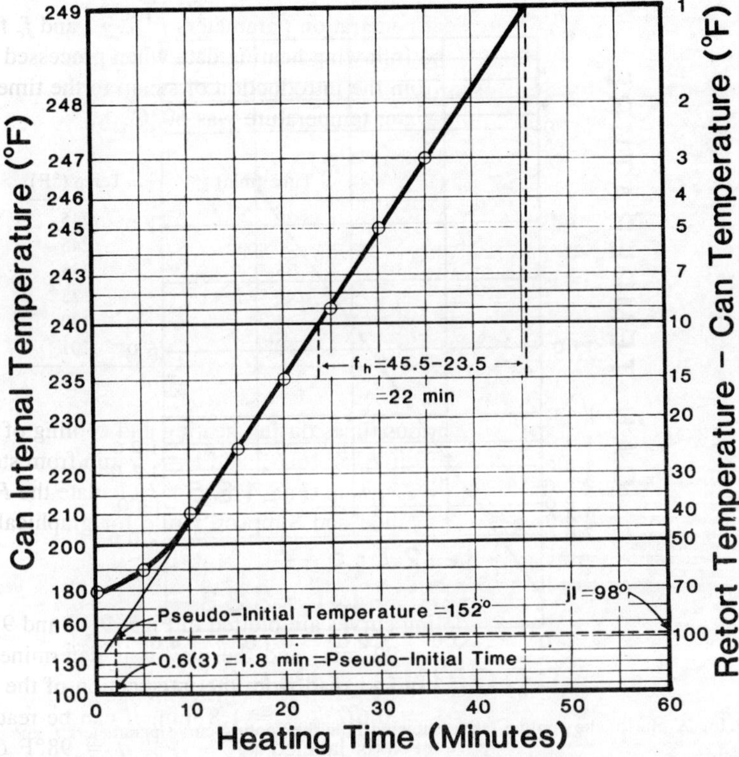

Fig. 9.12. A plot of the heating curve showing how the heating curve parameters f_h and j are determined.

The temperature during heating and cooling at the same point within a similar-sized container can be determined for any retort temperature, initial can temperature, or cooling water temperature once the heating and cooling curve parameters are known. Equation 36 is used to determine the temperature during heating. The initial heating period is assumed to be constant at T_0 until the calculated value for T exceeds T_0. This assumption does not introduce any errors in the calculation of the lethality of heat received, since at this low temperature lethality is negligible. Exceptions are rare and apply to cases where a very high initial temperature exists.

As previously discussed, when heating is carried out under conditions where a come-up time exists, the first 60% of the come-up time is assumed to have no heating value; therefore, the time variable in equation 36 should be zero when $t \leq t_{\text{come-up}}$. If t used in equation 36 is based on the time after "steam on," then 60% of $t_{\text{come-up}}$ must be subtracted from it when used in equation 36.

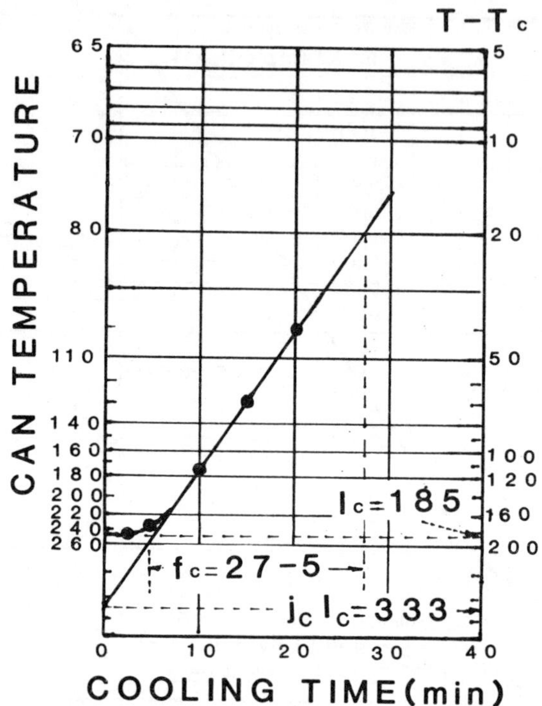

Fig. 9.13. A plot of the cooling curve showing how the cooling curve parameters f_c and j_c are obtained.

Let the exponential term in equation 36 = A.

$$A = 10^{-\{[t - 0.6(3)]/fh\}} = 10^{-[(t-1.8)/22]}$$

$$T = T_r - jIA = 251 - 1.4(251 - 160)A$$

Calculated temperatures are shown in Table 9.9.

During cooling, the curved portion of the cooling curve is constructed with the temperatures calculated using equations 39, 40, and 41. The time when the linear and curved portions of the cooling curve intersect is calculated using equation 39 and the previously calculated value of k of 0.96.

$$t_L = 22 \log\left(\frac{1.8}{0.96}\right) = 6.1 \text{ min}$$

Solving for parameter B in equation 38:

Table 9.9. Time and Temperature During Heating for an Example Problem

Time (min)	A	$T_r - jIa$	Temp (°F)
0	1.207	(97)	160
2	0.979	(126)	160
4	0.794	(150)	160
6	0.644	168	168
8	0.522	184	184
10	0.423	197	197
12	0.344	207	207
14	0.279	215.5	215.5
16	0.226	222	222
18	0.184	227.6	227.6
20	0.149	232	232
22	0.121	235.6	235.6
24	0.098	238.5	238.5
26	0.079	241.6	241.6
28	0.064	242.8	242.8
30	0.052	244.3	244.3

$$B = \frac{1}{6.1}\left[\arccos\left[\frac{\ln(1.8)(174.3) - (6.1/22)}{\ln(174.3)}\right]\right]$$

$$= \frac{1}{6.1}\arccos(0.99018) = \frac{1}{6.1}\left[(8.034°)\left[\frac{2\pi \text{ rad}}{360°}\right]\right]$$

$$= 0.02337$$

The temperature is then calculated using equation 39:

$$T = 70 + (174.3)^{\cos[(0.02337)(tc)]}$$

Let the exponential term $= E = \cos[(0.02337)(t_c)]$:

$$T = 70 + (174.3)^E$$

For the linear portion of the cooling curve, equation 38 is used. The temperature calculated using equation 30 will represent the actual can temperature only when $t_c > t_L$, since t_L is the intersection of the curved and linear portions of the cooling curve.

Equation 38: $T = 70 + 1.8(174.3)(10)^{(-tc/22)}$

Let $A = 10^{(-tc/22)}$; $\quad T = 70 + (313.74)^A$

The calculated temperatures for cooling, and the lethality and area elements for area calculation using Simpson's rule, are shown in Tables 9.10 and 9.11, respectively.

$$\text{Area} = F_0 = \left(\frac{2}{3}\right)(1.7297 + 5.3866 + 0.9698) = 5.39 \text{ min}$$

The lethality of the process expressed as an equivalent heating time at 250°F is 5.3 min. The complete procedure used in the evaluation of the lethality of a heating process in this example is the general method. The general method may be used on data obtained experimentally or on time-temperature data reconstructed from values of heating and cooling curve parameters calculated from experimental data.

Formula Methods for Thermal Process Evaluation. Formula methods are based on tabulated values for lethality expressed as the parameter f_h/U. These values were previously calculated for various conditions of heating and cooling when the unaccomplished temperature difference at the termination of the process is expressed as the parameter g. Two methods will be presented in this section: Stumbo's (1973) and Hayakawa's (1970). The purpose of presenting both methods is not to compare their accuracy but to provide a means for selecting the most convenient method to use for certain conditions.

Stumbo's f_h/U vs. g tables combine the lethalities of heating and cooling. A major assumption used in the calculation of lethality is that $f_h = f_c$. When actual conditions do not meet this assumption, Stumbo recommends using the general method. Hayakawa (1970) presents lethality of the heating and cooling stages of the process in separate tables, thus allowing the use of his method even under

Table 9.10. Time and Temperature During Cooling for an Example Problem

Time			Temp (°F)	
(t_c)	E	A	$70 + 174.3^E$	$70 + (313.74)^A$
0	1	1	244.3	(383.7)
2	0.998	0.811	243.3	(324.5)
4	0.993	0.658	240.4	(276.4)
6	0.984	0.534	235.7	(237.3)
8		0.433		199.4
10		0.351		180.2
12		0.285		159.4
14		0.231		142.5
16		0.187		128.8

Table 9.11. Time, Temperature, and Simpson's Rule Factors for Area Calculations for an Example Problem

Time (min)	Temp (°F)	L	$2L$	$4L$	L
Heating					
0	160	0			
2	160	0			0
4	160	0		0	
6	169	0	0		
8	184	0.0002		0.0008	
10	197	0.0011	0.0022		
12	207	0.0042		0.0168	
14	215.5	0.0121	0.0242		
16	222.2	0.0284		0.1136	
18	227.6	0.0571	0.114		
20	232	0.1005		0.402	
22	235.6	0.1588	0.3177		
24	238.5	0.2304		0.9215	
26	241.8	0.3414	0.6829		
28	242.8	0.3977		1.5908	
30	244.3	0.4899			0.4849
Cooling					
0	244.3	0.4849			0.4849
2	243.3	0.4244		1.6976	
4	240.4	0.2929	0.5858		
6	235.7	0.1605		0.6421	
8	199	0.0015	0.0030		
10	180	0.0001		0.0004	
12	159	0	0		
14	142	0		0	
16	129	0			0
Sums			1.7297	5.3866	0.9698

different rates of heating and cooling. Hayakawa's tabular values allow substitutions for different values of z, simplifying calculation of specific F_0^z values for different z values.

In this section, process calculations for products which exhibit simple heating curves will be discussed. Calculations for broken heating curves will be discussed in the section "Broken Heating Curves."

Both formula methods are based on the equation for the heating curve (equation 33). Let g = the unaccomplished temperature difference, $(T_r - T)$, at the termination of the heating period, and let B_b = the heating time at that point. B_b is the scheduled sterilization process. For products with simple heating curve, equation 33 becomes:

$$B_b = f_h[\log(jI) - \log(g)] \tag{44}$$

$B_b = t - 0.6t_{\text{come-up}}$. t = the time evaluated from steam introduction into the retort. In practice, B_b is timed from the point where the retort reaches the processing temperature to avoid the probability of errors arising from the need to correct for the come-up time.

g is obtained from the tables, using specified F_0 and z values. Stumbo's tables are simpler to use for thermal process determinations. Hayakawa's tables require an iterative procedure involving an assumption of a value of g, calculating the F_0, and repeating the calculations using a new value of g until the calculated F_0 matches the specified F_0.

The following parameters are used in the formula methods:

$U = F_0 F_i$ = time at T_r equivalent to F_0.

$$F_i = [10]^{250 - T/z} \tag{45}$$

Stumbo's Procedure. Stumbo has tabulated f_h/U vs. g with j_c as a parameter. j_c strongly influences the contribution of the cooling part of the process to the total lethality, as discussed in the section "Sterilizing Value of Processes Expressed as F_0." In general, j_c values are higher than j values. In the absence of j_c, j may be used and the error will be toward a longer process time or the safe side relative to spoilage. Condensed f_h/U vs. g tables for z values from 14 to 22 and from 30 to 45 are shown in Tables 9.12 and 9.13. Table 9.12 is used for microbial inactivation, and Table 9.13 is used for nutrient degradation. It is possible to interpolate between values in the table for other z values. Thermal process determinations can easily be made from specified F_0 values and product heat penetration parameters by solving for f_h/U, determining the corresponding value for g, and solving for B_b using equation 44.

Hayakawa's Procedure. Hayakawa's tables are shown in Table 9.14 for lethality of the heating part of the process and in Tables 9.15, 9.16, 9.17, 9.18, and 9.19 for lethality of the cooling part of the process. The tables are based on a z value of 20°F. The parameter g/K_s, with $K_s = z/20$, is tabulated against U/f_h. The latter is the reciprocal of Stumbo's f_h/U.

In Tables 9.15 to 9.19, for lethality of the cooling part of the process, j_c is used as a parameter. The parameter in the table for lethality of the cooling curve is $(T_g - T_c)/K_s = (T_r - g - T_c)/K_s$. U in the lethality tables for the cooling curve is the equivalent time of heating at T_g. Thus, it must be converted to U at T_r before being added to the U value obtained from Table 9.14 for heating. The conversion from U at T_g to the U at T_r is done using equation 46:

$$U = U'(10)^{-g/z} \tag{46}$$

where U is the process $U = F_0 F_i$ or the equivalent heating time at 250°F. U' is U obtained from Tables 9.15 to 9.19, the equivalent heating time at T_g for the lethality of the cooling part of the process.

Table 9.12. f_h/U vs. g Table Used for Thermal Process Calculations by Stumbo's Procedure

	z = 14		z = 18		z = 22	
$\frac{f_h}{U}$	g	$\frac{\Delta g}{\Delta j}$	g	$\frac{\Delta g}{\Delta j}$	g	$\frac{\Delta g}{\Delta j}$
0.2	0.000091	0.0000118	0.0000509	0.0000168	0.0000616	0.0000226
0.3	0.00175	0.00059	0.0024	0.00066	C.00282	0.00106
0.4	0.0122	0.0038	0.0162	0.0047	0.020	0.0067
0.5	0.0396	0.0111	0.0506	0.0159	0.065	0.0197
0.6	0.0876	0.0224	0.109	0.036	0.143	0.040
0.7	0.155	0.036	0.189	0.066	0.25	0.069
0.8	0.238	0.053	0.287	0.103	0.38	0.105
0.9	0.334	0.07	0.400	0.145	0.527	0.147
1.0	0.438	0.009	0.523	0.192	0.685	0.196
2.0	1.56	0.37	1.93	0.68	2.41	0.83
3.0	2.53	0.70	3.26	1.05	3.98	1.44
4.0	3.33	1.03	4.41	1.34	5.33	1.97
5.0	4.02	1.32	5.40	1.59	6.51	2.39
6.0	4.63	1.56	6.25	1.82	7.53	2.75
7.0	5.17	1.77	7.00	2.05	8.44	3.06
8.0	5.67	1.95	7.66	2.27	9.26	3.32
9.0	6.13	2.09	8.25	2.48	10.00	3.55
10	6.55	2.22	8.78	2.69	10.67	3.77
15	8.29	2.68	10.88	3.57	13.40	4.60
20	9.63	2.96	12.40	4.28	15.30	5.50
25	10.7	3.18	13.60	4.80	16.9	6.10
30	11.6	3.37	14.60	5.30	18.2	6.70
35	12.4	3.50	15.50	5.70	19.3	7.20
40	13.1	3.70	16.30	6.00	20.3	7.60
45	13.7	3.80	17.00	6.20	21.1	8.0
50	14.2	4.00	17.7	6.40	21.9	8.3
60	15.1	4.3	18.9	6.80	23.2	9.0
70	15.9	4.5	19.9	7.10	24.3	9.5
80	16.5	4.8	20.8	7.30	25.3	9.8
90	17.1	5.0	21.6	7.60	26.2	10.1
100	17.6	5.2	22.3	7.80	27.0	10.4
150	19.5	6.1	25.2	8.40	30.3	11.4
200	20.8	6.7	27.1	9.10	32.7	12.1

Source: Based on f_h/U vs. g tables in Stumbo, C. R. 1973. *Thermobacteriology in Food Processing*, 2nd ed. Academic Press, New York.

To use for values of j other than 1, solve for g_j as follows:

$$g_j = j_{j=1} + (j - 1)(\Delta g/\Delta j)$$

Example: g for $(f_h/U) = 20$ and $j = 1.4$ at $z = 18$.

$$g_{j=1.4} = 12.4 + (0.4)(4.28) = 14.11$$

Reprinted from: Toledo, R. T. 1980. Fundamentals of Food Process Engineering, 1st ed. AVI Pub. Co. Westport, CT.

Table 9.13. f_h/U vs. g Table Used for Thermal Process Calculation by Stumbo's Procedure

	z = 60		z = 70		z = 80		z = 90	
f_h/U	$g_{j=1}$	$\frac{\Delta g}{\Delta j}$	$g_{j=1}$	$\frac{\Delta g}{\Delta j}$	$g_{j=1}$	$\frac{\Delta g}{\Delta j}$	$g_{j=1}$	$\frac{\Delta g}{\Delta j}$
0.2	0.00018	0.00015	0.000218	0.000134	0.000253	0.00017	0.000289	0.00020
0.3	0.0085	0.000475	0.0101	0.0062	0.000253	0.00017	0.0134	0.0097
0.4	0.0583	0.032	0.0689	0.0421	0.0118	0.00775	0.0919	0.0661
0.5	0.185	0.1025	0.0219	0.0134	0.0802	0.0545	0.292	0.208
0.6	0.401	0.2225	0.474	0.292	0.255	0.17	0.632	0.452
0.7	0.699	0.3875	0.828	0.510	0.552	0.3675	1.101	0.791
0.8	1.064	0.595	1.263	0.777	0.963	0.6425	1.678	1.205
0.9	1.482	0.8325	1.76	1.08	1.469	0.9775	2.34	1.68
1.0	1.94	1.075	2.30	1.42	2.05	1.45	3.06	2.19
2.0	7.04	4.025	8.35	5.19	2.68	1.775	11.03	7.88
3.0	11.63	6.65	13.73	8.58	9.68	6.475	18.0	12.8
4.0	15.40	9.00	18.2	11.4	12.92	8.65	23.6	16.7
5.0	18.70	10.75	21.9	13.7	15.85	10.65	28.2	19.7
6.0	21.40	12.50	25.1	15.6	18.5	12.5		
7.0	23.80	13.75	27.9	17.2	20.9	14.0		
8.0	26.00	15.00	30.3	18.6	23.1	15.5		
9.0	27.90	16.00·	32.5	19.8	25.1	16.75		

Source: Based on f_h/U vs. g tables in Stumbo, C. R. 1973. *Thermobacteriology in Food Processing*, 2nd ed. Academic Press, New York.

EXAMPLE: For the example in the previous section, which was evaluated using the general method, calculate F_0 using Stumbo's and Hayakawa's procedures and calculate a process time needed to obtain an F_0 of 8 min. The following heating and cooling curve parameters were previously determined: $f_h = f_c = 22$ min; $j = 1.4$; $j_c = 1.8$.

Solution: The retort temperature, cooling water temperature, and process time are: $T_r = 251°F$; $T_c = 70°F$; and $t = 30 - 0.6(3) = 28.2$ min.

To use the formula methods to determine F_0, it is necessary to determine g from the process time and the heating curve parameters. The f_h/U vs. g table is then used to determine f_h/U, which corresponds to g, from which a value of U and F_0 can be calculated. Let T_g = can temperature at the termination of heating. Solving for T_g using equation 36:

$$T_g = 251 - (251 - 160)(10)^{-28.2/22} = 244.3°F$$

$$g = T_r - T_g = 251 - 244.3 = 6.66°F$$

Determination of F_0 *using Stumbo's procedure*. Table 9.12 is used to determine a value of f_h/U which corresponds to $g = 6.66$. Tabular parameters are: $z = 18$ and $j_c = 1.8$. It is necessary to interpolate. A value of $g = 6.66$

Table 9.14. g/K_s vs. U/f_h tables used for calculating the lethality of the heating part of a thermal process by Hayakawa's procedure

g/K_s (°F)	U/f_h	g/K_s (°F)	U/f_h	g/K_s (°F)	U/f_h
100.0000	0.4165(−06)	33.0000	0.2095(−02)	0.35000	0.1161(01)
98.0000	0.5152(−06)	32.0000	0.2413(−02)	0.30000	0.1226(01)
96.0000	0.6420(−06)	31.0000	0.2780(−02)	0.25000	0.1303(01)
94.0000	0.8051(−06)	30.0000	0.3205(−02)	0.20000	0.1397(01)
92.0000	0.1015(−05)	29.0000	0.3699(−02)	0.15000	0.1519(01)
90.0000	0.1284(−05)	28.0000	0.4272(−02)	0.10000	0.1693(01)
88.0000	0.1632(−05)	27.0000	0.4939(−02)	0.09000	0.1738(01)
86.0000	0.2079(−05)	26.0000	0.5715(−02)	0.08000	0.1789(01)
84.0000	0.2655(−05)	25.0000	0.6620(−02)	0.07000	0.1846(01)
82.0000	0.3398(−05)	24.0000	0.7677(−02)	0.06000	0.1913(01)
80.0000	0.4356(−05)	23.0000	0.8914(−02)	0.05000	0.1992(01)
78.0000	0.5593(−05)	22.0000	0.1036(−01)	0.04000	0.2088(01)
76.0000	0.7191(−05)	21.0000	0.1206(−01)	0.03500	0.2146(01)
74.0000	0.9256(−05)	20.0000	0.1407(−01)	0.03000	0.2212(01)
72.0000	0.1193(−04)	19.0000	0.1643(−01)	0.02500	0.2291(01)
70.0000	0.1539(−04)	18.0000	0.1922(−01)	0.02000	0.2388(01)
68.0000	0.1986(−04)	17.0000	0.2254(−01)	0.01500	0.2513(01)
66.0000	0.2567(−04)	16.0000	0.2648(−01)	0.01000	0.2688(01)
64.0000	0.3321(−04)	15.0000	0.3119(−01)	0.00900	0.2734(01)
62.0000	0.4300(−04)	14.0000	0.3684(−01)	0.00800	0.2785(01)
60.0000	0.5573(−04)	13.0000	0.4365(−01)	0.00700	0.2843(01)
58.0000	0.7229(−04)	12.0000	0.5191(−01)	0.00600	0.2909(01)
56.0000	0.9388(−04)	11.0000	0.6198(−01)	0.00500	0.2989(01)
54.0000	0.1220(−03)	10.0000	0.7435(−01)	0.00400	0.3085(01)
52.0000	0.1589(−03)	9.0000	0.8970(−01)	0.00350	0.3143(01)
50.0000	0.2070(−03)	8.0000	0.1090(00)	0.00300	0.3210(01)
49.0000	0.2364(−03)	7.0000	0.1335(00)	0.00250	0.3290(01)
48.0000	0.2701(−03)	6.0000	0.1652(00)	0.00200	0.3384(01)
47.0000	0.3087(−03)	5.0000	0.2073(00)	0.00150	0.3509(01)
46.0000	0.3529(−03)	4.0000	0.2652(00)	0.00100	0.3685(01)
45.0000	0.4036(−03)	3.5000	0.3029(00)	0.00090	0.3734(01)
44.0000	0.4618(−03)	3.0000	0.3490(00)	0.00080	0.3780(01)
43.0000	0.5286(−03)	2.5000	0.4067(00)	0.00070	0.3842(01)
42.0000	0.6053(−03)	2.0000	0.4816(00)	0.00060	0.3903(01)
41.0000	0.6934(−03)	1.5000	0.5839(00)	0.00050	0.3986(01)
40.0000	0.7947(−03)	1.0000	0.7367(00)	0.00040	0.4073(01)
39.0000	0.9113(−03)	0.9000	0.7777(00)	0.00035	0.4143(01)
38.0000	0.1045(−02)	0.8000	0.8241(00)	0.00030	0.4204(01)
37.0000	0.1200(−02)	0.7000	0.8773(00)	0.00025	0.4274(01)
36.0000	0.1378(−02)	0.6000	0.9395(00)	0.00020	0.4358(01)
35.0000	0.1584(−02)	0.5000	0.1014(01)	0.00015	0.4505(01)
34.0000	0.1821(−02)	0.4000	0.1106(01)	0.00010	0.4659(01)

Values in parentheses indicate powers of 10 by which tabulated values are to be multiplied; e.g., $U_{h/f}$ for g/K_s = 40°F is 0.0006646.

Source: Hayakawa, K., *Food Technol.* 24:1407, 1970. Corrected table courtesy of K. Hayakawa.

Table 9.15. g/K_s vs. U/f_c Tables used for Calculating the Lethality of the Cooling Part of a Thermal Process by Hayakawa's Procedure ($g/K_s \leq 200$)

I_c/K_s	U'/f_c for j_c = 0.40 to 1.90								
(°F)	0.40	0.60	0.80	1.00	1.20	1.40	1.60	1.80	1.90
200.00	0.9339(−2)	0.1086(−1)	0.1220(−1)	0.1976(−1)	0.7021(−1)	0.9440(−1)	0.1112	0.1243	0.1300
195.00	0.9585(−2)	0.1114(−1)	0.1253(−1)	0.2030(−1)	0.7114(−1)	0.9565(−1)	0.1126	0.1260	0.1318
190.00	0.9844(−2)	0.1145(−1)	0.1288(−1)	0.2086(−1)	0.7211(−1)	0.9695(−1)	0.1142	0.1277	0.1335
185.00	0.1012(−1)	0.1177(−1)	0.1325(−1)	0.2145(−1)	0.7312(−1)	0.9830(−1)	0.1158	0.1295	0.1354
180.00	0.1041(−1)	0.1212(−1)	0.1364(−1)	0.2208(−1)	0.7418(−1)	0.9972(−1)	0.1174	0.1313	0.1373
175.00	0.1072(−1)	0.1248(−1)	0.1405(−1)	0.2275(−1)	0.7529(−1)	0.1012	0.1192	0.1332	0.1394
170.00	0.1104(−1)	0.1287(−1)	0.1449(−1)	0.2346(−1)	0.7645(−1)	0.1027	0.1210	0.1353	0.1415
165.00	0.1139(−1)	0.1328(−1)	0.1496(−1)	0.2422(−1)	0.7767(−1)	0.1044	0.1229	0.1374	0.1437
160.00	0.1176(−1)	0.1372(−1)	0.1546(−1)	0.2503(−1)	0.7895(−1)	0.1061	0.1249	0.1396	0.1460
155.00	0.1216(−1)	0.1418(−1)	0.1599(−1)	0.2589(−1)	0.8029(−1)	0.1078	0.1270	0.1420	0.1485
150.00	0.1258(−1)	0.1469(−1)	0.1657(−1)	0.2682(−1)	0.8172(−1)	0.1097	0.1292	0.1444	0.1510
145.00	0.1304(−1)	0.1523(−1)	0.1719(−1)	0.2781(−1)	0.8322(−1)	0.1117	0.1315	0.1470	0.1538
140.00	0.1353(−1)	0.1582(−1)	0.1785(−1)	0.2889(−1)	0.8481(−1)	0.1138	0.1340	0.1498	0.1566
135.00	0.1407(−1)	0.1645(−1)	0.1858(−1)	0.3005(−1)	0.8651(−1)	0.1160	0.1366	0.1527	0.1597
130.00	0.1465(−1)	0.1714(−1)	0.1936(−1)	0.3131(−1)	0.8831(−1)	0.1184	0.1393	0.1558	0.1629
125.00	0.1528(−1)	0.1789(−1)	0.2022(−1)	0.3268(−1)	0.9025(−1)	0.1209	0.1423	0.1591	0.1663

120.00	0.1598(−1)	0.1872(−1)	0.2116(−1)	0.3418(−1)	0.9232(−1)	0.1236	0.1454	0.1626	0.1700
115.00	0.1675(−1)	0.1963(−1)	0.2219(−1)	0.3583(−1)	0.9456(−1)	0.1265	0.1488	0.1663	0.1739
110.00	0.1760(−1)	0.2065(−1)	0.2334(−1)	0.3766(−1)	0.9698(−1)	0.1296	0.1524	0.1703	0.1781
105.00	0.1857(−1)	0.2175(−1)	0.2462(−1)	0.3970(−1)	0.9961(−1)	0.1329	0.1563	0.1747	0.1826
100.00	0.1969(−1)	0.2296(−1)	0.2606(−1)	0.4199(−1)	0.1025	0.1366	0.1605	0.1794	0.1875
95.00	0.2104(−1)	0.2433(−1)	0.2769(−1)	0.4463(−1)	0.1057	0.1405	0.1651	0.1845	0.1929
90.00	0.2276(−1)	0.2589(−1)	0.2955(−1)	0.4779(−1)	0.1093	0.1449	0.1702	0.1901	0.1987
85.00	0.2414(−1)	0.2768(−1)	0.3170(−1)	0.5103(−1)	0.1132	0.1498	0.1757	0.1962	0.2051
80.00	0.2576(−1)	0.2976(−1)	0.3420(−1)	0.5460(−1)	0.1176	0.1552	0.1819	0.2031	0.2122
75.00	0.2768(−1)	0.3221(−1)	0.3715(−1)	0.5878(−1)	0.1227	0.1612	0.1888	0.2107	0.2202
70.00	0.2999(−1)	0.3512(−1)	0.4069(−1)	0.6373(−1)	0.1285	0.1682	0.1967	0.2193	0.2291
65.00	0.3280(−1)	0.3865(−1)	0.4499(−1)	0.6967(−1)	0.1353	0.1762	0.2057	0.2291	0.2394
60.00	0.3626(−1)	0.4300(−1)	0.5032(−1)	0.7687(−1)	0.1434	0.1855	0.2161	0.2405	0.2511
55.00	0.4061(−1)	0.4846(−1)	0.5703(−1)	0.8575(−1)	0.1532	0.1966	0.2284	0.2539	0.2650
50.00	0.4616(−1)	0.5546(−1)	0.6564(−1)	0.9687(−1)	0.1652	0.2101	0.2432	0.2698	0.2814
45.00	0.5340(−1)	0.6462(−1)	0.7692(−1)	0.1111	0.1803	0.2268	0.2613	0.2892	0.3014
40.00	0.6302(−1)	0.7688(−1)	0.9197(−1)	0.1295	0.1997	0.2478	0.2840	0.3132	0.3261
35.00	0.7607(−1)	0.9362(−1)	0.1124	0.1539	0.2251	0.2750	0.3129	0.3437	0.3573
30.00	0.9415(−1)	0.1170	0.1408	0.1868	0.2591	0.3108	0.3507	0.3834	0.3978
25.00	0.1197	0.1501	0.1808	0.2322	0.3056	0.3593	0.4014	0.4361	0.4515

Values in parentheses are powers of 10 by which tabulated value should be multiplied.
Source: Hayakawa, K., *Food Technol*. 24:1407, 1970.

Table 9.16. g/K_s vs. U/f_c Tables Used for Calculating the Lethality of the Cooling Part of a Thermal Process by Hayakawa's Procedure ($g/K_s \leq 200.00$)

I_c/K_s (°F)	U'/f_c for j_c = 2.0 to 2.8								
	2.00	2.10	2.20	2.30	2.40	2.50	2.60	2.70	2.[illegible]
200.00	0.1353	0.1403	0.1449	0.1492	0.1533	0.1572	0.1609	0.1645	0.1[illegible]
195.00	0.1371	0.1421	0.1468	0.1512	0.1553	0.1593	0.1630	0.1666	0.1[illegible]
190.00	0.1390	0.1440	0.1487	0.1532	0.1574	0.1614	0.1652	0.1689	0.1[illegible]
185.00	0.1409	0.1460	0.1508	0.1553	0.1596	0.1636	0.1675	0.1712	0.1[illegible]
180.00	0.1429	0.1481	0.1530	0.1575	0.1619	0.1660	0.1699	0.1736	0.1[illegible]
175.00	0.1450	0.1503	0.1552	0.1599	0.1643	0.1684	0.1724	0.1762	0.1[illegible]
170.00	0.1472	0.1526	0.1576	0.1623	0.1667	0.1710	0.1750	0.1788	0.1[illegible]
165.00	0.1495	0.1550	0.1600	0.1648	0.1693	0.1736	0.1777	0.1816	0.1[illegible]
160.00	0.1520	0.1575	0.1626	0.1675	0.1721	0.1764	0.1806	0.1845	0.1[illegible]
155.00	0.1545	0.1601	0.1653	0.1703	0.1749	0.1794	0.1836	0.1876	0.1[illegible]
150.00	0.1572	0.1629	0.1682	0.1732	0.1780	0.1825	0.1867	0.1908	0.1[illegible]
145.00	0.1600	0.1658	0.1712	0.1763	0.1811	0.1857	0.1901	0.1942	0.1[illegible]
140.00	0.1630	0.1689	0.1744	0.1796	0.1845	0.1891	0.1936	0.1978	0.2[illegible]
135.00	0.1661	0.1721	0.1777	0.1830	0.1880	0.1928	0.1973	0.2016	0.2[illegible]
130.00	0.1695	0.1756	0.1813	0.1867	0.1918	0.1966	0.2012	0.2056	0.2[illegible]
125.00	0.1730	0.1793	0.1851	0.1906	0.1958	0.2007	0.2054	0.2099	0.2[illegible]
120.00	0.1768	0.1832	0.1892	0.1948	0.2001	0.2051	0.2099	0.2145	0.2[illegible]
115.00	0.1809	0.1874	0.1935	0.1992	0.2046	0.2098	0.2147	0.2193	0.2[illegible]
110.00	0.1853	0.1919	0.1982	0.2040	0.2095	0.2148	0.2198	0.2246	0.2[illegible]
105.00	0.1900	0.1968	0.2032	0.2092	0.2148	0.2202	0.2253	0.2302	0.2[illegible]
100.00	0.1951	0.2021	0.2086	0.2148	0.2206	0.2261	0.2313	0.2363	0.2[illegible]
95.00	0.2006	0.2078	0.2146	0.2209	0.2268	0.2325	0.2378	0.2429	0.2[illegible]
90.00	0.2067	0.2141	0.2210	0.2275	0.2337	0.2395	0.2450	0.2502	0.2[illegible]
85.00	0.2134	0.2210	0.2281	0.2348	0.2412	0.2471	0.2528	0.2582	0.2[illegible]
80.00	0.2207	0.2286	0.2360	0.2429	0.2494	0.2556	0.2615	0.2670	0.2[illegible]
75.00	0.2290	0.2371	0.2447	0.2519	0.2587	0.2651	0.2711	0.2769	0.2[illegible]
70.00	0.2382	0.2467	0.2546	0.2620	0.2690	0.2757	0.2820	0.2879	0.2[illegible]
65.00	0.2488	0.2576	0.2658	0.2735	0.2808	0.2877	0.2943	0.3005	0.3[illegible]
60.00	0.2610	0.2701	0.2787	0.2868	0.2944	0.3015	0.3084	0.3149	0.3[illegible]
55.00	0.2752	0.2848	0.2937	0.3022	0.3101	0.3176	0.3248	0.3316	0.3[illegible]
50.00	0.2922	0.3022	0.3116	0.3204	0.3288	0.3367	0.3442	0.3513	0.3[illegible]
45.00	0.3127	0.3232	0.3331	0.3424	0.3512	0.3595	0.3674	0.3750	0.3[illegible]
40.00	0.3380	0.3491	0.3596	0.3694	0.3787	0.3876	0.3959	0.4039	0.4[illegible]
35.00	0.3700	0.3818	0.3929	0.4033	0.4132	0.4226	0.4315	0.4400	0.4[illegible]
30.00	0.4113	0.4239	0.4357	0.4468	0.4574	0.4674	0.4769	0.4860	0.4[illegible]
25.00	0.4659	0.4793	0.4920	0.5040	0.5153	0.5261	0.5363	0.5460	0.5[illegible]

Source: Hayakawa, K., *Food Technol.* 24:1407, 1970.

is not obtainable directly from Table 9.12, since a tabular entry for g is available only for a value of j_c = 1. Under the column "z = 18" in Table 9.12, a g value of 5.4 is listed which corresponds to $f_h/U = 5$. Also listed next to this value is an interpolating factor, $\Delta g/\Delta j = 1.59$. The value of g for $f_h/U = 5$ and for $j = 1.8$ is:

$$g_{j=1.8} = 5.4 + 0.8(1.59) = 6.672$$

The value $g = 6.672$ exceeds 6.66, the specified g; therefore, a lower value of $f_h/U = 4$ is chosen, a corresponding g value for $j_c = 1.8$ is calculated, and by interpolation, a value of f_h/U which corresponds to $g = 6.66$ is calculated. For $f_h/U = 4$, $g_{j=1} = 4.41$; $\Delta g/\Delta j = 1.34$; and $g_{j=1.8} = 4.41 + 0.8(1.34) = 5.482$. Interpolating:

$$\left(\frac{f_h}{U}\right)_{g=6.66} = 4 + \left(\frac{1}{6.672 - 5.482}\right)(6.66 - 5.482) = 4.99$$

$$U = \frac{f_h}{(f_h/U)_{g=6.66}} = \frac{22}{4.99} = 4.41$$

$$\text{At } 251°\text{F, } F_i = (10)^{-1/18} = 0.8799$$

$$F_0 = \frac{U}{F_i} = \frac{4.41}{0.8799} = 5.01$$

This value compares with 5.49 min calculated using the general method in the previous section.

Determination of process time for $F_0 = 8$ *using Stumbo's procedure.* B_b is required for $F_0 = 8.0$ min. A value of g is now required, and this value is obtained from Table 9.12 to correspond to a value of f_h/U. U is calculated as $U = F_0 F_i$. F_i at 251°F was calculated previously to be $(10)^{-1/18} = 0.8799$. Thus, $U = 8(0.8799) = 7.0392$. Solving for f_h/U:

$$\frac{f_h}{U} = \frac{22}{7.0392} = 3.1253$$

Table 9.12 is now used to determine g for $f_h/U = 3.1253$, $z = 18$, and $j_c = 1.8$ by interpolation. From Table 9.12, $z = 18$ and $f_h/U = 3$:

$$g_{j=1} = 3.26; \qquad \frac{\Delta g}{\Delta j} = 1.05; \qquad g_{j=1.8} = 3.26 + 0.8(1.05) = 4.10$$

For $f_h/U = 4$:

$$g_{j=1} = 4.41; \qquad \frac{\Delta g}{\Delta j} = 1.34; \qquad g_{j=1.8} = 4.41 + 0.8(1.34) = 5.482$$

Interpolating to obtain g for $f_h/U = 3.1253$:

Table 9.17. g/K_s vs. U/f_c **Tables Used for Calculating the Lethality of the Cooling Part of a Thermal Process by Hayakawa's Procedure** ($200 < g/K_s \leq 400$)

I_c/K_s	U'/f_c for j_c = 0.40 to 1.90								
(°F)	0.40	0.60	0.80	1.00	1.20	1.40	1.60	1.80	1.90
400.00	0.4642(−2)	0.5348(−2)	0.5964(−2)	0.9644(−2)	0.4919(−1)	0.6616(−1)	0.7794(−1)	0.8721(−1)	0.9124(−1)
395.00	0.4700(−2)	0.5416(−2)	0.6041(−2)	0.9769(−2)	0.4951(−1)	0.6658(−1)	0.7844(−1)	0.8777(−1)	0.9182(−1)
390.00	0.4760(−2)	0.5486(−2)	0.6120(−2)	0.9897(−2)	0.4983(−1)	0.6702(−1)	0.7895(−1)	0.8833(−1)	0.9241(−1)
385.00	0.4822(−2)	0.5558(−2)	0.6201(−2)	0.1003(−1)	0.5016(−1)	0.6746(−1)	0.7947(−1)	0.8892(−1)	0.9302(−1)
380.00	0.4886(−2)	0.5632(−2)	0.6284(−2)	0.1016(−1)	0.5049(−1)	0.6791(−1)	0.8000(−1)	0.8951(−1)	0.9364(−1)
375.00	0.4951(−2)	0.5708(−2)	0.6370(−2)	0.1030(−1)	0.5083(−1)	0.6837(−1)	0.8054(−1)	0.9011(−1)	0.9428(−1)
370.00	0.5017(−2)	0.5786(−2)	0.6458(−2)	0.1045(−1)	0.5118(−1)	0.6884(−1)	0.8109(−1)	0.9073(−1)	0.9492(−1)
365.00	0.5086(−2)	0.5866(−2)	0.6548(−2)	0.1059(−1)	0.5154(−1)	0.6931(−1)	0.8165(−1)	0.9136(−1)	0.9558(−1)
360.00	0.5157(−2)	0.5949(−2)	0.6641(−2)	0.1074(−1)	0.5190(−1)	0.6980(−1)	0.8223(−1)	0.9200(−1)	0.9625(−1)
355.00	0.5229(−2)	0.6033(−2)	0.6737(−2)	0.1090(−1)	0.5227(−1)	0.7030(−1)	0.8281(−1)	0.9266(−1)	0.9694(−1)
350.00	0.5304(−2)	0.6121(−2)	0.6835(−2)	0.1106(−1)	0.5265(−1)	0.7081(−1)	0.8341(−1)	0.9333(−1)	0.9764(−1)
345.00	0.5381(−2)	0.6211(−2)	0.6937(−2)	0.1122(−1)	0.5304(−1)	0.7133(−1)	0.8403(−1)	0.9401(−1)	0.9836(−1)
340.00	0.5460(−2)	0.6303(−2)	0.7041(−2)	0.1139(−1)	0.5344(−1)	0.7187(−1)	0.8466(−1)	0.9472(−1)	0.9909(−1)
335.00	0.5542(−2)	0.6398(−2)	0.7149(−2)	0.1157(−1)	0.5384(−1)	0.7241(−1)	0.8530(−1)	0.9543(−1)	0.9984(−1)
330.00	0.5626(−2)	0.6497(−2)	0.7260(−2)	0.1175(−1)	0.5426(−1)	0.7297(−1)	0.8595(−1)	0.9617(−1)	0.1006
325.00	0.5713(−2)	0.6598(−2)	0.7374(−2)	0.1193(−1)	0.5468(−1)	0.7354(−1)	0.8663(−1)	0.9692(−1)	0.1014
320.00	0.5802(−2)	0.6703(−2)	0.7493(−2)	0.1213(−1)	0.5512(−1)	0.7413(−1)	0.8731(−1)	0.9769(−1)	0.1022
315.00	0.5895(−2)	0.6811(−2)	0.7615(−2)	0.1232(−1)	0.5556(−1)	0.7472(−1)	0.8802(−1)	0.9847(−1)	0.1030

310.00	0.5990(−2)	0.6923(−2)	0.7741(−2)	0.1253(−1)	0.5602(−1)	0.7534(−1)	0.8874(−1)	0.9928(−1)	0.1039
305.00	0.6089(−2)	0.7038(−2)	0.7871(−2)	0.1274(−1)	0.5648(−1)	0.7597(−1)	0.8948(−1)	0.1001	0.1047
300.00	0.6191(−2)	0.7157(−2)	0.8006(−2)	0.1296(−1)	0.5696(−1)	0.7661(−1)	0.9024(−1)	0.1010	0.1056
295.00	0.6297(−2)	0.7281(−2)	0.8146(−2)	0.1319(−1)	0.5746(−1)	0.7727(−1)	0.9102(−1)	0.1018	0.1065
290.00	0.6406(−2)	0.7409(−2)	0.8291(−2)	0.1342(−1)	0.5796(−1)	0.7795(−1)	0.9182(−1)	0.1027	0.1075
285.00	0.6519(−2)	0.7541(−2)	0.8441(−2)	0.1367(−1)	0.5848(−1)	0.7865(−1)	0.9264(−1)	0.1036	0.1084
280.00	0.6636(−2)	0.7679(−2)	0.8596(−2)	0.1392(−1)	0.5901(−1)	0.7937(−1)	0.9348(−1)	0.1046	0.1094
275.00	0.6758(−2)	0.7821(−2)	0.8758(−2)	0.1418(−1)	0.5956(−1)	0.8010(−1)	0.9435(−1)	0.1055	0.1104
270.00	0.6884(−2)	0.7969(−2)	0.8925(−2)	0.1445(−1)	0.6012(−1)	0.8086(−1)	0.9524(−1)	0.1065	0.1115
265.00	0.7015(−2)	0.8123(−2)	0.9099(−2)	0.1473(−1)	0.6070(−1)	0.8164(−1)	0.9616(−1)	0.1076	0.1125
260.00	0.7152(−2)	0.8283(−2)	0.9281(−2)	0.1503(−1)	0.6130(−1)	0.8244(−1)	0.9710(−1)	0.1086	0.1136
255.00	0.7293(−2)	0.8449(−2)	0.9469(−2)	0.1533(−1)	0.6192(−1)	0.8327(−1)	0.9807(−1)	0.1097	0.1148
250.00	0.7441(−2)	0.8623(−2)	0.9666(−2)	0.1565(−1)	0.6255(−1)	0.8412(−1)	0.9908(−1)	0.1108	0.1159
245.00	0.7595(−2)	0.8803(−2)	0.9870(−2)	0.1599(−1)	0.6320(−1)	0.8500(−1)	0.1001	0.1120	0.1171
240.00	0.7755(−2)	0.8992(−2)	0.1008(−1)	0.1633(−1)	0.6388(−1)	0.8591(−1)	0.1012	0.1132	0.1184
235.00	0.7923(−2)	0.9188(−2)	0.1031(−1)	0.1669(−1)	0.6458(−1)	0.8685(−1)	0.1023	0.1144	0.1197
230.00	0.8098(−2)	0.9394(−2)	0.1054(−1)	0.1707(−1)	0.6530(−1)	0.8782(−1)	0.1034	0.1157	0.1210
225.00	0.8281(−2)	0.9609(−2)	0.1079(−1)	0.1747(−1)	0.6604(−1)	0.8882(−1)	0.1046	0.1170	0.1224
220.00	0.8472(−2)	0.9835(−2)	0.1104(−1)	0.1788(−1)	0.6682(−1)	0.8985(−1)	0.1058	0.1184	0.1238
215.00	0.8673(−2)	0.1007(−1)	0.1131(−1)	0.1832(−1)	0.6762(−1)	0.9093(−1)	0.1071	0.1198	0.1253
210.00	0.8884(−2)	0.1032(−1)	0.1159(−1)	0.1878(−1)	0.6845(−1)	0.9204(−1)	0.1084	0.1212	0.1268
205.00	0.9106(−2)	0.1058(−1)	0.1189(−1)	0.1926(−1)	0.6931(−1)	0.9320(−1)	0.1097	0.1228	0.1284

Values in parentheses are powers of 10 by which tabulated value should be multiplied.
Source: Hayakawa, K., *Food Technol.* 24:1407, 1970.

Table 9.18. g/K_s vs. U/f_c Tables Used for Calculating the Lethality of the Cooling Part of a Thermal Process by Hayakawa's Procedure ($200 < g/K_s \leq 400$)

I_c/K_s (°F)	U'/f_c for j_c = 2.00 to 2.80								
	2.00	2.10	2.20	2.30	2.40	2.50	2.60	2.70	2.80
400.00	0.9497(−1)	0.9844(−1)	0.1017	0.1048	0.1077	0.1105	0.1131	0.1156	0.1180
395.00	0.9557(−1)	0.9907(−1)	0.1024	0.1054	0.1084	0.1112	0.1138	0.1164	0.1188
390.00	0.9619(−1)	0.9971(−1)	0.1030	0.1061	0.1091	0.1119	0.1145	0.1171	0.1195
385.00	0.9682(−1)	0.1004	0.1037	0.1068	0.1098	0.1126	0.1153	0.1179	0.1203
380.00	0.9747(−1)	0.1010	0.1044	0.1075	0.1105	0.1134	0.1161	0.1186	0.1211
375.00	0.9813(−1)	0.1017	0.1051	0.1083	0.1113	0.1141	0.1168	0.1194	0.1219
370.00	0.9880(−1)	0.1024	0.1058	0.1090	0.1120	0.1149	0.1176	0.1203	0.1228
365.00	0.9948(−1)	0.1031	0.1065	0.1097	0.1128	0.1157	0.1184	0.1211	0.1236
360.00	0.1002	0.1038	0.1073	0.1105	0.1136	0.1165	0.1193	0.1219	0.1245
355.00	0.1009	0.1046	0.1080	0.1113	0.1144	0.1173	0.1201	0.1228	0.1254
350.00	0.1016	0.1053	0.1088	0.1121	0.1152	0.1182	0.1210	0.1237	0.1263
345.00	0.1024	0.1061	0.1096	0.1129	0.1161	0.1190	0.1219	0.1246	0.1272
340.00	0.1031	0.1069	0.1104	0.1138	0.1169	0.1199	0.1228	0.1255	0.1281
335.00	0.1039	0.1077	0.1113	0.1146	0.1178	0.1208	0.1237	0.1265	0.1291
330.00	0.1047	0.1085	0.1121	0.1155	0.1187	0.1217	0.1246	0.1274	0.1301
325.00	0.1055	0.1094	0.1130	0.1164	0.1196	0.1227	0.1256	0.1284	0.1311
320.00	0.1064	0.1102	0.1139	0.1173	0.1206	0.1237	0.1266	0.1294	0.1321
315.00	0.1072	0.1111	0.1148	0.1183	0.1215	0.1247	0.1276	0.1305	0.1332
310.00	0.1081	0.1120	0.1157	0.1192	0.1225	0.1257	0.1287	0.1315	0.1343
305.00	0.1090	0.1130	0.1167	0.1202	0.1236	0.1267	0.1297	0.1326	0.1354
300.00	0.1099	0.1139	0.1177	0.1212	0.1246	0.1278	0.1308	0.1337	0.1365
295.00	0.1109	0.1149	0.1187	0.1223	0.1257	0.1289	0.1319	0.1349	0.1377
290.00	0.1118	0.1159	0.1197	0.1234	0.1268	0.1300	0.1331	0.1360	0.1389
285.00	0.1128	0.1170	0.1208	0.1244	0.1279	0.1312	0.1343	0.1373	0.1401
280.00	0.1139	0.1180	0.1219	0.1256	0.1290	0.1323	0.1355	0.1385	0.1414
275.00	0.1149	0.1191	0.1230	0.1267	0.1302	0.1336	0.1367	0.1398	0.1427
270.00	0.1160	0.1202	0.1242	0.1279	0.1315	0.1348	0.1380	0.1411	0.1440
265.00	0.1171	0.1214	0.1254	0.1292	0.1327	0.1361	0.1393	0.1424	0.1454
260.00	0.1183	0.1226	0.1266	0.1304	0.1340	0.1374	0.1407	0.1438	0.1468
255.00	0.1194	0.1238	0.1279	0.1317	0.1354	0.1388	0.1421	0.1452	0.1483
250.00	0.1207	0.1251	0.1292	0.1331	0.1367	0.1402	0.1435	0.1467	0.1498
245.00	0.1219	0.1264	0.1305	0.1344	0.1381	0.1417	0.1450	0.1482	0.1513
240.00	0.1232	0.1277	0.1319	0.1359	0.1396	0.1432	0.1466	0.1498	0.1529
235.00	0.1245	0.1291	0.1333	0.1373	0.1411	0.1447	0.1482	0.1514	0.1546
230.00	0.1259	0.1305	0.1348	0.1389	0.1427	0.1463	0.1498	0.1531	0.1563
225.00	0.1274	0.1320	0.1363	0.1404	0.1443	0.1480	0.1515	0.1548	0.1580
220.00	0.1288	0.1335	0.1379	0.1421	0.1460	0.1497	0.1532	0.1566	0.1599
215.00	0.1301	0.1351	0.1396	0.1438	0.1477	0.1515	0.1551	0.1585	0.1617
210.00	0.1320	0.1368	0.1413	0.1455	0.1495	0.1533	0.1569	0.1604	0.1637
205.00	0.1336	0.1385	0.1430	0.1473	0.1514	0.1552	0.1589	0.1624	0.1657

Values in parentheses are powers of 10 by which tabulated value should be multiplied.
Source: Hayakawa, K., *Food Technol.* 24:1407, 1970.

Table 9.19. g/K_s vs. U/f_c Tables Used for Calculating the Lethality of the Cooling Part of a Thermal Process by Hayakawa's Procedure ($g/K_s \leq 400$)

I_c/K_s	U'/f_c for		I_c/K_s	U'/f_c for	
(°F)	$j_c = 2.90$	$j_c = 3.00$	(°F)	$j_c \doteq 2.90$	$j_c = 3.00$
400.00	0.1204	0.1226	200.00	0.1711	0.1742
395.00	0.1211	0.1234	195.00	0.1733	0.1765
390.00	0.1219	0.1242	190.00	0.1757	0.1789
385.00	0.1227	0.1250	185.00	0.1781	0.1813
380.00	0.1235	0.1258	180.00	0.1806	0.1839
375.00	0.1243	0.1266	175.00	0.1833	0.1866
370.00	0.1252	0.1275	170.00	0.1860	0.1894
365.00	0.1260	0.1284	165.00	0.1889	0.1923
360.00	0.1269	0.1293	160.00	0.1919	0.1954
355.00	0.1278	0.1302	155.00	0.1951	0.1986
350.00	0.1287	0.1311	150.00	0.1985	0.2020
345.00	0.1297	0.1321	145.00	0.2020	0.2056
340.00	0.1306	0.1331	140.00	0.2057	0.2094
335.00	0.1316	0.1341	135.00	0.2096	0.2134
330.00	0.1326	0.1351	130.00	0.2138	0.2177
325.00	0.1337	0.1361	125.00	0.2183	0.2222
320.00	0.1347	0.1372	120.00	0.2230	0.2270
315.00	0.1358	0.1383	115.00	0.2281	0.2321
310.00	0.1369	0.1394	110.00	0.2335	0.2377
305.00	0.1380	0.1406	105.00	0.2393	0.2436
300.00	0.1392	0.1417	100.00	0.2456	0.2500
295.00	0.1404	0.1430	95.00	0.2525	0.2570
290.00	0.1416	0.1442	90.00	0.2600	0.2646
285.00	0.1428	0.1455	85.00	0.2683	0.2730
280.00	0.1441	0.1468	80.00	0.2774	0.2823
275.00	0.1455	0.1481	75.00	0.2876	0.2926
270.00	0.1468	0.1495	70.00	0.2991	0.3043
265.00	0.1482	0.1509	65.00	0.3120	0.3174
260.00	0.1497	0.1524	60.00	0.3269	0.3325
255.00	0.1511	0.1539	55.00	0.3442	0.3501
250.00	0.1527	0.1555	50.00	0.3645	0.3707
245.00	0.1543	0.1571	45.00	0.3889	0.3954
240.00	0.1559	0.1587	40.00	0.4187	0.4256
235.00	0.1576	0.1604	35.00	0.4557	0.4631
230.00	0.1593	0.1622	30.00	0.5028	0.5107
225.00	0.1611	0.1640	25.00	0.5641	0.5725
220.00	0.1630	0.1659			
215.00	0.1649	0.1679			
210.00	0.1669	0.1699			
205.00	0.1689	0.1720			

Source: Hayakawa, K., *Food Technol.* 24:1407, 1970.

$$g = 4.10 + \left(\frac{5.482 - 4.10}{1}\right)(3.1253 - 3.0) = 4.273$$

B_b is calculated using equation 44:

$$B_b = 22[\ln(1.4)(251 - 160) - \ln(4.273)] = 32.4 \text{ min}$$

Determination of F_0 *using Hayakawa's procedure.* For a process time of 30 min from "steam on," the value of g was previously calculated to be g = 6.66. Table 9.14 is used to determine the lethality of the heating portion of the process. Tabular entry requires the parameter g/K_s. $K_s = 18/20 = 0.900$; $g/K_s = 7.398$.

From Table 9.14:

$$\frac{g}{K_s} = 7; \qquad \frac{U}{f_h} = 0.1252$$

$$= 8; \qquad = 0.1020$$

Interpolating to obtain U/f_h for $g/K_s = 7.398$:

$$\frac{U}{f_h} = 0.1252 - \left(\frac{0.1252 - 0.1020}{1}\right)(7.398 - 7)$$

$$= 0.11597$$

$$U = 22(0.11597) = 2.551$$

Tables 9.15 to 9.19 are now used to determine the lethality of the cooling part of the process. $T_g = 251 - 6.66 = 244.3$; $I_c = 244.3 - 70 = 174.3$; $I_c/K_s = 174.3/0.900 = 193.71$. The appropriate table is Table 9.15, since $I_c/K_s < 200$ and $j_c < 1.9$. Values of $I_c/K_s = 190$ and 195 can be read in Table 9.15. A value for U/f_c corresponding to $I_c/K_s = 193.71$ is obtained by interpolation. For $j_c = 1.8$, $I_c/K_s = 190$; $U'/f_c = 0.1277$; $I_c/K_s = 195$; and $U'/f_c = 0.1260$. Interpolating:

$$\frac{U'}{f_c} = 0.1277 - \left(\frac{0.1277 - 0.1260}{5}\right)(193.71 - 190)$$

$$= 0.1264$$

Solving for U' for the cooling part of the process:

$$U' = 22(0.1264) = 2.7816$$

U' is converted to U using equation 46: $U = 2.7816(10)^{-6.66/18} = 1.187$ for the cooling part of the process. Total U is the sum of U for heating and U for cooling.

$$U = 2.551 + 1.187 = 3.738$$

The process F_0 is then determined using $U = F_0 F_i$. F_i was previously calculated at 251°F to be 0.8799. Therefore, $F_0 = U/F_i = 3.738/0.8799 = 4.248$ min.

Determination of B_b *using Hayakawa's procedure.* A value of g is first assumed. Let $g = 3.7$°F. Since $K_s = 18/20 = 0.9$, $g/K_s = 3.7/0.9 = 4.111$°F. For the heating part of the process, Table 9.14 is used. The value of U/f_h corresponding to $g/K_s = 4.111$ will be obtained by interpolation.

From Table 9.14, for $g/K_s = 4$ and $U_h/f_h = 0.2514$; for $g/k_s = 5$ and $U_h/f_h = 0.1958$:

$$\frac{U}{f_h} = 0.2514 - \left(\frac{0.2514 - 0.1958}{1}\right)(4.1111 - 4)$$

$$= 0.2452$$

Solving for U, $U = 0.2452(22) = 5.39$ for the heating portion of the process.

For the cooling curve: $I_c = 251 - 3.7 - 70 = 177.3$. $I_c/K_s = 177.3/0.9 = 197$. The appropriate table to use is Table 9.15, since $I_c/K_s < 200$ and $j_c < 1.9$. From Table 9.15, for $j_c = 1.8$, U'/f_c; for $I_c/K_s = 197$ will be obtained by interpolation.

$$\frac{I_c}{K_s} = 195; \qquad \frac{U_g}{f_c} = 0.1260$$

$$= 200; \qquad = 0.1243$$

$$\frac{U'}{f_c} = 0.1260 - \left(\frac{0.1260 - 0.1243}{5}\right)(197 - 195) = 0.1253$$

Since $f_c = 22$ min, $U' = 0.1243(22) = 2.756$. U' is converted to U using equation 46: $U = 2.756(10)^{-3.7/18} = 1.716$ for the cooling portion of the process. The total U is the sum of U for the heating portion and U for the cooling portion.

$$U = 5.394 + 1.716 = 7.11 \text{ min} = F_0 F_i$$

F_i was previously determined to be 0.8799. $F_0 = U/F_i = 7.11/0.8799 = 8.08$ min. This is close to the required F_0 value of 8.0 min; therefore, the required value of g is 3.7°F. If the calculated F_0 for the assumed g is not close enough to the specified F_0, it will be necessary to assume another value of g and to repeat the calculations. When selecting another value of g, keep in mind that a smaller g will result in a larger calculated F_0 value.

The selected g of 3.7°F, which resulted in an F_0 value close to the specified F_0 of 8.0 min, is used to solve for the process time. Solving for B_b using equation 44:

$$B_b = 22[\log(1.4)(251 - 160) - \log(3.7)] = 33.8 \text{ min}$$

Evaluation of the Probability of Spoilage from a Given Process. This procedure is used to determine if a process which deviated from specifications will give a safe product. The procedures discussed in this section are also useful in cases of spoilage outbreaks where a spoilage organism is isolated, its heat resistance is determined, and it is desired to determine if a process schedule adjustment is necessary to prevent future spoilage. Another useful application of these procedures is in the conversion of standard F_0 values to F_0^z values for specific microorganisms.

Constant Temperature Processes. In this type of problem, a process time at a constant retort temperature and an initial can temperature are given. The procedure is similar to the one used in the example in the preceding section for calculating f_0. A value for g is calculated using equation 33. Table 9.12 or Tables 9.14 to 9.18 are then used to determine U at a specified z value, from which F_0^z is calculated. The probability of spoilage is then calculated by substituting F_0^z for t in equation 11.

EXAMPLE: The following data represent the heating characteristics of a canned product. $f_h = f_c = 22.5$ min; $j = j_c = 1.4$. If this product is processed for 25 min at 252°F from an initial temperature of 100°F, calculate (a) the F_0 and (b) the probability of spoilage if an organism with a D_0 value of 0.5 min and a z value of 14°F is present with an initial spore load of 10/can.

Solution: g is determined from the process time, using equation 33:

$$\begin{aligned} g &= [10]^{\log(jI) - t/f_h} \\ &= [10]^{\{\log[(1.4)(252 - 100)] - 25/22.5\}} \\ &= (10)^{1.2168} = 16.5°\text{F} \end{aligned}$$

(a) Since $f_h = f_c$, Stumbo's procedure is used. The F_0 is determined using Table 9.12 for $z = 18$°F and $j_c = 1.4$. Inspection of Table 9.12 reveals that to obtain $g = 16.5$ when $j = 1$, f_h/U has to be between 40 and 45. However, the interpolating factor $\Delta g / \Delta j$ is about 6; therefore, since $j_c = 1.4$, g in the table will increase by 0.4(6) or 2.4. Thus, the entry for $f_h/U = 30$ will be considered, and after calculating g at $j_c = 1.4$, the other entry to be used in the interpolation will be selected.

$$\frac{f_h}{U} = 30; \qquad g_{j=1.4} = 14.60 + 0.4(5.3) = 16.72$$

The value of g is greater than 16.5; therefore, the next lower value of f_h/U in the tables will be used to obtain the other value of g to use in the interpolation.

$$\frac{f_h}{U} = 25; \qquad g_{j=1.4} = 13.6 + 0.4(4.8) = 15.52$$

Interpolating between $g = 16.72$ and $g = 15.52$ to obtain f_h/U corresponding to $g = 16.5$:

$$\frac{f_h}{U} = \frac{25 + 5(16.5 - 15.52)}{(16.72 - 15.52)} = 29.1$$

$$U = \frac{f_h}{(f_h/U)_{g=16.5}} = \frac{22.5}{29.1} = 0.7736 = F_0 F_i$$

$$F_i = (10)^{-2/18} = 0.774$$

$$F_0 = \frac{0.773}{0.774} = 0.999 \text{ min}$$

(b) To evaluate lethality to the organism with $z = 14°\text{F}$, F_0^{14} must be determined. Using Table 9.12 for $z = 14°\text{F}$:

$$\frac{f_h}{U} = 50; \qquad g_{j=1.4} = 14.2 + 0.4(4.00) = 15.8$$

$$\frac{f_h}{U} = 60; \qquad g_{j=1.4} = 15.1 + 0.4(4.3) = 16.82$$

For $g = 16.5$:

$$\frac{f_h}{U} = \frac{50 + 10(16.5 - 15.8)}{(16.82 - 15.8)} = 56.9$$

$$U = \frac{f_h}{(f_h/U)_{g=16.5}} = \frac{22.5}{56.9} = 0.395 = F_0 F_i$$

$$f_i = (10)^{-2/14} = 0.7197$$

$$F_0^{14} = \frac{U}{F_i} = \frac{0.395}{0.7197} = 0.549 \text{ min}$$

The number of survivors will be:

$$N = 10[10]^{-Fo/Do} = 10(10)^{-0.549/0.5} = 0.798$$

The probability of spoilage is 79.80%.

Process Temperature Change. When the process temperature changes, errors in the formula method are magnified since evaluation of f_h and j is based on the original uniform initial temperature, while the starting temperature distribution with in-process temperature deviations is no longer uniform. The most accurate way of evaluating the effect of process temperature changes is to use finite difference methods for evaluation of temperature at the critical point and the general method for determining process lethality. If process deviation occurs before the temperature at the critical point exceeds 200°F, and the deviation simply involves a step change in processing temperature at $t = t_1$ from T_{r1} to T_{r2} and remains constant for the rest of the process, lethality may be approximated by the formula methods. The part of the process before the step temperature change is considered to have negligible lethality (if the temperature at the critical point did not exceed 200°F), and the temperature at t_1 is considered the initial temperature for a process at T_{r2}. Procedures for evaluation are the same as in the previous example.

BROKEN HEATING CURVES

Broken heating curves are those which exhibit a break in the continuity of the heating rate at some point in the heating process. Thus, two or more line segments are formed when the heat penetration data are plotted on semilogarithmic graphing paper. This type of heating behavior occurs when the product inside the can undergoes a physical change which alters the heat transfer characteristics. A typical broken heating curve is shown in Fig. 9.14. The slope indices of the curve are designated f_{h1} for the first line segment and f_{h2} for the second line segment. The retort–can temperature difference at the point of intersection of the first and second line segments is designated g_{bh}. The rest of the parameters of the heating curve are the same as for a simple heating curve.

The equation of the first line segment is:

$$\log\left(\frac{jI}{g_{bh}}\right) = \frac{t_{bh}}{f_{h1}} \tag{47}$$

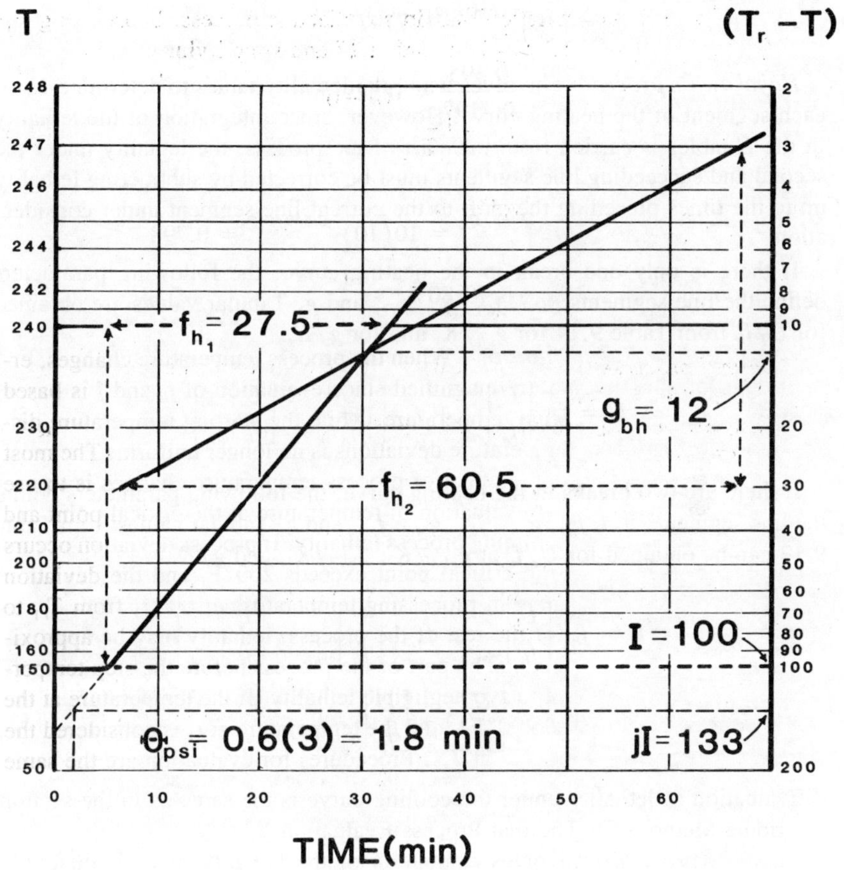

Fig. 9.14. Diagram of a broken heating curve showing the heating curve parameters.

The equation of the second line segment is:

$$\log\left(\frac{g_{bh}}{g}\right) = \frac{t - t_{bh}}{f_{h2}} \tag{48}$$

Combining equations 46 and 47:

$$t = f_{h1}\log\left(\frac{jI}{g_{bh}}\right) + f_{h2}\log\left(\frac{g_{bh}}{g}\right) \tag{49}$$

Equation 49 is used to calculate a process time to obtain g. An expression for g can be obtained by rearranging equation 49.

$$g = [10]^{1/f_{h2}[f_{h1}\log(jI) - (f_{h1} - f_{h2})\log(g_{bh}) - t]} \tag{50}$$

Hayakawa's procedure involves using the lethality tables to determine U for each segment of the heating curve. However, since integration of the lethality in the U tables is carried from the start of the process, the lethality under the second and succeeding line segments must be corrected by subtracting lethality up to the times preceding the shift to the current line segment under consideration.

If there is only one break in the heating curve, the following parameters define the line segments: j, I, f_{h1}, g_{bh}, f_{h2}, and g. Tabular values are obtained for U/f_h from Table 9.14 for g_{bh}/K_s and for g/K_s.

$$U = f_{h1}\left[\frac{U}{f_h}\right]_{g_{bh}} + f_{h2}\left[\left[\frac{U}{f_h}\right]_{g} - \left[\frac{U}{f_h}\right]_{g_{bh}}\right] \tag{51}$$

If there are two breaks in the heating curve, the following parameters define the line segments: j, I, f_{h1}, g_{bh1}, f_{h2}, g_{bh2}, f_{h3}, and g. Tabular values from Table 9.13 can be obtained for U/f_h at g_{bh1}/K_s, g_{bh2}/K_s, and g/K_s.

$$U = f_{h1}\left[\frac{U}{f_h}\right]_{g_{bh1}} + f_{h2}\left[\left[\frac{U}{f_h}\right]_{g_{bh2}} - \left[\frac{U}{f_h}\right]_{g_{bh1}}\right]$$
$$+ f_{h3}\left[\left[\frac{U}{f_h}\right]_{g} - \left[\frac{U}{f_h}\right]_{g_{bh2}}\right] \tag{52}$$

Evaluation of lethality under the cooling curve is the same as in the section "Formula Methods for Thermal Process Evaluation."

Stumbo's procedure involves evaluation of the lethality of individual segments of the heating curve. Since the f_h/U vs. g tables were developed to include the lethality of the cooling part of the process, a correction needs to be made for the lethality of cooling attributable to the first line segment of the heating curve, which does not exist. The r parameter was used to express the fraction of the total process lethality attributed to the heating part of the process.

For the first line segment which ends when $(T_r - T) = g_{bh}$:

$$U_1 = r\frac{f_{h1}}{\left[f_h/U\right]_{g_{bh}}} \tag{53}$$

The second line segment begins when $(T_r - T) = g_{bh}$ and ends when $(T_r - T) = g$. The lethality from the f_h/U tables considers the heating process with the same f_h value starting from time zero; therefore, the effective lethality up to $(T_r - T) = g_{bh}$ must be subtracted from the total.

$$U_2 = \frac{f_{h2}}{(f_h/U)_g} - r\frac{f_{h2}}{(f_h/U)_{gbh}} \tag{54}$$

Thus, the total U for the process is:

$$U = \frac{f_{h2}}{(f_h/U)_g} + \frac{r(f_{h1} - f_{h2})}{(f_h/U)_{gbh}} \tag{55}$$

The denominators $(f_h/U)_g$ or $(f_h/U)_{gbh}$ in equations 52 to 55 represent tabular values for f_h/U corresponding to g or g_{bh}. The parameter r is a function of g. Figure 9.15 can be used to obtain r corresponding to g.

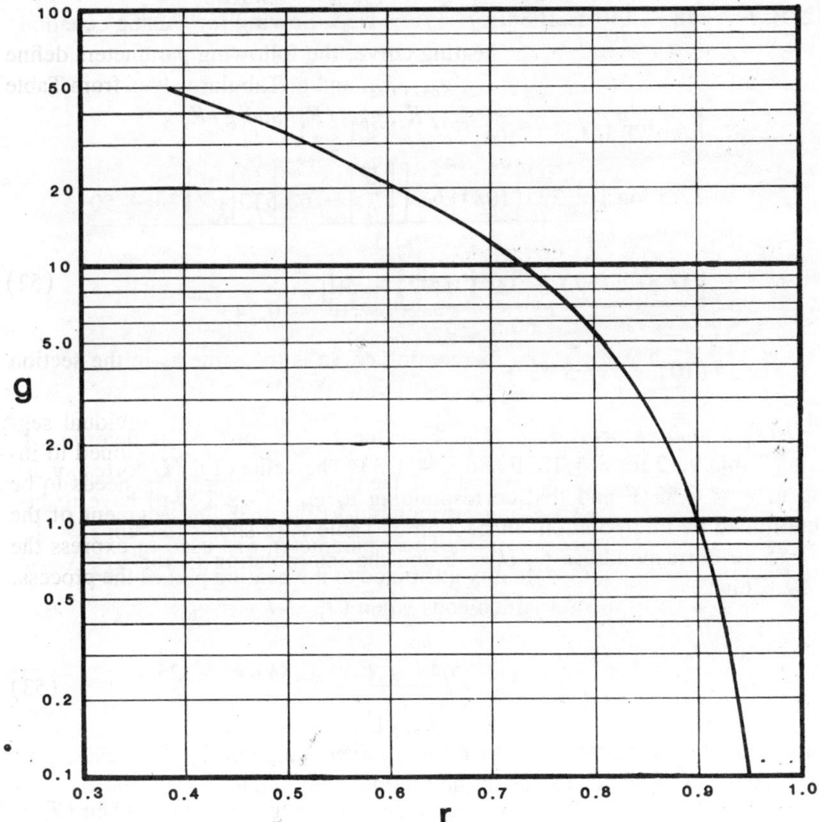

Fig. 9.15. Values of the parameter r corresponding to values of g. (Source: Anonymous 1952. Calculation of process for canned food. American Can Company, Technical Services Division, Maywood, Ill.)

EXAMPLE: A product exhibited the heating curve shown in Fig. 9.14. Assume that $f_c = f_{h2}$ and $j_c = j$ for this heating curve, since a cooling curve is not shown. The heat penetration parameters are used to evaluate a probability of spoilage resulting from a 50-min process at 248°F with an initial temperature of 140°F. Calculate F_0 and the probability of spoilage from an organism having a D_0 value of 1.5 min and a z value of 16°F in cans undergoing this process. The initial spore load is 100/can. The cooling water temperature is 60°F.

Solution: Note that the conditions under which the product is processed are different from those under which the heat penetration parameters were derived. Figure 9.14 is simply used to determine the heat penetration parameters, and these parameters are utilized in the specific process to evaluate the process lethality. From Fig. 9.14: $f_{h1} = 27.5$ min; $f_{h2} = 60.5$ min; $g_{bh} = 12$°F; $I = 248 - 140 = 108$°F; $j = j_c = 1.33$. Solving for g using equation 50:

$$\frac{1}{f_{h2}}\left[f_{h1}\log(jI) - (f_{h1} - f_{h2})\log(g_{bh}) - t\right]$$

$$= \frac{27.5\log[(1.33)(108)] - (27.5 - 60.5)\log(12) - 50}{60.5}$$

$$= \frac{[27.5(2.157) + 33(1.0792) - 50]}{60.5} = 0.743$$

$$g = (10)^{0.743} = 5.53\text{°F}$$

Using Stumbo's procedure. The F_0 value for the process is determined using Table 9.12 for $z = 18$°F and $j_c = 1.33$. The value of f_h/U corresponding to $g = 5.53$°F and that corresponding to $g_{bh} = 12$°F will have to be determined by interpolation. Inspection of Table 9.12 shows that for $j_c = 1$, $g = 5.40$ corresponds to $f_h/U = 5$ and the interpolating factor $\Delta g/\Delta j = 1.59$. Thus:

$$\frac{f_h}{U} = 5; \qquad g_{j=1.33} = 5.40 + 1.59(0.33) = 5.925$$

This value is higher than 5.53; therefore, a lower value of f_h/U is picked for the second set of values to use in the interpolation. Thus:

$$\frac{f_h}{U} = 4; \qquad g_{j=1.33} = 4.41 + 1.34(0.33) = 4.852$$

The value of f_h/U corresponding to $g = 5.53$ is calculated by interpolating between the above two tabular values of g:

$$\left(\frac{f_h}{U}\right)_{g=5.53} = 4 + \frac{(5-4)(5.53-4.852)}{(5.925-4.852)}$$

$$= 4.632$$

Next, the value of f_h/U corresponding to g_{bh} needs to be evaluated. g_{bh} is entered into Table 9.12 as a value of g, and the corresponding f_h/U is determined.

Inspection of Table 9.12 shows that for $j_c = 1$, $z = 18$, and $f_h/U = 15$ corresponds to $g = 10.88$ with the interpolating factor $\Delta g/\Delta j = 3.57$. Thus:

$$\frac{f_h}{U} = 15; \qquad g_{j=1.33} = 10.88 + 0.33(3.57) = 12.058$$

The value of g is greater than 12; therefore, the next lower tabular entry is selected as the other set of values used in the interpolation.

$$\frac{f_h}{U} = 10; \qquad g_{j=1.33} = 8.78 + 0.33(2.69) = 9.668$$

Interpolating between these two values to obtain f_h/U corresponding to g = 12:

$$\left(\frac{f_h}{U}\right)_{gbh} = \left(\frac{f_h}{U}\right)_{g=12}$$

$$= 10 + \frac{5(12-9.668)}{(12.058-9.668)} = 14.88$$

Equation 55 is used to determine U from values of $(f_h/U)_g$ and $(f_h/U)_{gbh}$. r needs to be evaluated from Fig. 9.15 to correspond to $g_{bh} = 12$. From Fig. 9.15, $r = 0.71$. Using equation 55:

$$U = \frac{60.5}{4.622} + \frac{0.71(27.5-60.5)}{14.88}$$

$$= 13.09 - 1.575 = 11.52$$

Using equation 45:

$$F_i = (10)^{2/18} = 1.291$$

$$F_0 = \frac{U}{F_i} = \frac{11.51}{1.291} = 8.91 \text{ min}$$

The probability of spoilage. Since the organism has a z value of 16°F, F_0 cannot be used in equation 11 to determine the probability of spoilage. t is necessary to calculate F_0^{16} for the conditions used in the process. Since the process is the same, g calculated in the first part of this problem is the same; $g = 5.53$. Evaluating f_h/U corresponding to this value of g, however, should be done using $z = 16$°F in Table 9.12. A double interpolation needs to be done. Values for f_h/U corresponding to $g = 5.53$ are determined for $z = 14$ and $z = 18$, and the two values are interpolated to obtain f_h/U at $z = 14$.

$$\frac{f_h}{U} = 6; \qquad g_{j=1.33, z=14} = 4.63 + 0.33(1.56) = 5.145$$

$$= 6.65 + 0.33(1.82) = 6.850$$

$$\frac{f_h}{U} = 6; \qquad g_{j=1.33, z=16} = 5.145 + \frac{(18 - 16)(6.850 - 5.145)}{(18 - 14)}$$

$$= 5.997$$

$$\frac{f_h}{U} = 5; \qquad g_{0.33, z=14} = 4.02 + 0.33(1.32) = 4.455$$

$$\frac{f_h}{U} = 5; \qquad g_{j=1.33, z=18} = 5.40 + 0.33(1.59) = 0.925$$

$$\frac{f_h}{U} = 5; \qquad g_{j=1.33, z=16} = 4.455 + \frac{(18 - 16)(5.925 - 4.455)}{(18 - 14)}$$

$$= 5.190$$

Interpolating between the two values of f_h/U at $z = 16$ which straddles $g = 5.53$:

$$\left(\frac{f_h}{U}\right)_{g=5.53, z=16} = 5 + \frac{(6 - 5)(5.53 - 5.190)}{(5.997 - 5.190)}$$

$$= 5.42$$

The same procedure is used to determine f_h/U corresponding to g_{bh}:

$$\frac{f_h}{U} = 20; \qquad g_{j=1.33,z=14} = 9.63 + 0.33(2.96) = 10.61$$

$$\frac{f_h}{U} = 20; \qquad g_{j=1.33,z=18} = 12.4 + 0.33(4.28) = 13.81$$

$$\frac{f_h}{U} = 20; \qquad g_{j=1.33,z=16} = 10.61 + \frac{(16 - 14)(13.81 - 10.61)}{(18 - 14)} = 12.21$$

$$\frac{f_h}{U} = 15; \qquad g_{j=1.33,z=18} = 10.88 + 0.33(3.57) = 12.06$$

$$\frac{f_h}{U} = 15; \qquad g_{j=1.33,z=14} = 8.29 + 0.33(2.68) = 9.17$$

$$\frac{f_h}{U} = 15; \qquad g_{j=1.33,z=16} = 9.7 + \frac{(16 - 14)(12.06 - 9.17)}{(18 - 14)} = 11.14$$

Interpolating between the two values of f_h/U at $z = 16$ which correspond to values of g which straddle $g_{bh} = 12$:

$$\left(\frac{f_h}{U}\right)_{g_{bh}} = 15 + \frac{(12 - 11.14)(20 - 15)}{(12.21 - 11.14)}$$

$$= 19.02$$

Since r is dependent only on g_{bh}, the same value as before is obtained from Fig. 9.15. For $g_{bh} = 12$, $r = 0.71$. U is obtained using equation 55:

$$U = \frac{60.5}{5.41} + \frac{0.71(27.5 - 60.5)}{19.02}$$

$$= 11.183 - 1.232 = 9.95$$

$$F_i = (10)^{2/16} = 1.333$$

$$F_0^{16} = \frac{U}{F_i} = \frac{9.95}{1.333} = 7.464$$

The number of survivors can now be calculated by substituting F_0^{16} for t in equation 11. Using equation 11:

$$\ln\left(\frac{N}{100}\right) = \frac{-F_0}{D_0} = \frac{-7.465}{1.5} = -4.977$$

$$N = 10^{-4.977} = 0.0011$$

The probability of spoilage is 11 in 10,000.
Hayakawa's procedure. The solution for F_0 will not be presented here. The procedure will be the same as for the determination of F_0^z, which is shown below. As an exercise, the reader can determine F_0. The calculated value is 8.16 min. The probability of spoilage is determined by calculating F_0^{16}. For the heating part of the process, U is calculated using Table 9.14 to obtain $(f_h/U)_g$ and $(f_h/U)_{g_{bh}}$, which is substituted in equation 51. The parameter for the table entry in Table 9.14 is $g/K_s = 5.53/0.8 = 6.91$. g was previously calculated for this problem to be 5.53°F and $K_s = 16/20 = 0.8$. $g_{bh}/K_s = 12/0.8 = 15.00$.

From Table 9.14: $(f_h/U)_{g/K_s=6} = 0.1555$ and $(f_h/U)_{g/K_s=7} = 0.1252$. f_h/U for $g/K_s = 6.91$ is obtained by interpolating between the above values.

$$\left(\frac{f_h}{U}\right)_g = 0.1555 - \frac{(0.1555 - 0.1252)(6.91 - 6)}{(7 - 6)}$$

$$= 0.1279$$

From Table 9.14 for $g_{bh}/K_s = 15.00$:

$$\left(\frac{f_h}{U}\right)_{gbh} = 0.02706$$

There is only one break in the heating curve; therefore, equation 51 is used to determine U for the heating portion of the process.

$$U = 27.5(0.02706) + 60.5(0.1279 - 0.02706) = 6.84 \text{ min}$$

The lethality of the cooling curve is determined using Tables 9.15 to 9.19. Tabular entry is done using I_c/K_s. Since $g = 5.53$°F, $T_g = 248 - 5.53 = 242.47$. $I_c = 242.47 - 60 = 182.47$; $I_c/K_s = 182.47/0.8 = 228$. Table 9.17 is the appropriate table to use, since $I_c/K_s > 200$ and $j_c < 1.9$. From Table 9.17:

$$\left(\frac{U'}{f_c}\right)_{I_c/K_s=225, j_c=1.2} = 0.06604$$

$$\left(\frac{U'}{f_c}\right)_{I_c/K_s=225, j_c=1.4} = 0.08862$$

Interpolating for $j_c = 1.33$:

$$\frac{U'}{f_c} = 0.06604 + \frac{(0.08862 - 0.06604)(1.22 - 1.2)}{0.2}$$

$$= 0.8072$$

From Table 9.17:

$$\left(\frac{U'}{f_c}\right)_{I_c/K_s = 230, j_c = 1.2} = 0.06530$$

$$\left(\frac{U'}{f_c}\right)_{I_c/K_s = 230, j_c = 1.4} = 0.08782$$

Interpolating for $j_c = 1.33$:

$$\frac{U'}{f_c} = 0.06530 + \frac{(0.08782 - 0.06530)(1.33 - 1.2)}{0.2}$$

$$= 0.07994$$

Solving for $(U'/f_c)_{I_c/K_s = 228, j_c = 1.33}$: $= 0.07994 + (0.08782 - 0.07994)(230 - 228)/5 = 0.0831$

$$U' = f_c\left[\left(\frac{U}{f_c}\right)_{I_c/K_s = 228, j_c = 1.33}\right]$$

$$= 60.5(0.0831) = 5.027$$

U' is then converted to U of the cooling part of the process using equation 46.

$$U = 5.027[10]^{-5.53/16} = 5.027(0.4512) = 2.27 \text{ min}$$

The total U is the sum of U for heating and U for cooling.

$$U = 2.27 + 6.86 = 9.13 \text{ min}$$

$$F_0^{16} = 9.13[10]^{248-250/16} = 6.85 \text{ min}$$

The number of survivors is calculated using equation 11 by substituting F_0^{16} for t:

$$\ln\left(\frac{N}{100}\right) = \frac{-6.85}{1.5} = -4.567$$

$$N = 100(10)^{-4.567} = 0.0027$$

The probability of spoilage = 27/10,000.

Hayakawa's procedure results in a higher probability of spoilage than Stumbo's procedure. F_0 calculated using Hayakawa's procedure is lower than that calculated using Stumbo's procedure, which in turn is lower than that calculated using the general method, as shown in the examples in the sections "Determination of f_c and j_c" and "Formula Methods for Thermal Process Evaluation." The safety factor built into the formula methods is primarily responsible for the success with which these thermal process calculation techniques have served the food industry over the years in eliminating the botulism hazard from commercially processed canned foods.

QUALITY FACTOR DEGRADATION

Quality factor degradation has to be evaluated on the basis of integrated lethality throughout the container. Unlike microbial inactivation, which leaves practically no survivors at regions in the can near the surface, there is substantial nutrient retention in the same regions. Quality factor degradation can be determined by separating the container into incremental cylindrical shells, calculating the temperature at each shell at designated time increments, determining the extent of degradation, and summing the extent of degradation at each incremental shell throughout the process. At the termination of the process, the residual concentration is calculated by integrating the residual concentration at each incremental cylindrical shell throughout the container. The procedure is relatively easy to perform using a computer, but the calculations can be onerous if done by hand.

For cylindrical containers, Stumbo (1973) derived an equation for the integrated residual nutrient based on the following observations on the temperature profiles for conduction heat transfer in cylinders:

1. In a container, an isotherm exists where the j value at that point, designated j_v, is half of the j value at the center. The g value at that point at any time, designated g_v, is half of the g value at the center and the volume enclosed by that isotherm is 19% of the total volume.
2. If v is the volume enclosed by the isotherm, and if the F value at the isotherm and at the critical point are F_v and F, respectively, then the difference, $(F_v - F)$, is proportional to $\ln(1 - v)$.

The following expression was then derived:

$$\bar{F} = F + D \log \left(\frac{D + 10.92(F_v - F)}{D} \right) \tag{56}$$

Equation 56 is an expression for the integrated lethal effect of the heating process ($\bar{F}$) on nutrients based on the lethality at the critical point, F, and the lethality, F_v, evaluated at the point where $g_v = 0.5g$ and $j_v = 0.5j$. Equation 54 has been found to be adequate for estimating nutrient retention in cylindrical containers containing foods which heat by conduction.

EXAMPLE: A food product has a j value of 1.2, a j_c value of 1.4, and $f_h = f_c = 35$ min. This product is processed at 255°F from an initial temperature of 130°F to an F_0 of 5.5. Calculate the residual ascorbic acid remaining in this product after the process if the initial concentration was 22 μg/g. The D_0 value for ascorbic acid in the product is 248 min, and the z value is 90°F.

Solution: Stumbo's procedure is used. The F_0 value is used to determine a value of g using the f_h/U tables for $z = 18$ and $j_c = 1.4$. From this value of g, a new f_h/U is determined for a z value of 91°F. The F_0^{91} obtained is the value of F in equation 56. F_v is determined from f_h/U which corresponds to $g_v = 0.5g$ and $j_v = 0.5j_c$.

$$U = F_0 F_i = 5.5(10)^{-5/18} = 2.901 \text{ min}$$

$$\frac{f_h}{U} = \frac{35}{2.901} = 12.06$$

From Table 9.12 for $z = 18$:

$$\frac{f_h}{U} = 10; \qquad g_{jc=1.4} = 8.78 + 2.69(0.4) = 9.956$$

$$\frac{f_h}{U} = 15; \qquad g_{jc=14.} = 10.88 + 3.57(0.4) = 12.308$$

For $f_h/U = 12.06$:

$$g = 9.956 + \frac{(12.308 - 9.956)(12.06 - 10)}{(15 - 10)}$$

$$= 10.925$$

For $g = 10.925$, using Table 9.13, $z = 90°F$:

$$\frac{f_h}{U} = 1; \qquad g_{j_c=1.4} = 3.06 + 0.4(2.19) = 3.936$$

$$\frac{f_h}{U} = 2; \qquad g_{j_c=1.4} = 11.03 + 0.4(7.88) = 14.182$$

f_h/U for $g = 10.925$:

$$= 1 + \frac{(2 - 1)(10.925 - 3.936)}{(14.182 - 3.936)}$$

$$= 1.682$$

U at the geometric center $= 35/1.682 = 20.81$ min.
$F_i = 10^{-5/90} = 0.8799$.
F at the geometric center $= U/F_i = 23.65$ min.
At a point where $g_v = 0.5g$ and $j_{vc} = 0.5j_c$, $g = 0.5(10.925) = 5.463$; $j_{vc} = 0.5(1.4) = 0.7$.
Using Table 9.13 again for $z = 90$:

$$\frac{f_h}{U} = 1; \qquad g_{j_c=0.7} = 3.06 - 0.3(2.19) = 2.403$$

$$\frac{f_h}{U} = 2; \qquad g_{j_c=0.7} = 11.03 - 0.3(7.88) = 8.666$$

$\frac{f_h}{U}$ for $g_v = 5.463$:

$$= 1 + \frac{(2 - 1)(5.463 - 2.403)}{(8.666 - 2.403)}$$

$$= 1.488$$

U at point where $g_v = 0.5g = 35/1.488 = 23.52$ min.
$F_v = U/F_i = 23.52/0.8799 = 26.73$ min.
Substituting in equation 56:

$$\bar{F} = 23.65 + 248 \log \left[\frac{248 + 10.92(26.73 - 23.65)}{248} \right]$$

$$= 23.65 + 248(0.055233) = 37.348$$

$$\log\left(\frac{C}{C_0}\right) = -\frac{F}{D} = \frac{-37.348}{248} = -0.1506$$

$$\frac{C}{C_0} = (10)^{-0.1506} = 70.6$$

The percent retention of ascorbic acid is 70.6%.
The residual ascorbic acid content is $0.706(22) = 15.55\ \mu g/g$.

Hayakawa's tables may also be used to determine F and F_t, to use in equation 56. U for the heating and cooling portions of the process has to be evaluated separately, as was done for microbial inactivation. When values of z are not the same as the tabulated values in Table 9.13, use of Hayakawa's tables is recommended since interpolation across large values of z in Table 9.13 may introduce too much of an error.

PROBLEMS

1. Calculate the D value of an organism which shows 30 survivors from an initial inoculum of 5×10^6 spores after 10 min at 250°F.
2. What level of inoculation of PA 3679 ($D_0 = 1.2$ min) is required such that a probability of spoilage of 1 in 100 attributed to PA 3679 is equivalent to 12 D inactivation of *C. botulinum*? Assume the same temperature process and the same z value for both organisms. The D_0 value of *C. botulinum* is 0.22 min.
3. Calculate the length of a holding tube in high-temperature processing in an aseptic packaging system that is necessary to provide a 5 D reduction of spores of PA 3679 ($D_{250} = 1.2$ min) at 280°F. Use a z value of 20°F. The rate of flow is 30 gal/min, density is 65 lb/ft^3, and viscosity is 10 cp. The tube has a 1.5-in. outside diameter and a wall thickness of 0.064 in.
4. If the same system were used on another fluid having a density of 65 lb/ft^3 and a viscosity of 100 cp, calculate the probability of spoilage when the process is carried out at 280°F ($z = 20$). The initial inoculum is 100 spores per can ($D_{250} = 1.2$ min). The rate of flow is 30 gal/min on a 1.5-in. outside diameter tube (wall thickness 0.064 in).
5. If an initial inoculum of 10 spores per gram of produce ($D_{250} = 1.2$ min) and a spoilage rate of 1 can in 100,000 is desired, calculate an F value for the process that will give the desired level of inactivation. Calculate the F_{280} for a z value of 18°F.
6. If an organism has a D value of 1.5 at 250°F and a z value of 15°F, calculate the F_{240} for a probability of spoilage of 1 in 10,000 from an initial inoculum of 100 spores per can.
7. Figure 9.16 shows an air sterilization system that supplies sterile air to a process. Calculate the length of the holding tube necessary to sterilize the air. The most heat-resistant organism that must be avoided requires 60 min of heating at 150°C for sterilization and has a z value of 70°C. The inside diameter of the holding tube is 0,695 in. Assume plug flow ($V_{max} = V_{avg}$).

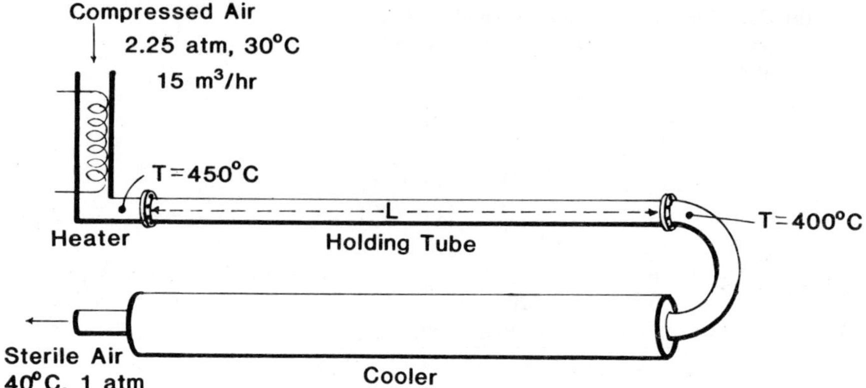

Fig. 9.16. Diagram of an air sterilization system by heat (for Problem 7).

8. (a) A food product in a 303 × 407 can has f_h = 5 and $j = j_c$ = 0.8. For an initial temperature of 80°F and a retort temperature of 250°F, calculate the process time B_b. Use F_0 = 4 min and z = 18°F.

(b) The product in part (a) is processed in a stationary retort, and it takes 4 min for the retort to reach 250°F from the time the steam is turned on. How many minutes after being turned on should the steam be turned off?

(c) In one of the retorts where the cans were processed, there was a misprocess and the record on the retort temperature chart showed the following:

Time (min)	Retort Temperature (°F)
0	70
3	210
10	210
Sudden jump from 210° to 250°F at 10 min	
15	250
16	Steam off, cooling water on

What is the F_0 of this process? The can temperature at time 0 was 80°F.

9. In a given product, PA 3679 has a D value of 3 min at 250°F and a z value of 20°F. If the process was calculated at 280°F for a z value of 18°F, how many minutes of heating is required for a 5 D process? What is the probability of spoilage of PA 3679 if N_0 is 100 spores per can?

10. The following heat penetration data were obtained on chili con carne processed at 250°F in a retort with a come-up time of 3 min. Assume that $j = j_c$.

(a) Calculate the f_h and j values and processes at:
250°F, z = 18, F_0 = 8 (initial temperature = 120°F)
240°F, z = 18, F_0 = 8 (initial temperature = 120°F)
260°F, z = 18, F_0 = 8 (initial temperature = 120°F)

(b) Calculate the probability of spoilage from FS 1518 that might occur from the process calculated at 240°F, $z = 18$, $F_0 = 8$ if FS 1518 has a D value at 250 of 4 min and a z value of 22°F for an initial spore load of 50/can.

Time (min)	Temp. (°F)	Time (min)	Temp. (°F)
0	170	35	223
5	170	40	228
10	180	45	235
15	187	50	236
20	200		
25	209		
30	216		

11. A canned food having $f_h = 30$ and $j = j_c = 1.07$ contains a spore load of 56 organisms per can and the organism had a D_{250} value of 1.2 min. A process with an F_0 of 6 min was calculated for this product at a retort temperature of 250°F and an initial temperature of 150°F.

Subsequent analysis revealed that the spores actually have a z value of 14°F instead of 18°F. If the same time used in the above process was used at a retort temperature of 248°F, calculate the probability of spoilage.

12. A process for a pack of sliced mushrooms in 303 × 404 cans on file with the Food and Drug Administration specifies processing at 252°F for 26 min from an initial temperature of 110°F. A spoiled can from one pack was analyzed microbiologically and was found to contain spore-forming organisms. Data on file for similar products show j values ranging from 0.98 to 1.15 and f_h values ranging from 14 to 18 min.

(a) Would the filed process be adequate to provide at least a 12 D reduction in spores of *C. botulinum* ($D = 0.25$ min; $z = 14$°F)?

(b) If the spores of the spoilage organism have a D_0 value of 1.1 min and a z value of 16°F, what would have been the initial number of organisms in the can to result in a probability of spoilage of 1 in 10,000 after the process?

(c) If you were evaluating the process, would you recommend a recall of the pack for inadequate processing? Would you call for additional technical data on the process before making a recommendation? Explain your action and provide as much detail as possible to convince nontechnical persons (i.e., lawyers and judges) that your action is the correct way to proceed.

13. The following data were collected in a heat penetration test on a canned food for thermal process determination.

Time (Min)	Temp. (°F)	Time (Min)	Temp. (°F)
0	128	35 (cool)	245
3	128	40	243
5	139	45	240
10	188	50	235
15	209	55	185
20	229	60	145
25	238	65	120
30	242	70	104

The processing temperature was 250°F and the retort come-up time was 2 min. Cooling water temperature was 60°F.

Calculate:

(a) The values of f_h, f_c, j_h, and j_c.

(b) If this product is processed from an initial temperature of 150°F at 248°F, how long after the retort temperature reaches 248°F must the process be carried out before the cooling water is turned on if the final can temperature at the time of cooling must be within 2°F of the retort temperature?

(c) If the process is to be carried out at 252°F from an initial temperature of 120°F, calculate a process time such that an organism with a D_0 value of 1.2 min and a z value of 18°F will have a probability of spoilage of 1 in 10,000 from an initial spore load of 100/can.

14. Beef stew is being formulated for canning. The marketing department of the company wants large chunks of meat in the can, and they stipulate that the meat should be 5-cm cubes.

The present product utilizes 2-cm cubes of all vegetables (carrots and potatoes) and meat, and the process time used is 50 min at 250°F from an initial temperature of 150°F. Marketing thinks that the change can be made without major alteration of the present process.

Heat penetration data for the current product obtained from the files did not specify if the thermocouple was embedded in a particle during the heating process. The f_h value was reported to be 35 min and the j value of heating was 1.55. No data were available on the j value of cooling. However, an inoculated pack was done where an inoculum of 1000 spores of an organism having a D_0 value of 1.2 min and a z value of 18°F injected into a single meat particle in each can resulted in a spoilage rate of 3 cans in 1000.

(a) Calculate the F_0 of the process used on the present product based on the available heat penetration data.

(b) Calculate the F_0 of the process based on the survivors from the inoculum. Does the inoculated pack data justify the assumption that the process for the current product is safe?

(c) Is it likely that the heat penetration data were obtained with the thermocouple inside a particle, or was it simply located in the fluid inside the can? Explain your answer.

(d) Calculate the most likely value for f_h if the thermocouple was located in the center of a particle, based on the inoculated pack data, assuming that the j value for a cube is the theoretical j of 2.02.

(e) If f_h varies directly with the square of the cube size and j remains the same at 2.02, estimate the process time for the 5-cm cube such that the F_0 value for the process will be similar to that based on the inoculated pack data for the present product.

15. A biological indicator unit (BIU) which consists of a vial containing a spore suspension and installed at the geometric center of a can was installed to check the validity of a thermal process for a canned food. The canned food has an f_h value of 30 min and a j value of 1.8. Of the 1000 spores originally in the BIU, an analysis after the process showed a survival of 12 spores. The spores in the BIU have a D_0 of 2.3 min and a z value of 16°F. The process was carried out at 248°F from an initial temperature of 140°F. Calculate:

(a) The F_0 value at the geometric center of the can.
(b) The process time.
(c) The sterilizing value of the process, expressed as a number of decimal reductions ($\ln N_0/N$) of *C. botulinum* having a D_0 value of 0.21 min and a z value of 18°F.

16. A canned food with an f_h of 30 min and a j value of 1.2 is to be given a process at 250°F with an F_0 of 8 min and a z of 18°F. To verify the adequacy of the process, an inoculated pack is to be processed using an organism with a D_0 of 1.5 min and a z of 22°F. If the process to be used on the inoculated pack is at 250°F from an initial temperature of 130°F, calculate the number of spores that must be inoculated per can such that a spoilage rate of 10 in 100 will be equivalent in lethality to the process F_0 desired. Assume that the j values of heating and cooling are the same.

SUGGESTED READING

Aiba, S., Humphrey, A. E., and Millis, N. F. 1965. *Biochemical Engineering*. Academic Press, New York.

Anonymous. 1952. Calculation of process for canned food. American Can Company, Technical Service Division Memo, Maywood, Ill.

Ball, C. O., and Olson, F. C. W. 1957. *Sterilization in Food Technology*. McGraw-Hill Book Co., New York.

Carson, V. R. 1969. *Aseptic Processing*, 3rd ed. Technical Digest CB 201, Cherry Burrel Corp., Chicago.

Charm, S. E. 1971. *Fundamentals of Food Engineering*, 2nd ed. AVI Publishing Co., Westport, Conn.

Cleland, A. C., and Robertson, G. L. 1986. Determination of thermal process to ensure commercial sterility of food in cans. In: *Developments in Food Preservation*, S. Thorn (ed.). Vol. 3. Elsevier Applied Science Publishers, New York.

Edgerton, E. R., and Jones, V. A. 1970. Effect of process variables on the holding time in an ultra high temperature steam injection system. *J. Dairy Sci.* 53:1353–1357.

Hayakawa, K. 1970. Experimental formulas for accurate estimation of transient temperature of food and their application to thermal process evaluation. *Food Technol.* 24(12):1407–1417.

Leniger, H. A. and Beverloo, W. A. 1975. Food Process Engineering. D. Riedel Publishing Co., Boston, MA.

NFPA 1968. *Laboratory Manual for Food Canners and Processors*. Vol. 1. National Food Processors Association, Washington D. C. Published by: AVI Pub. Co., Westport, CT.

Patasnik, M. 1953. A simplified procedure for thermal process evaluation. Food Technol. 7(1):1–6.

Ruthfus, R. R., Archer, D. H., Klimas, I. C., and Sikchi, K. G. 1957. Simplified flow calculations for tubes and parallel plates. *AIChE J.* 3:208.

Stumbo, C. R. 1973. *Thermobacteriology in Food Processing*, 2nd ed. Academic Press, New York.

10

Refrigeration

Cooling is a fundamental operation in food processing and preservation. Removal of heat may involve either the transfer of heat from one fluid to another or from a solid to a fluid, or it may be accomplished by vaporization of water from a material under adiabatic conditions. Knowledge of the principles of heat transfer is an essential prerequisite to the understanding of the design and operation of refrigeration systems.

Maintaining temperatures lower than ambient inside a system requires both the removal of heat and the prevention of incursion of heat through the system's boundaries. The rate of heat removal from a system necessary to maintain the temperature is the *refrigeration load*. Refrigeration systems must be sized to handle the refrigeration load adequately. When heat has to be removed from a system continuously, at temperatures below ambient and for prolonged periods, a mechanical refrigeration system acts as a pump that extracts heat at low temperatures and transfers it to another part of the system, where it is eventually dissipated to the surroundings at a higher temperature. The operation requires energy, and a well-designed system allows the maximum removal of heat at minimum energy cost.

MECHANICAL REFRIGERATION SYSTEM

Principle of Operation. The second law of thermodynamics mandates that heat will flow only in the direction of decreasing temperature. In a system that must be maintained at a temperature below ambient, heat must be made to flow in the opposite direction. A refrigeration system may be considered as a pump that conveys heat from a region of low temperature to another region of high temperature.

The low-temperature side of a refrigeration system is maintained at a lower temperature than the system it is cooling to allow spontaneous heat flow into the refrigeration system. The high-temperature side must have a temperature higher than ambient to allow dissipation of the absorbed heat to the surroundings. In some instances, this absorbed heat is utilized as a heat source for heating processes.

Both a high temperature and a low temperature can be maintained in a refrigeration system by the use of a refrigerant fluid which is continuously recirculated through the system. A liquid's boiling or condensation temperature is a function of absolute pressure. By reducing the pressure, a low boiling temperature is made possible, allowing for absorption of heat in the form of the refrigerant's heat of vaporization as it is vaporized at the low pressure and temperature. The vapors, when compressed to a high pressure, will condense at the high temperature, and the absorbed heat will be released from the refrigerant as it condenses back into liquid at the high temperature and pressure. Figure 10.1 shows the vapor pressure vs. temperature of commonly used refrigerants. It also illustrates how this pressure–temperature relationship is utilized for cooling and heating.

The Refrigeration Cycle. Figure 10.2 shows a schematic diagram of a mechanical refrigeration system. The heat of the system is the *compressor*.

When the compressor is operating, refrigerant gas is drawn into it continuously. Low pressure is maintained at the suction side, and because of the low pressure, the refrigerant can vaporize at a low temperature. In the compressor, the refrigerant gas is compressed, increasing in both pressure and temperature during the process. The hot refrigerant gas then flows into a heat exchange coil called the *condenser*, where heat is released in the process of condensation at constant pressure and temperature. From the condenser the liquid refrigerant flows into a liquid refrigerant holding tank. In small systems there may be no holding tank, and the refrigerant cycles continuously through the system.

When there is a demand for cooling, the liquid refrigerant flows from the holding tank to the low-pressure side of the refrigeration system through an *expansion valve*. The drop in pressure that occurs as the refrigerant passes through the expansion valve does not change the heat content of the refrigerant. However, the temperature drops to the boiling temperature of the liquid at the low pressure. The cold liquid refrigerant then flows to another heat exchange coil called the *evaporator*, where the system performs its cooling function, and heat is absorbed by the refrigerant in the process of vaporization at constant temperature and pressure. From the evaporator the cold refrigerant gas is drawn into the suction side of the compressor, thus completing the cycle.

A refrigeration system is usually equipped with low-pressure and high-pressure cut-off switches which interrupt power to the compressor when either the high-pressure set point is exceeded (this occurs if the cooling capacity of the condenser is inadequate) or the pressure drops below the low-pressure set point (this occurs when the compressor is running but the cooling demand is much less than the capacity of the refrigeration system). The low-pressure set point can be used to control the evaporator temperature. In some systems, liquid

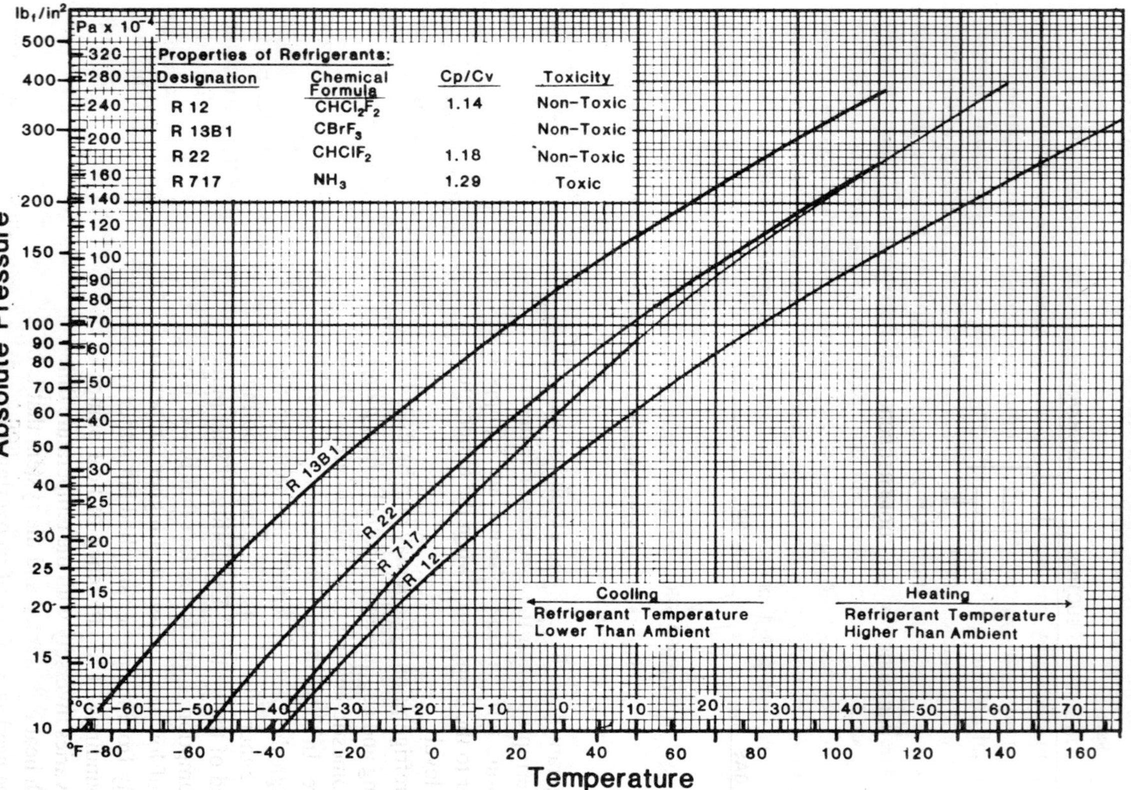

Fig. 10.1. Vapor pressure of commonly used refrigerants as a function of temperature.

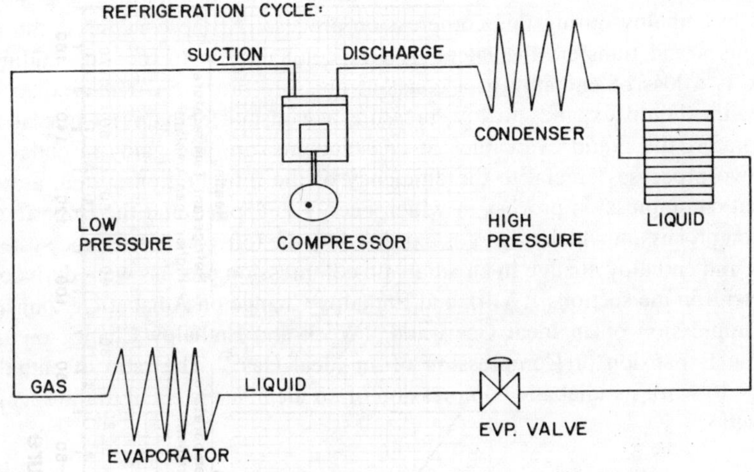

Fig. 10.2. Schematic diagram of a refrigeration system.

refrigerant flow through the expansion valve is thermostatically controlled and is interrupted when the evaporator temperature is lower than the set point. In addition, refrigeration systems are equipped with thermostats that interrupt power to the compressor when the temperature of the refrigerated room reaches a set level.

A refrigeration system may also be used for heating. A system that alternates heating and cooling duty is called a *heat pump*. These units are used extensively for domestic heating or cooling in areas where winter temperatures are not too severe. In these units, either a low-pressure or a high-pressure refrigerant goes through the heat exchange coils that constitute the evaporator and condenser; thus, either may act as a heating or cooling coil, depending upon the duty expected of the system. The ability of heat pumps to deliver heat with low power consumption is due to the fact that power is used only to pump energy from an area of low temperature to a higher temperature. The heat is not derived completely from the power supplied but rather is extracted from cooler air from the surroundings. The efficiency of heat pumps in transforming applied power to heat varies inversely with the temperature differential between the medium from which heat is extracted and the system that is being heated. Heat pump systems can be used as a means of recovering energy from low-temperature heat sources for use in low-temperature heat applications such as dehydration.

The Refrigeration Cycle as a Series of Thermodynamic Processes. Starting from the compressor in Fig. 10.2, low-pressure gas is compressed adiabatically to high pressure, which should allow condensation at ambient temperature. Work is required to carry out this process, and this energy is supplied in the form of electrical energy to drive the compressor motor. The gas also gains in enthalpy during this compression process. At the condenser, the gas condenses and transfers the latent heat of condensation to the surroundings. There is a loss of enthalpy in a constant pressure process. At the expansion valve, the liquid expands to low pressure at constant enthalpy, while at the evaporator, the liquid evaporates at constant pressure and gains in enthalpy. The two processes crucial to the efficiency of the refrigeration system are the adiabatic compression process in which energy is applied and the isobaric expansion process in which energy is extracted by the refrigerant from the system. Work and enthalpy change in an adiabatic compression process were derived in Chapter 4 in the sections "Work and Enthalpy Change on Adiabatic Expansion or Compression of an Ideal Gas" and "Work and Enthalpy Change on Isothermal Expansion or Compression of an Ideal Gas." The ratio of enthalpy change to work in adiabatic compression of an ideal gas is equal to the specific heat ratio.

$$\frac{\Delta H}{W} = -\gamma; \qquad W = \frac{-\Delta H}{\gamma} \tag{1}$$

The negative sign indicates that energy is being used on the system.

THE REFRIGERATION CYCLE ON THE PRESSURE-ENTHALPY DIAGRAM FOR A GIVEN REFRIGERANT

The thermodynamic properties of refrigerants when plotted on a pressure-enthalpy diagram result in a plot that is useful in determining capacities and power requirements for a refrigeration system. Figure 10.3 is a schematic diagram of

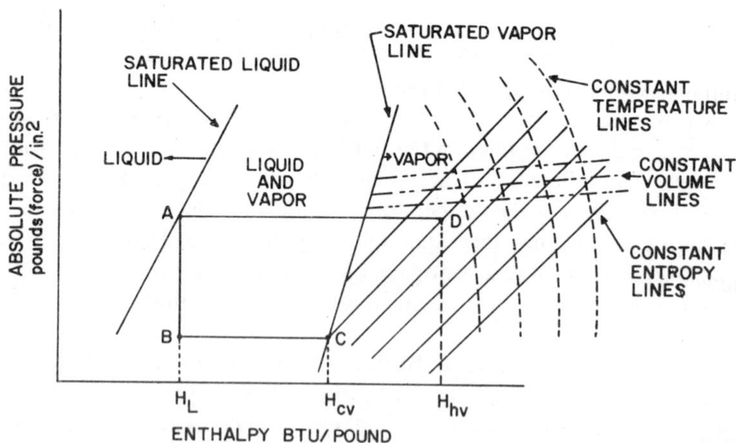

Fig. 10.3. Refrigeration cycle on a pressure-enthalpy diagram.

a refrigeration cycle on a pressure-enthalpy diagram. The diagram consists of lines representing the vapor and liquid pressure–enthalpy relationship, lines representing the change in enthalpy with pressure during adiabatic compression (constant entropy), and, in some charts, lines representing specific volumes at various pressures and enthalpies.

Point A in Fig. 10.3 represents the liquid refrigerant at high pressure entering the expansion valve. The refrigerant expands at constant enthalpy (H_L) as it goes through the expansion valve and leaves the valve as a mixture of liquid and vapor at a lower pressure, represented by point B. As the refrigerant absorbs heat in the evaporator, it gains in enthalpy, represented by the line BC. The refrigerant leaves the evaporator as saturated vapor (represented by point C) having the enthalpy H_{cv}. The compressor raises the pressure, and the change is represented by the line CD, which parallels the lines of constant entropy. As the compressed refrigerant gas leaves the compressor at point D, it has an enthalpy represented by H_{hv}. At the condenser, heat is dissipated, resulting in a drop in enthalpy, represented by the line AD. The liquid refrigerant leaves the condenser with the pressure and enthalpy represented by point A.

The cooling capacity of the refrigeration system is represented by the line BC.

$$\text{Cooling capacity} = H_{cv} - H_L \tag{2}$$

The condenser heat exchange requirement or condenser duty is represented by the line AD.

$$\text{Condenser duty} = H_{hv} - H_L \tag{3}$$

The change in enthalpy due to compression ΔH_c is:

$$\Delta H_c = H_{hv} - H_{cv} \tag{4}$$

From equation 1, the work required for compression is:

$$W = \frac{-\Delta H_c}{\gamma} = \frac{-(H_{hv} - H_{cv})}{\gamma} \tag{5}$$

If M is the mass of refrigerant recirculated through the system per unit time, the power requirement can be calculated as follows:

$$P = \frac{W}{\text{time}} = M\frac{-(H_{hv} - H_{cv})}{\gamma} \tag{6}$$

The negative sign on the work and power indicates that work is being added to the system.

The efficiency of a refrigeration system is also expressed in terms of a *coefficient of performance* (COP), which is a ratio of the cooling capacity over the gain in enthalpy due to compression.

$$COP = \frac{H_{cv} - H_L}{H_{hv} - H_{cv}} \tag{7}$$

The power requirement (P) can be expressed in terms of the coefficient of performance using equations 6 and 7.

$$P = \frac{H_{cv} - H_L}{\gamma(COP)} \cdot M \tag{8}$$

The refrigeration capacity is expressed in tons of refrigeration, the rate of heat removal sufficient to freeze 1 ton (2000 lb) of water in 24 h. Since the heat of fusion of water is 144 BTU/lb, this rate of heat removal is equivalent to 12,000 BTU/h. The refrigeration capacity in tons is:

$$(\text{tons})_r = \frac{(H_{cv} - H_L)(M)}{12{,}000} \tag{9}$$

Substituting equation 9 in equation 8:

$$P = \frac{(\text{tons})_r(12{,}000)}{\gamma(COP)} \left[\frac{\text{BTU}}{\text{h}(\text{ton})_r}\right]$$

Expressing the power requirement in horsepower (HP)/$(\text{ton})_r$:

$$\frac{\text{HP}}{(\text{ton})_r} = \frac{12{,}000}{\gamma(COP)} \left| \frac{\text{BTU}}{\text{h}(\text{ton})_r} \right| \frac{1\ \text{HP}}{2545\ \text{BTU/h}}$$

$$= \frac{4.715}{\gamma(COP)} \tag{10}$$

Since there is a certain degree of slippage of refrigerant past the clearance between the cylinder and the piston, particularly at high pressure and also because some frictional resistance occurs between the piston and the cylinder, the actual work expended would be higher than that determined using equation 10. The ratio between the theoretical horsepower as calculated from equation 10 and the actual horsepower expended is the efficiency of compression. The efficiency of compression depends upon the ratio of high-side to low-side pressure across the compressor and will be discussed further in the section "Refrigeration Load."

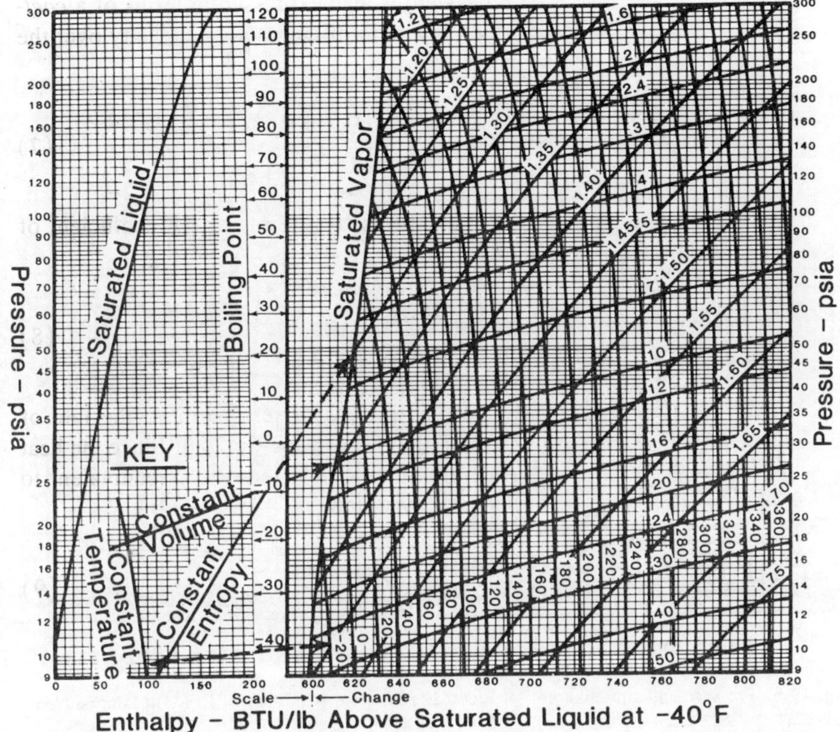

Fig. 10.4. Pressure-enthalpy diagram for refrigerant 717 (NH_3). (Adapted from *ASHRAE Guide and Data Book, Fundamentals and Equipment*, 1965.)

Pressure-enthalpy diagrams for ammonia, Freon 12, and Freon 22 are shown in Figs. 10.4, 10.5 and 10.6.

EXAMPLE PROBLEMS ON THE USE OF REFRIGERANT CHARTS

EXAMPLE 1: A refrigerated room is to be maintained at 40°F (4.44°C) and 90% relative humidity. The contents of the room vaporize moisture, requiring removal of moisture from the air to maintain the desired relative humidity. Assuming that air flow over the evaporator coil is such that the bulk mean air temperature reaches to within 2°F (1.11°C) of the coil temperature, determine the low-side pressure for an ammonia refrigeration system such that the desired humidity will be maintained.

Solution: From a psychrometric chart, the dew point of air at 40°F and 80% relative humidity is 33°F (0.56°C). When room air is cooled to 33°F in

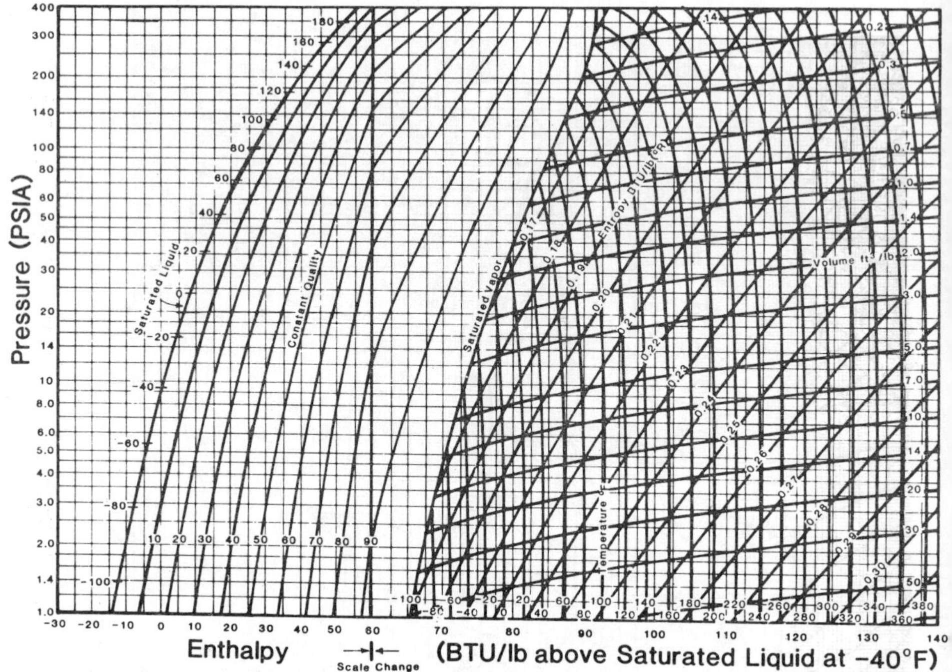

Fig. 10.5. Pressure-enthalpy diagram for Freon 12 (CCl_2F_2). (Courtesy of E. I. Du Pont de Nemours, Inc.)

passing over the evaporator coils, any moisture present in excess of the saturation humidity at this temperature will condense. If air flow over the coil is sufficient to allow enough moisture removal through condensation to equal the rate of moisture vaporization into the air, then the desired relative humidity will be maintained. The coil temperature should be 31°F (−0.56°C). From Fig. 10.1 for ammonia, the pressure corresponding to a temperature of 31°F is 65 psia (446 kPa absolute).

EXAMPLE 2: A heat pump is proposed as a means of heating a cabinet drier. Air is continuously recycled through the system, passing through the condenser coil at the air inlet to the drying chamber, where it is heated. Prior to recycling, the hot, moist air leaving the drying chamber is passed through the evaporator coil, where moisture is removed by condensation. Inlet air to the drying chamber is at 140°F (60°C) and 4% relative humidity. Assuming that the evaporator and condenser coil design and air flow are such that air temperature approaches 5°F (2.78°C) of the evaporator coil temperature and the log mean ΔT between condenser coil and air is 20°F (5.56°C), determine

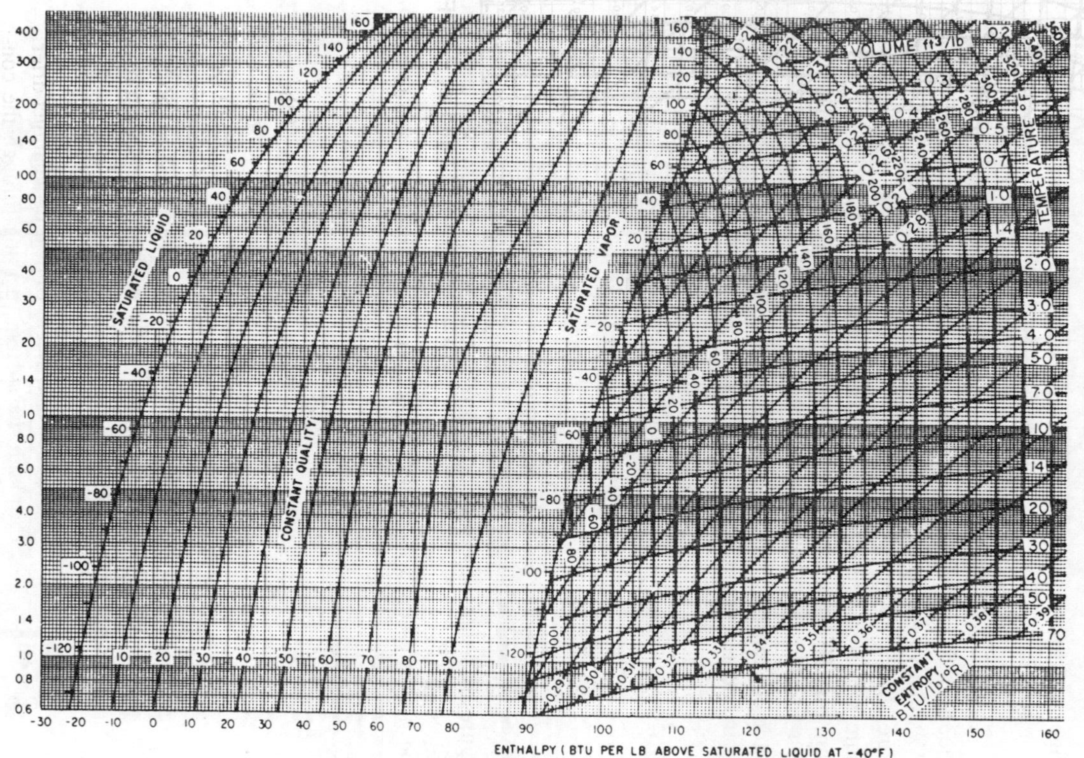

Fig. 10.6. Pressure-enthalpy diagram for Freon 22 ($CHClF_2$). (Courtesy of E. I. Du Pont de Nemours. Inc.)

the high- and low-side pressures for the refrigeration system using Freon 22 as the refrigerant.

Solution: From a psychrometeric chart, the dew point of air at 140°F (60°C) and 4% relative humidity is 38°F (3.33°C). Thus, the evaporator coil temperature should be at 33°F (0.56°C). The condenser coil temperature and pressure are calculated from the specified log mean ΔT.

From Fig. 10.1 for Freon 22, the low-side pressure corresponding to 33°F (0.56°C) is 77 psia (530 kPa absolute).

At the condenser, air enters at 38°F and leaves at 140°F. If the system is set up in countercurrent flow, the log mean ΔT can be calculated as follows:

$$\overline{\Delta T_1} = \frac{(T_1 - 38) - (T_g - 140)}{\ln (T_1 - 38)/(T_g - 140)}$$

where T_1 and T_g are the temperature of refrigerant liquid and gas, respectively.

Figure 10.7 shows the pressure-enthalpy diagram for the system under the conditions given.

The high pressure will be determined by trial and error from Fig. 10.6.

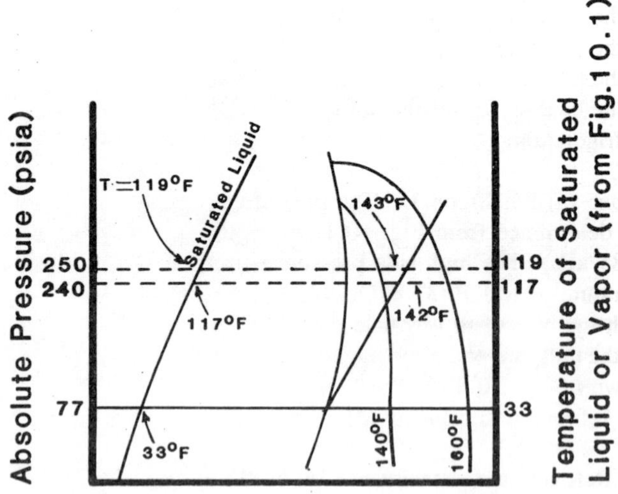

Fig. 10.7. Pressure-enthalpy diagram of the heat pump system in Example 2.

Following the constant entropy line from a pressure of 77 psia, a high pressure of 250 psia will give a hot refrigerant gas temperature of 143°F and a liquid temperature of 119°F. The log mean ΔT is:

$$\overline{\Delta T_1} = \frac{(119 - 38)(143 - 140)}{\ln 92/3} = 23.6$$

This is higher than the 20°F ΔT specified. At a pressure of 240 psia, the hot refrigerant gas temperature is 142°F and the liquid temperature is 117°F.

$$\overline{\Delta T_1} = \frac{(117 - 38) - (142 - 140)}{\ln 79/2} = 20.9$$

This calculated value for the log mean ΔT is almost equal to the specified 20°F; therefore, the high-side pressure for the system should be at 250 psia (1650 kPa).

EXAMPLE 3: A refrigeration system is to be operated at an evaporator coil temperature of −30°F (−34.4°C) and a condenser temperature of 100°F (37.8°C) for the liquid refrigerant. For (a) Freon 12 and (b) refrigerant 717, determine:

1. the high-side pressure,
2. the low-side pressure,
3. the refrigeration capacity per unit weight of refrigerant,
4. the coefficient of performance,
5. the theoretical horsepower of compressor per ton of refrigeration, and
6. the quantity of refrigerant circulated through the system per ton of refrigeration.

Solution: (a) For Freon 12: The pressure at the high- and low-sides (1 and 2) are determined from Fig. 10.1. At −30°F (−34.4°C), pressure is 12.3 psia (85 kPa). The high-side pressure corresponding to a refrigerant liquid temperature of 100°F (37.8°C) is 133 psia (910 kPa).

With the high- and low-side pressures known, Fig. 10.5 is used to construct the pressure-enthalpy diagram for the refrigeration cycle. This diagram is shown in Fig. 10.8.

The refrigeration capacity (3) is calculated using equation 2:

$$\begin{aligned}\text{Refrigeration capacity} &= (74 - 32) = 42 \text{ BTU/lb} \\ &= (17.2 - 7.4) \times 10^4 = 98{,}000 \text{ J/kg}\end{aligned}$$

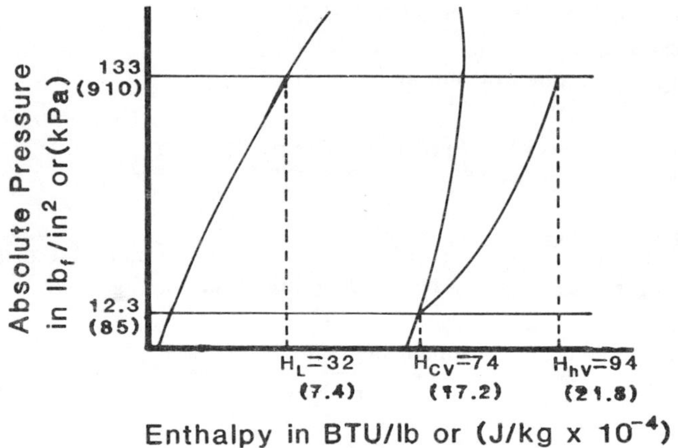

Fig. 10.8. Pressure-enthalpy diagram of refrigerant 12 for the problem in Example 3.

The coefficient of performance (4) is calculated using equation 7:

$$COP = \frac{74 - 32}{94 - 74} = 2.1$$

The theoretical horsepower per ton of refrigeration (5) is calculated using equation 10:

$$\frac{C_P}{C_v} \text{ for Freon 12} = \gamma = 1.14$$

$$\frac{\text{HP}}{(\text{ton})_r} = \frac{4.715}{(\gamma)(COP)} = \frac{4.715}{(1.14)(2.1)} = 1.97$$

One ton of refrigeration (6) is equivalent to 12,000 BTU/h or 3517 W. Weight of refrigerant circulated per hour is:

$$\text{Weight} = \frac{\text{cooling capacity/ton of refrigeration}}{\text{cooling capacity/unit weight of refrigerant}}$$

$$= \frac{12{,}000 \text{ BTU/h}}{42 \text{ BTU/lb}} = 286 \text{ lb refrigerant per hour}$$

$$= \frac{3517 \text{ J/s}}{98{,}000 \text{ J/kg}} = 0.0359 \text{ kg/s} = 129 \text{ kg/h}$$

Solution: (b) For refrigerant 717: The high- and low-side pressures (1 and 2) corresponding to $-30°F$ ($-34.4°C$) and 100°F (37.8°C) are determined from Fig. 10.1. The low-side pressure is 13.7 psia (94 kPa). The high-side pressure is 207 psia (142 kPa).

The pressure-enthalpy diagram (3) is then constructed, knowing the high- and low-side pressures using Fig. 10.4. Figure 10.9 shows this diagram.

$$\text{Refrigeration capacity} = 600 - 155 = 445 \text{ BTU/lb}$$
$$= (139 - 36) \times 10^4 = 1{,}030{,}000 \text{ J/kg}$$

The coefficient of performance (4) is:

$$COP = \frac{600 - 155}{784 - 600} = 2.42$$

The theoretical horsepower per ton of refrigeration (5) is calculated using a C_p/C_v ratio of 1.29.

$$\frac{\text{HP}}{(\text{ton})_r} = \frac{4.715}{1.29\,(2.42)} = 1.51$$

Weight of refrigerant circulated per hour $(\text{ton})_r$ (6) is:

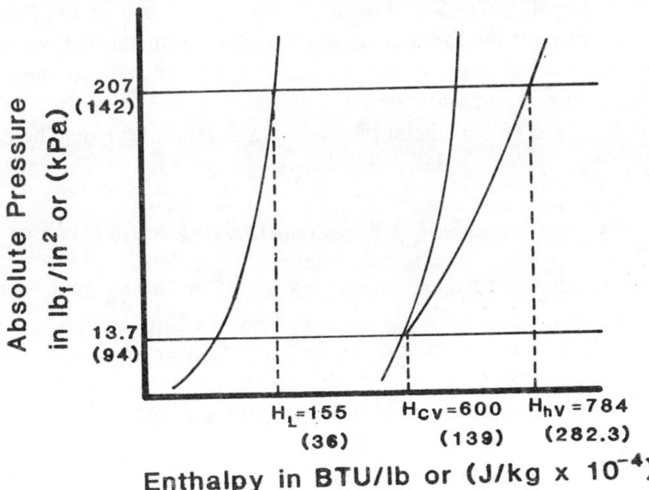

Fig. 10.9. Pressure-enthalpy diagram of ammonia for the problem in Example 3.

$$\text{Weight} = \frac{12{,}000\ \text{BTU}/(\text{h})(\text{ton})_r}{445\ \text{BTU}/\text{lb}} = 26.97\ \text{lb}/\text{h}(\text{ton})_r$$

$$= \frac{3517\ \text{J}/\text{s}(\text{ton})_r}{1{,}030{,}000\ \text{J}/\text{kg}} = 0.00341\ \text{kg}/\text{s}(\text{ton})_r$$

$$= 12.3\ \text{kg}/\text{h}(\text{ton})_r$$

THE CONDENSER AND THE EVAPORATOR

Most refrigeration systems transfer heat between the refrigerant and air. Since heat transfer coefficients to air are usually very low, the air film resistance controls the rate of heat transfer. Very large heat transfer areas would be required to achieve the necessary heat transfer rate. To achieve the necessary heat transfer rate and still have equipment that is reasonably sized, the heat transfer surface area of the tubes that comprise the evaporator or condenser coil is increased by the use of fins.

Finned heat exchange units are sized in terms of an effective heat transfer surface area which is the sum of the area of the bare tube A_t and the effective area of the fin, which is the product of the fin surface area A_f and the fin efficiency η.

$$A_{\text{eff}} = \eta A_f + A_t$$

The fin efficiency, η, decreases as the base area of the fin decreases and as the height of the fin increases. Furthermore, as the ratio of the heat transfer coefficient to air over the thermal conductivity of the metal that constitutes the fin (h_0/k) approaches zero, η approaches unity. Figure 10.10 shows fin efficiency for the most common systems. The fins in these units often consist of a stack of very thin sheets of metal pierced by a bundle of tubes. The equivalent fin surface area, A_f, is calculated as follows:

$$A_f = 2\pi(r_f^2 - r^2)(n_t)(n_f)$$

where n_t = number of tubes and n_f = number of fins along the length of each tube.

The capacity of a condensing or evaporating unit depends upon the surface area available for heat transfer and the mean temperature difference between the surfaces of the unit and the air going through the unit.

EXAMPLE 1: A condensing unit of a refrigeration system consists of 13 rows of 38.1-cm-long, 1.27-cm outside diameter copper tubes with 1.651-mm walls stacked four deep on 2.54-cm centers. The fin consists of 0.0254-cm-thick

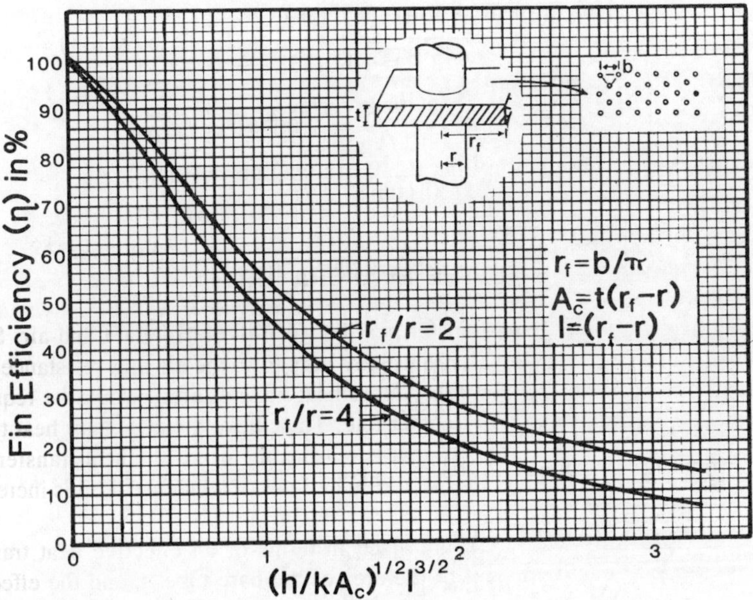

Fig. 10.10. Fin efficiency of fins consisting of a stack of sheet metal, each sheet of thickness l, pierced by tubes of radius r. The tubes are arranged with centers on an equilateral triangle having distance b between their centers.

aluminum sheets spaced 10 per 2.54 cm. If the rate of air flow through this condenser is such that the outside heat transfer coefficient, h, is 9 BTU/(h · ft^2 · °F) or 51.1 W/(m^2 · K), calculate the effective heat transfer surface area for this unit.

Solution: Refer to Fig. 10.10. Since the tubes are spaced on 2.54-cm centers, $b = 2.54$ cm. The thermal conductivity of aluminum (from Appendix Table A.9) is 206 W/(m · K). Figure 10.10 can now be used to determine the fin efficiency, η.

$$\text{Tube radius, } r = 0.00635 \text{ m}$$

$$\text{Equivalent fin radius, } r_f = \frac{b}{\sqrt{\pi}} = \frac{0.0254}{1.772} = 0.014334 \text{ m}$$

$$\text{Fin cross-sectional area, } A_c = t(r_f - r)$$

$$= 0.000254\,(0.014334 - 0.00635)$$

$$= 2.0279 \times 10^{-6}$$

$$\text{Height of the fin, } l = (r_f - r)$$
$$= (0.014334 - 0.00635)$$
$$= 0.00798 \text{ m}$$

Values for the abscissa in Fig. 10.10 can now be calculated:

$$\left[\frac{h}{kA_c}\right]^{0.5} = (l)^{1.5} = \left[\frac{51.1}{206(2.0279 \times 10^{-6})}\right]^{0.5} (0.007984)^{1.5}$$
$$= (349.7)(0.000713) = 0.2495$$
$$\frac{r_f}{r} = \frac{0.014334}{0.00635} = 2.26$$

From Fig. 10.10, $\eta = 0.9$. The equivalent fin area is calculated as follows:

$$n_t = 13(4) = 52 \text{ tubes}$$
$$n_f = 38.1 \text{ cm}\left[\frac{10 \text{ fins}}{2.54 \text{ cm}}\right] = 150$$
$$A_f = 2\pi(r_f^2 - r^2)^{n_t n_f}$$
$$= 2\pi[(0.014334)^2 - (0.00635)^2](52)(150) = 8.094 \text{ m}^2$$

The area of the bare tube A_t is the total tube area minus the area covered by the fins. Let l_t = length of the tube.

$$A_t = (2\pi r l_t)(n_t) - (2\pi r)(t)(n_t)(n_f)$$
$$t = \text{fin thickness} = 0.000254 \text{ m}$$
$$A_t = 2\pi(0.00635)(0.381)(52) - 2\pi(0.00635)(0.000254)(52)(150)$$
$$= 0.790 - 0.0790 = 0.711 \text{ m}^2$$

$$\text{Effective area } A_{eff} = 0.91(8.094) + 0.711$$
$$= 8.0765 \text{ m}^2$$

EXAMPLE 2: If the conditions are such that a mean temperature difference ΔT of 45°F (25°C) exists in the condensing system and the refrigeration unit has a coefficient of performance of 2, how many tons of refrigeration can be supplied by the refrigeration system fitted with the condensing unit in Example 1? Assume that the outside coefficient of heat transfer controls the heat transfer rate, $U = h_0 = 51.1\ W/(\mathrm{m}^2 \cdot \mathrm{K})$.

Solution: The rate of heat transferred through the unit is:

$$q = UA_{eff}\Delta T$$
$$= 51.1(8.0765)(25) = 10{,}318\ \mathrm{W}$$

From equation 7, let R_r = refrigeration capacity.

$$COP = \frac{R_r}{H_{hv} - H_{cv}}$$

where

$$R_r = H_{cv} - H_L$$
$$H_L = H_{cv} - R_r$$

$$\text{Condenser load, } C = H_{hv} - H_L = H_{hv} - H_{cv} + R_r$$
$$(H_{hv} - H_{cv}) = C - R_r$$

The coefficient of performance can be expressed in terms of the condenser load C and the refrigeration capacity R_r.

$$COP = \frac{R_r}{C - R_r}$$

Solving for R_r:

$$R_r = \frac{C(COP)}{1 + COP}$$

Substituting values for COP and C:

$$R_r = \frac{10{,}318(2)}{3} = 6878\ \mathrm{W}$$

$$(\text{tons})_r = \frac{6878\ \mathrm{W}}{3515.88\ \mathrm{W/Tm}} = 1.956\ \text{tons}$$

A similar procedure may be used to evaluate evaporators. The examples illustrate the dependence of the capacity of a refrigeration system on the heat exchange capacity of the condensing or evaporator unit in the system.

THE COMPRESSOR

The compressor must be adequate to compress the required amount of refrigerant per unit time between the high-side and low-side pressures to provide the necessary refrigeration capacity. Compressor capacities are determined by the displacement and the volumetric efficiency of the unit. The displacement is the volume displaced by the piston or rotary unit per unit time. For a reciprocating compressor, the displacement is calculated as follows:

$$\text{Displacement} = (\text{area of bore})(\text{stroke})(\text{cycles}/\text{min})$$

Theoretically, the quantity of gas passing through the unit per unit time should be independent of the pressure and should be a function only of the displacement of the unit. However, at high pressures, rotary units allow a certain amount of slip, and reciprocating units leave a certain amount of gas at the space between the cylinder head and the piston at the end of each stroke. Thus, the actual amount of gas passed through the unit per cycle is less than the displacement. The volumetric efficiency of a unit is the ratio of the actual volume delivered to the displacement. The volumetric efficiency decreases with increasing ratios of high-side to low-side pressures.

Large units that operate at very low pressures at the evaporator, e.g., units used for freezing, often are of a multistage type, where the refrigerant is compressed to an intermediate pressure and then partially cooled before it enters a second stage, where it is compressed to the next higher pressure, etc., until the desired high-side pressure is reached. The staging is designed such that equal work is accomplished at each stage and the volumetric efficiency remains high because of the low ratio of high-side to low-side pressure at each stage.

Figure 10.11 shows a comparison between the power requirement per ton of refrigeration between one-stage and two-stage units operating at various suction-side pressures. For an ammonia compressor operating at a suction side pressure equivalent to a temperature of −30°F, the motor horsepower per ton of refrigeration is 3.2 for a one-stage unit and only 2.5 for a two-stage unit. In Example 3, part b, in the section "Example Problems on the Use of Refrigerant Charts," the theoretical horsepower per ton of refrigeration under these conditions is only 1.51. Thus, the efficiency of the compressors represented by the graph in Fig. 10.11 is only 47% for a one-stage unit and 60% for a two-stage unit.

The presence of air in the refrigerant line reduces the efficiency of the compressor if expressed in terms of horsepower per ton of refrigeration, since the

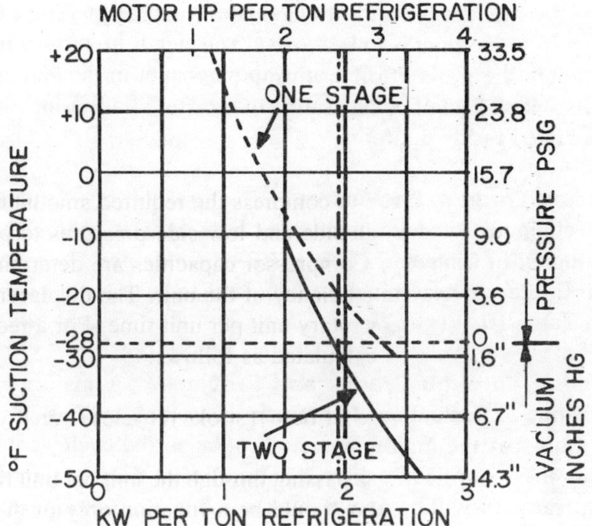

Fig. 10.11. Motor horsepower and kW/ton of refrigeration for one- and two-stage compressors. (From *Food Eng.* 41(11): 91, 1969. Used with permission.)

same amount of power is required to compress the gases to the required pressure. Yet, less refrigerant is passed through because part of the gas compressed is air.

REFRIGERATION LOAD

The amount of heat that must be removed by a refrigeration system per unit of time is the refrigeration load. The load may be subdivided into two categories: the *unsteady-state* load, which is the rate of heat removal necessary to reduce the temperature of the material being refrigerated to the storage temperature within a specific period of time, and the *steady-state* load, which is the amount of heat removal necessary to maintain the storage temperature. For food products, the temperature has to be reduced to the storage temperature in the shortest time possible to prevent microbiological spoilage and quality deterioration. Therefore, if large quantities need to be introduced into a refrigerated storage room intermittently, these materials are usually precooled to the storage temperature in a smaller precool or chill room or by other means prior to introduction into the large storage warehouse. This practice prevents the necessity of installing oversized refrigeration units in a large storage warehouse which would operate at full capacity only during the short periods when materials are being introduced into the warehouse.

The unsteady-state load includes the sensible heat of the product, the heat of respiration if the product is fresh produce, and the heat of fusion of the water in the product if the product is to be frozen. The steady-state load includes heat incursion through the walls of the enclosure, through cracks and crevices, and through doors; latent heat of condensation of moisture infiltrating into the room; and heat generated inside the room.

Heat Incursion Through Enclosures. Heat transfer rate through composite solids and techniques for determination of heat transfer coefficients were discussed thoroughly in Chapter 7. The *ASHRAE Guide and Data Book* for 1965 recommends the design heat transfer coefficients to be used for various surfaces, as shown in Table 10.1.

Heat Incursion Through Cracks and Crevices. A majority of the heat transferred is due to fluctuations in pressure caused by cycling of the temperature in the room. At the high-temperature point in the cycle, cold air will be expelled from the room. When the pressure drops at the low-temperature point in the cycle, warm air will be drawn into the room. The amount of air admitted into the room because of temperature cycling can be calculated using the principles discussed in Chapter 4.

Heat Incursion Through Open Doors. Opening doors allows entry of warm outside air and expulsion of cold air. The rate of heat incursion is dependent upon the size of the door and the temperature differential between the inside and outside. Data on rate of heat transfer through doors of refrigerated rooms have been determined empirically. The equation for rate of heat loss calculated

Table 10.1. Heat Transfer Coefficient to Air from Various Surfaces under Various Conditions

Surface	Heat Transfer Coefficients $\frac{\text{BTU}}{\text{h(ft}^2\text{)(°F)}}$	Heat Transfer Coefficients $\frac{\text{W}}{\text{m}^2\text{(K)}}$
Inside wall (still air)	1.5	8.5
Outside wall or roof 15 miles/hr or 24 km/h wind	5.9	33.5
Outside wall or roof 7.5 miles/hr or 12 km/h wind	4.0	22.7
Horizontal surface, still air, upward heat flow	1.7	9.7
Horizontal surface, still air, downward heat flow	1.1	6.25

Source: Adapted from ASHRAE, 1965. Design heat transmission coefficient. *ASHRAE Guide and Data Book. Fundamentals and Equipment for 1985 and 1986.* American Society of Heating, Refrigeration, and Air Conditioning Engineers, Atlanta, GA.

from data published in *Food Engineering* [41(11):91, 1969] for values of ΔT between 40° and 120°F (22.2° and 66.7°C) is:

$$q = 2126 W[e]^{0.0484(\Delta T)}(h)^{1.71} \tag{11}$$

where q = rate of heat incursion into the room in watts, W is the width of the door in m, ΔT is the temperature difference in °C inside and outside the room, and h is the height of the door in m.

EXAMPLE: The door to a refrigerated room is 3.048 m high and 1.83 m wide. It is opened and closed at least five times each hour and remains open for at least 1 min at each opening. Calculate the refrigeration load due to the door opening if the room is maintained at 0°C and ambient temperature is 29.4°C.

Solution: h = 3.048 m, W = 1.83 m, ΔT = 29.4°C. Substituting into equation 11:

$$q = 2126(1.83)(e)^{0.0484(29.4)}(3.048)^{1.71}$$
$$= 108.6 \text{ KW}$$

The total time the door was opened in 1 h is 300 s. Refrigeration load = 108.6 KW (300 s) = 35.58 MJ.

Heat Generation. Motors inside a refrigerated room generate heat at the rate of 1025.5 W/hp. This rate drops to 732.48 W/hp if only the motor is inside and the load it drives is outside the refrigerated room. Workers inside a room generate approximately 293 W of heat per person. The rate of heat generation by personnel increases with decreasing room temperature. The heat dissipated by light bulbs equals the wattage of the lamp.

Fruits and vegetables respire, and the heat of respiration adds to the refrigeration load. This heat of respiration is a function of temperature and can be calculated using the expression:

$$q = a(e)^{bT} \tag{12}$$

q = rate of heat generation (mW/kg in SI). Values for a and b for various fruits and vegetables are shown in Table 10.2.

During cooling, heat of respiration decreases as temperature decreases. An average heat of respiration may be calculated between two temperatures, T_1 and T_2, knowing the time Δt for the temperature to change between T_1 and T_2 and assuming that the temperature change during this period is linear.

Table 10.2. Heat of Respiration of Fruits and Vegetables in Air. Values of the Constants a and b in the Equation $q = a(e)^{bT}$

	T in °F, q in $\frac{\text{BTU}}{\text{ton (24 hr)}}$		T in °C, q in $\frac{\text{mW}}{\text{kg}}$	
Product	a	b	a	b
Apples	213	0.06	19.4	0.108
Asparagus	2779	0.048	173.0	0.086
Beans (green or snap)	829	0.064	86.1	0.115
Beans (lima)	376	0.071	48.9	0.128
Beets (topped)	1054	0.031	38.1	0.056
Broccoli	854	0.067	97.7	0.121
Brussels sprouts	1845	0.045	104.0	0.081
Cabbage	337	0.041	16.8	0.074
Cantaloupes	128	0.07	16.1	0.126
Carrots (topped)	498	0.046	29.1	0.083
Celery	237	0.058	20.3	0.104
Corn (sweet)	2465	0.043	131.0	0.077
Grapefruit	171	0.051	11.7	0.092
Lettuce (head)	416	0.049	26.7	0.088
Lettuce (leaf)	1188	0.041	59.1	0.074
Onions	89	0.055	6.92	0.099
Oranges	151	0.059	13.4	0.106
Peaches	104	0.074	14.8	0.133
Pears	42	0.096	12.1	0.173
Peas (green)	1249	0.059	111.0	0.106
Peppers (sweet)	796	0.040	33.4	0.072
Spinach	473	0.073	65.6	0.131
Strawberries	568	0.059	50.1	0.106
Sweet potatoes	796	0.034	31.7	0.061
Tomatoes	159	0.057	13.2	0.103
Turnips	590	0.037	25.8	0.067

Source: ASHRAE, 1974. Approximate rates of evolution of heat by certain fruits and vegetables when stored at temperatures indicated. In: *ASHRAE Handbook and Product Directory—1974 Applications*. American Society of Heating, Refrigeration, and Air Conditioning Engineers, Atlanta, GA.

[1] q calculated using the constants a and b are maximum values in the range reported. Minimum values average 67% of the maximum.

$$q = \frac{1}{\Delta t}\int_0^{\Delta t} a(e)^{bT}\, dt \tag{13}$$

Assuming a linear temperature change:

$$T - T_1 = \frac{T_1 - T_2}{\Delta t}\, t \tag{14}$$

Substituting equation 14 for T in equation 13:

$$q = \frac{a}{\Delta t} \int_0^{\Delta t} [e]^{bT_1} - \frac{T_1 - T_2}{\Delta t} bt\, dt$$

Integrating and substituting limits:

$$q = \frac{ae^{bT_1}}{b(T_1 - T_2)} [1 - (e)^{-b(T_1 - T_2)}] \qquad (15)$$

Equation 15 is the average rate of heat generation over the time Δt. The units are the same as the units of a, which in SI is mW/kg, as tabulated in Table 10.2. Multiplying q in equation 15 by Δt gives the total heat generated for the time period under consideration.

Although the temperature at any given point in a material during the process of cooling usually changes exponentially with time, materials that are cooled in bulk (e.g., head lettuce or boxed products) have a temperature gradient with the interior parts at a higher temperature than the external sections exposed to the cooling medium. Thus, if the bulk mean temperature is considered, the deviation from the linear temperature change assumption is minimal.

EXAMPLE 1: It is desired to cool cabbage from 32.2°C to 4.44°C in 4 h. Calculate the heat of respiration during this cooling period. Assume a linear temperature change.

Solution: T_1 = 32.2°C; T_2 = 4.44°C; Δt = 4 h = 14,400 s. From Table 10.2 for cabbage, a = 16.8 and b = 0.074 in SI units. Examination of equation 12 reveals that a will have the same units as q and b will have units of 1/°C.

$$q = \frac{16.8e^{0.074(32.2)}}{(0.074)(32.2 - 4.44)} [1 - (e)^{-0.074(32.2 - 4.44)}]$$

$$= 77.253 \text{ mW/kg}$$

Heat generated over the time period of 14,400 s is Q:

$$Q = (77.253 \times 10^{-3} \text{ W/kg})(14{,}400 \text{ s}) = 1112 \text{ J/kg}$$

EXAMPLE 2: Calculate the refrigeration load due to the heat of respiration of spinach at a constant temperature of 3.33°C:

Solution: T = 3.33°C. From Table 10.2, a = 65.6 mW/kg and b = 0.131°C^{-1}.

$$q = 65.6e^{0.131(3.35)}$$
$$= 101.74 \text{ mW/kg}$$

The Unsteady-State Refrigeration Load. Procedures for calculating the heat capacity and sensible heat gain or loss are discussed in Chapter 5. When a change in phase is involved, the latent heat of fusion of the water must be considered. The heat of fusion of ice is 144 BTU/lb, 80 cal/g, or 0.335 MJ/kg.

EXAMPLE: Calculate the refrigeration load when 100 kg/h of peas needs to be frozen from 30°C to −40°C. The peas have a moisture content of 74%.

Solution: Let the refrigeration load = Q_r; m = mass; C_1 = specific heat above freezing; C_s = specific heat below freezing; T_f = freezing point; H_f = enthalpy at the freezing point based on $H = 0$ at the reference temperature of 227.6 K, where all the water is in the form of ice; T_1 = initial temperature and T_2 = final temperature.

$$Q_r = mC_1(T_1 - T_f) + m(H_f - H_{T2})$$

The mass fraction of water is 0.74; thus, $M = 0.74$.
Using equation 7, Chapter 5:

$$C_1 = 3349(0.74) + 837.36 = 3315.6 \text{ J/(kg} \cdot \text{K)}$$

Using equation 20, Chapter 5:

$$T_f = 287.56 - 49.19(0.74) + 37.07(0.74)^2 = 271.5 \text{ K}$$

Using equation 22, Chapter 5:

$$H_f = 9792.46 + 405{,}096(0.74) = 309{,}563 \text{ J/kg}$$

Using equation 17, Chapter 5:

$$a = 0.362 + 0.0498(0.74 - 0.73) - 3.465(0.74 - 0.73)^2$$
$$= 0.3621$$

Using equation 18, Chapter 5:

$$b = 27.2 - 129.04(0.3621 - 0.23) - 481.46(0.3621 - 0.23)^2$$
$$= 1.752$$

$$T_r = \frac{(-40 + 273 - 227.6)}{(271.5 - 227.6)} = 0.123$$

Using equation 23, Chapter 5:

$$H_{T2} = H_f[(0.3621)(0.123) + (1 - 0.3621)(0.123)^{1.752}]$$
$$= H_f(0.06077)$$

$$T_f = 271.5\text{ K} - 273 = -1.5°\text{C}$$

$$Q_r = 3315.6(30 + 1.5) + 309{,}563(1 - 0.06077) = 395{,}193\text{ J/kg}$$

Total Q_r for 100 kg/h = 395,193 J/kg (100 kg/h) = 3.95193×10^7 J/h or 10.9776 kW.

Another approach, which is more general, is to use the freezing point depression principle to determine the fraction of unfrozen water remaining at any temperature below the freezing point. Let y_{12} = fraction of liquid water at temperature T_2; M = mass of fraction of total ice and liquid water in the material. The enthalpy at T_2, based on zero enthalpy at T = 227.6 K when all water is in the solid phase, is the sum of the heat of fusion of the fraction of water still liquid at T_2, the sensible heat of ice from the reference temperature to T_2, and the sensible heat of the solids fraction from the reference temperature to T_2.

$$H_{T2} = y_{12}(\Delta H_1) + Cs(M)(T_2 - 227.6)$$
$$+ (1 - M)(836.36)(T_2 - 227.6)$$

$$Cs = \text{specific heat of ice}$$
$$= 2093.4\text{ J/(kg} \cdot \text{K)}$$

The mass fraction liquid water in the material at T_2 is evaluated from the molality of a solution which gives a freezing point depression equal to the difference between T_2 and 273 K, the freezing point of pure water.

The freezing point depression of a solution is given by: $\Delta T_f = k_f M_s$, where k_f = cryoscopic constant = 1.9 for water and carbohydrate solutes and M_s = molality = moles of solute per 1000 g water.

At T_2, the freezing point depression is $(0 - T_2)$ and the equivalent molality which gives that freezing point depression is $M_{s2} = (0 - T_2)/1.9$.

The freezing point can be used to determine the original amount of solute. $M_s = (0 - T_f)/1.9$. Let n = number of moles of solute in 1 kg material.

$$n = \frac{M_s(M)}{1000}$$

$$M_{s2} = \frac{1000n}{y_{12}}$$

$$y_{12} = \frac{1000n}{M_{s2}}$$

In the case of peas, the freezing point, $T_f = -1.5°C$. The molality needed to reduce the freezing point 1.5°C is:

$$M_s = \frac{1.5}{1.9} = 0.789$$

$$n = \frac{0.789(0.74)}{1000} = 5.838 \times 10^{-4}$$

At −40°C, the molality of a solution which would depress the freezing point to a point where water is liquid at this temperature is:

$$M_{s2} = \frac{40}{1.9} = 21$$

$$y_{12} = \frac{1000(5.838 \times 10^{-4})}{21} = 0.0278$$

The heat of fusion of water is 334,944 J/kg.

$$H_{T2} = 0.0278(334{,}944) + 2093.4(0.74)(233 - 227.6)$$
$$+ 0.26(836.36)(233 - 227.6) = 18{,}851 \text{ J/kg}$$

The enthalpy at the freezing point is based on all water being in the liquid state.

$$H_f = M(334{,}944) + M(2093.4)(T_f - 227.6)$$
$$+ (1 - M)(836.36)T_f - 227.6$$
$$= 0.74(334{,}944) + 0.74(2093.6)(271.5 - 227.6)$$
$$+ 0.26(836.36)(271.5 - 227.6)$$
$$= 325{,}417 \text{ J/kg}$$

The refrigeration load is the sensible heat loss from the initial temperature plus the enthalpy difference from the freezing point to the final temperature.

$$Q_r = 100 \text{ kg/h}\left[(3315.6)(30 - (-1.5)) + (325{,}417 - 18{,}851)\right]$$

$$= 4.11 \times 10^7 \text{ J/h} \qquad \text{or } 11.42 \text{ kW}$$

The discrepancy between these calculations based on freezing point depression and Chang and Tao's (1981) correlations, discussed in Chapter 5, is only 3.8%.

COMMODITY STORAGE REQUIREMENTS

Most food products benefit from a reduction in the storage temperature provided that no freezing occurs. Reduced temperature reduces the rate of chemical reactions that cause the product to deteriorate and reduce microbiological activity. Freezing damages the cellular structure of fruits and vegetables and severely reduces their acceptability. Meat pigments darken irreversibly upon freezing. Acceleration of the development of a strong fishy flavor has been observed when fish are frozen at or near the freezing point and thawed. Thus, for perishable foods not specifically prepared for frozen storage, the freezing temperature should be the lowest acceptable one for storage. Some fruits and vegetables are susceptible to chill injury at temperatures above the freezing point. The lowest safe temperatures have been defined for these commodities. Table 10.3 shows the temperature and humidity recommended for storage of various fruits and vegetables.

CONTROLLED ATMOSPHERE STORAGE

Respiration. Fruits and vegetables continue their metabolic activity after harvest. Maintenance of this metabolic activity is essential in the preservation of quality. Metabolic activity is manifested by respiration, a process whereby the material consumes oxygen and emits carbon dioxide. A major nutrient metabolized during respiration is carbohydrate. Since respiration depletes nutrients in a product after harvesting, the key to prolonging the shelf life of fruits and vegetables is reducing the rate of respiration. Early studies on the relationship between respiration and quality established that storage life is inversely related to the rate of respiration and that for two products respiring at different rates, when the total quantity of CO_2 emitted by both products is equal, the products would have reached a comparable stage in their storage life.

Table 10.3. Recommended Storage Conditions for Fruits and Vegetables

(1) Storage temperature 30°-32°F (−1.11°-0°C). Highest freezing point 28°F (−2.22°C). 90-95% RH.

apricots	pears
cherries	peaches
grapes	plums

(2) Storage temperature 32°F (0°C)

65% RH	90% RH	95%	95% RH	95% RH
garlic	mushrooms	artichokes	sweet corn	carrots
onions	oranges	asparagus	endives[1]	lettuce[1]
	tangerines	lima beans	escarole[1]	parsnips
		beets	leafy greens	rutabagas
		broccoli	parsley	turnips
		brussels sprouts	green peas	
		cabbage	radishes	
		cauliflower	rhubarb	
		celery[2]	spinach[1]	

(3) Storage temperature 36°F (2.22°C). 95% RH.

apples

(4) Storage temperature 45°F (7.22°C). 90% RH. Subject to chill injury at temperature below 45°F.

green or snap beans
ripe tomatoes

(5) Storage temperature 50°F (10°C). Subject to chill injury at temperatures below 50°F.

85% RH	90-95% RH
melons	cucumbers
potatoes	eggplants
pumpkin	sweet peppers
squash	okra
green tomatoes	

(6) Storage temperature 58°-60°F (14.4°-15.6°C). 85-90% RH.

bananas
grapefruit
lemons

Source: ASHRAE. 1974. *ASHRAE Handbook and Product Directory—1974. Applications*. American Society of Heating, Refrigeration, and Air Conditioning Engineers, New York.
[1]Highest freezing point 31.9°F (−0.06°C).
[2]Highest freezing point 31.1°F (−0.5°C).
All others have freezing points below 31°F (−0.56°C).

Reduction of temperature is an effective means of reducing the rate of respiration. However, for some products subject to chill injury, the respiration rate may still be quite high even at the lowest safe storage temperature. Controlled atmosphere storage has been developed as a supplement to refrigeration in prolonging the storage life of actively respiring fruits and vegetables.

The reaction involved in respiration is primarily the oxidation of carbohydrates.

$$C_6H_{12}O_6 + 6O_2 = 6CO_2 + 6\ H_2O$$

The respiratory quotient (RQ), defined as the ratio of CO_2 produced to O_2 consumed, is 1 for the reaction. Most products show respiratory quotients close to 1. The heat evolved during respiration has also been shown to be related to the quantity of CO_2 produced in the same manner as that released during the combustion of glucose (10.7 J/mg CO_2). The rate of oxygen consumption and CO_2 evolution can be calculated from the heat of respiration and vice versa.

Controlled atmosphere (CA) storage is based on the premise that increasing the CO_2 level and decreasing the O_2 level in the storage atmosphere will result in a reduction of the rate of respiration. Indeed, calorimetric determination of the heat of respiration of products in continuously flowing CA of the optimum composition for the given product has shown that CA can reduce the respiration rate to approximately one-third of the respiration rate in air at the same temperature.

EXAMPLE: One pound (0.454 kg) of head lettuce is packaged in an airtight container with a volume of 4 L. The product occupies 80% of the volume, the rest being air. If the product is at a constant temperature of 4°C, calculate how long it will take for the oxygen content in the package to drop to 2.5%. Assume an RQ of 1.

Solution: From Table 10.2, the constants *a* and *b* for the heat of respiration of head lettuce (in mW/kg) are 26.7 and 0.888, respectively.

$$q = 26.7(e)^{0.088(4)} = 38\ \text{mW/kg}$$

$$\frac{\text{mg CO}_2}{\text{h}} = (38 \times 10^{-3})\left|\frac{\text{J}}{\text{s (kg)}}\right|\frac{1\ \text{mg CO}_2}{10.7\ \text{J}}\left|3600\ \frac{\text{s}}{\text{h}}\right.$$

$$(0.454\ \text{kg}) = 5.8$$

$$\frac{\text{g moles CO}_2}{\text{h}} = \frac{\text{g moles O}_2\ \text{depleted}}{\text{h}} = \frac{5.8 \times 10^{-3}}{44}$$

$$= 1.318 \times 10^{-4}$$

Air is approximately 21% O_2 and 79% N_2. The total number of moles of air originally in the container is:

$$n = \frac{PV}{RT} = \frac{(1\ \text{atm})(4)(0.2)1}{0.08206(1\ \text{atm/g mole K})(273 + 4)\text{K}}$$

$$= 0.0352\ \text{mole}$$

Since the RQ is 1, there will be no net change in the total number of moles of gases inside the container. The number of moles of oxygen when the concentration is 2.5% is:

$$n_{O_2} = 0.025(0.0352) = 0.00088$$

The original number of moles of O_2 is:

$$n_{O_2} = 0.21(0.0352) = 0.007392\ \text{mole}$$

The number of moles of O_2 that must be depleted by respiration is:

$$n_{O_2}\ \text{depletion} = 0.007392 - 0.00088 = 0.006512$$

The time required to deplete O_2 to the desired level is:

$$\text{Time} = \frac{0.006512\ \text{mole}}{0.0001318\ \text{mole/h}} = 49.4\ \text{h}$$

CA Gas Composition. CA storage has been used successfully with apples and pears. Experimental results with CA storage of cabbage, head lettuce, broccoli, and brussels sprouts have been very encouraging. Atmospheric modification of product containers during transit of vegetables is also gaining wider acceptance. CA is recommended when the transit time is 5 days or more.

Reduction of the O_2 concentration in the storage atmosphere to 3% or below has been shown to be most effective in reducing the respiration rate, with or without CO_2. Too low an O_2 concentration, however, leads to anaerobic respiration, resulting in the development of off-flavors in the product. The O_2 concentration for the onset of anaerobic respiration ranges from 0.8% for spinach to 2.3% for asparagus. The optimum CA composition varies from one product to another and among the varieties of a given product. CA compositions found to be effective for some products are shown in Table 10.4.

To be effective, CA storage rooms must be reasonably airtight. To test for airtightness, the room should be pressurized to a positive pressure of 1 in. water gage (wg) (249 Pa gage). A room is considered airtight if, at the end of 1 h, the pressure does not drop below 0.2 in. wg (49.8 Pa). This requirement is equivalent to an air incursion rate of 0.2% of the gas volume in the room per hour at a constant pressure differential between the inside and outside of 0.5 in.

Table 10.4. CA Storage Conditions Suitable for Use with Various Products

Product	% CO_2	% O_2
Apples	2–5	3
Asparagus	5–10	2.9
Brussels sprouts	2.5–5	2.5–5
Beans (green or snap)	5	2
Broccoli	10	2.5
Cabbage	2.5–5	2.5–5
Lettuce (head or leaf)	5–10	2
Pears	5	1
Spinach	11	1
Tomatoes (green)	0	3

Percentages are volume or mole percent. Balance is nitrogen.
Reprinted from Toledo, R. T. 1980. *Fundamentals of Food Process Engineering*. AVI Pub. Co., Westport, Conn.

wg (124 Pa). Current standards for refrigerated tractor-trailers call for a maximum air incursion rate of 2 ft^3/min (3.397 m^3/h) at a pressure differential of 0.5 in. water (124 Pa) between the inside and outside of the trailer. This rate is approximately 15 times the specification for airtightness.

CA in storage warehouses is maintained at the proper composition by introduction of CO_2 and/or N_2 if product respiration is insufficient to raise CO_2 levels or lower O_2 levels. In a full, tightly sealed warehouse, ventilation with fresh air and scrubbing of CO_2 from the storage atmosphere may be necessary to maintain the correct CA composition and prevent anaerobic respiration. Successful operation of a CA storage facility necessitates regular monitoring and analysis of gas composition and making appropriate corrections when necessary. Figures 10.12 and 10.13 are schematic diagrams of CA storage systems for warehouses and for in-transit systems.

EXAMPLE 1: The rate of flow of fluids through narrow openings is directly proportional to the square root of the pressure differential across the opening. If the pressure inside a chamber of volume V changes exponentially with time from 1 in. wg (249 Pa) at time 0 to 0.2 in. wg (49.8 Pa) at 1 h, calculate the mean volumetric rate of flow of gases out of the chamber and the mean pressure at which this rate of flow would be expected if the pressure is held constant.

Solution: The expression for the volumetric rate of flow Q is:

$$Q = k(\Delta P)^{1/2}$$

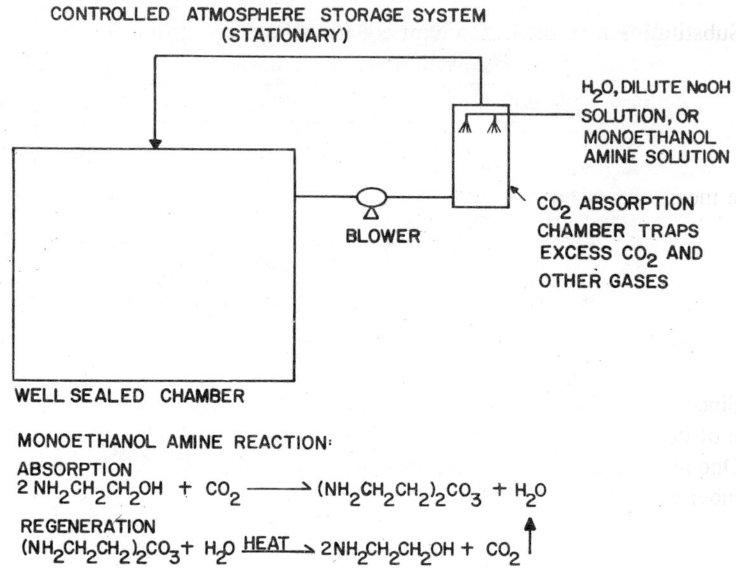

Fig. 10.12. Schematic diagram of a controlled atmosphere storage system for warehouses.

The volume dQ expelled over a period of time dt is:

$$dQ = k(\Delta P)^{1/2}\,dt$$

Since P is exponential with time, substituting the boundary conditions, $\Delta P = 1$ at $t = 0$, and $\Delta P = 0.2$ at $t = 1$, will give the equation for ΔP as a function of t as follows:

$$\Delta P = (e)^{-1.6094t}$$

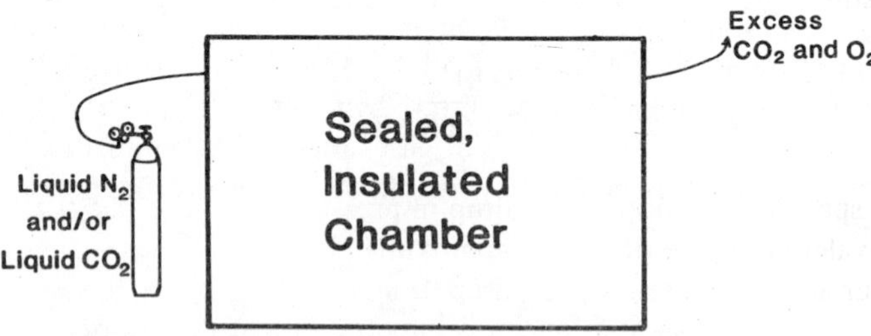

Fig. 10.13. Schematic diagram of a transportable system for generating and maintaining controlled atmospheres.

Substituting P in the differential equation and integrating:

$$Q = \int_0^1 \left[k(e)^{0.5(-1.6094)t}\right] dt$$

The mean rate of flow in 1 h, $\overline{q} = Q/t$, is:

$$\overline{q} = \frac{1}{1}\left[\left(\frac{k}{-0.8047}\right)(e)^{-0.8047t}\right]_0^1$$

$$= 0.687\ k$$

Since $\overline{q} = k(\overline{\Delta P})^{1/2}$, the mean pressure $\overline{\Delta P}$ that would give the equivalent rate of flow as $\overline{q}$ is $(0.687)^2$ or 0.472.

One atmosphere is 406.668 in. of water. Using the ideal gas equation, the number of moles gas in the chamber at pressures P_1 and P_2 is:

$$n_1 = \frac{P_1 V}{RT} \quad \text{and} \quad n_2 = \frac{P_2 V}{RT}$$

The number of moles lost with a drop in pressure from P_1 to P_2 is:

$$n_1 = \frac{V}{RT}(P_1 - P_2)$$

Expressing the quantity expelled as volume at standard atmospheric pressure:

$$V_1 = n_1 \frac{RT}{P} = \frac{(P_1 - P_2)V}{P}$$

Substituting $P_1 = 1$ in., $P_2 = 0.2$ in., and $P = 406.668$:

$$V_1 = \frac{0.8V}{408.668} = 0.00196V$$

The specified conditions of a drop in pressure from 1 to 0.2 in. wg in 1 h is equivalent to a rate of loss of approximately 0.2% of the chamber volume per hour at an average pressure drop of approximately 0.5 in. wg.

EXAMPLE 2: A refrigerated tractor-trailer is loaded with head lettuce such that 20% of its total volume is air space. The trailer has a total volume of 80

m^3 Assuming that the trailer meets the specifications for the maximum air incursion rate of 2 ft^3/min (3.397 m^3/h) at a pressure differential of 0.5 in. wg (124 Pa), what is the rate of addition of N_2 and CO_2 necessary to maintain the atmosphere inside the truck at 2% O_2 and 5% CO_2 when the truck is traveling at the rate of 50 miles/h (80 km/h)? The average temperature of the product is 1.5°C, and the trailer contains 18,000 kg of lettuce.

Solution: If the atmosphere inside the truck has been preequilibrated at the point of origin to the desired CA composition, the amount of N_2 and CO_2 that must be added during transit is that necessary to displace the air infiltrating into the trailer. Let x = rate of N_2 addition and y = rate of CO_2 addition. If n_{ai} = rate of air infiltration, n_{rc} = the rate of CO_2 generation by the product, and n_{ro} = the rate of oxygen consumption by the product during respiration, the following material balance may be set up for CO_2 and O_2, assuming that the expelled gases are of the same composition as those in the interior of the trailer.

O_2 balance (air is 21% O_2):

$$n_{ai}(0.21) = (n_{ai} + x + y)(0.02) + n_{ro}$$

CO_2 balance:

$$y + n_{rc} = (n_{ai} + x + y)(0.05)$$

The heat of respiration of head lettuce in mW/kg from values of a = 26.7 and b = 0.088 from Table 9.2 and for T = 1.5°C:

$$q = 26.7(e)^{1.5(0.088)} = 30.47 \text{ mW/kg}$$

$$\frac{\text{mg } CO_2}{\text{kg} \cdot \text{h}} = \frac{30.47 \times 10^{-3} \text{ J}}{\text{kg} \cdot \text{s}} \left| \frac{\text{mg } CO_2}{10.7 \text{ J}} \right| \frac{3600 \text{ s}}{\text{h}} = 10.25$$

$$\frac{\text{g moles } CO_2}{\text{kg} \cdot \text{h}} = \frac{\text{g moles } O_2}{\text{kg} \cdot \text{h}} = \frac{10.2 \times 10^{-3}}{44} = 0.233 \times 10^{-3}$$

$$n_{rc} = (18{,}000 \text{ kg}) \frac{0.233 \times 10^{-3} \text{ g moles}}{\text{kg} \cdot \text{h}} = 4.194 \frac{\text{g moles}}{\text{h}}$$

$$n_{ro} = 4.194 \text{ g moles/h}$$

The rate of air infiltration must now be calculated at a velocity of 80 km/h. The velocity head of a flowing fluid, $\Delta P/\rho$, is $V^2/2$. ρ is the fluid density.

$$\Delta P = \frac{(80{,}000)^2 \text{ m}^2}{\text{h}^2}(0.5)\frac{\rho \text{ kg}}{\text{m}^3}\frac{1 \text{ h}^2}{(3600)^2 \text{ s}^2} = 247 \rho \text{ Pa}$$

The specified infiltration rate of 3.397 m^3/h at a pressure of 124 Pa can be converted to the rate at the higher pressure by using the relationship that the rate of flow is directly proportional to the square root of the pressure.

$$Q = 3.397 \frac{247\rho^{1/2}}{124\rho} = 4.79 \text{ m}^3/\text{h}$$

Using the ideal gas equation, $n_{ai} = PV/RT$:

$$n_{ai} = \frac{(4.79 \text{ m}^3/\text{h})(1000 \text{ 1}/\text{m}^3)(1 \text{ atm})}{0.08206\ (1)(\text{atm})/\text{K}(274.5 \text{ K})} = 212.7 \text{ g moles/h}$$

Substituting values for n_{ai}, n_{ro}, and n_{rc} in the O_2 and CO_2 balance equations:

$$212(0.21) = (212 + x + y)(0.02) + 4.176$$

$$x + y = \frac{44.52 - 4.176 - 4.24}{0.02} = 1805$$

$$x = 1805 - y$$

$$y + 4.176 = (212 + x + y)(0.05)$$

$$0.95y = 0.05x + 6.424$$

$$0.95y - 0.05(1805 - y) = 6.424$$

$$1.0y = 6.424 + 90.25$$

$$y = 96.67 \text{ g moles } CO_2/\text{h}$$

$$x = 1708.3 \text{ g moles } N_2/\text{h}$$

Note that because of air infiltration, it would not be possible for product respiration to maintain the necessary CA composition inside the trailer. Furthermore, the respiration rate for this particular product is so low that flushing of the trailer with N_2 and CO_2 would have to be done initially to bring the CA to the desired composition within a short period of time.

Modified Atmosphere Packaging. Modified atmosphere (MA) and CA are similar, and the terms are sometimes used interchangeably. A feature of CA is some type of control which maintains constant conditions within the storage atmosphere. MA packaging involves changing the gaseous atmosphere surrounding the product at the time of packaging and using barrier packaging materials to prevent gaseous exchange with the environment surrounding the package. The use of gas-permeable packaging materials tailored to promote gas exchange to maintain relatively constant conditions inside the package may be considered a form of CA packaging.

Reduction of respiration rates of live organisms by lowering O_2 and elevating CO_2 was discussed in the previous section. MA packaging, as now practiced, is primarily designed to control the growth of microorganisms and, in some cases, enzyme activity in packaged uncooked or cooked food. The atmosphere used, if the product itself is nonrespiring, is usually 75% CO_2, 15% N_2, and 10% O_2. Bakery products, which are generally of lower moisture content than most foods, can be preserved longer in 100% CO_2. Dissolution of CO_2 and O_2 in the aqueous phase of the product is responsible for microbial inhibition. High CO_2 is inhibitory to most microorganisms. Although CO_2 lowers the pH, the inhibitory effect results from more than the pH reduction alone. The presence of O_2 has a slightly inhibitory effect on anaerobes, resulting in slower growth compared to a 100% CO_2 atmosphere.

Other gas combinations and the products on which they have been used are: 80% CO_2 and 20% N_2 for luncheon meats; 100% N_2 for cheese; 30% CO_2 and 70% N_2 for fresh poultry and fish; and 30% O_2, 30% CO_2, and 40% N_2 for red meats.

Typically, MA packaging involves drawing a vacuum from the product in a high-barrier packaging material, displacing the vacuum with the modified atmosphere, and sealing the package.

Another approach which is suitable for engineering modeling involves the use of gas-permeable films as the primary packaging material and enclosing the packages in a larger high-barrier bag filled with the desired gaseous mixture. The gas mixture composition and the package film permeability are balanced to achieve the desired constant MA inside the package. This is a rapidly growing area in packaging research. So far there are no satisfactory models which can adequately predict the shelf life of MA packaged products.

PROBLEMS

1. An ammonia refrigeration unit is used to cool milk from 30° to 1°C (86° to 33.8°F) by direct expansion of refrigerant in the jacket of a shell-and-tube heat exchanger. The heat exchanger has a total outside heat transfer surface area of 14.58 m^2 (157 ft^2). To prevent freezing, the temperature of the refrigerant in the heat exchanger jacket is maintained at −1°C (31.44°F).
   (a) If the average overall heat transfer coefficient in the heat exchanger is 1136 $W/m^2 \cdot K$ (200 $BTU/h \cdot ft^2 \cdot °F$) based on the outside area, calculate the rate at which milk with a specific heat of 3893 $J/kg \cdot K$ (0.93 $BTU/lb \cdot °F$) can be processed in this unit.
   (b) Determine the number of tons of refrigeration required for the refrigeration system.
   (c) The high-pressure side of the refrigeration system is at 1.72 MPa (250 psia). Calculate the horsepower of the compressor required for the refrigeration system, assuming a volumetric efficiency of 60%.

2. A single-stage compressor in a Freon 12 refrigeration system has a volumetric efficiency of 90% at a high-side pressure of 150 psia (1.03 MPa) and a low-side pressure of 50 psia (0.34 MPa). Calculate the volumetric efficiency of this unit if it is operated at the same high-side pressure but the low-side pressure is reduced to 10 psia (68.9 kPa). Assume that Freon 12 is an ideal gas.

3. Calculate the number of tons of refrigeration for a unit that will be installed in a cooler maintained at 0°C (32°F) given the following information on its construction and operation:

The cooler is inside a building.

Dimensions: 4 × 4 × 3.5 m (13.1 × 13.3 × 11.5 ft)
Wall and ceiling construction:
3.175-mm (1/8-in.)-thick polyvinyl chloride sheet inside (k of PVC = 0.173 W/m · K or 0.1 BTU/h · ft · °F)
15.24-cm (6-in.) fiberglass insulation
5.08-cm (2-in.) corkboard
3.17-mm (1/2-in.) PVC outside
Floor construction:
3.175-mm (1/8-in.)-thick floor tile (k = 0.36 W/(m · k) or 0.208 BTU/(h · ft · °F)
10.16-cm (4-in.) concrete slab
20.32-cm (8-in.) air space
Concrete surface facing the ground at a constant temperature of 15°C (59°F)
Door:
1 m wide × 2.43 m high (3.28 × 8 ft)
Design for door openings that average four per hour at 1 min per opening
Air infiltration rate:
1 m³/h (35.3 ft³/h) at atmospheric pressure and ambient temperature
Ambient conditions:
32°C (89.6°F)
Product cooling load:
Design for a capability to cool 900 kg of product (C_p = 0.76 BTU/lb · °F or 3181 J/kg · K) from 32° to 0°C (89.6° to 32°F) in 5 h. The freezing point of the product is −1.5°C (29.3°F)

4. The *stack effect* due to a difference in temperature between the inside and outside of a cooling room is often cited as the major reason for air infiltration. In this context, ΔP is positive at the lowest section of a cooler and negative at the highest section, with a zone, called the *neutral zone*, at approximately the center of the room, where ΔP is zero. If the area of the openings at the lowest sections where ΔP is positive equals the area of the openings in the highest sections where ΔP is negative, air will enter at the top and escape at the openings in the bottom at the same volumetric rate of flow (assuming no pressure change inside the room). If the room allows air leakage at the rate of 2% of the room volume per minute at a ΔP of 0.5 in. wg (124 Pa), determine the rate of air infiltration that can be expected in a room that is 2 m (6.56 ft) high to the neutral zone if the interior of the room is at −20°C (−4°F) and ambient temperature is 30°C (86°F). The rate of gas flow through the cracks is

proportional to the square root of ΔP. Assume that air is an ideal gas. ΔP due to a column of air of height h at different temperatures $= g(\rho_1 - \rho_2)h$, where ρ_1 and ρ_2 are the densities of the columns of air.

5. For a 1-ton refrigeration unit (80% volumetric efficiency) using Freon 12 at a high-side pressure of 150 psia (1.03 MPa) and a low-side pressure of 45 psia (0.31 MPa), operating at an ambient temperature of 30°C (86°F), determine the effect of the following on refrigeration capacity and on HP/(ton)$_r$. Assume the same compressor displacement in each case.
   (a) Reducing the evaporator temperature to −30°C (−22°F). High-side pressure remains at 150 psia (150 MPa).
   (b) Increasing ambient temperature to 35°C (95°F). Low-side pressure remains at 45 psia (0.31 MPa). The high-side pressure is to change such that ΔT between the hot refrigerant gas and ambient air remains the same as in the original set of conditions.
   (c) Air in the line such that the vapor phase of the refrigerant always contains 10% air and 90% refrigerant by volume. Assume that the condensation temperature of hot refrigerant gas and the temperature of cold refrigerant gas are the same as in the original set of conditions (partial pressures of refrigerant gas at the low- and high-side pressures are the same as in the original set of conditions, 45 and 150 psia or 0.31 and 1.03 MPa, respectively). Use $R = 1.987$ BTU/(lbmole · °R) or 8318 J/(kg · K). The specific heat ratio C_p/C_v for air is 1.4.
   (d) Oil trapped in the vapor return line such that ΔP across the constriction is 10 psi (68.9 kPa). Assume that evaporator temperature = 0°C.

SUGGESTED READING

ASHRAE. 1965. *ASHRAE Guide and Data Book. Fundamentals and Equipment for 1965 and 1966.* American Society of Heating, Refrigeration and Air Conditioning Engineers, Atlanta, GA.

ASHRAE. 1966. *ASHRAE Guide and Data Book. Applications for 1966 and 1967.* American Society of Heating, Refrigeration, and Air Conditioning Engineers, Atlanta, GA.

Charm, S. E. 1971. *Fundamentals of Food Engineering*, 2nd ed. AVI Publishing Co., Westport, Conn.

Ciobanu, A., Lascu, G., Bercescu, V., and Niculescu, L. 1976. *Cooling Technology in the Food Industry.* Abacus Press, Tunbridge Wells, England.

Hougen, O. A., and Watson, K. M. 1946. *Chemical Process Principles.* Part II. *Thermodynamics.* John Wiley & Sons, New York.

Perry, R. H., Chilton, C. H., and Kirkpatrick, S. D. 1963. *Chemical Engineers Handbook*, 4th ed. McGraw-Hill Book Co., New York.

Watson, E. L., and Harper, J. C. 1989. *Elements of Food Engineering*, 2nd ed. Van Nostrand Reinhold, New York.

11

Evaporation

The process of evaporation is employed in the food industry primarily as a means of bulk and weight reduction for fluids. It is used extensively in the dairy industry to concentrate milk; in the fruit juice industry to produce fruit juice concentrates; in the manufacture of jams, jellies, and preserves to raise the solids content necessary for gelling; and in the sugar industry to concentrate sugar solutions for crystallization. Evaporation can also be used to raise the solids of dilute solutions prior to spray or freeze drying.

Evaporation is used to remove water from solutions with or without insoluble suspended solids. If the liquid contains only suspended solids, dewatering can be achieved by either centrifugation or filtration. The process of evaporation involves the application of heat to vaporize water at the boiling point. The simplest form is atmospheric evaporation, in which liquid in an open container is heated and the vapors driven off are simply dispersed into the atmosphere. Atmospheric evaporation is simple, but it is slow and inefficient in the utilization of energy. Furthermore, since most food products are heat sensitive, prolonged exposure to high temperature during atmospheric evaporation causes off-flavors or general quality degradation. Evaporators used on food products remove water at low temperatures by heating the product in a vacuum. Efficient energy utilization can be designed into a system by using heat exchangers to extract heat from the vapors to preheat the feed or by using multiple effects where the vapors produced from one effect are used to provide heat in the succeeding effects.

Problems in evaporation involve primarily heat transfer and material and energy balances, the principles of which have been discussed.

SINGLE-EFFECT EVAPORATORS

Figure 11.1 is a schematic diagram of a single-effect evaporator. The system consists of a vapor chamber where water vapor separates from the liquid, a heat exchanger to supply heat for vaporization, a condenser to draw out the vapors

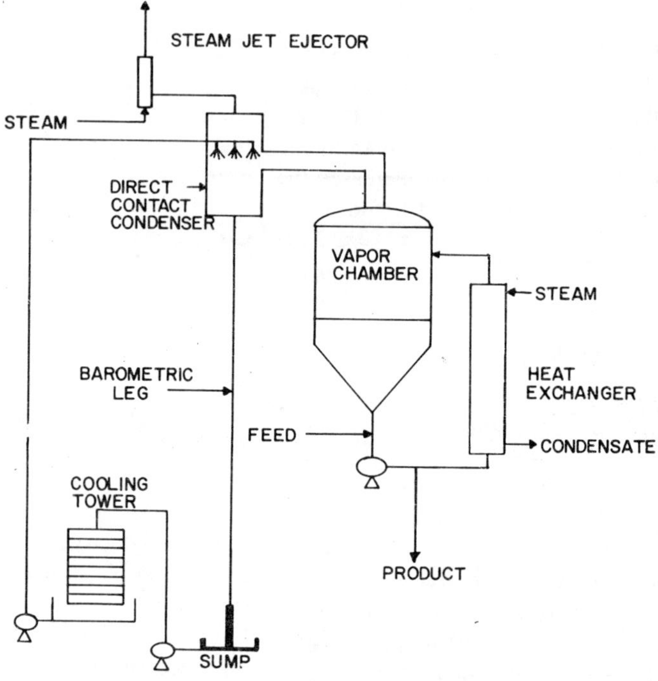

Fig. 11.1. Schematic diagram of a single-effect evaporator.

from the vapor chamber as rapidly as they are formed, and a steam jet ejector to remove noncondensible gases from the system. Each vapor chamber is considered an effect.

The Vapor Chamber. This is usually the largest and most visible part of the evaporator. Its main function is to allow separation of vapor from liquid and prevent carryover of solids by the vapor. It is also a reservoir for the product. The temperature inside an evaporator is determined by the absolute pressure in the vapor chamber. The vapor temperature is the temperature of saturated steam at the absolute pressure inside the chamber. When the liquid is a dilute solution, vapor and liquid temperatures will be the same. However, concentrated solutions exhibit a boiling point rise, resulting in a higher boiling temperature than that of pure water. Thus vapors leaving the liquid will be superheated steam at the same temperature as the boiling liquid. Depending upon the extent of heat loss to the surroundings around the vapor chamber, the vapor may be saturated at the absolute pressure within the vapor chamber or superheated steam at the boiling temperature of the liquid.

In most food products, the soluble solids are primarily organic compounds and the boiling point rise can be expressed as follows:

$$\Delta T_b = 0.51m \tag{1}$$

where ΔT_b in °C is the increase in the boiling point of a solution with molality m above the boiling point of pure water at the given absolute pressure. In addition to the boiling point rise due to the presence of solute, the pressure at the bottom of a liquid pool is higher than the absolute pressure of the vapor, and this difference in pressure also increases the boiling point of the liquid. The pressure exerted by a column of liquid of height h and density ρ is:

$$P = \rho(h)\,\frac{g}{g_c} \tag{2}$$

where $g_c = 1$ when using SI units.

EXAMPLE: Calculate the boiling temperature of liquid containing 30% soluble solids at a point 5 ft (1.524 m) below the surface inside an evaporator maintained at 20 in. Hg vacuum (33.8 kPa absolute). Assume that the soluble solids are hexose sugars and that the density of the liquid is 62 lb/ft^3 (933 kg/m^3). Atmospheric pressure is 30 in. Hg (101.5 kPa).

Solution: The absolute pressure in lb$_f$/in.^{2} corresponding to 20 in. Hg vacuum is (30 − 20)(0.491) = 4.91 psia. From the steam tables, the temperature corresponding to 4.91 psia is (by interpolation) 161.4°F (71.9°C). The molecular weight of a hexose sugar is 180. The molality of 30% soluble solids will be:

$$m = \frac{\text{moles solute}}{\text{1000 g solvent}}$$

$$= \frac{0.3/180}{0.7/1000} = 2.38$$

Using equation 1:

$$\Delta T_b = 0.51(2.38) = 1.21°\text{C or } 2.2°\text{F}$$

The absolute pressure at the level considered is the sum of the absolute pressure of the vapor and the pressure exerted by the column of liquid. Expressed in lb$_f$/in.^{2}, this pressure is:

$$P = 4.91 + \rho h\,\frac{g}{g_c}$$

$$= 4.91 + 62\,\frac{\text{lb}_m}{\text{ft}^3}\,(5\text{ ft})\cdot\frac{\text{lb}_f}{\text{lb}_m}\cdot\frac{1\text{ ft}^2}{144\text{ in.}^2}$$

$$= 7.06\text{ psia}$$

From the steam tables, the boiling temperature corresponding to 7.06 psia is (by interpolation) 175.4°F (79.7°C).

The boiling temperature of the liquid will be 175.4 + 2.2 = 177.6°F (80.9°C).

The significance of the boiling point rise is that the liquid leaving the evaporator will be at the boiling point of the liquid rather than at the temperature of the vapor. The boiling temperature at a point submerged below a pool of liquid will have the effect of reducing the ΔT available for heat transfer in the heat exchanger if the heat exchange unit is submerged far below the fluid surface.

The Condenser. Two general types of condensers are used. A surface condenser is used when the vapors need to be recovered. This type of condenser is actually a heat exchanger cooled by refrigerant or by cooling water. The condensate is pumped out of the condenser. It has a high first cost and is expensive to operate. For this reason, it is seldom used if an alternative is available. Condensers used on essence recovery systems fall into this category.

The other type of condenser is one in which cooling water mixes directly with the condensate. This condenser may be a barometric condenser in which vapors enter a water spray chamber on top of a tall column. The column full of water is called a *barometric leg*, and the pressure of water in the column balances the atmospheric pressure to seal the system and maintain a vacuum. The temperature of the condensate-water mixture should be about 5°F (2.78°C) below the temperature of the vapor in the vapor chamber to allow continuous vapor flow into the condenser. The barometric leg must be high enough to provide sufficient positive head at the base to allow the condensate-cooling water mixture to flow continuously out of the condenser at the same rate it enters. A jet condenser is one in which part of the cooling water is sprayed into the upper part of the unit to condense the vapors and the rest is introduced down the throat of a venturi at the base of the unit to draw the condensed vapor and cooling water out of the condenser. The jet condenser uses considerably more water than the barometric condenser, and the rate of water consumption cannot be easily controlled.

The condenser duty q_c is the amount of heat that must be removed to condense the vapor.

$$q_c = V(h_g - h_{fc}) \tag{3}$$

where V = the quantity of vapor to be condensed, h_g = the enthalpy of the vapor in the vapor chamber of the evaporator, and h_{fc} = the enthalpy of the liquid condensate.

For direct contact condensers, the amount of cooling water required per unit of vapor condensed can be determined by a heat balance:

$$W(h_{fc} - h_{fw}) = V(h_g - f_{fc}) \tag{4}$$

$$\frac{W}{V} = \frac{h_g - h_{fc}}{h_{fc} - h_{fw}} \tag{5}$$

where W = quantity of cooling water required and h_{fw} = enthalpy of cooling water entering the condenser.

The enthalpy of the condensate-water mixture, h_{fc}, should be evaluated, in the case of barometric condensers, at a temperature 5°F (2.7°C) lower than the vapor temperature.

EXAMPLE: Calculate the ratio of cooling water to vapor for a direct contact barometric condenser for an evaporator operating at a vapor temperature of 150°F (65.55°C). What is the minimum height of the water column in the barometric leg for the evaporator to operate at this temperature? Cooling water is at 70°F (21.1°C). Atmospheric pressure is 760 mm Hg.

Solution: At a temperature of 150°F, the absolute pressure of saturated steam is 3.7184 psia (25.6 kPa). h_g = 1126.1 BTU/lb or 2.619 MJ/kg. The condensate-cooling water mixture must be at 150 − 5 = 145°F (62.78°C). h_{fc} = 112.95 BTU/lb or 0.262 MJ/kg. The enthalpy of the cooling water, h_{fw} = 38.052 BTU/lb or 0.066 MJ/kg. Basis: V = one unit weight of vapor. Using equation 5:

$$\frac{W}{V} = \frac{1126.1 - 112.95}{112.95 - 38.052} = 13.52$$

The atmospheric pressure is 760 mm Hg or 101.3 kPa. From the steam tables, the density of water at 145°F is $1/V$ = 61.28 lb/ft³ or 981.7 kg/m³. The pressure that must be counteracted by the column of water in the barometric leg is the difference between the barometric pressure and the absolute pressure in the system.

$$\Delta P = 101.3 - 25.6 = 75.7 \text{ kPa}$$

$$= \rho g h$$

$$h = \frac{75{,}700 \text{ kg} \cdot \text{m}}{\text{s}^2 \cdot \text{m}^2} \frac{1}{981.7 \text{ kg} \cdot \text{m}^{-3}} \frac{1}{9.80 \text{ m} \cdot \text{s}^{-2}}$$

$$= 7.868 \text{ m or } 25.8 \text{ ft}$$

Removal of Noncondensible Gases. A steam jet ejector is often used. Figure 11.2 is a schematic diagram of a single-stage ejector. High-pressure steam

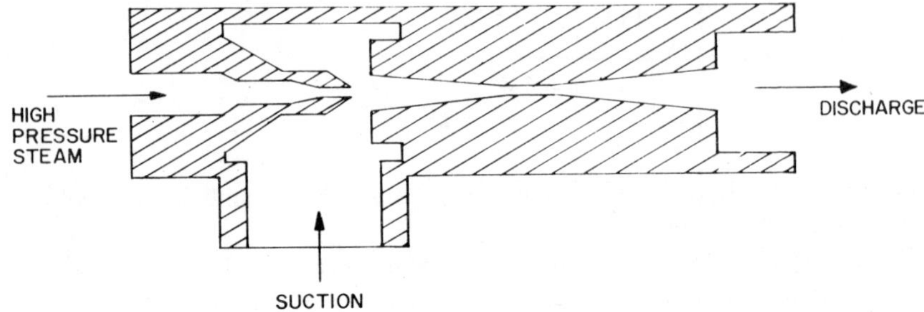

Fig. 11.2. Schematic diagram of a single-stage ejector.

is allowed to expand through a jet whose velocity increases. The movement of steam through the converging-diverging section at high velocity generates a zone of low pressure in the suction chamber, and noncondensible gases can be drawn into the ejector. The noncondensible gases mix with the high-velocity steam and are discharged into the atmosphere. Steam jet ejectors are more effective than vacuum pumps in that water vapor present in the noncondensible gases does not interfere with their operation. If suction absolute pressures are 4 in. Hg (1.352 kPa) or lower, multistage ejectors are used. The capacity of jet ejectors is dependent upon the design of the ejector, the pressure of the high-pressure steam, and the pressure differential between the suction and the discharge.

Capacity charts for steam jet ejectors are usually provided by their manufacturers. The capacity is expressed as weight of air evacuated per hour as a function of suction pressure and steam pressure.

The amount of noncondensible gases to be removed from a system depends upon the extent of leakage of air into the system and the amount of dissolved air in the feed and in the cooling water. In addition to removing the noncondensible gases, jet ejectors have to remove the water vapor present in the condenser. Air leakage has been estimated at 4 g air/h for every meter length of joints. The solubility of air in water at atmospheric pressure at various temperatures can be determined from Fig. 11.3. The amount of water vapor with the noncondensible gases leaving the condenser can be calculated as follows:

$$W_v = \frac{P_c(18)}{(P_v - P_c)29} \tag{6}$$

where W_v = kg water vapor/kg air, P_c = vapor pressure of water at the temperature of the condensate-cooling water mixture, and P_v = absolute pressure inside the evaporator.

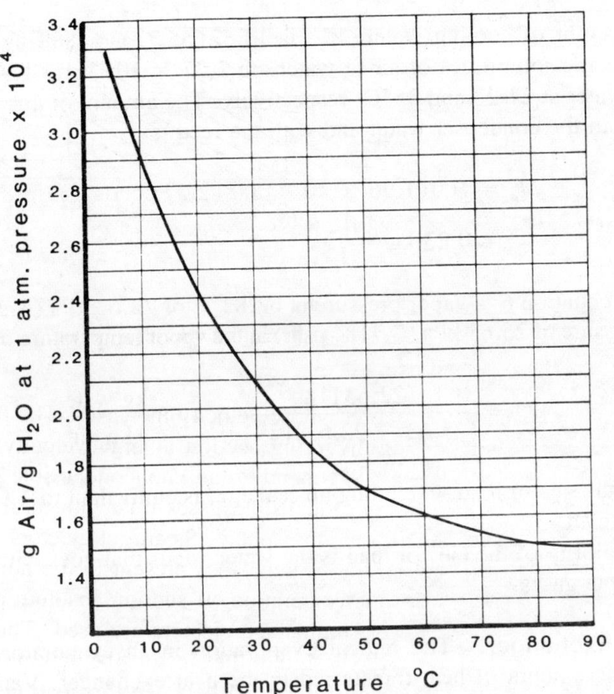

Fig. 11.3. Solubility of air in water at atmospheric pressure at various temperatures. (Data from Perry et al., 1963. *Chemical Engineers Handbook*, McGraw-Hill Book Co., New York.)

EXAMPLE: Calculate the ejector capacity required for an evaporator that processes 100 kg/h of juice from 12% to 35% solids. The evaporator is operated at 65.6°C (150°F). Cooling water is at 21.2°C (70°F). Product enters at 35°C (89.6°F). The condenser temperature is maintained at 2.78°C (5°F) below the vapor temperature.

Solution: From the example in the previous section, the ratio of cooling water to vapors for an evaporator operated under the conditions specified in this problem was determined to be 13.52 kg of water at 21.1°C. A material balance gives:

$$\text{Wt concentrate} = \frac{100(0.12)}{0.35} = 34.28 \text{ kg/h}$$

$$\text{Wt vapors} = 100 - 34.28 = 65.71 \text{ kg/h}$$

The weight of cooling water, W_c, is 13.52(65.71) = 888 kg/h. From Fig. 11.3 the solubilities of air in water are 2.37×10^{-4} and 1.96×10^{-4} air/kg water at 21.2° and 35°C, respectively. The amount of air, M_a, introduced with the condenser water and with the feed is:

$$M_a = 100(1.96 \times 10^{-4})888(2.37 \times 10^{-4})$$
$$= 0.23 \text{ kg/h}$$

Using equation 6, a vapor pressure at 62.82°C of 22.63 kPa (3.2825 psia) and a pressure of 25.63 kPa (3.7184 psia) at the vapor temperature of 65.6°C:

$$W_v = \frac{(22.63)18}{(25.63 - 22.63)(29)} = 4.68 \frac{\text{kg water}}{\text{kg air}}$$

The total ejector load, excluding air leakage, is equivalent to 4.68 + 0.23 or 4.91 kg/h.

The majority of the ejector load is the water vapor that goes with the noncondensible gases.

The Heat Exchanger. The rate of evaporation in an evaporator is determined by the amount of heat transferred in the heat exchanger. Variations in evaporator design can be seen primarily in the manner in which heat is transferred to the product. Considerations like the stability of the product to heat, fouling of heat exchange surfaces, ease of cleaning, and whether the product allows a rapid enough rate of heat transfer by natural convection dictate the design of the heat exchanger for use on a given product. The schematic diagram of an evaporator presented in Fig. 11.1 shows a long tube, vertical, forced circulation evaporator. This heat exchanger design is usually used when a single effect concentrates a material that becomes very viscous at a high solids content. Because of the forced circulation, heat transfer coefficients are fairly even at the high viscosity of the concentrate. In some evaporators, the heat exchanger is completely immersed in the fluid being heated inside the vessel that constitutes the fluid reservoir and vapor chamber. Heat is transferred by natural convection. This type of heat exchange is suitable when the product is not very viscous and is usually utilized in the first few effects of a multiple-effect evaporator.

Heat exchangers on evaporators for food products have the food flowing inside the tubes for ease of cleaning. In very-low-temperature evaporation such as in fruit juice concentration, the operation has to be stopped regularly to prevent microbiological buildup and to remove deposits of food product on the heat exchange surfaces.

The capacity of an evaporator is determined by the amount of heat transferred to the fluid by the heat exchanger. If q = the amount of heat transferred, P = mass of concentrated product, C_c = specific heat of the concentrate, V = mass of the vapor, h_g = enthalpy of the vapor, h_f = enthalpy of the water component of the feed that is converted to vapor, T_1 = feed inlet temperature, and T_2 = liquid temperature in the evaporator, a heat balance would give:

$$q = PC_c(T_2 - T_1) + V(h_g - h_f) \tag{7}$$

The rate of heat transfer can be expressed as:

$$Q = UA\Delta T \tag{8}$$

A material balance would give:

$$P = \frac{Fx_f}{x_p} \tag{9}$$

and

$$V = F\left(1 - \frac{x_f}{x_p}\right) \tag{10}$$

Equations 7 and 10 can be used to calculate the capacity of an evaporator in terms of a rate of feed, F, knowing the initial solids content, x_f, the final solids content, x_p, and the amount of heat transferred in the heat exchanger expressed in terms of the heat transfer coefficient, U, the area available for heat transfer, A, and the temperature difference between the boiling liquid in the evaporator and the heating medium, ΔT.

In evaporators where heat transfer to the product occurs by natural convection, products which tend to form deposits on the heat exchange surface foul this surface and reduce the overall heat transfer coefficient, U. When the evaporation rate slows down considerably, seriously affecting production, the operation is stopped and the evaporator is cleaned. In evaporators used on tomato juice, the temperatures are high enough so that microbiological buildup is not a factor, and shutdown for cleanup is usually done after about 14 days of operation. In some models of evaporators used on orange juice, on the other hand, operating time between cleanup is much shorter (2–3 days) because of microbiological buildup at the lower temperatures used.

Fouling of heat exchange surfaces is minimized by reducing ΔT across the heat exchange surface and by allowing the product to flow rapidly over this surface. Although forced recirculation through the heat exchanger results in rapid heat transfer, a disadvantage is the long residence time of the product inside the evaporator.

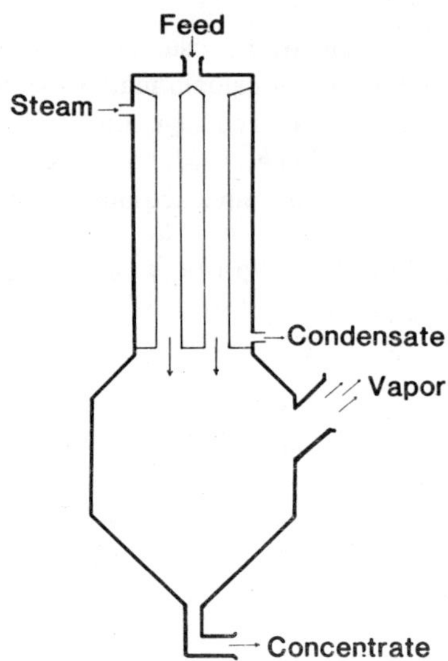

Fig. 11.4. Schematic diagram of a falling film evaporator.

For products that are heat sensitive and for which low temperature differentials are allowable in the heat exchanger, the falling-film heat exchanger is used. Figure 11.4 is a schematic diagram of this heat exchanger, which is used extensively in the concentration of orange juice. The product flows in a thin film down heated tubes, where heat is transferred and vapor is removed. The product passes through the heat exchange tube of one effect only once; this short contact with a hot surface minimizes heat-induced flavor or color changes and nutrient degradation.

The coefficient of heat transfer, U, in evaporator heat exchangers is on the order of 200 BTU/(h · ft^2 · °F) or 1136 W/(m^2 · K) for natural convection and 400 BTU/h(ft^2)(°F) or 2272 W/m^2 · K for forced convection. The effect of increased viscosity on heat transfer can be estimated by using this relationship: the heat transfer coefficient is proportional to the viscosity raised to the power −0.44. This, $U_1/U_2 = (\mu_1/\mu_2)^{-0.44}$, where U_1 is the heat transfer coefficient corresponding to viscosity μ_1 and U_2 is the heat transfer coefficient corresponding to viscosity μ_2. This relationship is useful in estimating how much of a change in evaporator performance would be expected with a change in operating conditions that would result in variations in product viscosity.

EXAMPLE: A fruit juice is to be concentrated in a single effect, forced recirculation evaporator from 10% to 45% soluble solids. The feed rate is 5500 lb/h or 2497 kg/h. Steam condensing at 250°F (121.1°C) is used for heating. The vapor temperature in the evaporator should be at 130°F (54.4°C).

Assume that the soluble solids are hexose sugars in calculating the boiling point rise. Use Siebel's formula in calculating the specific heat of the juice. The feed is at 125°F (51.7°C). The heat transfer coefficient, U, is 500 BTU/(h · ft^2 · °F) or 2839 W/(m^2 · K). Calculate the steam economy to be expected and the heating surface area required.

Solution: Equations 7 through 10 will be used. The steam economy is defined as the ratio of vapor produced to steam consumed. C_c is calculated using Siebel's formula. $C_c = 0.8(0.55) + 0.2 = 0.64$ BTU/lb(°F) or 2679 J/(kg · K).

$$h_g = \text{enthalpy of vapor at } 130°\text{F}$$

$$= 1117.8 \text{ BTU/(lb} \cdot °\text{F)}$$

$$= 2679 \text{ J/(kg} \cdot \text{K)}$$

$$h_f = \text{enthalpy of the water component of the feed at } 125°\text{F}$$

$$= 92.96 \text{ BTU/lb or } 0.216 \text{ MJ/kg}$$

The temperature of the concentrate leaving the evaporator, T_2, is the sum of the vapor temperature and the boiling point rise, ΔT_b. Using equation 1, m for 45% soluble solids is:

$$m = \frac{45/180}{55/1000} = 4.545 \text{ moles sugar/1000 g water}$$

$$\Delta T_b = 0.51(4.545) = 2.32°\text{C or } 4.2°\text{F}$$

$$T_2 = 130 + 4.2 = 134.2°\text{F or } 56.72°\text{C}$$

Substituting in equation 10:

$$V = 5500\left[1 - \frac{0.10}{0.45}\right] = 4278 \text{ lb/h or } 1942 \text{ kg/h}$$

Using equation 9:

$$P = 5500\left[\frac{0.10}{0.45}\right] = 1222 \text{ lb/h or } 555 \text{ kg/h}$$

Using equation 7:

$$q = 1222(0.64)(134.2 - 125) + 4278(1117.8 - 92.96)$$

$$= 4{,}391{,}500 \text{ BTU/h or } 1.2827 \text{ MW}$$

Equation 8 can be used to calculate the heat transfer surface area.

$$A = \frac{q}{U\Delta T} = \frac{4{,}391{,}500}{500(250 - 134.2)} = 75.8 \text{ ft}^2 \text{ or } 7.04 \text{ m}^2$$

Note that the liquid boiling temperature was used in determining the heat transfer ΔT rather than the vapor temperature. The enthalpy of vaporization of steam at 250°F is 945.5 BTU/lb or 2.199 MJ/kg. Steam required = q/h_{fg} = 4,391,500/945.5 = 4645 lb/h or 2109 kg/h. Steam economy = 4278/4645 = 0.92.

IMPROVING THE ECONOMY OF EVAPORATORS

Poor evaporator economy results from wasting the heat present in the vapors. Some of the techniques used to reclaim heat from the vapors include use of multiple effects such that vapors from the first effect are used to heat the succeeding effects, use of vapors to preheat the feed, and vapor recompression.

Vapor Recompression. Adiabatic recompression of vapor results in an increase in temperature and pressure. Figure 11.5 is a Mollier diagram for steam in the region involved in vapor recompression for evaporators. Recompression involves increasing the pressure of the vapor to increase its condensing temperature above the boiling point of the liquid in the evaporator. Compression of saturated steam results in superheated steam at high pressure. It is necessary to convert this vapor to saturated steam by mixing it with liquid water before introducing it into the heating element of the evaporator. Superheated steam in the heat exchanger can lower the overall heat transfer coefficient.

In Chapter 10, the work involved in adiabatic compression was found to be the difference in the enthalpy of the low-pressure saturated vapor and the high-pressure superheated vapor. The ratio between the latent heat of the saturated steam produced from the hot vapors and the work of compression is the coefficient of performance of the recompression system.

EXAMPLE: Using the example problem of the previous section, determine the coefficient of performance for a vapor recompression system used on the unit.

Solution: From Fig. 11.5, the initial point for compression is saturated vapor with an enthalpy of 1118 BTU/lb or (2.600 MJ/kg). Isentropic compression to the pressure of saturated steam at 250°F (121.1°C) gives a pressure of 29.84 psia (206 kPa), an enthalpy of 1338 BTU/lb (3.112 MJ/kg), and a temperature of 612°F (322°C). Figure 11.6 is a schematic diagram showing how these numbers were obtained from Fig. 11.5.

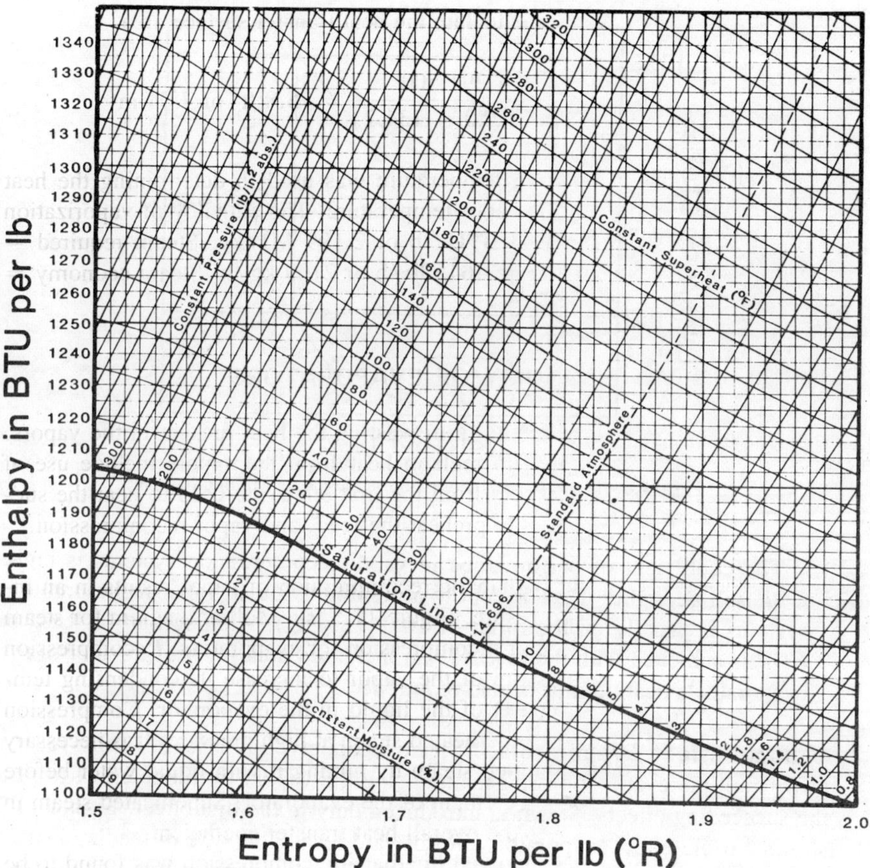

Fig. 11.5. Mollier diagram for steam in the region involved in vapor recompression for evaporators. (Diagram compliments of Combustion Engineering, Inc.)

If condensate from the heat exchanger at 250°F is used to mix with the superheated vapor after compression to produce saturated steam at 250°F, the amount of saturated steam produced will be:

$$\text{Wt saturated steam} = 1 + \frac{h_{g1} - h_{g2}}{h_{fg}}$$

$$= 1 + \frac{1338 - 1164}{945.5} = 1.184 \text{ lb } (0.537 \text{ kg})$$

$$COP = \frac{1.184(945.5)}{1(1338 - 1118)} = 5.09$$

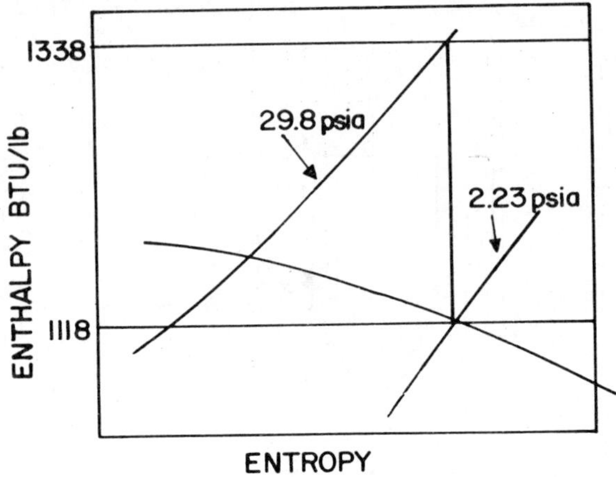

Fig. 11.6. The vapor recompression process on a Mollier diagram.

The coefficient of performance in vapor recompression systems is high. *COP* will be higher if ΔT is kept to a minimum. ΔT in vapor recompression systems is usually on the order of 10°F (5.6°C). Although an increase in *COP* is achieved with the low ΔT, an increased area for heat transfer in the evaporator's heating unit is also required.

Multiple-Effect Evaporators. Steam economy can also be improved by using multiple evaporation stages and using the vapors from one effect to heat the succeeding effects. Steam is introduced only in the first effect. Figure 11.7 shows a triple-effect evaporator with forward feed. This type of feeding is used when the feed is at a temperature close to the vapor temperature of the effect where it is introduced. If substantial amounts of heat are necessary to bring the feed temperature to the boiling temperature, other arrangements such as backward feeding may be used. In a backward feed arrangement, the flow of the feed is countercurrent to the flow of vapor.

Multiple-effect evaporators are often constructed with the same heat transfer surface areas in each effect. The governing equation for evaporation rate in such evaporators is the heat transfer equation (equation 8), as in single-effect evaporators. However, the ΔT in each effect of a multiple-effect evaporator is only a fraction of the total ΔT; therefore, for the same rate of evaporation and the same total ΔT, a multiple-effect evaporator with n effects requires approximately n times the heat exchange area needed for a single-effect evaporator. The saving in energy costs with the improvement in steam economy is achieved only with an increase in the required heat transfer surface area.

Only two parameters affecting heat transfer ΔT can be controlled in the operation of a multiple-effect evaporator: the temperature of the vapor in the last

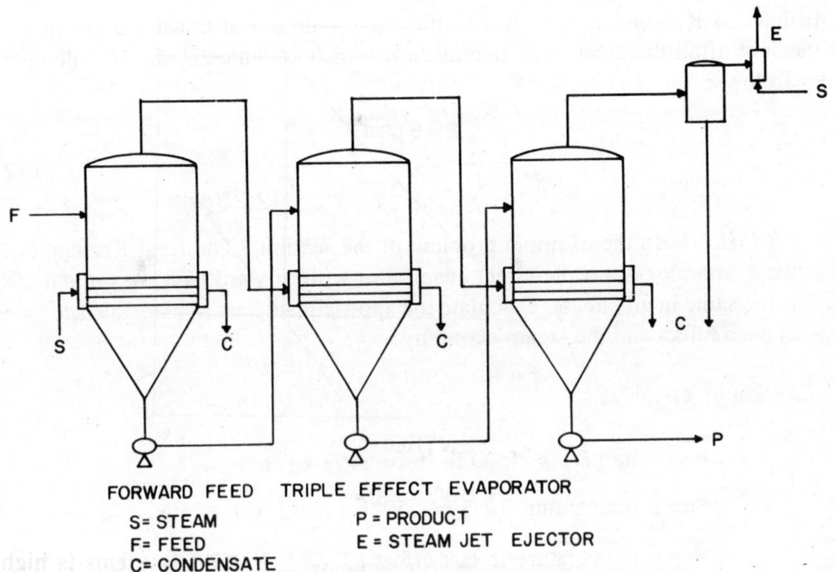

Fig. 11.7. Schematic diagram of a triple-effect evaporator with forward feed.

effect and the temperature of the steam in the first effect. The vapor temperature and pressure in the first and intermediate effects develop spontaneously according to the heat balance occurring within these various effects. The total ΔT in a multiple-effect evaporator is the difference between the steam temperature and the vapor temperature in the last effect. A boiling point rise would decrease the available ΔT.

$$\Delta T = T_s - T_{vn} - \Delta T_{b1} - \Delta T_{b2} - \ldots \Delta T_{bn} \tag{11}$$

where ΔT = total available temperature drop for heat transfer; T_s = steam temperature; T_{vn} = vapor temperature in the last effect; and ΔT_{b1}; ΔT_{b2} . . . ΔT_{bn} = boiling point rise in effects 1, 2, . . . n.

Multiple-effect evaporator calculations are done using a trial-and-error method. ΔT is assumed for each effect, and by making heat and material balances in each effect, the rate of heat transfer in each effect is compared with the heat input necessary to achieve the desired evaporation rate in each effect. Adjustments are then made on the assumed ΔTs until the heat input and the heat requirement for each effect are in balance. Multiple-effect evaporator calculations are tedious and time-consuming. They are best done on a computer.

Reasonable approximations can be made on the capacity of a given multiple-effect evaporator, knowing the heat transfer surface areas and the heat transfer coefficients by assuming equal evaporation in each effect and making a heat balance as if the evaporator is a single effect with a heat transfer area equal to the sum of all the areas. The overall heat transfer coefficient can be calculated as follows:

$$\frac{1}{U} = \frac{1}{U_1} + \frac{1}{U_2} + \cdots \frac{1}{U_n} \tag{12}$$

EXAMPLE: In the example problem of the section "The Heat Exchanger," the evaporator is a triple-effect evaporator with forward feed. Assume that U is the same in all effects. Calculate the approximate heat transfer surface areas in each effect and the steam economy.

Solution: Given:

Feed rate (F) = 5500 lb/h or 2497 kg/h

Steam temperature (T_s) = 250°F (121.1°C)

Vapor temperature in last effect (T_{v3}) = 130°F (54.4°C)

Feed temperature (T_f) = 125°F (51.7°C)

$$U_1 = U_2 = U_3 = 500 \frac{\text{BTU}}{\text{h} \cdot \text{ft}^2 \cdot {}^\circ\text{F}} \quad \text{or} \quad 2839 \frac{\text{W}}{\text{m}^2 \cdot \text{K}}$$

Solids content of feed (x_f) = 10%

Solids content of product (x_p) = 45%

Molecular weight of solids (M) = 180

First, determine the overall heat transfer, ΔT, using equation 11. It is necessary to determine the boiling point rise in each effect. Using equation 10:

$$\text{Total evaporation } (V) = 5500 \left[1 - \frac{0.10}{0.45} \right] = 4278 \text{ lb/h or } 1942 \text{ kg/h}$$

Assuming equal evaporation, $V_1 = V_2 = V_3$ = 1426 lb/h or 647.3 kg/h. The solids content in each effect can be calculated by rearranging equation 10. The subscripts n on V and F refer to the feed to and vapor from effect n.

$$x_n = \frac{x_f F_n}{F_n - V_n}$$

$$x_1 = \frac{0.10(5500)}{5500 - 1426} = 0.135$$

$$x_2 = \frac{0.135(5500 - 1426)}{5500 - 1426 - 1426} = \frac{0.135(4074)}{2648} = 0.208$$

$$x_3 = x_p = 0.45$$

The molalities are:

$$m_1 = \frac{0.135(1000)}{(1 - 0.135)(180)} = 0.067 \text{ molal}$$

$$m_2 = \frac{0.208(1000)}{(1 - 0.208)(180)} = 1.459 \text{ molal}$$

$$m_3 = \frac{0.45(1000)}{(1 - 0.45)(180)} = 4.545 \text{ molal}$$

Using equation 1, the boiling point rises are:

$$\Delta T_{h1} = 0.51(0.067) = 0.0342°\text{C or } 0.06°\text{F}$$

$$\Delta T_{h2} = 0.51(1.459) = 0.744°\text{C or } 1.34°\text{F}$$

$$\Delta T_{h3} = 0.51(4.545) = 2.32°\text{C or } 4.17°\text{F}$$

The total ΔT for heat transfer is calculated using equation 11:

$$\Delta T = 250 - 130 - 4.17 - 1.34 - 0.06$$
$$= 114.4°\text{F or } 63.57°\text{C}$$

Overall U is calculated using equation 12:

$$\frac{1}{U} = \frac{1}{500} + \frac{1}{500} + \frac{1}{500} = 0.006$$

$$U = 166.67 \text{ BTU/(h} \cdot \text{ft}^2 \cdot °\text{F) or } 946.3 \text{ W/(m}^2 \cdot \text{K)}$$

From the section "The Heat Exchanger," the required heat transfer rate for this evaporator was determined to be 4,391,500 BTU/h or 1.2827 MW. Using equation 8:

$$A = \frac{4{,}391{,}500}{166.67(114.4)} = 230.3 \text{ ft}^2 \text{ or } 21.4 \text{ m}^2$$

The heat transfer surface area for each effect is 230.3/3 ft^2 = 76.77 ft^2 or 7.13 m^2.

Calculations of steam economy can only be done using the trial-and-error procedure necessary to establish the ΔT and the vapor temperature of each effect. The steam economy of multiple-effect evaporators is a number slightly less than the number of effects.

ESSENCE RECOVERY

A major problem in concentration of fruit juices is the loss of essence during the evaporation process. With condensers in which the cooling water directly contacts the vapor, it is not possible to recover the flavor components that are vaporized from the liquid. In the past, the problem of essence loss was solved in the orange juice industry by concentrating the juice to a higher concentration than is desired and diluting the concentrate with fresh juice to the desired solids concentration. The essence in the fresh juice gave the necessary flavor to the concentrate.

One method for essence recovery consists of flashing the juice into a packed or perforated plate column maintained at a very low absolute pressure. Flash evaporation is a process whereby hot liquid is introduced into a chamber which is at an absolute pressure where the boiling point of the liquid is below the liquid temperature. The liquid boils immediately upon exposure to the low pressure, vapor is released, and the liquid temperature drops to the boiling point of the liquid at the given absolute pressure.

The feed is preheated to 120° to 150°F (48.9° to 65.6°C) and is introduced into a column maintained at an absolute pressure of approximately 0.5 psia (3.45 kPa). There is no heat input in the column; therefore, evaporative cooling reduces the temperature of the liquid. The vapors rise continually up the packed column, becoming richer in the volatile components as they proceed up the column. A surface condenser cooled by a refrigeration system traps the volatile components. The essence concentrate recovered is blended with the concentrated product.

In a multiple-effect evaporator, a backward feed arrangement is used and the vapors from the last effect are condensed using a surface condenser. The condensate containing the essence is flashed into the essence recovery unit.

PROBLEMS

1. A single-effect falling-film evaporator is used to concentrate orange juice from 14% to 45% solids. The evaporator utilizes a mechanical refrigeration cycle using ammonia as a refrigerant for heating and for condensing the vapors. The refrigeration cycle is operated at a high-side pressure of 200 psia (1.379 MPa) and a low-side pressure of 50 psia (344.7 kPa). The evaporator is operated at a vapor temperature of 90°F (32.2°C). Feed enters at 70°F (21.1°C). The ratio of insoluble to soluble solids in the juice is 0.09, and the soluble solids consist of glucose and sucrose in a 70:30 ratio. Consider the ΔT as the log mean-ΔT between the liquid refrigerant

temperature and the feed temperature at one point and the hot refrigerant gas temperature and the concentrated liquid boiling temperature at the other point. The evaporator has a heat transfer surface area of 100 ft^2 (9.29 m^2), and an overall heat transfer coefficient of 300 BTU/(h · ft^2 · °F) or 1073 W/(m^2 · K) may be expected.

Calculate:

(a) The evaporator capacity in weight of feed per hour.

(b) The tons of refrigeration capacity required for the refrigeration unit based on the heating requirement for the evaporator.

(c) The additional cooling required for condensation of vapors if the refrigeration unit is designed to provide all of the heating requirements for evaporation.

2. Condensate from the heating unit of one effect in a multiple-effect evaporator is flashed to the pressure of the heating unit in one of the succeeding effects. If the condensate is saturated liquid at 7.511 psia (51.79 kPa) and the heating unit contains condensing steam at 2.889 psia (19.92 kPa), calculate the total available latent heat that will be in the steam produced from a unit weight of the condensate.

3. A single-effect evaporator was operating at a feed rate of 10,000 kg/h, concentrating tomato juice at 160°F (71.1°C) from 15% to 28% solids. The ratio of insoluble to soluble solids is 0.168, and the soluble solids consist of hexose sugars. Condensing steam at 29.840 psia (205.7 kPa) was used for heating, and the evaporator was at an absolute pressure of 5.993 psia (41.32 kPa). It is desired to change the operating conditions to enable the efficient use of a vapor recompression system. The steam pressure is to be lowered to 17.186 psia (118.37 kPa). Assume that there is no change in the heat transfer coefficient because of the lowering of the heating medium temperature.

Calculate:

(a) The steam economy for the original operating conditions.

(b) The capacity in weight of feed per hour under the new operating conditions.

(c) The steam economy of the vapor recompression system.

Express the steam economy as the ratio of the energy required to concentrate the juice to the energy required to compress the vapor, assuming a mechanical efficiency of 50% for the compressor. Assume that condensate from the heating element is added to the superheated steam to reduce the temperature to saturation.

SUGGESTED READING

Bennet, C. O., and Myers, J. E. 1962. *Momentum, Heat, and Mass Transport*. McGraw-Hill Book Co., New York.

Charm, S. E. 1971. *Fundamentals of Food Engineering*, 2nd ed. AVI Publishing Co., Westport, Conn.

Foust, A. S., Wenzel, E. A., Clump, C. W., Maus, L., and Andersen, L. B. 1960. *Principles of Unit Operations*. John Wiley & Sons, New York.

Heldman, D. R. 1973. *Food Process Engineering*. AVI Publishing Co., Westport, Conn.

McCabe, W. L., and Smith, J. C. 1967. *Unit Operations of Chemical Engineering*, 2nd ed. McGraw-Hill Book Co., New York.

Peters, M. S. 1954. *Elementary Chemical Engineering*. McGraw-Hill Book Co., New York.

Perry, R. H., Chilton, C. H., and Kirkpatrick, S. D. 1963. *Chemical Engineers Handbook*, 4th ed. McGraw-Hill Book Co., New York.

12

Dehydration

Dehydration is an important method of food preservation. The reduced weight and bulk of dehydrated products and their dry shelf stability reduce product storage and distribution costs. As dehydration techniques that produce good-quality, convenient foods are developed, more dehydrated products will be commercially produced. At present, instant beverage powders, dry soup mixes, spices, and ingredients used in further processing are the major food products dehydrated.

WATER ACTIVITY

Dehydrated foods are preserved because water activity is at a level where no microbiological activity can occur and where deteriorative chemical and biochemical reaction rates are reduced to a minimum. Water activity (a_w) is measured as the equilibrium relative humidity (ERH), the percent relative humidity (RH) of an atmosphere in contact with a product at the equilibrium water content. a_w is also the ratio of the partial pressure of water in the headspace of a product (P) to the vapor pressure of pure water (P^0) at the same temperature.

$$a_w = ERH = \frac{P}{P^0} \tag{1}$$

The relationship between a_w and the rate of deteriorative reactions in food is shown in Fig. 12.1. Reducing a_w below 0.7 prevents microbiological spoilage. However, although microbial spoilage does not occur at $a_w = 0.7$, prevention of other deteriorative reactions needed to preserve a food product successfully by dehydration requires reduction of a_w to $\equiv 0.3$.

A food material may also be dehydrated only for weight or bulk reduction and finally preserved using other techniques.

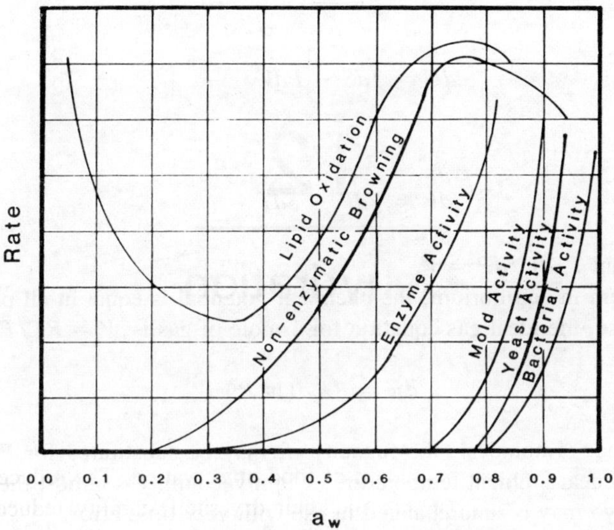

Fig. 12.1. Relationship between water activity and deteriorative reactions in foods. (From Labuza et al. *J. Food Sci.* 37:154, 1972. Used with permission.)

Thermodynamic Basis for Water Activity. The first and second laws of thermodynamics were discussed in the section "Thermodynamics." In addition to the thermodynamic variables of enthalpy (H), internal energy (E), work (W), and heat (Q) discussed under "Thermodynamics," other thermodynamic variables related to chemical changes will be discussed in this section.

F = free energy, another form of energy in a system which is different from PV work. This form of energy is responsible for chemical or electrical work and is responsible for driving chemical reactions.

$$F = H - TS \tag{2}$$

μ = chemical potential. It is directly related to the free energy. In any system undergoing change, $d\mu = dF$:

Since:

$$dF = dH - TdS - SdT$$

and:

$$dH = dU + PdV + VdP$$

and:

$$dU = TdS - PdV$$

Then:

$$dF = VdP - SdT$$

At constant T, $d\mu/dP = V$.

In a system in equilibrium, the chemical potential is equal in all phases. In the gas phase, the ideal gas equation for 1 mole of gas is: $V = RT/P$. Thus:

$$d\mu = RT\,d(\ln P) \tag{3}$$

If the vapor is nonideal, the fugacity, f, may be substituted for P. At low pressures of near ambient temperature or in a vacuum, f is almost the same as P and the two may be interchanged in use with very little error.

Thus, the chemical potential can be measured directly by the vapor pressure of a component. The activity of a component is defined as:

$$a = \frac{f}{f^0}$$

where f^0 is the fugacity of the pure component at the same temperature. If the component is water:

$$a_w = \frac{f}{f^0} \equiv \frac{P}{P^0} \tag{4}$$

The thermodynamic basis for the water activity is the chemical potential of water, its ability to participate in chemical reactions.

Osmotic Pressure. One consequence of reduced water activity is an increase in osmotic pressure, which interferes with food water and waste transport between a cell and its surroundings. The osmotic pressure is related to the activity coefficient as follows:

$$\pi = \frac{RT}{V} \ln x\gamma \tag{5}$$

π = osmotic pressure, atm; R = gas constant 82.06 mL · atm/mole · K; γ = activity coefficient; V = molar volume = 18 mL/mole for water; x = mole fraction of water; and T = absolute temperature.

Water Activity at High Moisture Contents. At high moisture contents, depression of the water vapor pressure by soluble solids is similar to the vapor pressure depression by solutes in solution. The presence of insoluble solids is ignored, and the solution phase determines the water activity.

For ideal solutions, the water activity is equal to the mole fraction of water, x_w.

$$a_w = x_w \tag{6}$$

For nonideal solutions, the activity coefficient, γ, corrects for deviation from nonideality.

$$a_w = \gamma x_w \tag{7}$$

The mole fraction of water, x_w, expressed in terms of the weight fraction water (x'_w) and solute (x'_s) having molecular weights of 18 and M_s, respectively, is:

$$\begin{aligned} x_w &= \frac{\text{moles water}}{\text{moles water + moles solute}} \\ &= \frac{x'_w/18}{x'_w/18 + x'_s/M_s} \end{aligned} \tag{8}$$

Equation 8 is Raoult's law. As equation 8, Raoult's law is often cited to govern the dependence of water activity on water and solute concentration in foods. Qualitatively, Raoult's law explains adequately the dependence of water activity on water and solute content in foods. Solutes with low molecular weights also provide a larger reduction of vapor pressure per unit weight of solute than those with high molecular weights. However, a_w calculated using equation 8 will differ from actual a_w because the solution phase in most foods is nonideal.

The deviation of a_w from Raoult's law is related to the heat of mixing involved in the dissolution of solute. Let ΔH_{m1} = partial molal heat of mixing, defined as the change in enthalpy of mixing with the addition of or removal of 1 mole of component 1 at constant temperature and pressure. Let N_1 = moles of component 1. Equation 9 can be derived (Perry et al., 1963) by assuming that the excess entropy of mixing is zero.

$$\overline{\Delta H}_{m1} = \frac{d(\Delta H_m)}{dN_1} = RT \ln \gamma_1 \tag{9}$$

Solutes in food systems exhibit negative heat of mixing with water (heat is released in mixing), and γ in equation 9 is less than 1. Solutions with negative heat of mixing will have a_w less than that calculated using Raoult's law.

The activity coefficient of a regular solution can be derived using Van Laar's approximations based on van der Waal's equation of state:

$$P = \frac{RT}{V - b} - \frac{a}{V^2} \tag{10}$$

The constant a in van der Waal's equation of state represents the magnitude of attractive forces between molecules, and the constant b represents the reduction in the volume as a result of these attractive forces. In a mixture containing N_1 and N_2 moles of components 1 and 2, respectively, a for the mixture will be determined by the constant a_1 for component 1 and a_2 for component 2. Since a is associated with attraction between molecules, the rules of permutation give N_1 ways in which N_1 molecules can interact. If each of these interactions yields a_1, the total yield of N_1^2 interactions will be $a_1 N_1^2$. The yield of all interactions between N_2 molecules of component 2 will be $a_2 N_2^2$. Molecules of components 1 and 2 also interact with each other, and the yield of each of these interactions is the geometric mean of a_1 and a_2. The number of ways in which N_1 molecules of component 1 and N_2 molecules of component 2 can interact with each other (excluding interactions between like molecules) is $2N_1 N_2$, and the yield of these interactions is $(a_1 a_2)^{1/2}(2N_1 N_2)$. Thus, for a mixture, the van der Waal's constant a can be expressed in terms of the constants of the components according to:

$$a = a_1 N_1^2 + a_2 N_2^2 + 2N_1 N_2 (a_1 a_2)^{1/2} \tag{11}$$

Constant b is associated with the volume of each molecule; therefore, b for a mixture can be expressed in terms of b_1 and b_2:

$$b = b_1 N_1 + b_2 N_2 \tag{12}$$

Van Laar's first approximation assumes the ratio a/b to be the internal energy of the mixture which approximates the heat of vaporization. Furthermore, in solutions, the volume changes on mixing are small, so that internal energy, E, and enthalpy, H, are equivalent. The heat of mixing is the difference between the sum of the internal energies of pure component 1 and pure component 2 and the internal energy of the mixture.

$$-\Delta H_m = E_1 + E_2 - E_{12} \tag{13}$$

E_1 is a_1/b_1, E_2 is a_2/b_2, and E_{12} is a/b for the mixture determined using equations 11 and 12. Equation 13 becomes:

$$-\Delta H_m = \frac{a_1}{b_1} + \frac{a_2}{b_2} - \frac{a_1 N_1^2 + a_2 N_2^2 + 2N_1 N_2 (a_1 a_2)^{1/2}}{N_1 b_1 + N_2 b_2}$$

Simplifying:

$$-\Delta H_m = \frac{N_1 N_2 b_1 b_2}{N_1 b_1 + N_2 b_2} \left[\frac{a_1^{1/2}}{b_1} - \frac{a_2^{1/2}}{b_2} \right]^2$$

If the values of b for the two components are equal:

$$-\Delta H_m = \frac{N_1 N_2}{(N_1 + N_2) b} (a_1^{1/2} - a_2^{1/2})^2 \qquad (14)$$

The partial molal heat of mixing can be determined by differentiating equation 14 with respect to N_1, keeping N_2 constant.

$$\overline{\Delta H_{m1}} = \frac{-d\Delta H_m}{dN_1} = \frac{N_2 (a_1^{1/2} - a_2^{1/2})^2}{b} \frac{d}{dN_1} \left[\frac{N_1}{N_1 + N_2} \right]$$

$$= \frac{N_2 (a_1^{1/2} - a_2^{1/2})}{b} \left[\frac{N_2}{(N_1 + N_2)^2} \right]$$

Since a_1, a_2, and b are all constants:

$$-\overline{\Delta H_{m1}} = k' \left[\frac{N_2}{N_1 + N_2} \right]^2$$

$$= k' x_2^2 \qquad (15)$$

Combining equations 9 and 15:

$$RT \ln \gamma_1 = -k' x_2^2 \qquad (16)$$

Since from equation 7 $\gamma_w = a_w / x_w$, and in a two-component system $x_2 = 1 - x_w$, at constant temperature equation 16 can be written as:

$$\log \frac{a_w}{x_w} = -k(1 - x_w)^2 \qquad (17)$$

Equation 17 shows that a plot of $\log a_w / x_w$ against $(1 - x_w)^2$ is linear with a negative slope. Equation 17 has been used successfully by Norrish (1966) to predict the water activity of sugar solutions. Values of the constant k for various solutes are shown in Table 12.1.

Table 12.1. Values of the Constant *k* for Various Solutes in Norrish's Equation for Water Activity of Solutions

Sucrose	2.7
Glucose	0.7
Fructose	0.7
Invert sugars	0.7
Sorbitol	0.85
Glycerol	0.38
Propylene glycol	−0.12
NaCl	15.8 ($x_2 < 0.02$)
	7.9 ($x_2 < 0.02$)
Citric acid	6.17
d-Tartaric acid	4.68
Malic acid	1.82
Lactic acid	−1.59

Sources: Norrish, R. S., *J. Food Technol.* 1:25, 1966; Toledo, R. T., *Proc. Meat Ind. Res. Conf.*, 1973; Chuang, L., M.S. thesis, University of Georgia, 1974; Chirife, J., and Ferro-Fontain, C., *J. Food Sci.* 45:802, 1980.

EXAMPLE: Calculate the water activity of a 50% sucrose solution.

Solution: From Table 12.1, the *k* value for sucrose is 2.7. Sucrose has a molecular weight of 342. The mole fraction of water is:

$$x_w = \frac{50/18}{50/18 + 50/342} = 0.95$$

Using equation 17:

$$\log a_w = \log x_w - 2.7(1 - x_w)^2$$

$$= \log 0.95 - 2.7(0.05)^2$$

$$a_w = 0.935$$

Gibbs-Duhem Equation. The chemical potential is a partial molal quantity which can be represented as the change in free energy with a change in the number of moles of that component, all other conditions being maintained constant. $\mu_1 = (dF/dN_1)$

$$dF = \left(\frac{dF}{dN_1}\right)dN_1 + \left(\frac{dF}{dN_2}\right)dN_2$$

$$= \mu_1\, dN_1 + \mu_2\, dN_2$$

But:

$$F = \mu_1 N_1 + \mu_2 N_2$$

$$dF = \mu_1\, dN_1 + N_1\, d\mu_1 + \mu_2\, dN_2 + N_2\, d\mu_2$$

and:

$$N_1\, d\mu_1 + N_2\, d\mu_2 = 0 \tag{18}$$

Equation 18 and similar expressions in equations 19 and 20 are different forms of the Gibbs-Duhem equation.

$$x_1 \left(\frac{d \ln a_1}{d_{x1}}\right) + x_2 \left(\frac{d \ln a_2}{dx_1}\right) = 0 \tag{19}$$

or:

$$x_1\, d \ln a_1 + x_2 d \ln a_2 = 0 \tag{20}$$

Let $(a_w)^0$ represent the activity of water in a system with only one solute and water in the mixture.

$$\ln (a_w)^0 = -\int \left(\frac{x_2}{x_1}\right) d \ln (a_2)$$

In a multicomponent system involving two solutes:

$$d \ln a_w = -\left(\frac{x_2}{x_1}\right) d \ln a_2 - \left(\frac{x_3}{x_1}\right) d \ln a_3$$

$$\ln a_w = -\int \left(\frac{x_2}{x_1}\right) d \ln a_2 - \int \left(\frac{x_3}{x_1}\right) d \ln a_3$$

$$a_w = (a_w)_2^0 (a)_3^0 \tag{21}$$

The water activity of a mixture involving several components can be determined from the product of the water activity of each component separately if all the water present in the mixture is mixed with individual components. Equation 21 was derived by Ross (*Food Technol.* 29(3):26, 1975).

EXAMPLE: Calculate the water activity of a fruit preserve containing 65% soluble solids, 2% insoluble solids, and the rest water. The soluble solids may be assumed to be 50% hexose sugars and 50 sucrose.

Solution: Basis: 100 g of fruit preserve.

$$\text{g hexose sugars} = 65(0.5) = 32.5 \text{ g}$$

$$\text{g sucrose} = 65(0.5) = 32.5 \text{ g}$$

$$\text{g water} = 33 \text{ g}$$

Consider that sucrose dissolves in all of the water present.

$$x_s = \frac{32.5/342}{33/18 + 32.5/342} = 0.0492$$

$$x_w = 1 - 0.0492 = 0.9507$$

Using equation 17 and a k value of 2.7 for sucrose:

$$\ln (a_{w1})^0 = \ln 0.9507 - 2.7(0.0492)^2$$

$$(a_{w1})^0 = 0.9365$$

Consider that hexose dissolves in all of the water present.

$$x_i = \frac{32.5/180}{32.5/180 + 33/18} = 0.0896$$

$$x_w = 1 - 0.0896 = 0.9103$$

Using equation 17 and a k value for hexose of 0.7:

$$\ln (a_{w2})^0 = \ln 0.9103 - 0.7(0.0896)^2$$

$$(a_{w2})^0 = 0.8986$$

Using equation 21, the a_w of the mixture is:

$$a_w = (a_{w1})^0(a_{w2})^0 = 0.9365(0.8986)$$

$$= 0.841$$

Other Equations for Calculating Water Activity

Bromley Equation. For salts and other electrolytes, Bromley's equation (Bromley, *AIChE J.* 19:313, 1973) accounts for ionic dissociation and non-ideality.

$$a_w = \exp\left(-0.018\ \Sigma m_i\ \phi\right) \tag{22}$$

ϕ = osmotic coefficient; m_i = moles of ionic species i per kilogram of solvent. $m_i = I(m)$, where I = number of ionized species per mole and m = molality. The osmotic coefficient, ϕ, is calculated using equation 23.

$$\phi = 1 + 2.303\left[F_1 + (0.06 + 0.6B)F_2 + 0.5BI\right] \tag{23}$$

The parameter f_1 in equation 23 is calculated using equation 24.

$$F_1 = F_{id}\left[-0.017zI^{0.5}\right] \tag{24}$$

The parameter F_{id} in equation 24 is calculated using equation 25.

$$F_{id} = 3I^{-1.5}\left[1 + I^{0.5} - \frac{1}{1 + I^{0.5}} - 2\ln\left(1 + I^{0.5}\right)\right] \tag{25}$$

The parameter F_2 in equation 23 is calculated using equation 26.

$$F_2 = \frac{z}{aI}\left[\frac{1 + 2aI}{(1 + aI)^2} - \frac{\ln(1 + aI)}{aI}\right] \tag{26}$$

The parameter I is the ionic strength, evaluated as half the sum of the product of the molality of dissociated ions and the square of their charge; for example, for $MgCl_2$, $I = 0.5[m(2)^2 + 2m(-1)^2] = 3m$; B and a are constants for each salt obtained from a regression of activity coefficient data against ionic strength. A comprehensive listing of a and B values for different salts is given by Bromley (1973). Chirife and Ferro-Fontan (*J. Food Sci.* 45:802, 1980) reported a value of B for sodium lactate of 0.050 kg/gmol. z is the charge number, the ratio of the sum of the product of the stoichiometric number of ions; for example, for $MgCl_2$, $z = [1(2) + 2(1)]/3 = 4/3 = 1.333$.

The values of the parameters z, I, B, and a for various salts are as follows:

For NaCl: $z = 1$; $I = m$; $B = 0.0574$; $a = 1.5$

For KCl: $z = 1$; $I = m$; $B = 0.0240$; $a = 1.5$

For KNO_3: $z = 1$; $I = m$; $B = -0.0862$; $a = 1$

For $MgCl_2$: $z = 1.333$; $I = 3m$; $B = 0.1129$; $a = 1.153$

Lang-Steinberg (J. Food Sci. 46:670, 1981) Equation. This equation is useful for mixtures of solids where there are no solute–insoluble solids interaction. The moisture content of each component X_i was considered to be linear when plotted against log $(1 - a_{wi})$, with a slope b_i and intercept a_i. a_{wi} is the water activity of component i at moisture content X_i. If X is the equilibrium moisture content of the mixture, g water/g dry matter, and S is the total mass of dry matter in the mixture, the a_w of the mixture can be calculated from the mass of dry matter in each component, S_i as follows:

$$\log(1 - a_w) = \frac{XS - \Sigma(a_i S_i)}{\Sigma(b_i X_i)} \tag{27}$$

Water Activity at Low Moisture Contents. If the water content of a food is plotted against the water activity at constant temperature, a sigmoid curve usually results. The curve is known as the *sorption isotherm* for that product. Figure 12.2 shows a sorption isotherm for raw beef at 25°C.

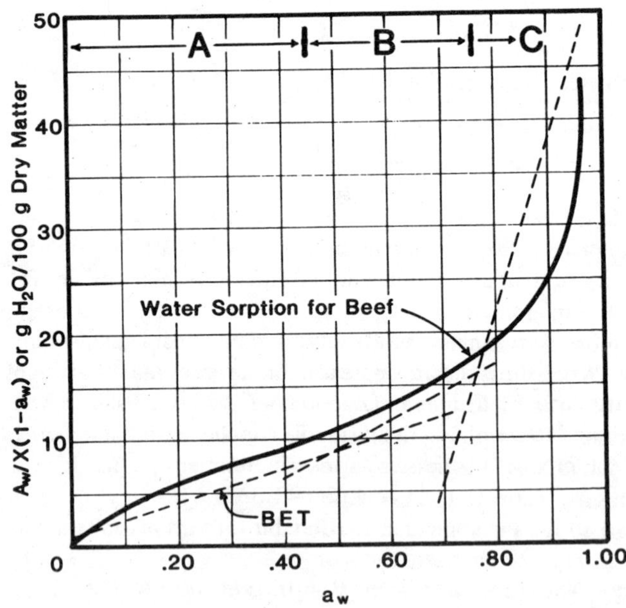

Fig. 12.2. Sorption isotherm of dried raw beef at 25°C, showing the three segments of the BET plot.

The sorption isotherm can be subdivided into three zones, each zone representing a different mechanism for water sorption. In zone C, the influence of insoluble solids on a_w is negligible. a_w is dependent upon the solute and water content of the solution phase and can be calculated using equations 17 and 21.

In zone B, the influence of insoluble solids on a_w becomes significant. The sorption isotherm flattens out, and very small changes in moisture content are reflected by very large changes in water activity. In this zone, water is held in the solid matrix by capillary condensation and multilayer adsorption. Some of the solutes may also be in the form of hydrates. Some of the water may still be in the liquid phase, but its mobility is considerably restricted because of attractive forces with the solid phase. The quantity of water present in the material that does not freeze at the normal freezing point usually is within this zone.

Zone A represents adsorption of water on the surface of solid particles. None of the water is in the liquid phase. The heat of vaporization of water in this zone is higher than the heat of vaporization of pure water, since both heat of vaporization and heat of adsorption must be supplied to remove the water molecules from the solid surface.

The relationship between water activity and moisture content in Zone A is best described by the Brunauer-Emmett-Teller (BET) equation. If x is the moisture content in grams of H_2O/per gram of dry matter:

$$\frac{a_w}{x(1 - a_w)} = \frac{1}{x_m C} + \frac{C - 1}{x_m C}(a_w) \tag{28}$$

x_m in equation 28 is the mass fraction of water in the material equivalent to a unimolecular layer of water covering the surface of each particle. C is a constant at constant temperature and is related to the heat of adsorption of water on the particles. C is temperature dependent.

A plot of $a_w/x(1 - a_w)$ against a_w is called the *BET plot*. The plot is linear, and the slope and intercept of the line can be used to determine the constant x_m, the moisture content at which the water molecules coat the surface of the solid particles in a monomolecular layer.

The region of maximum stability for a food product is usually in zone B in Fig. 12.2. When the moisture content in the product drops to a level insufficient to cover the solid molecules in a monomolecular layer, the rate of lipid oxidation increases. Determining the moisture content for the maximum shelf stability of a dehydrated material involves determining the sorption isotherm and calculating the value of x_m in equation 28 from a BET plot.

The dashed lines in Fig. 12.2 are BET plots of the sorption data for raw beef. The line in zone A represents monomolecular absorption and is used for calculating x_m. The slope and intercept of the line are 13.7 and 0.98, respectively. From equation 28, the intercept is $x_m C$. $x_m C = 1/0.98 = 1.02$. The slope is $(C - 1)/x_m C$.

$$\frac{C-1}{1.02} = 13.7$$

$$C = 14.974$$

$$x_m = \frac{1.02}{14.974} = 0.068 \text{ g } H_2O/\text{g dry matter}$$

GAB (Guggenheim-Anderson-de Boer) Equation. This equation gives a better fit than the BET over a wider range of moisture contents. Let X = moisture content on a dry solids basis, kg water/kg dry matter, and X_m = moisture content on a dry basis, equivalent to a monomolecular layer of water. Then:

$$X = \frac{X_m Cka_w}{(1 - ka_w)(1 - ka_w + Cka_w)} \tag{29}$$

The GAB equation is a three-parameter equation with k, C, and X_m as the constants. C and X_m are similar in significance to the BET equation. k is a third parameter which improves the fit to a wider range of moisture content than the BET.

Evaluation of X_m using equation 29 is more precise than that using the BET equation, since the fit of equation 29 to the data extends over a wide range of moisture contents. The narrow range of moisture contents to which equation 28 fits often presents some problems in the determination of X_m using the BET equation, particularly when data points are not close enough at the lower moisture contents.

Rearranging equation 29 by taking the reciprocal:

$$\frac{1}{X} = \frac{1}{X_m}\left[\frac{1}{Cka_w} - \frac{1}{C}\right]\left[\frac{1}{Cka_w} - \frac{1}{C} + 1\right]$$

$$= \frac{1}{X_m(Ck)^2}\frac{1}{(a_w)^2} + \left[\frac{1}{CkX_m} - \frac{2}{C^2kX_m}\right]\frac{1}{a_w} - \frac{1}{CX_m}$$

Thus, a polynomial nonlinear regression of $(1/X)$ against $(1/a_w)$ will give values for the coefficient of the quadratic term, α, the coefficient of the linear term, β, and the constant, σ.

$$\beta = \frac{1}{X_m(Ck)^2}; \qquad \alpha = \frac{1}{CkX_m} - \frac{2}{C^2kX_m}; \qquad \sigma = \frac{1}{CX_m}$$

These three equations can be used to evaluate C, k, and X_m.

Other equations which have been used to fit the sorption isotherms of foods are equations 30 to 39. Iglesias and Chirife (1982) fitted equations with the corresponding values of the equation parameters. Halsey's (equation 34), Henderson's (equation 35), Oswin's (equation 38), and Smith's (equation 39) equations were reported to fit more water sorption/desorption data among these equations.

Caurie Equation

$$\ln (C) = A - a_w \ln (r) \tag{30}$$

$C = (100 - \%\ \text{water})/\%\ \text{water}$; A and r are constants.

Chen Equation:

$$a_w = \exp [-K \exp (-bX)] \tag{31}$$

k, a, and b are constants.

Chong-Pfost Equation

$$\ln (a_w) = \frac{-A}{T} \exp (-BX) \tag{32}$$

A and B are constants.

Day-Nelson Equation

$$a_w = 1 - \exp [P_1 T^{P_2} X^{P_3} T^{P_4}] \tag{33}$$

P_1, P_2, P_3, and P_4 are constants; T = absolute temperature.

Halsey Equation

$$a_w = \exp \left[\frac{-A}{T} \Theta^b \right] \tag{34}$$

$\Theta = X/X_m$; a and b are constants.

Henderson Equation

$$1 - a_w = \exp [-aX^b] \tag{35}$$

a and b are constants.

Iglesias-Chirife Equation

$$X = B_1\left[\frac{a_w}{1 - a_w}\right] + B_2 \tag{36}$$

B_1 and B_2 are constants.

Kuhn Equation

$$X = \frac{A}{\ln(a_w)} + \mathrm{B} \tag{37}$$

Oswin Equation

$$X = A\left[\frac{a_w}{1 - a_w}\right]^B \tag{38}$$

Smith Equation

$$X = A - B\ln(1 - a_w) \tag{39}$$

MASS TRANSFER

During dehydration, water is vaporized only from the surface. The transfer of water vapor from the wet surface to a stream of moving air is analogous to convection heat transfer; therefore, a mass transfer coefficient is used. Moisture flux is proportional to the driving force which is the difference in vapor pressure on the surface and the vapor pressure of water in air surrounding the surface. At the same time that water is removed from the surface, water diffuses from the interior of a solid toward the surface. The latter is a general form of diffusion which is analogous to conduction heat transfer. The differential equations for conduction also apply to diffusion, but mass diffusivity is used in place of thermal diffusivity.

Mass Diffusion. For an infinite slab, equation 89 in the section "The Infinite Slab," Chapter 7, can be used for the dimensionless moisture change with time. Expressing this equation in terms of moisture content and mass diffusivity:

$$\Theta = \frac{X - X_m}{X_0 - X_m}$$

X = moisture content at any time, dry matter basis (kg water/kg dry matter), X_0 = initial moisture content and X_m = equilibrium moisture content. The equation for moisture content at any point y in the solid measured from the center, at any time t, when the moisture content X was originally uniform at X_0 is:

$$\Theta = 2 \sum_{n=0}^{\infty} \left[\frac{(-1)^n}{(n + 0.5)\pi} [e]^{-(n+0.5)^2 \frac{\pi^2 D_m t}{L^2}} \right] \left[\cos \left(\frac{(n + 0.5)\pi y}{L} \right) \right] \quad (40)$$

Assumptions used in the derivation of this equation are constant diffusivity and constant surface moisture at the moisture content in equilibrium with the drying air during the process. The value of X_m is determined from the sorption isotherm.

During the early stages of dehydration, moisture is transferred from the center toward the surface by capillary action. This mechanism is more rapid than diffusion, and the rate of surface evaporation controls the rate of drying. However, in the later stages of drying, diffusion controls the rate of moisture migration within the solid. Diffusivity may be constant if cells do not collapse and pack together. Firm solids such as grain may exhibit constant diffusivity, but high-moisture products such as fruits and vegetables may exhibit variable diffusivity with moisture content, depending upon the physical changes which occur as water is removed.

Mass diffusivity, D_m, has the same units as thermal diffusivity and can be used directly to substitute for α in heat transfer equations. Of interest in dehydration processes is the average moisture content in the slab at any time during the drying process. Let W and Z represent the width and length of the slab, respectively. The total moisture in the slab is:

$$\text{Total moisture} = \rho WZ \int_0^L X\, dy$$

The mean moisture, $\overline{X}$ = total moisture/total mass:

$$\overline{X} = \frac{1}{L} \int_0^L X\, dy$$

Substituting the expression for X and integrating:

$$\overline{\Theta} = \frac{\overline{X} - X_m}{X_0 - X_m}$$

$$\overline{\Theta} = 2 \sum_{n=0}^{\infty} \left[\frac{(-1)^n}{(n + 0.5)^2 \pi^2} [e]^{-(n+0.5)^2 \frac{\pi^2 D_m t}{L^2}} \right] [\sin (n + 0.5)\pi] \quad (41)$$

The squared term in the denominator of equation 41 results from the integration of the cosine function of equation 40.

Table 12.2. Dimensionless mean moisture ratio as a function of $D_m t/L^2$

$\frac{D_m t}{L^2}$	$\overline{\Theta}$	$\frac{D_m t}{L^2}$	$\overline{\Theta}$	$\frac{D_m t}{L^2}$	$\overline{\Theta}$
0	1.0000	0.10	0.6432	1.10	0.0537
0.01	0.8871	0.20	0.4959	1.20	0.0411
0.02	0.8404	0.30	0.3868	1.30	0.0328
0.03	0.8045	0.40	0.3021	1.40	0.0256
0.04	0.7743	0.50	0.2360	1.50	0.0200
0.05	0.7477	0.60	0.1844	1.60	0.0156
0.06	0.7236	0.70	0.1441	1.70	0.0122
0.07	0.7014	0.80	0.1126	1.80	0.0095
0.08	0.6808	0.90	0.0879	1.90	0.0074
0.09	0.6615	1.00	0.0687	2.00	0.0058

Table 12.2 shows the values of the dimensionless mean moisture ratio $\overline{\Theta}$ as a function of $(D_m t)/L^2$, calculated using equation 41. L is the half-thickness of the slab. For three-dimensional diffusion, $\overline{\Theta}_{L1}$ is obtained for $(D_m t/L_1^2)$, $\overline{\Theta}_{L2}$ is obtained for $(D_m t/L_2^2)$, and $\overline{\Theta}_{L3}$ is obtained for $(D_m t/L_3^2)$. The composite $\overline{\Theta}$ is the product of $\overline{\Theta}_{L1}$, $\overline{\Theta}_{L2}$, and $\overline{\Theta}_{L3}$.

When $(D_m t)/L^2 > 0.1$, the first three terms in the series in equation 41 are adequate for series convergence. Taking the logarithms of both sides of equation 41:

$$\log(\overline{\Theta}) = \frac{\pi^2 D_m}{L^2} t \log(e) + B$$

$$B = \log(2) + \log A_1(e)^{0.25} + \log A_2(e)^{2.25} + \log A_3(e)^{6.25}$$

$$A_1, A_2, A_3 = \frac{(-1)^n \sin\left[(n + 0.5)\pi\right]}{(n + 0.5)^2 \pi^2} \quad \text{for } n = 0, 1, 2 \tag{42}$$

Equation 42 shows that a semilogarithmic plot of the dimensionless mean moisture ratio against the time of drying is linear if D_m is constant. The mass diffusivity can be calculated from the slope.

EXAMPLE: The desorption isotherm of apples was reported by Iglesias and Chirife (1982) to best fit Henderson's equation with the constants $a = 4.471$ and $b = 0.7131$. Experimental drying data for apple slices showed that when the average moisture content was 1.5 kg water/kg dry matter, the drying rate, which may be assumed to be diffusion controlled, was 8.33×10^{-4} kg water/(kg dry matter · s). The apple slices were 1.5 cm thick and 2.5 cm wide and were long enough so that diffusion could be considered to occur from two dimensions. The drying air has a relative humidity of 5%.

(a) Calculate the mass diffusivity of water at this stage of drying.
(b) Calculate the drying rate and moisture content 1 h into the drying process from the time when the moisture content was 1.5 kg water/kg dry matter.

Solution: Henderson's equation is used to calculate the equilibrium moisture content, X_m. Solving for X_m by substituting X_m for X in Henderson's equation:

$$\ln(1 - a_w) = -aX^b; \qquad X_m = \left[\frac{-\ln(1 - 0.05)}{4.471}\right]^{1/0.7131}$$

$$X_m = 0.0019$$

Equation 42 for diffusion from two dimensions is:

$$\log(\overline{\Theta}) = \left[\frac{1}{(L_1)^2} + \frac{1}{(L_2)^2}\right]\pi^2 D_m \log(e)t + B \tag{43}$$

Differentiating equation 43:

$$\frac{d}{dt}[\log(\overline{\Theta})] = \left[\frac{1}{(L_1)^2} + \frac{1}{(L_2)^2}\right]\pi^2 D_m \log(e) \tag{44}$$

Differentiating the expression for Θ with respect to t:

$$\frac{d}{dt}[\log(\overline{\Theta})] = \frac{1}{(\overline{X} - X_m)[\ln(10)]}\frac{d\overline{X}}{dt} \tag{45}$$

Combining equations 44 and 45 and solving for D_m:

$$D_m = \frac{d\overline{X}/dt}{(\overline{X} - X_m)\left[\frac{1}{(L_1)^2} + \frac{1}{(L_2)^2}\right]\pi^2} \tag{46}$$

Substituting known quantities:

$$D_m = \frac{8.33 \times 10^{-4}}{(1.5 - 0.0019)[(1/(0.0075)^2 + (1/(0.0125)^2]\pi^2}$$

$$= 2.3302 \times 10^{-9}\ \text{m}^2/\text{s}$$

(b) Solving for $D_m t/L^2$ for $t = 3600$ s:

$$L = 0.0075; \left(\frac{D_m t}{L^2}\right) = 0.149$$

$$L = 0.0125; \left(\frac{D_m t}{L^2}\right) = 0.054$$

From Table 12.2:
When $(D_m t/L^2) = 0.149$, by interpolation:

$$\overline{\Theta} = \frac{0.6432 - (0.6432 - 0.4959)(0.049)}{0.1} = 0.571$$

When $(D_m t/L^2) = 0.054$, by interpolation:

$$\overline{\Theta} = \frac{0.7477 - (0.7477 - 0.7236)(0.004)}{0.01} = 0.738$$

For diffusion from two directions:

$$\overline{\Theta} = 0.571(0.738) = 0.4214$$

$$\overline{X} = 0.4214(1.5 - 0.0019) + 0.0019$$

$$= 0.633 \text{ kg water/kg DM}$$

The drying rate is $d\overline{X}/dt$. Using equation 46:

$$\frac{d\overline{X}}{dt} = D_m(\overline{X} - X_m)\left[\frac{1}{L_1^2} + \frac{1}{L_2^2}\right]\pi^2$$

$$= \left[2.330 \times 10^{-9}(0.633 - 0.0019)(\pi^2)\right]\left[\frac{1}{(0.0075)^2} + \frac{1}{(0.0125)^2}\right]$$

$$= 0.000351 \text{ kg water/(s} \cdot \text{kg dry matter)}$$

Mass Transfer from a Surfaces to Flowing Air. When air flows over a wet surface, water is transferred from the surface to air. The equations governing the rate of mass transfer are similar to those for heat transfer. By analogy, the driving force for mass transfer is a concentration difference, and the proportionality constant between the mass flux and the driving force is the mass transfer coefficient.

$$\frac{dW_w}{A\,dt} = k_g M_w(a_{ws} - a_{wa}) \tag{47}$$

W_w = mass of water vapor transferred from the surface to the moving air, A = surface area exposed to air, M_w = molecular weight of water, a_{ws} = water activity on the surface, and a_{wa} = water activity of the drying air. Using the dimensionless a_w difference as the driving force results in the mass transfer coefficient, k_g, having units in a general form widely used in the literature, kg-moles/($m^2 \cdot s$).

Determination of the mass transfer coefficient is analogous to that in heat transfer, which involves the use of dimensionless groups. The equivalent of the Nusselt number in mass transfer is the Sherwood number (Sh).

$$Sh = \frac{k_g D}{D_{wm}} \tag{48}$$

where D = the diameter or characteristic length and D_{wm} = the diffusivity expressed in kg-mole/(m · s).

The equivalent of the Prandtl number in mass transfer is the Schmidt number, expressed as either mass diffusivity, D_m, m^2/s, or D_{wm}, kg-mole/(m · s). Physical property terms in the Schmidt number for dehydration processes are those for air. M_a = the molecular weight of air, 29 kg/kg-mole. D_m = diffusivity of water in air = 2.2×10^{-5} m^2/s.

$$Sc = \frac{\mu}{\rho D_m} = \frac{\mu}{M_a D_{wm}} \tag{49}$$

Correlation equations for the mass transfer coefficient are similar to those for heat transfer. Typical expressions are as follows:

Gilliland-Sherwood Equation

$$Sh = 0.023 Re^{0.81} Sc^{0.44} \tag{50}$$

Equation 50 was derived for vaporization from a water film flowing down a vertical tube to air flowing upward through the tube. The Reynolds number for flowing air ranged from 2000 to 35,000, and the pressure ranged from 0.1 to 3 atm.

Colburn j Factor

$$j = \frac{k_g}{G} Sc^{0.666} = 0.023 Re^{-0.2} \tag{51}$$

G = the molar flux of air, kg-moles/($m^2 \cdot s$) = PV/RT. P = pressure in Pa; V = velocity, m/s; R = 8315 N · m/(kgmole · K); and T = absolute temperature. Equation 51 may be used for mass transfer into air flowing through particles in a packed bed. This equation would be suitable for air flowing parallel to the surface of a bed of particles.

Ranz-Marshall Equation

$$Sh = 2 + 0.6Re^{0.5}Sc^{0.33} \tag{52}$$

Equation 52 is suitable for mass transfer from surfaces of individual particles such as those in fluidized beds.

In dehydration, mass transfer is not the rate-limiting mechanism, particularly at the high air velocities needed to maintain low humidity in the drying air.

EXAMPLE: Calculate the rate of dehydration expected from mass transfer when 1-cm carrot cubes having a density of 1020 kg/m^3 are dried in a fluidized bed with air at 2% relative humidity flowing at the rate of 12 m/s. Atmospheric pressure is 101 kPa, and air temperature is 80°C. Express the dehydration rate as kg/(s · kg dry matter) when moisture content is 5 kg water/kg dry matter.

Solution: At 80°C, the viscosity of air from the *Handbook of Chemistry and Physics* is 0.0195 cP. The characteristic length of a cube may be calculated as the diameter of a sphere having the same surface area.

$$D = L\left(\frac{6}{\pi}\right)^{0.5} = 1.382L$$

The mass flux is calculated using the ideal gas equation by substituting air velocity, $\overline{V}$, for volume: $G = P\overline{V}M_a/RT$. M_a is the molecular weight of air, 29 kg/kg-mole.

$$G = \frac{101{,}000(12)(29)}{8315(353)} = 11.975\ \frac{\text{kg}}{\text{m}^2 \cdot \text{s}}$$

The Reynolds number is $DG/\mu = 1.382LG/\mu$.

$$Re = \frac{1.382(0.01)(11.975)}{0.0195(0.001)} = 8487$$

The density, ρ, is obtained using the ideal gas equation.

$$\rho = \frac{PM_a}{RT} = \frac{101{,}000(29)}{8315(353)} = 0.998\ \frac{\text{kg}}{\text{m}^3}$$

$$Sc = \frac{\mu}{\rho D_m} = \frac{0.0195(0.001)}{0.998(2.2 \times 10^{-5})} = 0.888$$

$$Sh = 2 + 0.6(8487)^{0.5}(0.888)^{0.33} = 55.15$$

Equation 49:

$$D_{wm} = \left(\frac{\rho D_m}{M_a}\right) = \frac{0.998(2.2 \times 10^{-5})}{29}$$

$$= 7.57 \times 10^{-7}\ \text{kgmole}/(\text{m} \cdot \text{s})$$

Equation 48:

$$k_g = Sh\left(\frac{D_{wm}}{D}\right)$$

$$= \frac{55.150(7.57 \times 10^{-7})}{0.01(1.382)} = 0.00302\ \frac{\text{kgmole}}{\text{m}^2 \cdot \text{s}}$$

Express area as m^2/kg dry matter (DM).

$$A = \frac{6(0.01)^2}{(0.01)^3(1020)(1/6)} = 3.529\ \frac{\text{m}^2}{\text{kg DM}}$$

Equation 47:

$$\frac{dW}{dt} = [3.529][0.003021(18)(1 - 0.02)] = 0.188\ \frac{\text{kg}}{\text{s} \cdot \text{kg DM}}$$

The drying rate, based on surface mass transfer, is 0.188 kg water/(s · kg DM).

PSYCHROMETRY

Carrying Capacity of Cases for Vapors. The mass of any component of a gas mixture can be calculated from the partial pressure using the ideal gas equation. The ideal gas equation can be written for the whole mixture of air and water or for a single component as follows:

$$P_tV = n_tRT \quad \text{for the whole mixture}$$

or:

$$P_aV = n_aRT \quad \text{for air} \tag{53}$$

where P_a and n_a are the partial pressure and number of moles of air in the mixture, respectively. If the mixture consists only of air and water, and P_w and n_w represent the partial pressure and number of moles of water vapor, respectively:

$$P_wV = n_wRT$$

$$P_t = P_a + P_w; \qquad P_w = P_t - P_a \tag{54}$$

$$(P_t - P_a)(V) = n_wRT \tag{55}$$

Dividing equation 53 by equation 55:

$$\frac{P_a}{(P_t - P_a)} = \frac{n_a}{n_w}$$

If M_a is the molecular weight of air and M_w is the molecular weight of water, the mass ratio of the two components can be determined from the partial pressure as follows:

$$\frac{W_w}{W_a} = \frac{n_wM_w}{n_aM_a} = \frac{P_w}{(P_t - P_w)}\frac{M_w}{M_a} \tag{56}$$

The mass ratio of water to air is known as the *humidity* or *absolute humidity*. If P_w is equal to the vapor pressure of water at the given air temperature, the mass ratio of water to air is the saturation humidity. If P_w is less than the vapor pressure, the ratio P_w/P_s, where P_s is the saturation partial pressure or the vapor pressure, is the saturation ratio. This ratio, expressed as a percentage, is the *relative humidity*.

> *EXAMPLE:* Dry air is passed through a bed of solids at the rate of 1 m^3/s at 30°C and 1 atm. If the solids have an equilibrium relative humidity of 80%, and assuming that the bed is deep enough so that equilibrium is attained between the solids and the air before the air leaves the bed, determine the amount of water removed from the bed per hour. Use 29 for the average molecular weight of air. Atmospheric pressure is 101.325 kPa.

Solution: At equilibrium, the partial pressure of water in the air should be 80% of the vapor pressure of water at 30°C. From Appendix Table A.3, the vapor pressure of water at 30°C is 4.243 kPa. The partial pressure of water in the air is 80% of 4.243, or 3.394 kPa. The molecular weight of water is 18.

The absolute humidity can be calculated using equation 56:

$$\frac{W_w}{W_a} = \frac{3.394}{(101.325 - 3.394)}\frac{18}{29} = 0.0215\ \frac{\text{kg water}}{\text{kg dry air}}$$

In order to calculate the total amount of water removed in 1 h, the total amount of dry air that passed through the bed per hour must be calculated. For the dry air component:

$$PV = nRT = \frac{W_a}{M_a}RT$$

$$W_a = \frac{P_a M_a}{RT}V$$

Substituting values for P_a, M_a, R, and T and using $R = 8315$ N · m/(kg-mole · K):

$$W_a = \frac{(101{,}325\ \text{N} \cdot \text{m}^{-2})(29)(\text{kg})(\text{kg-mole})^{-1}}{8315\ \text{N} \cdot \text{m}\ (\text{kg-mole})^{-1}(\text{K})^{-1}(30 + 273)\text{K}}\left[1\ \frac{\text{m}^3}{\text{s}}\right]$$

$$= 1.166\ \text{kg/s of dry air.}$$

The weight of water removed per hour is:

$$W_w = 1.166\ \frac{\text{kg dry air}}{\text{s}}\ 0.0215\ \frac{\text{kg water}}{\text{kg dry air}}\ \frac{3600\ \text{s}}{\text{h}}$$

$$= 90.24\ \text{kg/h of water}$$

The Psychrometric Chart. Psychrometry is the study of the behavior of mixtures of air and water. In the preceding section, the determination of saturation humidities from the vapor pressure and the determination of the humidity from the partial pressure of water in air were discussed. A graph of humidity as a function of temperature at varying degrees of saturation forms the main body of a psychrometric chart. For processes that involve loss or gain or moisture by air at room temperature, the psychrometric chart is very useful for determining changes in temperature and humidity.

Another main feature of a psychrometric chart is the wet bulb temperature. When a thermometer is fitted with a wet sock at the bulb and placed in a stream of air, evaporation of water from the sock cools the bulb to a temperature lower than that which would register if the bulb were dry. The difference is known as the *wet bulb depression* and is a function of the relative humidity of the air. More humid air allows less vaporization, resulting in a lower wet bulb depression.

The various quantities that can be determined from a psychrometric chart are as follows:

Humidity (absolute humidity, H), the mass ratio of water to dry air in the mixture.

Relative humidity (% RH), the ratio of the partial pressure of water in the air to the vapor pressure of water, expressed as a percentage.

Dry bulb temperature (T_{db}), air temperature measured using a dry temperature sensing element.

Wet bulb temperature (T_{wb}), air temperature measured using a wet sensing element that allows cooling by evaporation of water.

Dew point (T_{dp}), temperature to which a given air-water mixture needs to be cooled to start condensation of water. At the dew point, the air is saturated with water vapor. The dew point is also that temperature at which the vapor pressure of water equals the partial pressure of water in the air.

A psychrometric chart has for its axes temperature on the abscissa and humidity on the ordinate. Two parameters are necessary to establish a point on the chart that represents the condition of air. These parameters could be any two of the following: relative humidity, dry bulb temperature, wet bulb temperature, and dew point or absolute humidity. Dew point and absolute humidity are not independent; knowing one establishes the other.

Psychrometric charts in metric and English units are presented in the Appendix (Figs. A6 and A7). Figure 12.3 is a diagrammatic representation of how quantities are read from a psychrometric chart. Figure 12.3A shows how dew point is determined from the humidity. Figure 12.3B shows how the wet bulb temperature is determined from the humidity and the dry bulb temperature. Figure 12.3C shows how the percentage of relative humidity is determined knowing the dry bulb and the humidity, and how dry bulb temperature is determined from humidity and percent relative humidity. Figure 12.3D shows how humidity and dew point are determined from the wet and dry bulb temperatures.

EXAMPLE 1: Air has a dew point of 40°C and a relative humidity of 50%. Determine (a) the absolute humidity, (b) the dry bulb temperature, and (c) the wet bulb temperature.

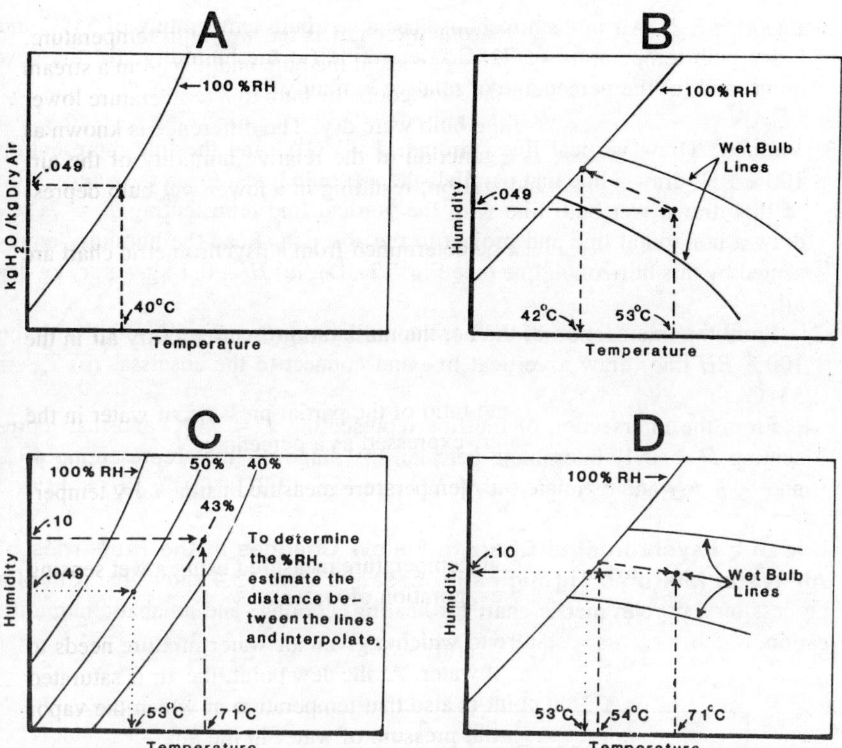

Fig. 12.3. Diagram showing the use of a psychrometric chart for determining properties of air-water mixtures. (Courtesy of Proctor and Schwartz, Inc. Copyrighted by Proctor and Schwartz, Inc.)

Solutions: Using a psychrometric chart, draw a vertical line through T = 40°. At the intersection with the line representing 100% relative humidity (*% RH*), draw a horizontal line and read the absolute humidity represented by this horizontal line at the abscissa (see Fig. 12.3A). (a) H = 0.049 kg water/kg dry air.

Extend the horizontal line representing the humidity until it intersects the diagonal line representing 50% RH. Draw a vertical line through this intersection and read the dry bulb temperature representing the vertical line at the abscissa (see Fig. 12.3C). (b) Dry bulb temperature = 53°C.

From the point represented by H = 0.049 and T_{tb} = 53°C, draw a line that parallels the wet bulb lines. Project this line to its intersection with the 100% RH line. Draw a vertical line at this intersection, project to the abscissa, and read the wet bulb temperature at the abscissa (see Fig. 12.3B). (c) T_{db} = 42°C.

EXAMPLE 2: Air in a smokehouse has a wet bulb temperature of 54°C and a dry bulb temperature of 71°C. Determine (a) the humidity, (b) the dew point, and (c) the percentage of relative humidity.

Solution: Draw vertical lines through T = 54°C and the line representing 100% RH; draw a line that parallels the wet bulb lines. From the intersection of this drawn wet bulb line with the vertical line representing T = 71°C, draw a horizontal line and project to the abscissa. Read the humidity represented by this horizontal line (see Fig. 12.3D). (a) H = 0.1 kg water/kg dry air.

From the intersection of the horizontal line representing H = 0.1 with the 100% *RH* line, draw a vertical line and connect to the abscissa. (b) T_{dp} = 53°C.

From the intersection of the line representing T = 71°C and that representing H = 0.1, interpolate between the diagonal lines representing 40% and 50% *RH* and estimate the % *RH* (see Fig. 12.3C). (c) % *RH* = 43%.

Use of a Psychrometric Chart to Follow Changes in the Properties of Air-Water Mixtures Through a Process. Figure 12.4 shows the path of a process on a psychrometric chart for heating, cooling, and adiabatic humidification.

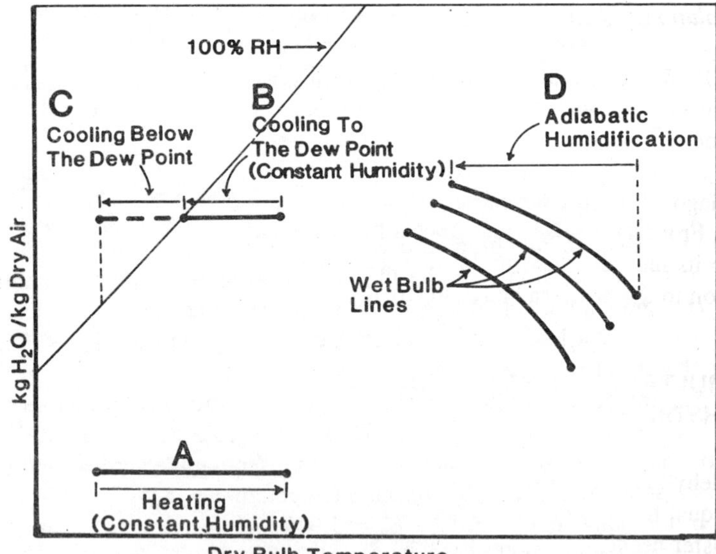

Fig. 12.4. Diagram showing paths of cooling, heating, condensation, and adiabatic humidification on a psychrometric chart.

When the temperature of air is increased at constant pressure and no water is added to or removed from the air, the process is a constant humidity process. T_{db} and T_{wb} increases but % *RH* decreases. The process is represented by *A* in Fig. 12.4.

When the temperature of air is decreased above the dew point, the process is a constant humidity process represented by B in Fig. 12.4. Both T_{db} and T_{wb} decrease and % *RH* increases. Cooling below the dew point results in condensation of water and humidity drops. The % *RH* remains at 100%, and the temperature and humidity drop following the line representing 100% *RH*. This is shown by C in Fig. 12.4.

Adiabatic humidification is a process whereby water is picked up by air and the heat required to vaporize the added water comes from the sensible heat loss that results from a reduction of the air temperature. The path traced by the temperature and humidity of the air is represented by D in Fig. 12.4, and this path parallels the wet bulb lines. The wet bulb temperature of the air remains constant, humidity and % *RH* increase, and T_{db} decreases. This is the process that occurs when air is passed over or through a bed of wet solids in drying or when air is passed through water sprays in a water cooling tower.

EXAMPLE: Ambient air at 25°C and 50% RH is heated to 175°C. Determine (a) the % *RH* and (b) the wet bulb temperature of the heated air.

Solution: Use the psychrometric chart and draw a vertical line representing T_{db} = 25°C. Project this line until it intersects the diagonal line representing 50% *RH*. Draw a horizontal line and project to the ordinate to determine the absolute humidity. H = 0.0098 kg water/kg dry air. Project the horizontal line representing H = 0.0098 until it intersects the vertical line representing T_{db} = 175°C. At the intersection, the % *RH* can be interpolated between the diagonal lines representing 1.5% *RH* and 2% *RH*. (a) % *RH* = 1.8%.

From the intersection, draw a line parallel to the wet bulb line and project to its intersection with the line representing 100% *RH*. Project the intersection to the abscissa and read (b) T_{wb} = 45°C.

SIMULTANEOUS HEAT AND MASS TRANSFER IN DEHYDRATION

In dehydration, moisture is removed by evaporation. Heat must be transferred to equal the heat of vaporization. If the rate of mass transfer exceeds the heat transfer needed to supply the heat of vaporization, the temperature of the material will drop. This process is called *evaporative cooling*. The mass transfer equation (equation 47) can be written for a material undergoing dehydration in the form:

$$\frac{dW_w}{Adt} = k_{gw}(H_i - H) \tag{57}$$

dW_w/dt = the mass of water transferred, H_i = the humidity at the interface where water is vaporized, the H = the humidity of the drying air. H_i is the equilibrium humidity at the interface. When the temperature at the interface drops, H_i also drops and the rate of mass transfer slows down according to equation 57.

The rate of heating needed to vaporize, dW_w/dt, is:

$$q = \frac{dW_w}{dt} h_{fg}$$

where h_{fg} = the enthalpy of vaporization at the temperature of the interface. Heat transfer needed for vaporization is:

$$q = k_{fg}(A)(H_i - H)(h_{fg}) \tag{58}$$

At equilibrium, the heat for vaporizing the mass transferred (equation 58) equals the rate of heat transfer (equation 59). The temperature and humidity at the interface are T_s and H_s, respectively.

$$q = hA(T - T_s) \tag{59}$$

Equating equations 58 and 59:

$$(h_{fg})k_{gw}(A)(H_s - H) = hA(T - T_s)$$

$$H = H_s - \frac{hT_s}{k_{gw}(h_{fg})} + \frac{h}{k_{gw}(h_{fg})} T \tag{60}$$

Equation 60 is the equation of a wet bulb line, the relationship between equilibrium temperature and humidity when a thermometer bulb is wrapped with a wet sock and exposed to a flowing stream of air. The line will go through the point T_s and H_s, the wet bulb temperature and the saturation humidity at the wet bulb temperature, respectively.

Food products during drying follow the relationship between heat and mass transfer as expressed in equations 58, 59, and 60.

If the temperature of a product during dehydration is higher than the wet bulb temperature, then the rate of heat transfer (equation 59) exceeds evaporative cooling by mass transfer (equation 58) and mass transfer controls the drying rate. This phenomenon is often observed in the later stages of drying when the interface from vaporization is removed from the surface requiring vapor to flow through the pores of the dried material close to the surface before mixing with the drying air at the surface.

The air temperature drops as it undergoes adiabatic humidification. Vaporization of water to increase the humidity requires energy, which comes from a drop in the sensible heat of air. The heat balance is as follows.

Heat of vaporization, q_v, to increase humidity from H to H_s is:

$$q_v = (H_s - H)(h_{fg}) \tag{61}$$

The loss in sensible heat, q_s, for air with a specific heat, C_p, is:

$$q_s = C_p(T - T_s) \tag{62}$$

Equating equations 61 and 62:

$$C_p(T - T_s) = H_s - H(h_{fg})$$

$$H = H_s - \frac{C_p}{h_{fg}} T_e + \frac{C_p}{h_{fg}}(T) \tag{63}$$

Equation 63 is the adiabatic humidification line, the change in the humidity of air as it drops in temperature during adiabatic humidification. Equation 63 will be exactly equal to equation 60 if the ratio of the heat transfer coefficient to the mass transfer coefficient, h/k_{gw}, equals the specific heat of air, C_p. Indeed, this has been shown to be true for water vaporizing into air at 1 atm and at moderate temperatures. Thus, in air drying, wet bulb lines on the psychrometric chart coincide exactly with the adiabatic humidification lines.

EXAMPLE: Room air at 80°F (26.7°C) and 50% RH is heated to 392°F (200°C) and introduced into a spray drier, from which it leaves at a temperature of 203°F (95°C). Determine the humidity and relative humidity of the air leaving the drier. Assume adiabatic humidification in the drier.

Solution: Using a psychrometric chart, locate the point that represents T = 80°F and 50% RH. The humidity at this point is 0.011. When air is heated, the temperature increases at constant humidity. Air leaving the heater will have a humidity of 0.011 and a temperature of 392°F (200°C). Starting from T = 392°F (200°C) and H = 0.011, draw a curve that approximates the closest wet bulb curve until a temperature of 203°F (95°C) is reached. The humidity at this point is 0.055 and the relative humidity is 10%.

THE STAGES OF DRYING

Drying usually occurs in a number of stages, characterized by different dehydration rates in each of the stages. Figure 12.5 shows the desorption isotherm and the rate of drying of apple slices in a cabinet dryer with air flowing across the trays containing the slices at 0.3 m/s, 150°F (65.6°C), and 20% RH. The drying rate curve shows a constant drying rate (line AB) from a starting moisture content of 8.7 g H_2O/g dry matter to a moisture content of 6.00 kg H_2O/kg dry matter. At this stage in the drying cycle, vaporization is occurring at the product surface and free water with an a_w of 1 is always available at the surface to vaporize. This stage of drying is the constant rate stage. The rate of drying is limited by the rate at which heat is transferred to the material from air. The product temperature is usually at the wet bulb temperature of the drying air.

With a moisture content of 6.00 g H_2O/g dry matter, the drying rate decreased linearly with the moisture content. Point B is the critical moisture content, and line BC represents the first falling rate period. The first falling rate period is characterized by a slight increase in the temperature of the product, although this temperature may not be very much higher than the wet bulb temperature. Free moisture is no longer available at the surface, and the rate of drying is controlled by moisture diffusion toward the surface. Most of the water in the material is still free water with an a_w of 1. However, diffusion toward the surface is necessary for vaporization to occur.

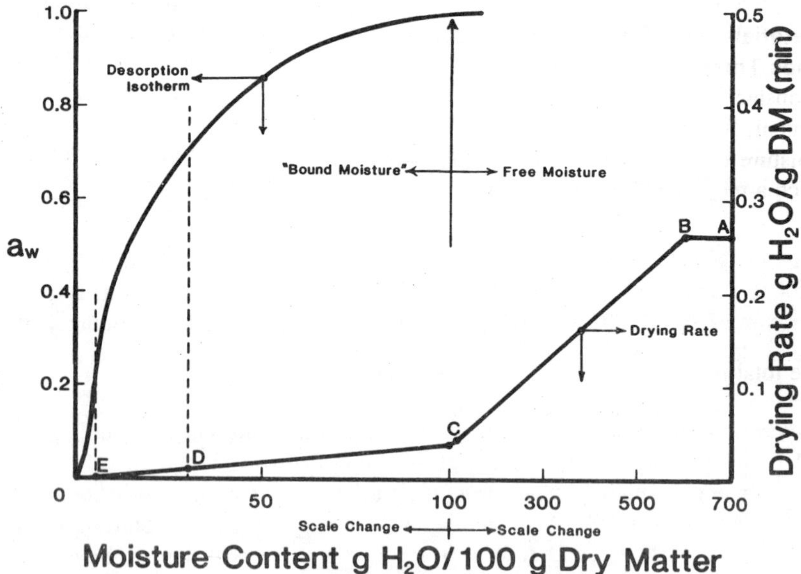

Fig. 12.5. Desorption isotherm of raw apple slices and drying rates at different moisture contents.

With a moisture content of 1.20 g H_2O/g dry matter, the slope of the drying rate against moisture content changes and the rate of drying goes through a second falling rate period. The second falling rate period starts at point C, where the equilibrium relative humidity for the material begins to drop below 100%. The moisture content at point C is the bound water capacity of the material. During the second falling rate period, dehydration proceeds through the portion of the sorption isotherm where water in the material is held by multimolecular adsorption and capillary condensation. The heat of vaporization of water at this stage of the dehydration process is higher than the heat of vaporization of pure water, since the heats of adsorption and vaporization must be provided. Vaporization at this stage occurs in the interior rather than at the surface, and water vapor has to diffuse to the surface before it mixes with the flowing stream of air.

At the moisture content of 0.30 g H_2O/g dry air, the drying rate goes through another falling rate zone. This stage of drying corresponds to the region of the sorption isotherm where water is held in mono- or multimolecular layers. Dehydration should be terminated at any point along the line DE. Point E represents the equilibrium moisture content where dehydration stops because the a_w values of the surface and air are equal.

PREDICTION OF DRYING TIMES FROM DRYING RATE DATA

Materials with One Falling Rate Stage Where the Rate of Drying Curve Goes Through the Origin. The drying rate curve for these materials shows a constant rate, R_c, from the initial moisture content, X_0, to the critical moisture content, X_c. The drying rate then falls in a linear relationship with decreasing moisture content until it becomes 0 at $X = 0$. The drying times required to reach a moisture content X in either of the stages of drying are:

$$\text{Constant rate:} \quad -\frac{dX}{dt} = R_c \tag{64}$$

The total time, t_c, for the constant rate stage is:

$$t_c = \frac{X_0 - X_c}{R_c} \tag{65}$$

$$\text{At the falling rate period:} \quad -\frac{dX}{dt} = \frac{R_c}{X_c}(X) \tag{66}$$

$$\int_{t_c}^{t} dt = \frac{X_c}{R_c} \int_{X_c}^{X} \frac{dX}{X}$$

$$t - t_c = \frac{X_c}{R_c} \ln \frac{X_c}{X} \tag{67}$$

The total time from X_0 to X in the falling rate stage can be calculated by substituting t_c in equation 65 into equation 67.

$$t = \frac{X_0 - X_c}{R_c} + \frac{X_c}{R_c} \ln \frac{X_c}{X} \tag{68}$$

Equation 68 shows that if a material exhibits only one falling rate stage of drying and the drying rate is 0 only at $X = 0$, the drying time required to reach a desired moisture content can be determined from the constant rate, R_c, and the critical moisture content, X_c. For these materials, X_c is usually the moisture content when a_w starts to drop below 1 in a desorption isotherm.

EXAMPLE: A material shows a constant drying rate of 0.15 kg H_2O/(min · kg dry matter) and has an a_w of 1 at moisture contents above 1.10 kg H_2O/kg dry matter. How long will it take to dry this material from an initial moisture content of 75% (wet basis) to a final moisture content of 8% (wet basis)?

Solution: Converting the final and initial moisture contents from a wet to a dry basis:

$$X_0 = \frac{0.75 \text{ kg water}}{0.25 \text{ kg dry matter}} = 3.0 \text{ kg water/kg dry matter}$$

$$X = \frac{0.08 \text{ kg water}}{0.92 \text{ kg dry matter}} = 0.0869 \text{ kg water/kg dry matter}$$

$$X_c = 1.10 \text{ kg water/kg dry matter}$$

$$R_c = 0.15 \text{ kg water/(min} \cdot \text{kg dry matter)}$$

Using equation 68:

$$t = \frac{3.0 - 1.10}{0.15} + \frac{1.10}{0.15} \ln \frac{1.10}{0.0869}$$

$$= 12.7 + 18.6 = 31.3 \text{ min}$$

Materials with More Than One Falling Rate Stage. Most food solids exhibit this drying behavior. The drying rate curve shown in Fig. 12.5 for apple slices is a typical example. The drying time in the constant rate follows equation 65. However, since the rate of drying against the moisture content plot no longer goes to the origin from the point (X_c, R_c), equation 66 cannot be used for the falling rate stage. If the rate vs. moisture content line is extended to the abscissa, the moisture content where the rate is 0 may be designated as the residual moisture content, X_r, and for the first falling rate period:

$$\frac{d(X - X_{r1})}{dt} = \frac{R_c}{X_{c1} - X_{r1}}(X - X_{r1}) \tag{69}$$

Integrating equation 69 and using equation 65 for drying time in the constant rate zone, drying to a moisture content X in the first falling rate stage takes:

$$t = \frac{X_0 - X_{c1}}{R_c} + \frac{X_{c1} - X_{r1}}{R_c} \ln \frac{X_{c1} - X_{r1}}{X - X_{r1}} \tag{70}$$

where X_{c1} and X_{r1} represent the critical moisture content and the residual moisture content for the first falling rate stage of drying, respectively. The time required to dry to moisture content X in the second falling rate stage is:

$$t = \frac{X_0 - X_{c1}}{R_c} + \frac{X_{c1} - X_{r1}}{R_c} \ln \left[\frac{X_{c1} - X_{r1}}{X_{c2} - X_{r1}}\right] + \left[\frac{X_{c1} - X_{r1}}{R_c}\right]\left[\frac{X_{c2} - X_{r2}}{R_c}\right] \ln \left[\frac{X_{c2} - X_{r2}}{X - X_{r2}}\right] \tag{71}$$

EXAMPLE: Figure 12.6 shows the drying curve for apple slices blanched in 10% sucrose solution and dried in a cabinet drier using air in parallel flow at a velocity of 3.65 m/s; T_{db} = 170°F (76.7°C) and T_{wb} = 100°F (37.8°C) for the first 40 min, and T_{db} = 160°F (71.1°C) and T_{wb} = 110°F (43.3°C) for the rest of the drying period. Calculate the drying time needed to reach a moisture content of 0.15 kg water/kg dry matter.

Solution: The drying rate curve was obtained by drawing tangents to the drying curve at the designated moisture contents and determining the slopes of the tangents. The drying time required to obtain a moisture content of 0.15 kg water/kg dry matter is:

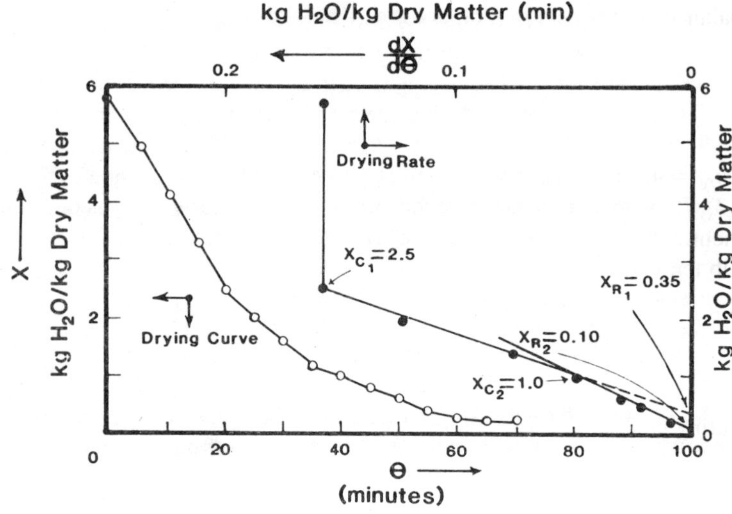

Fig. 12.6. Drying curve and drying rate as a function of the moisture content of blanched apple slices, showing several breaks in the drying rate.

$$t = \frac{5.3 - 2.5}{0.163} + \frac{2.5 - 0.35}{0.163} \ln\left[\frac{2.5 + 0.35}{1.0 - 0.35}\right]$$

$$+ \left[\frac{2.5 - 0.35}{0.163}\right]\left[\frac{1 - 0.1}{0.163}\right] \ln\left[\frac{1 - 0.1}{0.15 - 0.35}\right]$$

$$= 20.2 + 15.8 + 52.7 = 88.7 \text{ min}$$

The drying curve in Fig. 12.6 shows a drying time of 90 min at $X = 0.15$.

The Constant Drying Rate. The constant drying rate, R_c, is controlled by heat transfer and can be calculated using a heat balance. Let ρ_s = the dry solids density, kg dry solids/m^3 of wet material. ρ_s = wet solids density × mass fraction dry solids in the wet material.

If L is the depth of the innermost section of the material from the drying surface (if drying occurs from both sides, L will be half the total thickness of the solid), ρ_s is the dry matter density of the material (kg dry matter/m^3 of material), M_s is the mass of dry solids, and A is the area of the top surface of the solid, the volume V of material will be:

$$V = \text{surface area (depth)} = A(L) = \frac{M_s}{\rho_s}$$

$$\frac{A}{M_s} = \frac{1}{L(\rho_s)} \tag{72}$$

Heat balance: latent heat of evaporation = heat transferred.

$$\frac{dX}{dt} M_s h_{fg} = hA(T_a - T_s) \tag{73}$$

where h_{fg} = latent heat of vaporization at the surface temperature of the material, T_s, T_a = the dry bulb temperature of the air, and h is the heat transfer coefficient. The surface temperature during the constant rate period is also the wet bulb temperature (T_{wb}) of the air. $T_s = T_{wb} \cdot dX/dt = R_c$.

$$R_c = \frac{h(T_a - T_s)}{h_{fg}} \cdot \frac{A}{M_s} \tag{74}$$

Substituting equation 72 in equation 74:

$$R_c = \frac{h(T_a - T_s)}{h_{fg} L \rho_s} \tag{75}$$

Equation 75 can be used to calculate the constant rate of drying from the heat transfer coefficient and the wet and dry bulb temperatures of the drying air for a bed of particles with drying air flowing parallel to the surface.

Expressions similar to equation 75 may be derived using the same procedure as above for cubes with sides L evaporating water at all sides:

$$R_c = \frac{6h(T_a - T_s)}{h_{fg} L \rho_s} \tag{75a}$$

For a brick-shaped solid with sides a and $2a$ and thickness L:

$$R_c = \frac{h(T_a - T_s)}{h_{fg} \rho_s .}\left[\frac{3}{a} + \frac{2}{L}\right] \tag{75b}$$

The heat transfer coefficient can be calculated using the following correlation equations (Sherwood, *Ind. Eng. Chem.* 21:976, 1029):

If air flow is parallel to the surface:

$$H = 0.0128G^{0.8} \tag{76}$$

where h = heat transfer coefficient in BTU/(h · ft^2 · °F) and G = mass rate of flow of air, lb$_m$/(h · ft^2). In SI units, equation 76 is:

$$h = 14.305G^{0.8} \tag{77}$$

where h is in W/(m^2 · K) and G is in kg/(m^2 · s). If flow is perpendicular to the surface:

$$h = 0.37G^{0.37} \tag{78}$$

where h is in BTU/(h · ft^2 · °F) and G is in lb$_m$/(ft^2 · h). In SI units, equation 78 is:

$$h = 413.5G^{0.37} \tag{79}$$

where h is in W/(m^2 · K) and G is in kg/(m^2 · s).

When air flows through the bed of solids, the Ranz-Marshall equation (*Chem. Eng. Prog.* 48 (3):141, 1956; Appendix Table A.12 for particles in a gas stream) may be used for determining the heat transfer coefficient.

EXAMPLE: Calculate the constant drying rate for blanched apple slices dried with air flowing parallel to the surface at 3.65 m/s. The initial moisture content was 85.4% (wet basis), and the slices were in a layer 0.5 in. (0.0127 m) thick. The wet blanched apples had a bulk density of approximately 35 lb/ft^3 (560 kg · m^3) at a moisture content of 87% (wet basis). Dehydration proceeds from the top and bottom surfaces of the tray. Air is at T_{db} = 76.7°C (170°F) and T_{wb} = 37.8°C (100°F)

$$\rho_s = \frac{560 \text{ kg}}{\text{m}^3}\,\frac{0.13 \text{ kg DM}}{\text{kg}} = 72.8\frac{\text{kg DM}}{\text{m}^3} \text{ or } 4.55\,\frac{\text{lb DM}}{\text{ft}^3}$$

$$V = 3.65 \text{ m/s or } 12.0 \text{ ft/s}$$

Solution: Using the ideal gas equation: R = 8315 N m/kgmole · K; T = 76.7°C; P = 1 atm = 101.3 kPa; and M = 29 kg/kgmole.

$$\frac{\text{kg air}}{\text{m}^3} = \frac{p(M)}{R(T)} = \frac{(101{,}300)(29)}{8315(349.7)} = 1.01\,\frac{\text{kg}}{\text{m}^3} \text{ or } 0.063\,\frac{\text{lb}}{\text{ft}^3}$$

$$G = \frac{\text{kg air}}{\text{m}^3} \times \text{velocity} = 1.01\,\frac{\text{kg}}{\text{m}^3}\,3.65\,\frac{\text{m}}{\text{s}}$$

$$= 3.687 \text{ kg/(m}^2 \cdot \text{s) or } 2713 \text{ lb/(ft}^2 \cdot \text{h)}$$

Using equation 77:

$$h = 14.305(3.687)^{0.8}$$

$$= 40.6 \text{ W/(m}^2 \cdot \text{K) or } 7.15 \text{ BTU/(h} \cdot \text{ft}^2 \cdot {}^\circ\text{F)}$$

$$T_a - T_s = 76.7 - 37.8 = 38.9°C \text{ or } 70°F$$

$$h_{fg} = \text{heat of vaporization at } 37.8°C\ (100°F)$$

$$= 1037.1 \text{ BTU/lb or } 2.4123 \text{ MJ/kg}$$

Since drying occurs on the top and bottom surfaces, $L = 0.0127/2 = 0.00635$. Using equation 75:

$$R_c = \frac{40.6(38.9)}{2.4123 \times 10^6(0.00635)(72.8)} = 0.001416 \frac{\text{kg water}}{\text{s} \cdot \text{kg DM}}$$

$$= 5.098 \text{ kg water}/(\text{h} \cdot \text{kg DM})$$

SPRAY DRYING

Spray drying is a process whereby a liquid droplet is rapidly dried as it comes in contact with a stream of hot air. Figure 12.7 is a schematic diagram of a spray drier in which the atomized feed travels concurrently with the drying air. The small size of the liquid droplets allows very rapid drying, and the residence time of the material inside the spray drier is on the order of seconds. The dried

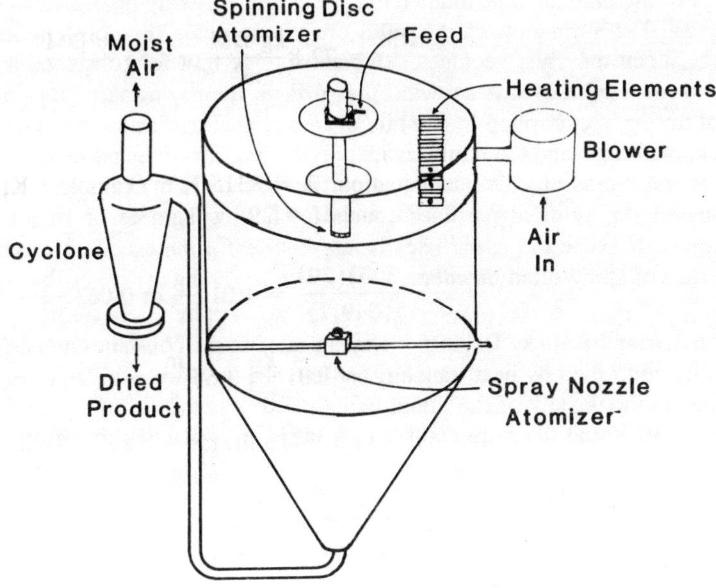

Fig. 12.7. Schematic diagram of a spray drier.

material, in powder form, is separated from the air in a cyclone separator. The powder is continuously withdrawn and cooled. Heat may damage the product if contact with the high-temperature drying air is prolonged.

While the droplets are drying, the temperature remains at the wet bulb temperature of the drying air. For this reason drying air at very high temperatures can be tolerated in a drier with a minimum of damage to the heat-sensitive components. Furthermore, the rate of degradative reactions in foods slows down at low moisture contents. Thus, the portion of the drying process in which product temperatures rise above the wet bulb temperature does not result in severe heat damage to the product.

A major requirement of successful spray drying is the reduction of the moisture content of the liquid droplets to a level that prevents the particles from sticking to a solid surface as the particles impinge on that surface. The rate of drying of the particles must be such that from the time the particles leave the atomizer to the time they impinge on the walls of the spray drier, the particles are dry. The trajectory and velocity of the particles determine the available drying time. The rate of drying and the time required to dry depend on the temperature of the drying air, the heat transfer coefficient, and the diameter of the droplets being dried.

A constant rate and a falling rate drying stage are also manifested in a spray drying process. As the wet droplets leave the atomizer, the surface rapidly loses water. Solidified solute and suspended solids rapidly form a solid crust on the surface of the particle. The diameter of the particle usually decreases as drying proceeds. The formation of the solid crust constitutes the constant stage of drying. When the crust becomes sufficiently thick to offer considerable resistance to movement of water toward the surface, the drying rate drops and the rate of drying is controlled by the rate of mass transfer. The temperature of the particle increases, and the liquid trapped in the interior of the particle vaporizes and generates pressure. Eventually, a portion of the crust breaks and the vapor is released. Spray-dried particles consist of hollow spheres or fragments of spheres. This shape of the particles is responsible for the excellent rehydration properties of spray-dried powders.

Drying Times in Spray Drying. Drying rates in the constant rate period are generally controlled by heat transfer, and the heat balance is given in equation 73. If ρ_L is the density of the liquid being dried, r is the radius of the droplet, and X_0 is the initial moisture content (dry basis, kg water/kg dry matter):

$$M_s = \frac{4\pi r^3(\rho_L)}{3(1 + X_0)}$$

and equation 73 becomes:

$$\frac{dX}{dt} \cdot \frac{4\pi r^3(\rho_L)(h_{fg})}{3(1 + X_0)} = h(\pi r^2)(T_a - T_s)$$

$$\frac{dX}{dt} = \frac{3(1 + X_0)(h)(T_a - T_s)}{4r\rho_L h_{fg}} \tag{80}$$

The constant drying rate in equation 75 can be determined if the heat transfer coefficient, h, and the radius of the liquid droplets are known. T_s during the constant drying stage is the wet bulb temperature of the drying air. Integrating equation 80, the constant rate drying time, t_c, is:

$$t_c = \frac{4(X_0 - X_c)(r)(\rho_L)h_{fg}}{3(1 + X_0)(h)(T_a - T_s)} \tag{81}$$

For water vaporizing from a very small spherical particle in slow-moving air, the following relationship has been derived for the limiting case of very small Reynolds numbers in Froessling's boundary layer equations for a blunt-nosed solid of revolution:

$$\frac{hr}{k_f} = 1.0; \quad h = \frac{k_f}{r} \tag{82}$$

k_f in equation 82 is the thermal conductivity of the film envelope around the particle. In spray drying, k_f may be assumed to be the thermal conductivity of saturated air at the wet bulb temperature. Substituting equation 82 in equation 81:

$$t_c = \frac{4(X_0 - X_c)(r^2)(\rho_L)h_{fg}}{3k_f(1 + X_0)(T_a - T_s)} \tag{83}$$

t_c represents the critical time in spray drying which must be allowed in a particle's trajectory before it impinges on a solid surface in the drier. The drying time in the falling rate period derived by Ranz and Marshal is:

$$t_f = \frac{h_{fg}(\rho_s)(r_c^2)(X_c - X)}{3k_f(\overline{\Delta T})} \tag{84}$$

where r_c is the radius of the dried particle and $\overline{\Delta T}$ is the mean temperature between the drying air and the surface of the particle during the falling rate period. $\overline{\Delta T}$ may be considered as log mean between the wet bulb depression and the difference between the exit air and product temperatures. ρ_s is the dry solids density, weight solids/volume dry matter. For liquids that contain a high concentration of suspended solids or crystallizable solutes, there is very little change in the droplet diameter during spray drying. Both r and r_c in equations 83 and 84, therefore, may be approximated to be the diameter of the liquid droplet leaving the atomizer.

For centrifugal atomizers, the diameter of the droplets as a function of the peripheral speed of the atomizer is shown in Fig. 12.8. For pneumatic atomizers, a graph of drop diameter as a function of atomizing air pressure at different liquid flow rates is shown in Fig. 12.9.

EXAMPLE: Calculate the drying time for a liquid atomized in a centrifugal atomizer at a feed rate of 15 lb/min (6.8 kg/min) at a peripheral speed of 200 ft/s. Base the drying time on a particle size representing a diameter larger than that of 90% of the total droplets produced. Assume that there is no change in droplet diameter with drying. The liquid originally has a density

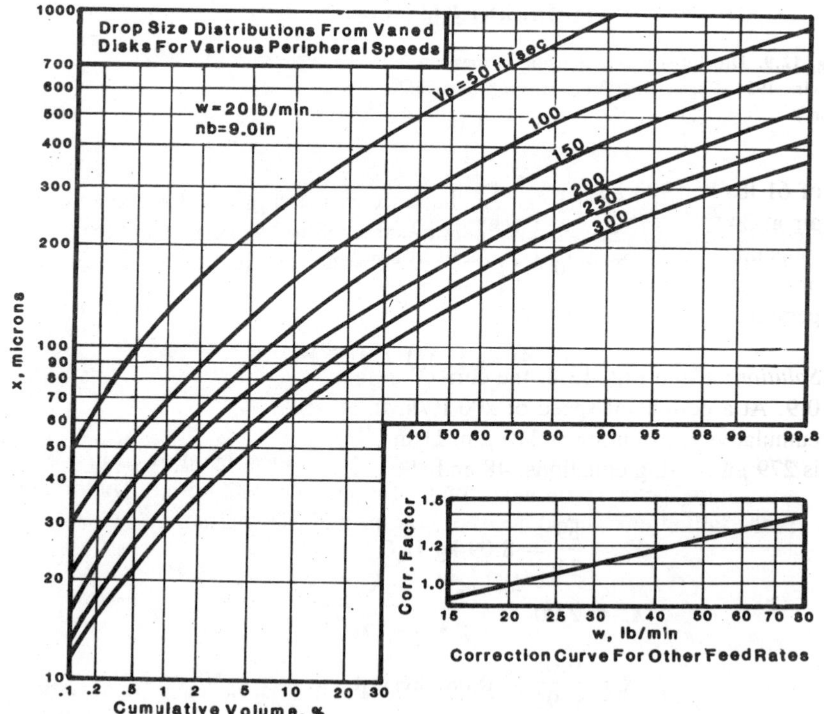

Fig. 12.8. Droplet size as a function of peripheral speed of a centrifugal atomizer. (From Marshall, W. R., Jr., 1954. *Chem. Eng. Prog. Monogr. Ser.* 50(2):71. AIChE, New York. Used with permission.)

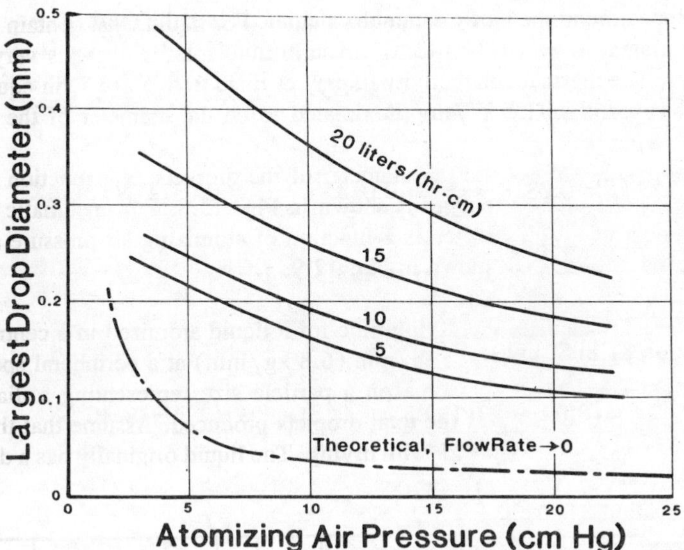

Fig. 12.9. Droplet size as a function of pressure in a pneumatic atomizer. (From Marshall, W. R., Jr., 1954. *Chem. Eng. Prog. Monogr. Ser.* 50(20):79. AIChE, New York. Used with permission.)

of 61 lb/ft^3 (993 kg/m^3) and a moisture content of 8% (wet basis), using air at 347°F (175°C) and a humidity of 0.001 H_2O/dry air. The critical moisture content is 2.00 g H_2O/g dry matter. The dried solids have a density of 0.3 g/cm^3. Exit air temperature is 220°F (104.4°C). Product exit temperature is 130°F (54.4°C).

Solution: From Fig. 12.8, the correction factor for a 15 lb/min feed rate is 0.9. At a peripheral speed of 200 ft/s, the diameter corresponding to 90% cumulative distribution is 310 μm. Using the correction factor, the diameter is 279 μm. Using equations. 48 and 49:

$$X_0 = \frac{89}{11} = 8.09 \text{ kg } H_2O/\text{kg dry matter}$$

$$X_c = 2.00$$

$$X = \frac{0.08}{0.92} = 0.087 \text{ kg } H_2O/\text{kg dry matter}$$

From a psychrometric chart, $T_s = T_{wb} = 109°F$ (43°C). h_{fg} at 109°F is 1031.4 BTU/lb or 2.3999 MJ/kg. The thermal conductivity of the gas film envelope around the particle can be calculated. It is the thermal conductivity of saturated air at 109°F (43°C). The humidity is 0.056 kg H_2O/kg dry matter. The thermal conductivity of dry air is 0.0318 W/m · K, and that of water vapor is 0.0235 W/m · K.

$$k_f = 0.0318\frac{1}{1.056} + 0.0235\frac{0.056}{1.056} = 0.0314 \text{ W/m} \cdot \text{K}$$

$$r = \frac{279}{2} \times 10^{-6} \text{ m} = 193.5 \times 10^{-6} \text{ m}$$

Equation 81 in SI units:

$$T_c = \frac{4(8.09 - 2)(139.5 \times 10^{-6})^2(993)(2.399 \times 10^6)}{3(0.0314)(1 + 8.09)(175 - 43)} = 10 \text{ s}$$

For equation 84, the $\overline{\Delta T}$ is:

$$\overline{\Delta T} = \frac{(175 - 43) - (104.4 - 54.4)}{\ln 132/50} = 84.5 \text{ K}$$

$$t_f = \frac{2.399 \times 10^6(0.3 \times 1000)(193.5 \times 10^{-6})^2(2.0 - 0.087)}{3(0.0314)(84.5)}$$

$$= 6.5 \text{ s}$$

FREEZE DRYING

Dehydration carried out at low absolute pressures allows the vaporization of water from the solid phase. Figure 12.10 shows the vapor pressure of water over ice at various temperatures below the freezing point of water. To carry out freeze drying successfully, the absolute pressure in the drying chamber must be maintained at a minimum of 620 Pa.

Figure 12.11 is a schematic diagram of a freeze drier. The absolute pressure inside the drying chamber is determined by the temperature at which the vapor trap is maintained. This pressure corresponds to the vapor pressure over ice at the vapor trap temperature. The vacuum pump is designed primarily to exhaust the vacuum chamber at the start of the operation and to remove noncondensing gases and whatever air leaked into the system. The volume of vaporized water at the low absolute pressure in freeze drying is very large; therefore, removal of the vapor by the vacuum pump alone requires a very large pump. Condensing the vaporized water to ice in the vapor trap is an efficient means of reducing the volume of gases to be removed from the system by the vacuum pump.

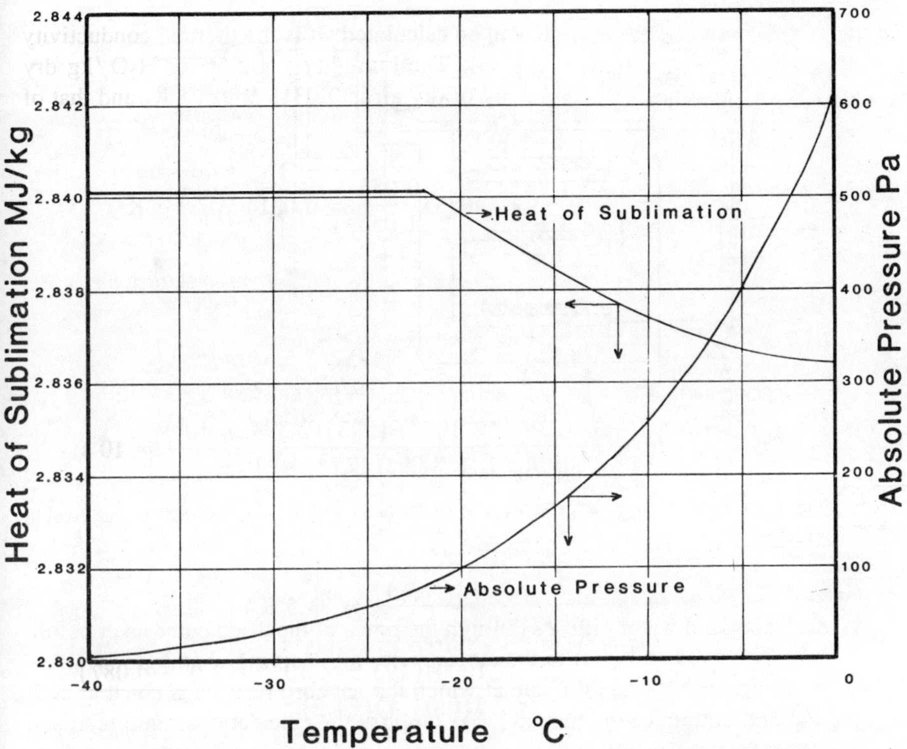

Fig. 12.10. Heat of sublimation (ΔH_s) and vapor pressure of water above ice (P_i). (Based on data from Charm, S. E. 1971. *Fundamentals of Food Engineering*, 2nd ed. AVI Publishing Co., Westport, Conn.)

Heat must also be supplied to the material being dried to provide the energy of vaporization. This is accomplished by the use of hollow shelves through which a heated liquid is circulated. The temperature of the shelves can be regulated by regulating either the temperature or the supply of the heat transfer medium. The material to be dried rests on top of the heated plates. Heat transfer occurs by conduction from the heated plates, by convection from the air inside the drying chamber to the exposed surfaces, and by radiation.

Drying Times for Symmetrical Drying. Analysis of freeze drying is different from that of conventional drying in that drying proceeds from the exposed surfaces toward the interior. The outer layers are completely dry as the ice core recedes. Vaporization of water occurs at the surface of the ice core. Heat of sublimation is conducted to the surface of the ice core through the dried outer

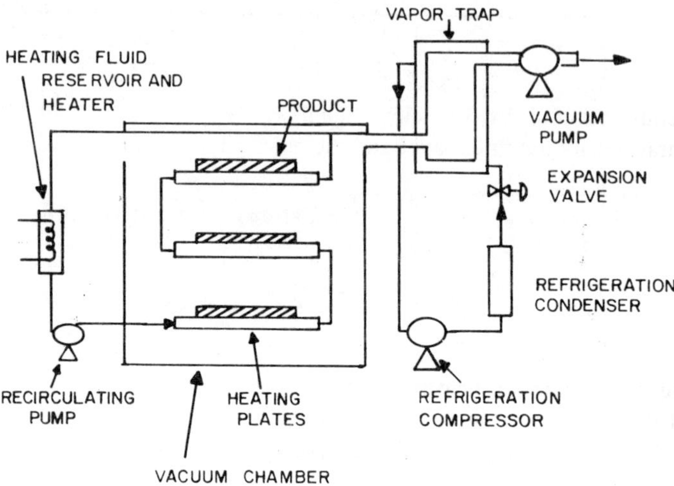

Fig. 12.11. Schematic diagram of a freeze drier showing a vapor trap external to the vacuum chamber.

layer. Vaporized water diffuses through the pores of the dried outer layer before it leaves the solid and goes to the atmosphere in the drying chamber. Symmetrical drying occurs when the rate at which the ice core recedes is equal at both the top and bottom of the material. To simplify the calculations, unidirectional heat transfer is assumed.

Let W = kg water/m^3 of wet material. If ρ is the density of the wet material and X_0 is the initial moisture content on a dry basis, kg water/kg dry matter, $W = (\rho)X_0/1 + X_0$.

If drying is symmetrical, the mass of water evaporated, M_c, expressed in terms of a dried layer, ΔL, is:

$$M_c = WA\,(\Delta L)(2)$$

Let X' = fraction of water evaporated = M_c/total water.

$$X' = \frac{W(A)(\Delta L)(2)}{W(A)(L)} = \frac{\Delta L(2)}{L}$$

$$X' \text{ is also } \frac{X_0 - X}{X_0}$$

where L = the thickness of the solid and X = moisture content, kg H_2O/kg dry matter.

$$\Delta L = \frac{LX'}{2}$$

The overall heat transfer coefficient calculated from the sum of the resistance to heat transfer by the dried layer and the heat transfer coefficient is:

$$\frac{1}{U} = \frac{1}{h} + \frac{\Delta L}{k} = \frac{1}{h} + \frac{LX'}{2k} \tag{85}$$

$$U = \frac{k}{k/h + LX'/2} \tag{86}$$

If drying is symmetrical, heat is transferred from both sides. Let T_a = temperature of the atmosphere in the drier and let T_f temperature of the frozen core surface. The heat transferred to the ice core is

$$q = UA\Delta T = \frac{k}{k/h + LX'/2}(2A)(T_a - T_f) \tag{87}$$

For simultaneous heat and mass transfer, heat transferred = heat of vaporization. ΔH_s is the heat of sublimation of ice at T_f.

$$\Delta H_s(W)(A)(L)\frac{dX'}{dt} = \frac{2A(T_a - T_f)(k)}{k/h + LX'/2} \tag{88}$$

Simplifying and integrating equation 88:

$$t = \frac{\Delta H_s(W)(L)}{2k(T_a - T_f)}\left[\frac{kX'}{h} + \frac{LX'^2}{4}\right] \tag{89}$$

W in equation 89 can be expressed in terms of either the density of the wet material, ρ, or the density of the dried material, ρ_s.

$$W = X_0\rho_s \quad \text{or} \quad W = (\rho)\frac{(X)_0}{1 + X_0}$$

$$t = \frac{\Delta H_s(X_0)(\rho_s)(L)}{2k(T_a - T_f)}\left[\frac{kX'}{h} + \frac{LX'^2}{4}\right] \tag{90}$$

or:

$$t = \frac{\Delta H_s(\rho)(X_0)(L)}{2k(1 + X_0)(T_a - T_f)}\left[\frac{kX'}{h} + \frac{LX'^2}{4}\right] \tag{91}$$

Either equation 90 or equation 91 can be used to calculate the time of drying, depending on whether ρ or ρ_s is known.

EXAMPLE: The density of a sample of beef is 60 lb/ft^3 (965 kg/m^3). How long will it take to dry a 1-in. (2.54-cm)-thick strip of this sample from an initial moisture of 75% to a final moisture content of 4% (wet basis)? Freeze drying is carried out at an absolute pressure of 500 μm of mercury. The air in the drying chamber is at 80°F (26.7°C). Assume symmetrical drying. The thermal conductivity of the dried meat is 0.0692 W/(m · K). Estimate the heat transfer coefficient by assuming that a 3-mm-thick layer of vapor [k_v of water = 0.0235 W/(m · K)] envelopes the surfaces where dying occurs and that the heat transfer coefficient is equivalent in resistance to the resistance of this vapor film.

$$h = \frac{k}{x} = \frac{0.0235}{0.003} = 7.833 \text{ W/m}^2\text{K}$$

The absolute pressure in the chamber is:

$$P = 500 \times 10^{-6} \text{ m Hg} \frac{133.3 \times 10^3 \text{ Pa}}{\text{m Hg}} = 66.65 \text{ Pa}$$

From Fig. 12.10, the temperature of ice in equilibrium with a pressure of 66.65 Pa is −24.5°C. The heat of sublimation H_s = 2.8403 MJ/kg. The final moisture content on a dry basis is: $x = 0.04/0.96 = 0.0417$ kg water/kg dry matter. The initial moisture content on a dry basis is: $X_0 = 0.75/0.25 = 3.00$. The fraction of water remaining at the completion of the drying period is $X' = X_0 - X/X_0 = 3 - 0.0417/3.0 = 0.986$.

Substituting in equation 91:

$$t = \left[\frac{2.8403 \times 10^6(965)(3.00)(0.0254)}{2(0.0692)(1+3)(26.7+24.5)}\right] \left[\frac{0.0692(0.986)}{7.833} + \frac{0.0254(0.986)^2}{4}\right]$$

$$= 109{,}673 \text{ s} = 30.46 \text{ h}$$

PROBLEMS

1. Pork has an a_w of 1 at a moisture content of 50% (wet basis) or higher. If pork is infused with sucrose and sodium chloride and dehydrated such that at the end of the dehydration process the moisture content is 60% (wet basis) and the concentrations

of sugar and sodium chloride are 10% and 3%, respectively, calculate the a_w of the cured product.

2. What concentration of sodium chloride in water would give the same water activity as a 20% solution of sucrose?
3. The following data were obtained on the dehydration of a food product: initial moisture content = 89.7% (wet basis).

Drying time (Min)	Net Weight (kg)
0	24.0
10	17.4
20	12.9
30	9.7
40	7.8
50	6.2
60	5.2
70	4.5
80	3.9
90	3.5

Draw the drying curve for this material, and construct a curve for the drying rate as a function of the moisture content.
(a) What is the critical moisture content for each of the falling rate zones?
(b) What is the constant drying rate?
(c) Determine the residual moisture content for each of the falling rate stages.
(d) The dehydration was conducted at an air flow rate of 50 m/s at a dry bulb temperature of 82°C and a wet bulb temperature of 43°C. The wet material has a density of 947 kg/m^3 and was dried in a layer 2.5 cm thick. If the same conditions were used but the initial moisture content was 91% (wb) and a thicker layer of material (3.5 cm) was used on the drying trays, how long will it take to dry this material to a final moisture content of 12% (wet basis)?
4. A continuous countercurrent drier is to be designed to dry 500 kg/h of food product from 60% (wet basis) moisture to 10% (wet basis) moisture. The equilibrium moisture content for the material is 5% (wet basis), and the critical moisture content is 30% (wet basis). The drying curve of the material in preliminary drying studies showed only one falling rate zone. Air at 66°C dry bulb temperature and 30°C wet bulb temperature will be used for drying. The exit air's RH is 40%. Assume adiabatic humidification of the air. The drying air is drawn from room temperature at 18°C and 50% RH. The wet material has a density of 920 kg/m^3. The drying tunnel should use trucks that hold a stack of 14 trays, each 122 cm wide, 76 cm deep along the length of the tunnel, and 5 cm thick. The distance between the trays on the stack is 10 cm. The drying tunnel has a cross-sectional area of 2.93 m^2. The material in the trays will be loaded to a depth of 12.7 mm. Calculate:
(a) The number of trays of product through the tunnel per hour.
(b) The rate of travel by the trucks through the tunnel. Assume that the distance between the trucks is 30 cm.
(c) The constant drying rate and the total time for drying.

(d) The length of the tunnel.
(e) If air recycling is used, the fraction of the inlet air to the drier that must come from recycled air.
(f) The capacity of the heater required for the operation with recycling.

5. A laboratory drier is operated with a wet bulb temperature of 115°F and a dry bulb temperature of 160°F. The air leaving the drier has a dry bulb temperature of 145°F. Assume adiabatic operation. Part of the discharge air is recycled. Ambient air at 70°F and 60% RH is heated and mixed with the recycled hot air. Calculate the proportion of fresh air and recycled hot air that must be mixed to achieve the desired inlet dry and wet bulb temperatures.
6. If it takes 8 h to dry a material in a freeze drier from 80% to 10% water (wet basis) at an absolute pressure of 100 μm and a temperature of 110°F (43.3°C), how long will it take to dry this material from 80% to 40% water if the dehydration is carried out at 500 μm and 80°F (26.7°C)? The material is 25 mm thick and has a density of 950 kg/m^3, and the thermal conductivity of the dried material is 0.35 W/(m · K). Thermal conductivity and heat transfer coefficients are independent of plate temperature and vacuum.
7. Calculate the constant rate of drying in a countercurrent continuous belt dehydrator that processes 200 lb/h (90.8 kg/h) of wet material containing 80% water to 30% water. Air at 80°F (26.7°C) and 80% RH is heated to 180°F (82.2°C) in an electric heater, enters the drier, and leaves at 10% RH. The critical moisture content of the material is 28%. The drier is 4 ft (1.21 m) wide, the belt is loaded to a depth of 2 in. (5.08 cm) of material, and the clearance from the top of the drier to the top of the material on the belt is 10 in. (25.4 cm). The density of the dry solids in the material is 12 lb/ft^3 (193 kg/m^3).
8. In a spray drying experiment, a sample containing 2.15% solids and 97.8% water was fed at the rate of 6.9 lb/h (3.126 kg/h) and was dried at 392°F (200°C) inlet air temperature. The exit air temperature was 200°F (93.3°C). The dried product was 94.5% solids, and the outside air was at 79°F (26.1°C) and 20% RH. Calculate:
   (a) The weight of water evaporated per hour.
   (b) The % RH of the exit air.
   (c) The mass flow rate of air through the drier in weight dry air per hour.
   (d) In this same drier, if the inlet air temperature is changed to 440°F (226.7°C) and the % RH of the exit air is kept the same as in (b), what weight of sample containing 5% solids and 98% water can be dried to 2% water in 1 h? (Air flow rate is the same as before.) What would be the exit temperature of the air from the drier under the conditions? Assume adiabatic drying.
9. A dehydrator, when operated in the winter where the outside air is 10°F (−12.2°C) and 100% RH (H = 0.001), can dry 100 lb (45.5 kg) of fruit per hour from 90% to 10% water. The inlet temperature of the air to the drier is 150°F (65.6°C) and leaves at 100°F (37.8°C). In the summer when the outside air is 90°F (32.2°C) and 80% RH, determine the moisture content of the product leaving the drier if the operator maintains the same rate of 100 lb (45.4 kg) of wet fruit per hour and the exit air from the drier has the same % RH as it does in the winter.
10. The desorption isotherm of water in carrots at 70°C is reported to fit the Iglesias-Chirife equation (equation 36) with the constants B_1 = 3.2841 and B_2 = 1.3923.

(a) Determine the moisture contents where a shift in drying rate may be expected in the dehydration of carrots.
(b) The following data represent the equilibrium water activity (a_w) for carrots at various moisture contents in kg water/kg dry matter (X): (a_w, X): (0.02,0.0045), (0.04,0.009), (0.06,0.0125), (0.08,0.016), (0.10,0.019), (0.12,0.0225), (0.14,0.025), (0.016,0.028), (0.18,0.031), (0.20,0.034). Fit these data to the BET isotherm and determine the moisture content for a unimolecular layer, X_m.
(c) Fit the data to the GAB equation and determine the constants.

11. The diffusivity of water in scalded potatoes at 69°C and 80% moisture (wet basis) has been determined to be 0.22×10^{-5} m^2/h. If 1-cm potato cubes are dried using air at 1.5 m/s velocity and 1% RH, calculate the dry bulb temperature of the air which can be used so that the diffusion rate from the interior to the surface will be equal to the surface dehydration rate. Assume that the air flows parallel to the cubes and that dehydration proceeds from all faces of each cube. The density of the potato cube is 1002 kg/m^3 at 80% moisture.

12. Puffing can be induced during dehydration of diced carrots if the dehydration rate at the constant rate period is on the order of 1 kg water/(min · kg DM). In a fluidized bed drier where the air contacts individual particles at a velocity of 12 m/s, calculate the minimum dry bulb temperature of the drying air which would induce this rate of drying at the constant rate period. Assume that drying air has a humidity of 0.001 kg water/kg dry air and that surface temperature under these conditions is 5°C higher than the wet bulb temperature. Calculate the mass transfer rate under these conditions. Is dehydration rate heat or mass transfer controlled?

SUGGESTED READING

Charm, S. E. 1971. *Fundamentals of Food Engineering*, 2nd ed. AVI Publishing Co., Westport, Conn.

Foust, A. S., Wenzel, L. A. Clump, C. W., Maus, L., and Andersen, L. B. 1960. *Principles of Unit Operations*. John Wiley & Sons, New York.

Goldblith, S. A., Rey, L., and Rothmayr, W. W. 1975. *Freeze Drying and Advanced Food Technology*. Academic Press, New York.

Heldman, D. R. 1973. *Food Process Engineering*. AVI Publishing Co., Westport, Conn.

Hildebrand, J., and Scott, R. L. 1962. *Regular Solutions*. Prentice-Hall, Englewood Cliffs, N.J.

Hougen, O. A., and Watson, K. M. 1946. *Chemical Process Principles*. Part II. *Thermodynamics*. John Wiley & Sons, New York.

Iglesias, H. A., and Chirife, J. 1982. *Handbook of Food Isotherms*. Academic Press, New York.

Leniger, H. A., and Beerloo, W. A. 1975. *Food Process Engineering*. D. Riedel, Boston.

Marshall, W. R., Jr. 1954. Atomization and spray drying. *Amer. Inst. Chem. Eng. Prog. Monogr. Ser*. 50(2).

McCabe, W. L., and Smith, J. C. 1967. *Unit Operations of Chemical Engineering*, 2nd ed. McGraw-Hill Book Co., New York.

McCabe, W. L., Smith, J. C., and Hariott, P. 1985. *Unit Operations of Chemical Engineering*, 4th ed. McGraw-Hill Book Co., New York.

Norrish, R. S. 1966. An equation for the activity coefficient and equilibrium relative humidity of water in confectionery syrups. *J. Food Technol*. 1:25–39.

Perry, R. H., Chilton, C. H., and Kirkpatrick, S. D. 1963. *Chemical Engineers Handbook*, 4th ed. McGraw-Hill Book Co., New York.

Peters, M. S. 1954. *Elementary Chemical Engineering*. McGraw-Hill Book Co., New York.

Rockland, L. B., and Stewart, G. F., eds. 1981. *Water Activity: Influences on Food Quality*. Academic Press, New York,

Ross, K. D. 1975. Estimation of a_w in intermediate moisture foods. *Food Technol*. 29(3):26–34.

Sandal, O. C., King, C. J., and Wilke, C. R. 1967. The relationship between transport properties and rates of freeze drying of poultry meat. *AICHE J*. 13(3):428–438.

Watson, E. L., and Harper, J. C. 1989. *Elements of Food Engineering*, 2nd Ed. Van Nostrand Reinhold, New York.

13

Physical Separation Processes

Food technology has evolved from the practice of preserving products in much the same form as they occur in nature to one where desirable components are separated and converted to other forms. Separation processes have been in use in the food industry for years, but sophistication in their use is a fairly recent occurrence. Current technology makes it possible to remove haze from wine and fruit juices or nectars, separate the proteins of cheese whey into fractions having different functional properties, separate foreign matter from whole or milled grains, and concentrate fruit juices without having to employ heat. Efficient separation processes have been instrumental in making economically viable the recovery of useful components from food processing wastes.

FILTRATION

Filtration is the process of passing a fluid containing suspended particles through a porous medium. The medium traps the suspended solids, producing a clarified filtrate. Filtration is employed when the valuable component of the mixture is the filtrate. Examples are clarification of fruit juices and vegetable oil. If the suspended material is the valuable component (e.g. recovery of precipitated proteins from an extracting solution) and rapid removal of the suspending liquid cannot be carried out without the addition of a filter aid, other separation techniques must be used.

Surface filtration is a process whereby the filtrate passes across the thickness of a porous sheet while the suspended solids are retained on the surface of the sheet. A sheet with large pores has low resistance to flow; therefore, filtrate flow is rapid. However, small particles may pass through, resulting in a cloudy filtrate. Surface filtration allows no cake accumulation. Flow stops when solids cover the pores. If the solids do not adhere to the filter surface, the filter may be regenerated by backwashing the surface. Filtration sterilization of beer using microporous filters is a form of surface filtration.

Depth filtration is a process whereby the filter medium is thick and solids penetrate the depth of the filter. Eventually solids block the pores and stop filtrate flow, or solids may break through the filter and contaminate the filtrate. Once filtrate flow stops or slows down considerably, the filter must be replaced. In depth filtration, particle retention may occur by electrostatic attraction in addition to the sieving effect. Thus, particles smaller than the pore size may be retained. Depth filters capable of electrostatic solids retention are ideal for rapid filtration. Cartridge, fiber and sand filters are the forms used for depth filtration. Depth filtration is not very effective when the suspended solids concentration is very high.

Filter aid filtration involves the use of agents which form a porous cake with the suspended solids. Filter aids used commercially are diatomaceous earth and perlite. Diatomaceous earth consists of skeletal remains of diatoms and is very porous. Perlite is milled and classified perlite rock, an expanded crystalline silicate. In filter aid filtration, the filter medium is a thin layer of cloth or wire screen which has little capacity for retention of the suspended solids and serves only to retain the filter aid. A layer of filter aid (0.5 to 1 kg/m^2 of filter surface) is precoated over the filter medium before the start of filtration. Filter aid, referred to as body feed is continuously added to the suspension during filtration. As filtration proceeds, the filter aid and suspended solids are deposited as a filter cake, which increases in thickness with increasing filtrate volume. The body feed filter aid concentration is usually in the range of one to two times the suspended solids concentration. A body feed concentration must be used which will produce a cake with adequate porosity for filtrate flow. Inadequate body feed concentration of the filter aid will result in a rapid decrease in filtrate flow and consequently will shorten the filtration cycles. Figure 13.1 is a diagram of a filtration system for filter aid filtration. The system is designed to make it easy to change from precoat operation to body feed filtration operation with the switch of a three-way valve. Some filters, such as the vertical leaf filter, can be back-

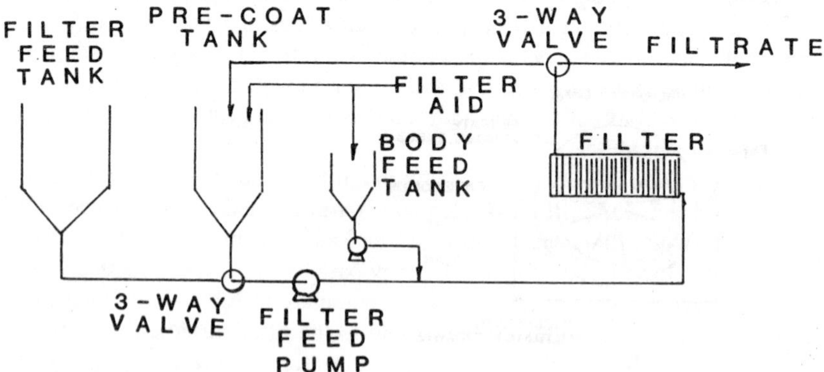

Fig. 13.1. Diagram of a system for filter aid filtration.

flushed to remove the filter cake. Others, like the plate and frame filter, are designed to be easily disassembled to remove the filter cake. The body feed may be a slurry of the filter aid which is metered into the filter feed, or the dry filter aid may be added to the filter feed tank using a proportional solids feeder. With continuous vacuum filters, precoating is done only once, at the start of the operation. Body feed is added directly to the vacuum filter pan. Figure 13.2 is a diagram of the most common filters used for filter aid filtration. Figure 13.2A is a vertical leaf filter, Fig. 13.2B is a plate and frame filter, and Fig. 13.2C is a continuous vacuum rotary drum filter.

Filtrate Flow Through Filter Cake. In filter aid filtration, filtrate flow through the pores in the cake is dependent upon the pressure differential across the cake and the resistance to flow. The total resistance increases with increasing cake thickness; thus, filtrate flow decreases with time of filtration. The resistance to filtrate flow across the filter cake is expressed as the *specific cake resistance*. The resistance of the filter medium and the precoated layer of filter aid is the *medium resistance*.

Figure 13.3 is a section of a filter showing the filter medium, precoat, and

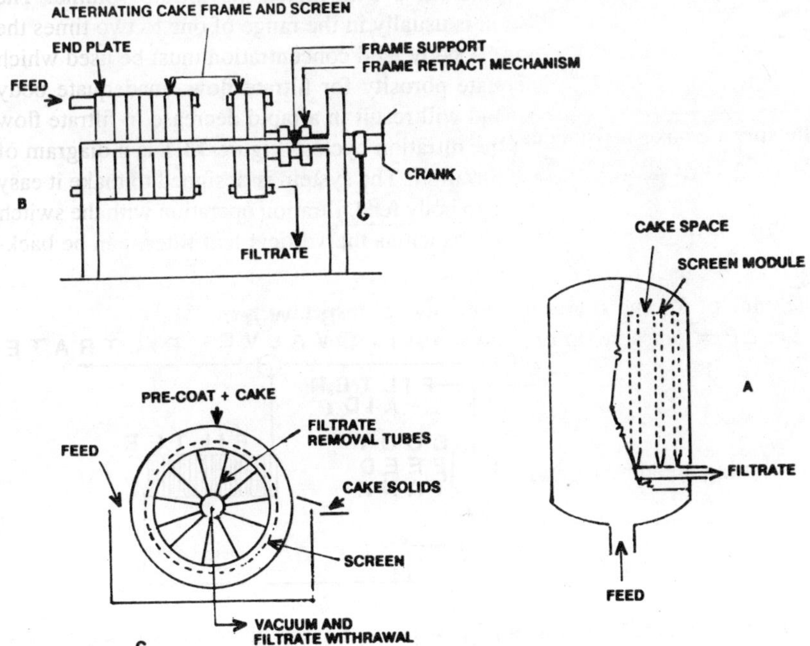

Fig. 13.2. Schematic diagram of a vertical leaf filter (A), a plate-and-frame filter (B), and a rotary drum filter (C).

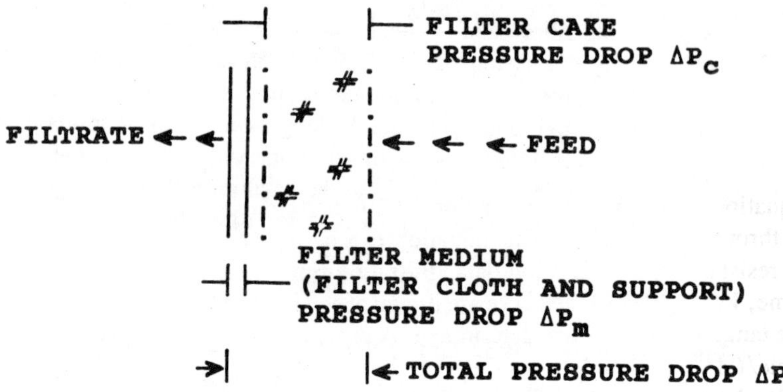

Fig. 13.3. Section through a filter and a filter cake.

filter aid. The total pressure drop across the filter is the sum of the pressure drop across the filter medium (the filter cloth and precoat), ΔP_m, and that across the cake, ΔP_c.

Let v = velocity of filtrate flow, μ = filtrate viscosity, A = filter area, and m = mass of filter cake. The medium resistance, R_m, is:

$$R_m = \frac{\Delta P_m}{\mu v} \tag{1}$$

The specific cake resistance, α, is:

$$\alpha = \frac{\Delta P_c}{\mu v(m/A)} \tag{2}$$

The units of R_m and α are m^{-1} and m/kg, respectively, in SI.

Let ΔP = pressure differential across the filter.

$$\Delta P = \Delta P_m + \Delta P_c \tag{3}$$

Substituting equations 1 and 2 in equation 3:

$$\Delta P = \alpha\mu v \frac{m}{A} + R_m \mu v \tag{4}$$

Let v = volume of filtrate and c = concentrate of cake solids in the suspension to be filtered. $m = Vc$. The filtrate velocity, $v = (1/A)\, dV/dt$.

Substituting for m and v in equation 4:

$$\Delta P = \frac{\mu(dV/dt)}{A}\left[\frac{\alpha Vc}{A} + R_m\right] \tag{5}$$

$$\frac{dt}{dV} = \frac{\mu}{A\Delta P}\left[\frac{\alpha Vc}{A} + R_m\right] \tag{6}$$

Equation 6 is the Sperry equation, the most widely used model for filtrate flow through filter cakes. This equation can be used to determine the specific cake resistance from filtration data. Filtration time, t, is plotted against filtrate volume, V; tangents to the curve are drawn at several values of V; and the slopes of the tangents, dt/dV, are determined. A plot of dt/dV vs. V has a slope equal to $\alpha c\mu/(A^2 \cdot \Delta P)$ and an intercept on the ordinate at $V = 0$ equal to $R_m\ \mu/(A \cdot \Delta P)$. An easier method for determining α and R_m will be shown in the next section.

Constant Pressure Filtration. When a centrifugal pump is used as the filter feed pump, the pressure differential across the filter, ΔP, is constant, and equation 6 can be integrated to give:

$$t = \frac{\mu}{\Delta P}\left[\frac{\alpha c}{2}\frac{V^2}{A^2} + R_m\frac{V}{A}\right] \tag{7}$$

Dividing equation 7 through by V:

$$\frac{t}{V} = \frac{\mu\alpha c}{2A^2\Delta P}V + \frac{\mu R_m}{A\Delta P} \tag{8}$$

Equation 8 shows that a plot of t/V against V is linear, and the values of α and R_m can be determined from the slope and intercept. A common problem with the use of either equation 6 or equation 8 is that negative values for R_m may be obtained. This may occur when R_m is much smaller than α; when finely suspended material is present which rapidly reduces medium porosity, even with a very small amount of cake solids deposited; or when α increases with the time of filtration such that the least squares method of curve fitting weighs the data heavily during the later stages of filtration relative to those at the early stages. To avoid having negative values for R_m, equation 7 may be used in the analysis. If filtration is carried out using only filtrate without suspended solids, $c = 0$ and equation 7 becomes:

$$t = \frac{\mu R_m}{\Delta PA}V \tag{9}$$

Since $R_m = \Delta P_m/(\mu \cdot v)$; and since during filtration with only the filter cloth and precoat the pressure differential is ΔP_m; and since Av = the filtrate volumetric rate of flow, q, the coefficient of V in equation 9 is $1/q$. Thus, equation 7 can be expressed as:

$$t = \frac{\mu \alpha c}{2 \Delta P A^2} V^2 + \left(\frac{1}{q}\right) V \tag{10}$$

q is evaluated separately as the volumetric rate of filtrate flow on the precoated filter at the pressure differential used in filtration with body feed. Since ΔP across the precoated filter is primarily due to the resistance of the deposited filter aid, q is primarily a function of the type of filter aid, the pressure applied, and the thickness of the cake. For the same filter aid and filtrate, q is proportional to the thickness and the applied pressure; therefore, the dependence can be quickly established. If q is known, α can be easily determined from the slope of a regression equation for $(t - V/q)$ against V^2 or from the intercept of a log-log plot of $(t - V/q)$ against V.

The use of filtration model equations permits determination of the filtration constants R_m and α on a small filter which can then be used to scale up to larger filtrations. A typical laboratory filtration module is shown in Fig. 13.4. This is a batch filter with a tank volume of 7.5 L and a filter area of 20 cm^2. An opening at the bottom allows draining of the precoat suspension at the beginning of a filtration cycle and charging of the tank after precoating. To precoat, a suspension of filter aid is made at such a concentration that filtration of 1000 mL gives the precoating level of 1.0 kg/m^2. For example, 2.0 g of dry filter aid in 1000 mL of water is equivalent to a precoat of 1.0 kg/m^2 on a filter area of 20 cm^2 if 1000 ml of filtrate is allowed to pass through the filter from this suspension. Depending upon the fineness of the filter aid used, filter paper or cloth may be used as the filter medium.

EXAMPLE: Data in Table 13.1 were collected in filtration clarification of apple juice. The apple juice was squeezed from macerated apples and treated with pectinase. After settling and siphoning of the clarified layer, the cloudy juice which remained in the tank was filter clarified using perlite filter aid. The cloudy juice contained 1.19 g solids/100 mL. Perlite filter aid was used which had a pure water permeability of 0.40 mL/(s · 1 cm^2 filter area · 1 atm pressure drop · 1 cm thick cake). Whatman no. 541 filter paper was used as the filter medium. The precoat was 0.1 g/cm^2 (1 kg/m^2). The filter shown in Fig. 13.4, which had a filtration area of 20 cm^2, was used. Flow of clear apple juice through the precoated filter was 0.6 mL/s at a pressure differential of 25 lb$_f$/in.^{2} (172.369 kPa). The filtrate had a viscosity of 1.6 centipoises.

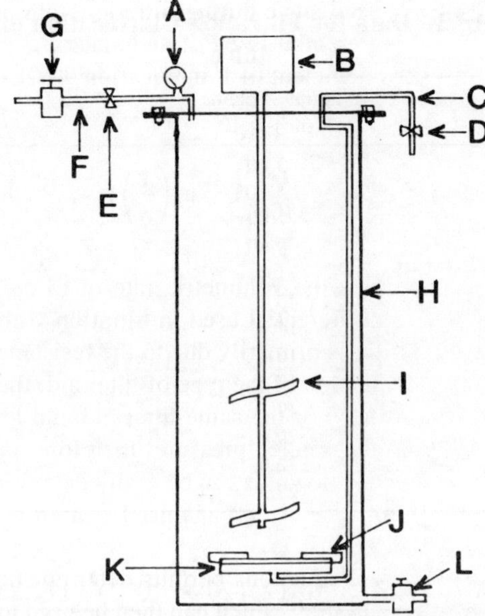

Fig. 13.4. Schematic diagram of a test filtration cell. (From Chang and Toledo, *J. Food Proc. Pres.* 12:353, 1969.) A, pressure gage; B, gear motor; C, filtrate exit tube; D, needle valve; E, needle valve; F, compressed air inlet tube; G, pressure regulator; H, pressure vessel; I, stirrer element; J, replaceable ring; K, filtration cell assembly; L, drain valve.

Calculate α and R_m. Calculate the average filtration rate in $m^3/(min \cdot m^2$ filter area) if the filtration time per cycle is 30 min and the pressure differential across the filter is 30 $lb_f/in.^2$ (206.84 kPa). Assume the same solids and body feed concentration.

The cumulative filtrate volume with filtration time at a filter aid concentration of 2.5 g/100 mL of suspension is shown in Table 13.1.

Solution: Using equation 8, a regression of (t/V) against V was done using the transformed data in Table 13.1.

The regression equation was:

$$\frac{t}{V} = 7.61 \times 10^9 V + 1684000$$

The coefficient of V is $\alpha\mu c/(2\Delta P A^2)$.

c = concentration of filter cake solids in the feed which is the sum of original suspended solids, 1.19 g/100 mL or 11.9 g/L, and the body feed filter aid concentration, which is 2.5 g/100 mL or 25 g/L.

Table 13.1. Data for Filtration Clarification of Apple Juice

Time t(s)	Filtrate Volume V (m^3) × 10^5	$t/V \times 10^{-6}$
60	3.40	1.7647
120	5.80	2.0689
180	7.80	2.3077
240	9.60	2.5000
300	11.4	2.6316
360	13.1	2.7586
420	14.7	2.8669
480	16.1	2.9721
540	17.6	3.0769
600	19.0	3.1496
660	20.6	3.2117
720	22.0	3.2727
780	23.3	3.3476

$$\alpha = \frac{7.61 \times 10^9(2)(20 \times 10^{-4})^2(172{,}369)}{1.6(0.001)(25 + 11.9)}$$

$$= 1.78 \times 10^{11} \text{ m/kg}$$

The intercept is $R_m\, \mu/(A\Delta P)$.

$$R_m = \frac{1.684 \times 10^6(20 \times 10^{-4})(172{,}369)}{0.001(1.6)}$$

$$= 3.63 \times 10^{11} \text{ m}^{-1}$$

Similar values for α and R_m were obtained using equation 10 by performing a regression on $(t - V/q)$ against V^2. q is 0.6×10^{-6} m^3/s. A slightly better fit was obtained using equation 10 with $R^2 = 0.994$ compared with $R^2 = 0.964$ using equation 8. The regression equation using equation 10 was:

$$\left(t - \frac{V}{q}\right) = 7.29 \times 10^9 V^2$$

Values for R_m and α were 3.59×10^{11} and 1.7×10^{11}, respectively, in SI units. These values are more accurate than those obtained using equation 8 because of the better R^2 value for the regression. However, the values are very close and can be considered the same for most practical purposes.

The average filtrate flow is calculated using equation 7. t = 30 min. = 1800 s.

$$\frac{1.6(0.001)}{206,840}\left[\frac{1.78 \times 10^{11}(36.9)}{2}\left[\frac{V}{A}\right]^2 + 3.63 \times 10^{11}\frac{V}{A}\right]$$

$$32.841 \times 10^{11}\left(\frac{V}{A}\right)^2 + 3.63 \times 10^{11}\left(\frac{V}{A}\right) = 2.32698375 \times 10^{11}$$

Dividing through by 32.841×10^{11}:

$$\left(\frac{V}{A}\right)^2 + 0.1105\left(\frac{V}{A}\right) - 0.070856 = 0$$

$$\left(\frac{V}{A}\right) = \frac{-0.1105 \pm [(0.1105)^2 + 4(0.070856)]^{0.5}}{2}$$

$$= 0.217 \text{ m}^3/(\text{m}^2 \text{ filter area})$$

This is the volume of filtrate which passes through a unit area of the filter for a filtration cycle of 30 min. The average rate of filtration is:

$$\frac{(V/A)_{avg}}{t} = \frac{0.217}{30}$$

$$= 0.00722 \text{ m}^3/(\text{min} \cdot \text{m}^2 \text{ filter area})$$

The average filtration rate can be used to size a filter for a desired production rate if the same filtration cycle time is used.

Two approaches were used to evaluate R_m and α in the preceding example. In both, the traditional graphical method for determining filtration rate by drawing tangents to the filtration curve was eliminated. In this example, the resistance of the filter medium was of the same magnitude as the specific cake resistance; therefore, the Sperry equation worked well in determining R_m. There are instances, particularly when $R_m << \alpha$, when a negative intercept is obtained using equation 8. Under these conditions, R_m is best determined by filtration of a solids-free filtrate to obtain q and, using equation 10, to determine α.

In the above example, use of the values of α and R_m calculated using equation 10 gives a filtrate volume per 30-min cycle of 0.221 m^3/m^2 of filter area, or an average filtration rate of 0.00737 $m^3/(\text{min} \cdot m^2$ filter area).

Filtration Rate Model Equations for Prolonged Filtration When Filter Cakes Exhibit Time-Dependent Specific Resistance. The Sperry equation (equation 6) and equations based on it (equation 7) assume constant specific cake resistance. These equations usually provide a good fit to experimental filtration data for short filtration times, as shown in the previous example. However, when the filtration time is extended, equation 8 usually overestimates the filtrate flow. Some filter cakes undergo compaction, or fine solids may migrate within the pores, blocking flow and increasing cake resistance as filtration proceeds.

Exponential Dependence of Rate on Filtrate Volume. The Sperry equation has been modified to account for changing resistance with increasing filtration time. De la Garza and Bouton (*Am. J. Enol. Vitic.* 35:189, 1984) assumed α to be constant and modified the Sperry equation such that the filtration rate is a power function of filtrate volume.

$$\frac{dt}{dV} = \frac{\mu}{\Delta PA}\left[\alpha c\left(\frac{V}{A}\right)^n + R_m\right] \tag{11}$$

Integration of equation 11 gives:

$$t = \frac{\mu}{\Delta PA}\left[\frac{\alpha c(V/A)^{n+1}}{n+1} + R_m V\right] \tag{12}$$

Bayindirly et al. (*J. Food Sci.* 54:1003, 1989) tested equation 12 on apple juice and discovered that a common n in equation 11 could not be found to describe all data at different body feed concentrations. Consequently, an alternative equation was proposed which combines the specific cake resistance and solids concentration into a parameter, k.

$$\frac{dt}{dV} = \frac{\mu}{A\Delta P} R_m [e]^{kV/A} \tag{13}$$

Integration of equation 13 gives:

$$t = \frac{\mu}{k\Delta P} R_m [e]^{kV/A} \tag{14}$$

Taking the natural logarithm of equation 14:

$$\ln(t) = \ln\left[\frac{\mu R_m}{k\Delta P}\right] + \frac{k}{A} V \tag{15}$$

A semilogarithmic plot of t against (V/A) is linear, with slope k. R_m is evaluated from the intercept.

Table 13.2. Filtration Data for Apple Juice

Time (s)	ln (t)	Filtrate Volume V (m^3) × 10^5	$t/V \times 10^{-5}$	$t - (V/q)$
84.3	4.353	9.74	8.659	14.78
112.5	4.723	12.3	9.141	24.59
140.6	4.946	14.4	9.793	38.06
196.9	5.283	17.4	11.291	72.33
281.2	5.639	20.0	14.062	138.39
421.9	6.045	22.6	18.697	260.70
759.4	6.632	27.2	27.939	565.24
1068.8	6.974	29.7	35.932	856.29
1575	7.362	32.3	48.750	1344.2
2250	7.719	24.9	64.522	2000.9
3825	8.249	39.5	96.867	3542.9

EXAMPLE: The data in Table 13.2 were obtained from Bayindirly et al. on filtration of apple juice through a filter with a 30.2-cm^2 area using a diatomaceous earth filter aid with an average particle size of 20 μm. The juice was said to have a viscosity just slightly greater than that of water (assume that μ = 1.0 centipoises), and suspended solids in the juice was 0.3% (3.0 kg/m^3). Precoating was 0.25 g/cm^2 (2.5 kg/m^2), and body feed was 0.005 g/mL (5.0 kg/m^3). The pressure differential was 0.65 atm. Calculate the filtration parameters k and R_m and the average filtration rate in m^3/(min · m^2 of filter area) if a filtration cycle of 60 min is used in the filtration process. Solution: Table 13.2 also shows the transformed data on which a linear regression was performed. Linear regression of (t/V) against V to fit equation 8 results in the following correlation equation:

$$\frac{t}{V} = 2.64 \times 10^{10} V - 3{,}107{,}386$$

The problem of a negative value for the term involving the medium resistance, R_m, is apparent in the analysis of these data. The correlation coefficient was 0.835, which may indicate a reasonable fit; however, if filtration time against filtrate volume is plotted, the lack of fit is obvious. Thus, the Sperry equation could not be used on these filtration data.

A linear regression of ln (t) against V according to equation 15 results in the following regression equation:

$$\ln(t) = 13155.92\,V + 3.70293$$

The correlation coefficient is 0.998, indicating a very good fit. A plot of filtration time against filtrate volume also shows very good agreement between the values calculated using the correlation equation and the experimental data. The exponential dependence of filtration rate on filtrate volume appropriately described the filtration data.

The value of k is determined from the coefficient of V in the regression equation.

$$\frac{k}{A} = 13155.92; \qquad k = 13155.92(30.2 \times 10^{-4})$$

$$k = 39.73 \text{ m}^{-1}$$

The constant is $\ln(\mu R_m / k \cdot \Delta P)$.

$$R_m = \frac{e^{3.0729}(39.73)(0.65)(101{,}300)}{0.001(1)}$$

$$= 5.652 \times 10^{10} \text{ m}^{-1}$$

The correlation equation is used to calculate the filtrate volume after a filtration time of 60 min.

$$V = \frac{\ln(3600) - 3.70293}{13155.92} = 0.000348 \text{ m}^3$$

Average filtration rate $= (V/A)_{\text{avg}}/t$.

$$\frac{(V/A)_{\text{avg}}}{t} = \frac{0.000341}{(30.2 \times 10^{-4})(60)}$$

$$= 0.001882 \text{ m}^3/(\text{min} \cdot \text{m}^2 \text{ filter area})$$

Model Equation Based on Time-Dependent Specific Cake Resistance. Chang and Toledo (*J. Food Proc. Pres.* 12:253, 1989) modified equation 10, derived from the Sperry equation, by assuming a linear dependence of the specific cake resistance with time. The model fitted experimental data on body feed filtration of poultry chiller water overflow for recycling, using perlite filter aid.

$$t = (k_0 + \beta t)V^2 + \left(\frac{1}{q}\right)V \tag{16}$$

The ratio $(t - V/q)/(V^2)$ is $(k_0 + \beta t)$ at t. A regression analysis of this ratio against t gives the values for k_0 and β. The filtrate volume at time t can be calculated from the positive root of equation 16.

$$V = \frac{-(1/q) + \left[(1/q)^2 + 4(k_0 + \beta t)t\right]^{0.5}}{2(k_0 + \beta t)} \qquad (17)$$

Filtration data may be analyzed using this procedure by working on the raw volume vs. time data; the filter area is not considered until the final analysis, i.e., the calculated value of filtrate volume or filtration rate is converted to a per unit area basis. Another approach is to convert filtrate volume data to volume per unit area of filter, in which case the final calculated value of the filtrate volume will already be on a per unit area basis. The former approach is used in the following example to avoid having to manipulate very small numbers.

The units of k_0 and β in equation 16 are $m^{-6} \cdot s$ and m^{-6}, respectively, since V is not expressed on a per unit area basis.

EXAMPLE: Table 13.3 shows data on filtration of poultry chiller water overflow using perlite filter aid which has a rated pure water permeability of 0.4 mL/(s · m² · cm cake atm pressure differential). The filter has an area of 20 cm², water at 2°C was the filtrate, precoating was equivalent to 1.0 kg/m² of filter area, and Whatman no. 541 filter paper was used as the filter medium. Body feed was 5 kg/m³ and suspended solids was 5 kg/m³. Pressure across the filter was 172 kPa. Filtrate flow across the precoated filter was 25.5 mL/s at a 172 kPa pressure differential. Calculate the parameters for the time dependence of specific cake resistance and determine the fit of equation 16 with experimental data. Calculate the average filtration rate if the cycle time is 20 min.

Solution: Table 13.3 also shows the values of $(t - V/q)/V^2$ on which a regression analysis was done against time to yield the following regression equation ($R^2 = 0.9883$):

Table 13.3. Data on Filtration of Poultry Chiller Water

Time (s)	Volume of Filtrate V (m³)	$t - V/q$	$R = (t - V/q)/V^2 \times 10^8$
60	0.000483	41.059	1.760
180	0.000823	147.7	2.181
300	0.001031	259.6	2.442
420	0.001173	374.0	2.718
540	0.001292	489.3	2.931
660	0.001385	605.7	3.156
780	0.001466	722.5	3.362
900	0.001538	839.7	3.550
1020	0.001601	957.1	3.734
1140	0.001658	1075.0	3.910
1200	0.001660	1134.9	4.118

$q = 25.5 \times 10^{-6}$ m³/s.

$$k_0 + \beta t = 191{,}347t + 1.83 \times 10^8 \quad (17a)$$

Table 13.4 shows the calculation of filtrate volume against filtration time using the expression for $k_0 + \beta t$ in equation 17a and equation 17. A graph of calculated filtrate volume against filtration time shows good agreement between the model and the experimental value.

For a filtration time of 20 min, the calculated value of V in Table 13.4 is 0.001659 m^3. The average filtration rate is:

$$\frac{(V/A)_{avg}}{t} = \frac{0.001659}{[(20)(20 \times 10^{-4})]}$$

$$= 0.0415 \text{ m}^3/(\text{m}^2 \text{ filter area} \cdot \text{s}) \text{ at 172}$$

kPa pressure differential across the filter

Optimization of Filtration Cycles. Filtration cycles are generally based on obtaining the fastest average filtrate flow. As filtrate flow drops with increasing cake thickness, a very long filtration time results in a lowered average filtration rate. However, the labor involved in disassembling the filter, removing the cake, and precoating is a major expense. Since precoating is needed before actual filtration, shortened filtration cycles necessitate the increased consumption of filter aid. Thus, optimum cycles based on maximum filtrate flow may not always be the ideal from the standpoint of economics. Filtrations with short cycle

Table 13.4. Calculation of Filtrate Volume from Cake Resistance Data

Time (s)	$k_0 + \beta t$ $\times 10^{-8}$	$1/q$	C $\times 10^{-10}$	Filtrate Volume (V) Calculated from Equation 17 (mL)
60	1.94	39215	4.82	464
180	2.17	39215	15.8	824
300	2.40	39215	29.0	1038
420	2.63	39215	44.4	1191
540	2.86	39215	62.0	1307
660	3.09	39215	81.8	1399
780	3.32	39215	104	1474
900	3.55	39215	128	1537
1020	3.78	39215	154	1591
1140	4.01	39215	183	1637
1200	4.13	39215	198	1659

$q = 2.55 \times 10^{-5}$; From equation 17: $C = (1/q)^2 + 4(k_0 + \beta t)$;
$V = [-(\text{column 2}) + (\text{column 3})^{0.5}]/[2(\text{column 1})]$.

times for maximum filtrate flow per unit filter area may have to be done using a continuous rotary vacuum filter to prevent excessive filter aid use.

The optimum cycle time for maximizing filtrate flow per cycle is derived as follows:

For filtrations which fit the Sperry equation (equation 7), a cycle is the sum of filtration time, t, and the time needed to disassemble, assemble and precoat the filter, t_{DAP}.

$$\frac{V}{\text{Cycle}} = \frac{V}{t + t_{DAP}}$$

Let $k_1 = \mu\alpha c/2\Delta P$. Equation 7 becomes:

$$t = k_1\left(\frac{V}{A}\right)^2 + k_2\left(\frac{V}{A}\right)$$

To obtain the optimum cycle time, the filtrate volume per cycle is maximized by taking the derivative and equating to zero as follows:

$$\frac{d}{dV}\left[\frac{V}{t + t_{DAP}}\right] = \frac{d}{dV}\left[\frac{V}{k_1(V/A)^2 + k_2(V/A) + t_{DAP}}\right]$$

The derivative is equated to zero to obtain V/A for the maximum V/cycle.

$$0 = \frac{[k_1(V/A)^2 + k_2(V/A) + t_{DAP}] - V[2k_1(V/A^2) + k_2/A]}{[k_1(V/A)^2 + k_2(V/A) + t_{DAP}]^2}$$

$$t_{DAP} - k_1\left(\frac{V}{A}\right)^2 = 0$$

$$\left(\frac{V}{A}\right)_{max} = \left[\frac{t_{DAP}}{k_1}\right]^{0.5}$$

The optimum cycle time is:

$$t_{opt} = k_1\left[\frac{t_{DAP}}{k_1}\right] + k_2\left[\frac{t_{DAP}}{k_1}\right]^{0.5}$$

$$= t_{DAP} + k_2\left[\frac{t_{DAP}}{k_1}\right]^{0.5} \qquad (18)$$

$$= t_{DAP} + R_m\left[\frac{2\mu t_{DAP}}{\alpha c\Delta P}\right]^{0.5}$$

When filtration data fit equation 14, the (V/A) for maximum filtrate flow derived using the above procedure is the root of equation 19:

$$\mu R_m A[e]^{k(V/A)}\left[\frac{k}{A^2} - \frac{V}{A}\right] - t_{DAP} = 0 \tag{19}$$

When filtration data fit equation 16, t as a function of k_0, V, and q only is determined as follows:

$$t = \frac{k_0 V^2 + q^{-1} V}{1 - \beta V^2}$$

The optimum V for one filtration cycle is the root of equation 20.

$$V^4(\beta k_0) + V^3(2\beta q^{-1}) + V^2(k_0 + 2\beta t_{DAP}) - t_{DAP} = 0 \tag{20}$$

EXAMPLE: Calculate the optimum filtration cycle in the example in the previous section for poultry chiller water overflow filtration to maximize filtrate flow, assuming 10 min for disassembly, assembly, and precoating. $k_0 = 1.83 \times 10^8$ s/m^6; $\beta = 191{,}347$ 1/m^6; $q = 25.5 \times 10^{-6}$ m^3/s.

Solution: The values of k_0, β, q, and $t_{DAP} = 600$ s are substituted in equation 20, which is then solved for V. A BASIC program is used to solve for the optimum filtrate volume, V, and the filtration time. Lines 10 to 50 in the program assign the known values of β, k_0, q, and t_{DAP}. Line 60 is equation 20. Line 70 solves for the t required for the filtrate volume, V, to flow through the filter.

```
10 FOR V = 0.00112 TO 0.00113 STEP 0.000001
20 B = 191347
30 K = 1.83e08
40 Q = 0.0000225
50 TP = 600
60 F = V ^ 4 * B * K + 2 * B * Q ^ -1 * V ^ 3
+ (K + 2 * B * TP) * v ^ 2 - TP
70 T = (K * V ^ 2 + Q ^ -1 * V) / (1 - B * V ^ 2)
70 PRINT F,V,T
80 NEXT V
```

The program, when run, gives $F = -1.04$ when $V = 0.001122$ and $F = 0.142$ when $V = 0.001123$. Thus, the optimum V for a cycle is 0.001123 m^3 or 1123 mL. The optimum cycle time is 370 s.

The filtration behavior shown in this example results in very short cycle times because of very rapid loss of filter cake porosity. Thus, use of a batch filter in carrying out this filtration results in excessive filter aid use. Use of a continuous rotary vacuum filter is indicated for this application.

Pressure-Driven Membrane Separation Processes. A form of filtration which employs permselective membranes as the filter medium is used to separate solute, macromolecules, and small suspended particles in liquids. A thin membrane with small pore size and with selectivity for passing solute or solvent is used. The solvent and small molecules pass through the membrane and other solutes; macromolecules or suspended solids are retained, either by repulsive forces acting on the membrane surface or by a sieving effect. Figure 13.5 shows a typical membrane filtration system. Cross-flow filtration is the most efficient configuration, as discussed later, and retentate recycling is needed to obtain both high cross-membrane fluid velocities and the desired final solids concentration in the product. The fluid crossing the membrane is the permeate, and the fluid retained on the feed side of the membrane is the *retentate*. The fluid entering the membrane is the feed.

Pressure-driven membrane separation processes include:

Microfiltration (MF)—the size of the particles retained on the membrane is in the range of 0.02 to 10 μm. Sterilizing filtration is an MF process.

Ultrafiltration (UF)—the size of the particles retained on the membrane is in the range of 0.001 to 0.02 μm. Concentration of cheese whey and removal of lactose is a UF process.

Reverse osmosis (RO)—solute molecules with a molecular size of less than 0.001 μm (molecular weight $<$ 1000 daltons) are retained on the membrane surface. In RO, solutes increase the osmotic pressure; therefore, the transmembrane pressure is reduced by the osmotic pressure as the driving force for solvent flux across the membrane. Concentration of apple juice and desalination of brackish water are RO processes.

Membranes are either isotropic or anisotropic. An isotropic membrane has uniform pores all the way across the membrane thickness, while an anisotropic

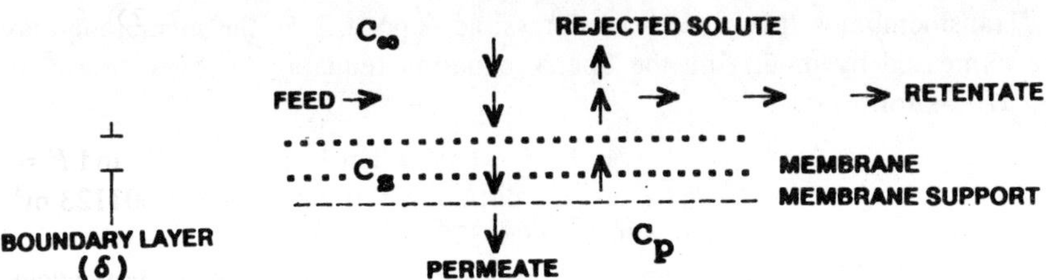

Fig. 13.5. Schematic diagram of a typical membrane filtration system.

membrane consists of a thin layer of permselective material on the surface and a porous backing. MF membranes are isotropic, while the permselective, high-flux membranes used in UF and RO are often anisotropic. Membranes are formed by casting sheets using a polymer solution in a volatile solvent followed by evaporation of the solvent. Removal of the solvent leaves a porous structure in the membrane. Larger-pore membranes are produced by subjecting a non-porous, polymeric membrane to high-energy radiation. The polymer is changed at specific points of entry of the radiation into the membrane. A secondary treatment to dissolve the altered polymer produces pores within the membrane. Dynamic casting may also be employed. A porous support is coated with the membrane material by pumping a solution containing the membrane material across its surface. As the solvent filters across the porous support, the membrane material is slowly deposited. Eventually, a layer with the desired permselectivity is produced.

Two major factors are important in the design and operation of pressure-driven membrane separation processes: transmembrane flux and solute rejection properties of the membrane. In general, transmembrane flux and rejection properties are dependent on the properties of the solute and the suspended solids and on the membrane characteristics of (1) mean pore size, (2) range of pore size distribution, (3) tortuous path for fluid or particle flow across the membrane thickness or, *tortuosity*, (4) membrane thickness, and (5) configuration of pores. Operating conditions also affect flux and rejection of solutes; this topic will be discussed in the succeeding sections.

Transmembrane Flux in Pressure-Driven Membrane Separation Processes (Polarization Concentration and Fouling). Qualitatively, similarities exist in ordinary filtration and membrane separation processes. The net transmembrane pressure, which drives solvent flow across the membrane, is the sum of the pressure drop across the filter medium, ΔP_m, and the pressure drop across the combined deposited solids on the surface and the boundary layer of concentrated suspension flowing across the membrane surface, ΔP_δ. The development of the equation for transmembrane flux is similar to that for filtration rate in filter aid filtrations. The pure water permeability of the membrane, usually given by the membrane manufacturer, is q in equation 10.

Transmembrane flux when a compressible deposit is on the membrane may be expressed by modifying the Sperry equation (equation 6) by letting $\alpha = \alpha_0 \Delta P^n$ as follows:

$$\frac{dV}{dt} = \frac{Aq\Delta P}{\mu \alpha_0 \Delta P^n cq(V/A) + A\Delta P} \tag{21}$$

n is a compressibility factor which is 0 for a completely incompressible solid deposit and 1 for a completely compressible solid deposit. q = the pure water permeability at the transmembrane pressure used in the filtration, ΔP.

If the second term in the denominator is much smaller than the first term, i.e., the resistance of the deposit and the fluid boundary layer at the member surface controls the flux, equation 21 becomes:

$$\frac{dV}{dt} = \frac{A^2 \Delta P^{1-n}}{\mu \alpha_0 c V} \tag{22}$$

Equation 22 shows that if n is 1, i.e., the solid deposit is completely compressible, the transmembrane flux is independent of the transmembrane pressure. This phenomenon has been observed in UF and RO operations. For example, in UF of cheese whey, transmembrane flux is proportional to pressure at transmembrane pressure between 30 and 70 $lb_f/in.^2$ gage (207 – 483 kPa) and is independent of pressure at higher pressures. The phenomenon of decreasing flux rates with increased permeate throughput, V, shown in equation 22, has also been observed. Solids deposition on the membrane surface is referred to as *fouling*.

Even in the absence of suspended solids, transmembrane flux with solutions is different from pure water permeability. This decrease in flux is referred to as *polarization concentration*. In simplistic terms, equation 22 qualitatively expresses polarization concentration in membrane filtrations. As solvent permeates through the membrane, the solids concentration, c, increases in the vicinity of the membrane. The transmembrane flux is inversely proportional to the solids concentration in equation 22. The concentration on the membrane surface increases with increasing transmembrane flux; therefore, the flux decrease due to polarization concentration is more apparent with high transmembrane flux. The flux decrease due to polarization concentration is minimized by increasing the flow across the membrane surface by cross-flow filtration. In the absence of suspended solids which form fouling deposits, flux is a steady state. In the presence of suspended solids which form compressible deposits, flux continually decreases with increasing throughput of permeate, V.

Flux decrease due to polarization concentration is a function of the operating conditions, and if flow rate on the membrane surface, solids concentration in the feed, and transmembrane pressure are kept constant, flux should remain constant. Flux decline under constant operating conditions can be attributed to membrane fouling.

Porter (1979) presented theoretical equations for evaluation of polarization concentration in ultrafiltration and reverse osmosis. The theoretical basis for the equations is the mass balance for solute or solids transport occurring at the fluid boundary layer at the membrane surface.

In cross-flow filtration, fluid flows over the membrane at a fast rate. A laminar boundary layer of thickness δ exists at the membrane surface. If C is the solids concentration in the boundary layer and C_∞ is the concentration in the liquid bulk, a mass balance of solids entering the boundary layer with the solvent and those leaving the boundary layer by diffusion will be:

$$\left(\frac{\partial V}{\partial t}\right)C = -D\left(\frac{\partial C}{\partial x}\right)$$

D = the mass diffusivity of solids. Integrating with respect to x, designating the transmembrane flux $\partial V/\partial t = J$, and using the boundary conditions $C = C_\infty$ at $x = \delta$:

$$\ln\left(\frac{C}{C_\infty}\right) = \frac{J}{D}(\delta - x) \tag{23}$$

At the surface, $x = 0$ and $C = C_s$.

$$\ln\left(\frac{C_s}{C_\infty}\right) = \frac{J}{D}\delta \tag{24}$$

Equation 24 shows that polarization concentration expressed as the ratio of surface to bulk solids concentration increases as the transmembrane flux and thickness of boundary layer increase. Equation 24 demonstrates that the polarization concentration can be reduced by reducing the transmembrane flux and decreasing the boundary layer thickness. Since flux must be maximized in any filtration process, reduced polarization concentration can be achieved at maximum flux if high fluid velocities can be maintained on the membrane surface to decrease the boundary layer thickness, δ.

Rearranging equation 24:

$$J = \frac{D}{\delta}\ln\left(\frac{C_s}{C_\infty}\right)$$

The ratio D/δ may be represented by a mass transfer coefficient for the solids, k_s. Thus:

$$J = k_s\left[\ln(C_s) - \ln(C_\infty)\right] \tag{25}$$

Equation 25 shows that a semilogarithmic plot of the bulk concentration, C_∞, against the transmembrane flux under conditions where C_s is constant is linear, with a negative slope of $1/k_s$. Thus, transmembrane flux decreases with increasing bulk solids concentration, and the rate of flux decrease is inversely proportional to the mass transfer coefficient for solids transport between the surface and the fluid bulk.

Equations for estimation of mass transfer coefficients have been applied to determine the effects of cross-flow velocity on transmembrane flux when polarization concentration alone controls flux. In Chapter 12, mass transfer coefficients were determined using the same equations used for heat transfer. The Sherwood number (Sh) corresponds to the Nusselt number, and the Schmidt number (Sc) corresponds to the Prandtl number.

The Dittus-Boelter equation can thus be used to estimate mass transfer in turbulent flow.

$$Sh = 0.023 Re^{0.8} Sc^{0.33}$$

Since $Sh = k_s d_h / D$, $Re = d_h v \rho / \mu$, and $Sc = \mu / (D \cdot \rho)$:

$$k_s = 0.023 \frac{D}{d_h} \frac{d_h^{0.8} v^{0.8} \rho^{0.8}}{\mu^{0.8}} \frac{\mu^{0.33}}{D^{0.33} \rho^{0.33}} = \frac{0.023 D^{0.67} v^{0.8} \rho^{0.47}}{d_h^{0.2} \mu^{0.47}} \qquad (26)$$

Although the Dittus-Boelter equation has been derived for tube flow, an analogy with pressure drops through noncircular conduits reveals that equations for tube flow can be applied to noncircular conduits if the hydraulic radius is substituted for the diameter of the tube. D = mass diffusivity, m^2/s; and d_h = hydraulic radius = 4 (cross-sectional area)/wetted perimeter.

If fluid flows in a thin channel between parallel plates with width W and channel depth $2b$:

$$d_h = \frac{4(2b)(W)}{(4b + 4W)} = \frac{2bW}{(b + w)}$$

If $b \ll W$, $d_h = 2b$ = the channel depth.

In laminar flow, the Sieder-Tate equation for heat transfer can be used as an analog to mass transfer:

$$Sh = 1.86 \left[Re \cdot Pr \cdot \left(\frac{d}{L} \right) \right]^{0.33} \qquad (27)$$

Substituting the expressions for Sh, Re, and Pr:

$$k_s = 1.86 \frac{v^{0.33} D^{0.66}}{d_h^{0.33} L^{0.33}} \qquad (28)$$

L is the length of the channel.

Equations 25 and 28 can be used to predict transmembrane flux when polarization concentration occurs if the diffusivity, D, of the molecular species involved and the surface concentration are known. The form of these equations agrees with experimental data. However, they are ineffective in predicting actual transmembrane flux by their inability to predict the concentration at the membrane surface, C_s. The value of these equations is in interpolating within experimentally observed values for flux and in extrapolating fluxes within reasonable limits when the data at one set of operating conditions are known. Porter (1979) suggests using the Einstein-Stokes equation (equation 29) to estimate D:

$$D = \frac{1.38 \times 10^{-23} T}{6\pi\mu r} \tag{29}$$

where: D is in m^2/s, T = °K, μ = medium viscosity in Pa · s, and r = molecular radius in meters.

EXAMPLE: The molecular diameter of β-lactoglobulin having a molecular weight of 37,000 daltons is 1.2×10^{-9} m. When performing ultrafiltration at 30°C at a membrane surface velocity of 1.25 m/s, a solids concentration in the feed of 12%, and a transmembrane pressure of 414 kPa, the flux was 6.792 L/(m² · h). Calculate the concentration at the membrane surface and the flux under the same conditions but at a higher cross-membrane velocity of 2.19 m/s. The membrane system was a thin channel with a separation of 7.6 mm. The flow path was 30 cm long. The viscosity of the solution at 30°C was 4.8 centipoises, and the density was 1002 kg · m³.

Solution: The diffusivity using equation 29 is:

$$D = \frac{1.38 \times 10^{-23}(30 + 273)}{6\pi(4.8)(0.001)(1.2 \times 10^{-9})} = 3.851 \times 10^{-11}\ m^2/s$$

The hydraulic radius, d_h = channel depth = 7.6×10^{-3} m. The Reynolds number is:

$$Re = \frac{7.6 \times 10^{-3}(1.25)(1002)}{[(4.8)(0.001)]} = 1983$$

For turbulent flow (equation 26):

$$k_s = \frac{0.023(3.851 \times 10^{-11})^{0.66}(1.25)^{0.33}(1002)^{0.47}}{(7.6 \times 10^{-3})^{0.2}[(4.8)(0.001)]^{0.47}}$$

$$= \frac{0.023(1.338 \times 10^{-7})(1.076)(25.72)}{0.376(0.081317)}$$

$$= 2.785 \times 10^{-6} \text{ m/s}$$

Given: $J = 6.792 \text{ L/(m}^2 + \text{h)} = 1.887 \times 10^{-6} \text{ m}^3/(\text{m}^2 \cdot \text{s})$:
Using equation 25:

$$C_s = C_\infty [e]^{J/k_s}$$

C_∞ is given as 0.12 g solids/g solutions.

$$C_s = 0.12[e]^{1.887 \times 10^{-6}/2.785 \times 10^{-6}}$$

$$= 0.12(1.969) = 0.236 \text{ g solute/g solution}$$

For $v = 2.19$ m/s:

$$k_s = 2.785 \times 10^{-6}\left[\frac{2.19}{1.25}\right]^{0.8} = 4.362 \times 10^{-6} \text{ m/s}$$

Using equation 25:

$$J = 4.362 \times 10^{-6}\left[\ln\left(\frac{0.236}{0.12}\right)\right]$$

$$= 2.95 \times 10^{-6} \text{ m}^3/(\text{m}^2 \cdot \text{s}) \text{ or } 10.62 \text{ L/(m}^2 \cdot \text{h)}$$

Solute Rejection. Solute rejection in membrane separations is dependent on the type of membrane and the operating conditions. A solute rejection factor, R, is used, defined as:

$$R = \frac{C_s - C_p}{C_s} \tag{30}$$

where C_s = concentration of solute on the membrane surface on the retentate side of the membrane and C_p = concentration of solute in the permeate.

Equation 30 shows that polarization concentration and increasing solute concentration in the feed decrease the solute rejection by membranes.

Rejection properties of membranes are specified by the manufacturer in terms of the *molecular weight cutoff*, an approximate molecular size which will be retained by the membrane with a rejection factor of 0.99 in very dilute solutions. The ideal membrane with a sharp molecular weight cutoff does not exist.

On a particular membrane, the curve for rejection factor against molecular weight is often sigmoidal with increasing molecular weight of the solute. Low molecular weight solutes completely pass through the membrane and $R = 0$. As the solute's molecular weight increases, small increases in R are observed until the molecular size becomes too great for the solute to pass through the majority of the pores. When the molecular size exceeds the size of all the pores, the rejection factor is 1.0. The pore size distribution, therefore, determines the distribution of molecular sizes which will be retained by the membrane.

Some solutes are retained because they are repelled by the membrane at the surface. This is the case with mineral salts on cellulose acetate membrane surfaces. This property of a membrane to repel a particular solute is beneficial under conditions where separation from the solvent is desired (e.g., concentration), since high rejection factors can be achieved with larger membrane pore size, allowing solute rejection at high transmembrane flux.

Interactions between solutes also affect the rejection factor. For example, the rejection factor for calcium in milk or whey is higher than for aqueous calcium solutions. A protein–calcium complex will bind the calcium, preventing its permeation through the membrane.

Sterilizing Filtrations. Sterilizing filtrations employ either depth cartridge filters or microporous membrane filters in a plate-and-frame or cartridge configuration. These filtrations are MF processes. Microporous membrane filters have pore sizes smaller than the smallest particle to be removed and perform a sieving process. They are preferred for sterilizing filtrations on liquids, since the pressure drop across the membrane is much smaller than in a depth filter and the possibility of microorganisms breaking through is minimal. The pore size of microporous membranes for sterilizing filtrations is less than 0.2 μm.

The filtration rate in sterilizing filtrations has been found by Peleg and Brown (*J. Food Sci.* 41:805, 1976) to consist of the following relationship:

$$\frac{dV}{dt} = k\frac{V^{-n}}{c} \tag{31}$$

where c = load of microorganisms and suspended material which is removed by the filter, and k and n are constants which are characteristic of the fluid being filtered, the filter medium, the suspended solids, and the fluid velocity across the membrane surface. k is also dependent on the transmembrane pressure. n is greater that 1; therefore, the filtration rate decreases rapidly with the increase in filtrate volume. No cake accumulation occurs in this type of filtration. The time needed for membrane replacement and presterilization must be included in the analysis of the optimum cycle time, and the procedures for optimization are similar to that in the section "Optimization of Filtration Cycles."

Cycle time can be increased when fluid is in cross-flow across the membrane surface at high velocities. This configuration minimizes solids deposition and membrane fouling. Fouled membrane surfaces may be rejuvenated by occasionally interrupting filtrations through reduction of transmembrane pressure while maintaining the same velocity of fluid flow across the membrane. This procedure is called *flushing*, as opposed to *backwashing*, in which the transmembrane pressure is reversed. Some membrane configurations are suitable for backwashing, while in others, membrane fragility may restrict removal of surface deposits to flushing. When rejuvenation is done by flushing or backflushing, the flushing fluid must be discarded or filtered through another coarser filter to remove the solids; otherwise, mixing with new incoming feed will result in rapid loss of filtration rate.

Presterilization of filter assemblies may be done using high-pressure steam or chemical sterilants such as hydrogen peroxide, iodophores, or chlorine solutions. Care must be taken to test for filter integrity after presterilization, particularly when heating of filter assemblies is used for presterilization. One method to test for filter integrity in line is the *bubble point test*, in which the membrane is wetted, sterile air is introduced, and the pressure needed to dislodge the liquid from the membrane is noted. Intact membranes require specific pressures to dislodge the liquid from the surface and any reduction in this bubble pressure is an indication of a break in the membrane.

Sterilizing filtrations are successfully employed in cold pasteurization of beer and wine, in the pharmaceutical industry for sterilization of injectable solutions, and in the biotechnological industry for sterilization of fermentation media and enzyme solutions.

Ultrafiltration. UF is widely employed in the increasingly important biotechnology industry for separation of fermentation products, particularly enzymes. Its largest commercial use is in the dairy industry for recovery of proteins from cheese whey and for preconcentration of milk for cheese making. UF systems also have potential for use as biochemical reactors, particularly in enzyme or microbial conversion processes where the reaction products have an inhibitory effect on the reaction rate.

A large potential use for UF is in the extraction of nectar from fruits which cannot be pressed for extraction of the juice. The fruit mash is treated with pectinase enzymes, and the slurry is clarified by UF. UF applications in recycling of food processing waste water and in recovery of valuable components of food processing wastes are currently under development.

More membranes are available for UF applications than for RO applications, indicating increasing commercial applications for RO in the food industry.

Reverse Osmosis. RO has potential for generating energy savings in the food processing industry when used as an alternative to evaporation in the concentration of products containing low molecular weight solutes. The early work on RO was devoted to desalination of brackish water, and a number of RO systems have now been developed which can purify waste water for reuse after an RO treatment. In the food industry, RO has potential for juice concentration without the need for application of heat.

In RO, the transmembrane pressure is reduced as a driving force for solvent flux by the osmotic pressure. Since the solutes separated by RO have low molecular weights, the influence of concentration on osmotic pressure is significant. Equation 5 in Chapter 12 can be used to determine the osmotic pressure. Since the product of the activity coefficient of water and the mole fraction water is the water activity, this equation can be written as:

$$\pi = -\frac{RT}{V} \ln (a_w) \tag{32}$$

where π = osmotic pressure, R = gas constant, T = absolute temperature, and V = molar volume of water. a_w = water activity.

The minimum pressure differential needed to force solvent across a membrane in RO is the osmotic pressure. Equations 17 and 21 in Chapter 12 can be used to determine the a_w of sugar solutions.

EXAMPLE 1: Calculate the minimum transmembrane pressure which must be used to concentrate sucrose solution to 50% sucrose at 25°C.

Solution: From the example problem in the section "Water Activity at High Moisture Contents" in Chapter 12, the water activity of a 50% sucrose solution was calculated to be 0.935.

R = 8315 Nm/(kgmole · °K), T = 303 °K, and the density of water at 25°C = 997.067 kg/m^3 from the steam tables.

$$V = \frac{18 \text{ kg}}{\text{kgmole}} \frac{1}{997.067 \text{ kg/m}^3} = 0.018053 \text{ m}^3/\text{kgmole}$$

$$\pi = \frac{-[8315 \text{ Nm/(kgmole} \cdot \text{K)}][303 \text{ °K}]}{0.018053 \text{ m}^3/\text{kgmole}} \ln (0.935)$$

$$= 9379 \text{ kPa}$$

A minimum of 9379 kPa transmembrane pressure must be applied just to overcome the osmotic pressure of a 50% sucrose solution.

This example illustrates the difficulty of producing very high concentrations of solutes in RO processes and the large influence of concentration polarization in reducing transmembrane flux in RO systems.

Transmembrane flux in RO systems is proportional to the difference between the transmembrane pressure and the osmotic pressure.

$$J = \left(\frac{k_w}{x}\right)(P - \Delta\pi) \tag{33}$$

where J = transmembrane flux, kg water/(s · m^2); k_w = mass transfer coefficient, s; x = thickness of the membrane, m; P = transmembrane pressure, Pa; and π = osmotic pressure, Pa.

EXAMPLE 2: The pure water permeability of a cellulose acetate membrane with a rejection factor of 0.93 for a mixture of 50% sucrose and 50% glucose is 34.08 kg/h of water at a transmembrane pressure of 10.2 atm (1034 kPa) for a membrane area of 0.26 m^2. This membrane system had a permeation rate of 23.86 kg/h when a 1% glucose solution was passed through it at a transmembrane pressure of 10.2 atm at 25°C.

Orange juice is to be concentrated using this membrane system. In order to prevent fouling, the orange juice was first centrifuged to remove the insoluble solids, and only the serum which contained 12% soluble solids was passed through the RO system. After centrifugation, the serum was 70% of the total mass of the juice.

Assume that the soluble solids are 50% sucrose and 50% glucose. The pump which feeds the serum through the system operates at a flow rate of 0.0612 kg/s (60 mL/s of juice having a density of 1.02 g/mL). The transmembrane pressure was 2758 kPa (27.2 atm).

Calculate the concentration of soluble solids in the retentate after one pass of the serum through the membrane system.

Solution: This problem illustrates the effects of transmembrane pressure, polarization concentration, and osmotic pressure on transmembrane flux and solute rejection in RO systems. The governing equations are equation 25 for polarization concentration and equation 33 for transmembrane pressure.

The reduction in pure water permeability when the glucose solution is passed through the membrane system can be attributed to the polarization concentration. Equation 25: $J = k_s \ln(C_s/C_\infty)$. Let N_w = water permeation rate in kg/h. $N_w = JA$ m^3/s (3600 s/h)(1000 kg/m^3).

$$J = \left(\frac{N_w}{A}\right)\left(\frac{1}{3.6}\right) \times 10^{-6}$$

The osmotic pressure at the membrane surface responsible for the reduction of transmembrane flux in the 1% glucose solution is calculated using equation 33.

$$N_w = \left(k_w \frac{A}{x}\right)(P - \pi)$$

The factor $(k_w A/x)$ can be calculated using the pure water permeability data, where $\pi = 0$:

$$\left(k_w \frac{A}{x}\right) = \frac{N_w}{P} = \frac{34.08}{10.2} = 3.34 \text{ kg/(h} \cdot \text{atm)}$$

For the 1% glucose solution, using equation 33:

$$N_w = \left(k_w \frac{A}{x}\right)(P - \pi)$$

$$23.86 = 3.34(10.2 - \pi)$$

$$\pi = 3.05 \text{ atm}$$

A 1% glucose solution has a water activity of practically 1.0; therefore, the osmotic pressure must be exerted at the membrane surface. Thus, C_s can be calculated as the glucose concentration which gives an osmotic pressure of 3.05 atm. Using equation 32, for $P = 3.05(101{,}325.) = 309{,}041$ Pa:

$$309{,}041 = -(8315)\left(\frac{298}{0.018}\right)\ln(a_w)$$

$$a_w = e^{-0.00225} = 0.99775$$

For very dilute solutions, $a_w = x_w$. Let C_s = solute concentration at the surface, g solute/g solution.

$$a_w = x_w = \frac{(1 - C_s)/18}{(1 - C_s)/18 - C_s/180}$$

$$a_w = \frac{(1 - C_s)(18)(180)}{18[180(1 - C_s) + 18C_s]} = 0.99775$$

$$321.271\,C_s + 3232.71 - 3232.71\,C_s = 3420 - 3420\,C_s$$

$$C_s = \frac{7.29}{330.561} = 0.022 \text{ g solute/g solution}$$

Solving for J from the permeation rate:

$$J = \left(\frac{23.86}{0.26}\right)\left(\frac{1}{3.6}\right) \times 10^{-6} = 2.55 \times 10^{-5} \text{ m/s}$$

Using equation 25:

$$k_s = \frac{2.55 \times 10^{-5}}{\ln(0.022/0.01)}$$
$$= 3.234 \times 10^{-5} \text{ m/s}$$

For the orange juice serum: $C_\infty = 0.12$ g solute/g solution. Solving for C_s using equation 25:

$$J = 3.234 \times 10^{-5} \ln\left(\frac{C_s}{0.12}\right)$$

Solving for N_w from J:

$$N_w = JA(3.6)(10^6)$$
$$= \left[3.234 \times 10^{-5} \ln\left(\frac{C_s}{0.12}\right)\right](0.26)(3.6 \times 10^6)$$
$$= 30.27 \ln\left(\frac{C_s}{0.12}\right)$$

Solving for N_w from the transmembrane pressure using equation 36:

$$N_w = \left(k_w \frac{A}{x}\right)(P - \pi) = 3.34(27.2 - \pi)$$

Equating the two equations for permeation rate:

$$30.27 \ln\left(\frac{C_s}{0.12}\right) = 3.34(27.2 - \pi)$$
$$\ln\left(\frac{C_s}{0.12}\right) = 0.1103(27.2 - \pi)$$

Solving for π in atm using equation 32:

$$\pi = -(8315)\left(\frac{298}{0.018}\right) \ln(a_w)[1 \text{ atm}/101{,}325 \text{ Pa}]$$
$$= -1358.59 \ln(a_w), \text{ atm.}$$

Substituting in equation 32 for C_s:

$$\ln\left(\frac{C_s}{0.12}\right) = 0.1103[27.2 + 1358.59 \ln(a_w)]$$

a_w can be solved in terms of C_s as follows:
For sucrose, x_{ws} = mole fraction of water in the sucrose fraction of the solution; a_{ws} = water activity due to the sucrose fraction.

$$x_{ws} = \frac{(1 - 0.5C_s)/18}{(1 - 0.5C_s)/18 + C_s/342}$$

$$a_{ws} = x_{ws}[10]^{-2.7(1 - x_{ws})^2}$$

For the glucose fraction of the mixture, let x_{wg} = mole fraction of glucose; a_{wg} = water activity due to the glucose fraction.

$$x_{wg} = \frac{(1 - 0.5C_s)/18}{(1 - 0.5C_s)/18 + C_s/180}$$

$$a_{wg} = x_{wg}[10]^{-0.7(1 - x_{wg})^2}$$

$$a_w = a_{ws} \cdot a_{wg}$$

C_s will be solved using a BASIC program. In this program, a wide range of C_s values from 0.13 to 0.2 was first substituted in line 10, with a larger step change of 0.01. The printout was then checked for the two values of C_s between which the function in line 70 crossed zero. This occurred between C_s = 0.19 and C_s = 0.20. A narrow range of C_s between 0.19 and 0.195 was later chosen with a narrower step change in order to obtain a value of C_s to the nearest third decimal place which satisfied the value of the function in line 70 to be zero.

```
10 For Cs = 0.19 to 0.195 step 0.001
20 XWG = ((1 - 0.5 * CS) / 18) / (((1 - 0.5 * CS) /
 18) + 0.5 * CS / 180)
30 AWG = XWG * 10 ^ ( -0.7 * (1 - XWG) ^ 2)
40 XWS = ((1 - 0.5 * CS) / 18) / (((1 - 0.5 * CS) /
 18) + 0.5 * CS / 342)
50 AWS = XWS * 10 ^ ( 2.7 * (1 - XWS) ^ 2)
60 AW = AWS * AWG
70 F = LOG(CS / 0.12) - 0.11034 * (27.2 + 1358.59
 * LOG(AW))
80 PRINT CS, F
90 NEXT CS
100 END
```

The following was the output for the above program as written when run:

```
0.19       -2.3738 × 10^-2
0.191      -3.1625 × 10^-3
0.192       1.7461 × 10^-2
0.193       3.7950 × 10^-2
0.194       0.05825
```

The value of C_s is 0.191. The rejection factor is used to calculate the permeate solute concentration.

$$R = \frac{(C_s - C_p)}{C_s}$$

$$C_p = C_s(1 - R) = 0.191(1 - 0.93)$$

$$C_p = 0.0134 \text{ g solute/g permeate}$$

Total mass balance:

$$F = P + R; \quad F = 0.0612 \text{ kg/s}(3600 \text{ s/h}) = 220.3 \text{ kg/h}$$

$$P = \frac{N_w}{(1 - C_p)}$$

$$N_w = 30.27 \ln\left(\frac{0.191}{0.12}\right) = 14.07 \text{ kg/h}$$

$$P = \frac{14.07}{(1 - 0.0134)} = 14.26 \text{ kg/h}$$

$$R = 220.3 - 14.26 = 206.04 \text{ kg/h}$$

Solute balance:

$$FC_f = RC_r + PC_p$$

$$0.12(220.3) = 206.04(C_r) + 14.26(0.0134)$$

$$C_r = 0.127 \text{ g solute/g retentate}$$

The effect of polarization concentration shown in this example problem demonstrates that increasing transmembrane pressure may not always result in a higher flux. A pressure will be reached beyond which polarization concentration will negate the effect of the applied pressure, and any further increase in transmembrane pressure will result in no further increase in flux.

Fouling remains a major problem in both RO and UF applications. Optimization of operating cycles using the principles discussed in the section "Optimization of Filtration Cycles" is an aspect of the operation to which an engineer can make contributions to successful operations. Application of theories and equations on polarization concentration as discussed in the preceding sections will help identify optimum operating conditions, but absolute values of transmembrane flux are better obtained empirically on small systems, and the results used for scale-up, than calculating theoretical values. Material balance calculations as discussed in Chapter 3, in the section "Multistage Processes" will also help in analyzing UF systems, particularly in relation to recycling and multistage operations to obtain the desired composition in the retentate.

Other Membrane Separation Processes. Membranes are also used to remove inorganic salts from protein solutions or to recover valuable inorganic compounds from process wastes. Dialysis and electrodialysis are two commonly used processes.

Dialysis is a diffusion rather than a pressure-driven process. A solute concentration gradient drives solute transport across the membrane. Dialysis is well known in the medical area for removing body wastes from the blood of persons with diseased kidneys. In the food and biochemical industries, dialysis is used extensively for removal of mineral salts from protein solutions by immersing the protein solution contained in a dialysis bag or tube in flowing water. Solute flux in dialysis is very slow.

Electrodialysis is a process whereby solute migration across the membrane is accelerated by the application of an electromotive potential. Anion- and cation-selective membranes are used. The electromotive potential forces the migration of ionic species across the membrane toward the appropriate electrode. Electrodialysis is used to demineralize whey, but food industry applications of the process are still rather limited.

SIEVING

Sieving is a mechanical size separation process. It is widely used in the food industry for separating fines from larger particles and for removing large solid particles from liquid streams prior to further treatment or disposal. Sieving is a gravity-driven process. Usually a stack of sieves are used when fractions of various sizes are to be produced from a mixture of particle sizes.

To assist in the sifting of solids in a stack of sieves, sieve shakers are used. The shakers may be in the form of an eccentric drive which gives the screens a gyratory or oscillating motion, or, they may take the form of a vibrator which gives the screens small-amplitude, high-frequency, up-and-down motion. When the sieves are inclined, the particles retained on the screen fall off at the lower end and are collected by a conveyor. Screening and particle size separation can thus be carried out automatically.

Standard Sieve Sizes. Sieves may be designated by the opening size, US-Sieve mesh, or Tyler Sieve mesh. The Tyler mesh designation refers to the number of openings per inch, while the US-Sieve mesh designation is the metric equivalent. The latter has been adopted by the International Standards Organization. The two mesh designations have equivalent opening size, although the sieve number designations are not exactly the same. Current sieve designations, unless specified, refer to the US-Sieve series. Particle sizes are usually designated by the mesh size which retains particles that have passed through the next larger screen size. A clearer specification of particle size by mesh number would be a plus sign before the mesh size which retains the particles and a negative sign the mesh size which passes the particles. If a mixture of different-sized particles is present, the designated particle size must be the weighted average of the particle sizes.

Table 13.5 shows US-Sieve mesh designations and the size of openings on the screen. A cumulative size distribution of powders can be made by sieving through a series of standard sieves and determining the mass fraction of the particles retained on each screen. The particles are assumed to be spherical, with a diameter equal to the mean of the sieve opening which passed the particles and that which retained the particles.

EXAMPLE: Table 13.6 shows the mass fraction of a sample of milled corn retained on each of a series of sieves. Calculate a mean particle diameter which should be specified for this mixture.

Table 13.5. Standard US-Sieve Sizes

US-Sieve Size (mesh)	Opening (mm)	US-Sieve Size (mesh)	Opening (mm)
2.5	8.00	35	0.500
3	6.73	40	0.420
3.5	5.66	45	0.354
4	4.76	50	0.297
5	4.00	60	0.250
6	3.36	70	0.210
7	2.83	80	0.177
8	2.38	100	0.149
10	2.00	120	0.125
12	1.68	140	0.105
14	1.41	170	0.088
16	1.19	200	0.074
18	1.00	230	0.063
20	0.841	270	0.053
25	0.707	325	0.044
30	0.595	400	0.037

Source: Perry, R. H., Chilton, C. H., and Kirkpatrick, S. D. 1963. *Chemical Engineers Handbook*, 4th ed. McGraw-Hill Book Co., New York.

Table 13.6. US-Sieve Size Distribution of Powder Sample and Calculation of Mean Particle Diameter

Mass Fraction Retained	US-Sieve Screen	Sieve Opening (mm)	Particle size (mm)	Fraction × Size
0	35	0.500		
0.15	45	0.354	0.437	0.065
0.35	60	0.250	0.302	0.1057
0.45	80	0.177	0.213	0.09585
0.05	120	0.125	0.151	0.00755
				Sum = 0.2741

Solution: The sieve opening is obtained from Table 13.6. The mean particle size retained on each screen is the mean opening size between the screen which retained the particle fraction and the one above it. The mass fraction is multiplied by the mean particle size on each screen, and the sum is the weighted mean diameter of the particle. From Table 13.5, the mean particle diameter is 0.2741 mm.

This example clearly shows that the choice of sieves used in classifying the sample into the different size fractions will affect the mean particle diameter calculated. The use of a series of sieves of adjacent sizes in Table 13.5 is recommended.

GRAVITY SEPARATIONS

This type of separation depends on the density differences between several solids in suspension or between the suspending medium and the suspended material. Generally, the force of gravity is the only driving force for the separation. However, the same principles apply when the driving force is increased by application of centrifugal force.

Force Balance on Particles Suspended in a Fluid. When a solid is in suspension, an *external force*, F_e, acts on it to move it in the direction of the force. This external force may be the gravitational force which will cause a particle to move downward, or it can be a centrifugal force which will cause movement towards the center of rotation. In addition, a *buoyant force*, F_b, exists which acts in a direction opposite to the external force. Another force, the *drag force*, F_d, is due to the motion of the suspended solid and acts in the direction opposite the direction of flow of the solid. Let F_t = the net force on the particle. The component of forces in the direction of solid motion will obey the following force balance equation:

$$F_t = F_e - F_d - F_b \qquad (34)$$

In applying equation 34, the sign on F_t will determine its direction based on the designation of the direction of F_e as the positive direction.

Buoyant Force. F_b is the mass of fluid displaced by the solid multiplied by the acceleration provided by the external force. Let a_e = acceleration due to the external force, m = mass of the particle, ρ_p = density of the particle, and ρ = density of the fluid.

$$\text{Mass of displaced fluid} = \text{volume of the particle/density} = \frac{m}{\rho_p}$$

The mass of displaced fluid $= m(\rho/\rho_p)$.

$$F_b = ma_e\left(\frac{\rho}{\rho_p}\right) \tag{35}$$

Drag Force. F_d is fluid resistance to particle movement. Dimensional analysis is used to derive an expression for F_d.

Assume that F_d/A_p, where A_p is the projected area perpendicular to the direction of flow, is a function of the particle diameter, d, the relative velocity between fluid and particle, v_r, the fluid density, ρ, and the fluid viscosity, μ. The general expression used in dimensional analysis is that the functionality is a constant multiplied by the product of the variables, each raised to a power.

$$\frac{F_d}{A_p} = C(v_r)^a \rho^b \mu^c d^e$$

Expressing in terms of the base units mass, M length, L, and time, t:

$$\frac{ML}{t^2}\frac{1}{L^2} = \frac{L^a}{t^a}\frac{M^b}{L^{3b}}\frac{M^c}{L^c t^c}L^e$$

Consider the coefficients of each base unit. To be dimensionally consistent, the exponents of each unit must match the opposite side of the equation.

Exponents of L:

$$-1 = a - 3b - c - e \tag{i}$$

Exponents of t:

$$2 = a + c; \quad a = 2 - c \tag{ii}$$

Exponents of M:

$$1 = b + c; \qquad b = 1 - c \tag{iii}$$

Combining equations (ii) and (iii) in (i):

$$-1 = 2 - c - 3(1 - c) - c + e$$
$$e = -c$$

Thus, the expression for drag force becomes:

$$\frac{F_d}{A_p} = CV^{2-c}\rho^{1-c}\mu^c d^{-c}$$

Combining terms with the same exponent:

$$\frac{F_d}{A_p} = C\left[\frac{\mu}{dv_r\rho}\right]^n (v_r)^2\rho \tag{36}$$

The first two terms on the right are replaced by a drag coefficient, C_d defined as: $C_d = C/Re^n$. Equation 36 is often written in terms of C_d as follows:

$$F_d = C_d\frac{(v_r)^2\rho}{2} \tag{37}$$

Consider a spherical particle settling within a suspending fluid. A sphere has a projected area of a circle: $A_p = \pi d^2/4$. If particle motion through the fluid results in laminar flow at the fluid interface with the particle, $C_d = 24/Re$. The equation for drag force becomes:

$$\begin{aligned} F_d &= \left(\frac{24}{Re}\right)\left(\frac{\pi d^2}{4}\right)(v_r)^2\rho \\ &= 3\pi\mu dv_r \end{aligned} \tag{38}$$

Terminal Velocity. The terminal velocity of the particle is the velocity when $dv_r/dt = 0$, i.e., setting will occur at a constant velocity. Substituting equations 35 and 38 into equation 34:

$$F_t = ma_e - ma_e\left(\frac{\rho}{\rho_p}\right) - 3\pi\mu dv_r$$

$F_t = ma_p = m(dv_r/dt)$. The mass of the particle is the product of the volume and the density. For a sphere: $m = \pi d^3\rho/6$. Thus, at the terminal velocity, v_t:

$$ma_e\left(1 - \frac{\rho}{\rho_p}\right) = 3\pi\mu v_t d$$

Substituting for m:

$$v_t = \frac{a_e(\rho_p - \rho)d^2}{18\ \mu} \qquad (39)$$

Equation 39 is *Stokes' law* for velocity of particle settling within the suspending fluid. For gravity settling, $a_e = g$. For centrifugation, $a_e = \omega^2 r_c$, where r_c = radius of rotation and ω = angular velocity of rotation.

Drag Coefficient. C_d used in the derivation of equation 39 is good only when v_r is small and laminar conditions exists at the fluid–particle interface. In general, C_d can be determined by calculating an index, K (dimensionless), as follows:

$$K = d\left[\frac{a_e\rho(\rho - \rho_p)}{\mu^2}\right]^{0.33} \qquad (40)$$

Data by Lapple and Shepherd (*Ind. Eng. Chem.* 32:605, 1940) show that the value of C_d can be determined for different values of K as follows:

$$\text{If } K > 0.33 \text{ or } Re < 1.9,\ C_d = \frac{24}{Re}$$

$$\text{If } 1.3 < K < 44 \text{ or } 1.9 < Re < 500,\ C_d = \frac{18.5}{Re^{0.6}}$$

$$\text{If } K > 44 \text{ or } Re > 500,\ C_d = 0.44$$

If the Reynolds number at the fluid–particle interface exceeds 1.9, equations 35 and 37 are combined in equation 34, and under conditions when the particle is at the terminal velocity:

$$ma_e\left[1 - \left(\frac{\rho}{\rho_p}\right)\right] = C_d A_p (v_r)^2 \frac{\rho}{2}$$

Substituting $A_p = \pi d^2/4$ and $m = \pi d^3\ \rho_p/6$ and solving for v_r:

$$v_r = 1.155\left[\frac{a_e d(\rho_p - \rho)}{\rho C_d}\right]^{0.5} \tag{41}$$

EXAMPLE: In the processing of soybeans for oil extraction, the hulls must be separated from the cotyledons so that the soy meal eventually produced will have a high value as feed or as a material for further processing for food proteins. The whole soybeans are passed between a cracking roll which splits the cotyledons and produces a mixture of cotyledons and hulls. Figure 13.6 shows a system that can be used to separate the hulls from the cotyledons by air classification. The projected diameters and densities of the cotyledons and hulls are 4.76 mm and 1003.2 kg/m^3 and 6.35 mm and 550 kg/m^3, respectively. The process is carried out at 20°C. Calculate the terminal velocity of the hulls and cotyledons in air. The appropriate velocity through the aperture will be between these two calculated terminal velocities.

Solution: The density of air at 20°C is calculated using the ideal gas equation. The molecular weight of air is 29; atmospheric pressure = 101.325 kPa; R = 8315 N · M/(kgmole · K).

$$\rho = \frac{P(29)}{RT} = \frac{101{,}325(29)}{8315(293)} = 1.206\ kg/m^3$$

The viscosity of air at 20°C is 0.0175 centipoise (from Perry's *Chemical Engineers' Handbook*). Using equation 40:

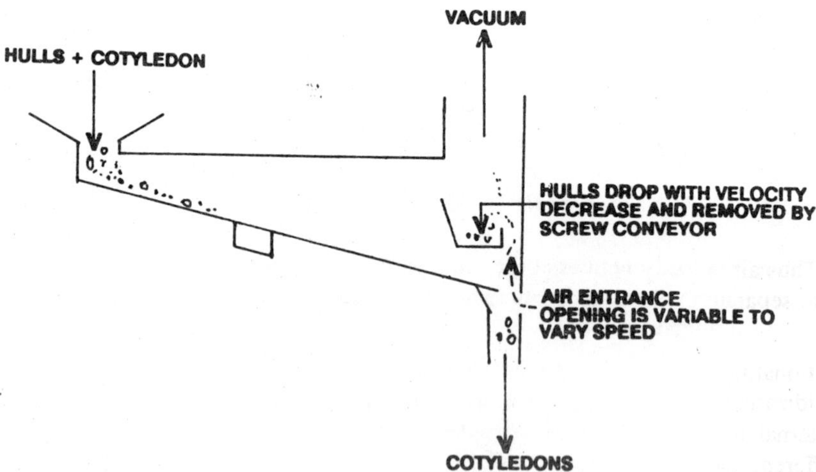

Fig. 13.6. Schematic diagram of a system for airflow separation of soybean hulls and cotyledons.

$$a_e = \text{acceleration due to gravity} = 9.8 \text{ m/s}^2$$

For the cotyledons:

$$K = 0.00476\left[\frac{9.8(1.206)(1003.2 - 1.206)}{[(0.0175)(0.001)]^2}\right]^{0.33}$$

$$= 145$$

Since $K > 44$, $C_d = 0.44$.
Using equation 41:

$$a_e = 9.8 \text{ m/s}^2.$$

$$v_r = 1.155\left[\frac{9.8(0.00476)(1.003.2 - 1.206)}{1.206(0.44)}\right]^{0.5}$$

$$= 10.84 \text{ m/s}$$

For the hulls:

$$K = 0.00476\left[\frac{9.8(1.206)(550 - 1.206)}{[0.0175(0.001)]^2}\right]^{0.33}$$

$$= 118.9$$

$K > 44$; therefore, $C_d = 0.44$.
Using equation 41:

$$V_r = 1.155\left[\frac{9.8(0.00635)(550 - 1.206)}{1.206(0.44)}\right]^{0.5}$$

$$= 9.266 \text{ m/s}$$

Thus air velocity between 9.266 and 10.84 m/s must be used on the system to separate the hulls properly from the cotyledons.

Equations 37 to 40 also apply in sedimentation, flotation, centrifugation, and fluidization of a bed of solids. In flotation separation, where gravity is the only external force acting on the particles, equation 39 shows that solids having different densities can be separated if the flotation fluid density is between the densities of the two solids. The solid with the lower density will float and the one with the higher density will sink. Air classification of solids with different densities, as shown in the example, can be achieved by suspending the solids in air of the appropriate velocity; the denser solid will fall, and the lighter solid will be carried off by the air stream. Air classification is a commonly used method for removal of lighter foreign material from grains and in fractionation of milled wheat into flour with different protein and bran contents. Density gradient separations are employed in fractionating beans with different degrees of maturity and in separating crab meat from scraps after manual removal of flesh.

In fluidization, the fluid velocity must equal the terminal velocity to maintain the particles in stationary motion in the stream of air. Fluidized bed drying and freezing are commercially practiced in the food and pharmaceutical industries.

PROBLEMS

1. Based on the data for apple juice filtration supplied in Table 13.2, calculate the filtration area of a rotary filter which must be used to filter 4000 L/h of juice having a suspended solids content of 50 kg/m^3. Assume that k is proportional to the solids content. The medium resistance increases with increasing precoat thickness. A 10-cm-thick precoat is usually used on rotary filters. Assume that the medium resistance is proportional to the precoat thickness. The rotational speed of the filter is 2 rev/min, and the diameter is 2.5 m. A filtration cycle on a rotary filter is the time of immersion of the drum in the slurry. In this particular system, 45% of the total circumference is immersed at any given time.
2. The following data were obtained in a laboratory filtration of apple juice using a mixture of rice hulls and perlite as the filter aid. The filtrate had a viscosity of 2.0 centipoises, the suspended solids in the juice is 5 g/L, and rice hulls and perlite were added, each at the same concentration as the suspended solids. The filter has an area of 34 cm^3. Calculate:
   (a) The specific cake resistance and the medium resistance.
   (b) The optimum filtration time per cycle and the average volume of filtration per unit area per hour at the optimum cycle time.

Time (s)	Volume (mL)	Time (s)	Volume (mL)
100	40	400	83
200	60	500	91
300	73		

3. The following data were reported by Slack (*Proc. Biochem.* 17(4):7, July/August 1982) on flux rates at different solids contents during ultrafiltration of milk. Test if the flux vs. solids content follows what might be predicted by the theory on polarization concentration. What might be the reasons for the deviation? All flow rates are the same, and the same membrane was used in the series of tests.

Mean solids content(%)	Flux $(L/(\mathrm{m}^2 \cdot \mathrm{h}))$
12.3	28.9
13.6	25.5
14.9	22.1
24.1	23.8

4. The following analysis has been reported for retentate and permeate in ultrafiltration of skim milk. Retentate: 0.15% fat, 16.7% protein, 4.3% lactose, 22.9% total solids. Permeate: 0% fat, 0% protein, 4.6% lactose, 5.2% total solids.
(a) Calculate the rejection factor for lactose by the membrane used in this process.
(b) If the membrane flux in this process follows the data in Problem 3, calculate the lactose content of skim milk which was subjected to a diafiltration process where the original milk containing 8.8% total solids, 4.5% lactose, 3.3% protein, and 0.03% fat was concentrated to 17% total solids, diluted back to 10% total solids, and reconcentrated by UF to 20% total solids.

5. Calculate the average particle diameter and total particle surface area assuming spherical particles, per kilogram of solids, for a powder which has a bulk density of 870 $\mathrm{kg/m^3}$ and the following particle size distribution:

+14,	−10 mesh	10%
+18,	−14 mesh	14.6%
+25,	−18 mesh	22.3%
+35,	−45 mesh	30.2%
+60,	−45 mesh	22.9%

SUGGESTED READING

Cheryan, M. 1986. *Ultrafiltration Handbook*. Technomic Publishing Co., Lancaster, Pa.
Poole, J. B., and Doyle, D. 1968. *Solid-Liquid Separation*. Chemical Publishing Co., New York.
Porter, M. C. 1979. Membrane filtration. In: *Handbook of Separation Techniques for Chemical Engineers*, P. A. Schweitzer, ed. McGraw-Hill Book Co., New York.
Sourirajan, S. 1970. *Reverse Osmosis*. Logos Press, New York.

14

Extraction

One area of food processing which is receiving increasing attention is extraction. This separation process involves two phases. The solvent is the material added to form a phase different from that in which material to be separated was originally present. Separation is achieved when the compound to be separated dissolves in the solvent while the rest of the components remain where they were originally. The two phases may be solid and liquid, immiscible liquid, immiscible liquid phases, or solid and gas. Solid-liquid extraction is also called *leaching*. Supercritical fluid extraction is a recent development which involves a solid and a gas at supercritical conditions. Extraction has been practiced in the oil industry for a long time. Oil from soybean, corn, and rice bran cannot be separated by mechanical pressing; therefore, solvent extraction is used for their recovery. Oil from peanuts is recovered by mechanical pressing and extraction of the pressed cake to remove the oil completely. One characteristic of solvent-extracted oilseed meal is the high quality of the residual protein, which is suitable for further processing into food grade powders. It may also be texturized to form food protein extenders.

Extraction of spice oils and natural flavor extracts is also practiced in the flavor industry.

Extraction is also used in the beet sugar industry to separate sugar from sugar beets. Sugar from sugar cane is separated by multistage mechanical expression, with water added between stages. This process may also be considered a form of extraction. Roller mills used for mechanical expression of sugar cane juice are capital intensive, and when breakdowns occur, the down time is usually very lengthy. The process is also energy intensive; therefore, modern cane sugar processing plants are installing diffusers, which use a water extraction process, instead of the multiple-roller mills previously used.

In other areas of the food industry, extraction with water or an organic solvent is used to remove caffeine from coffee beans, and water extraction is used to prepare coffee and tea solubles for freeze or spray drying. Supercritical fluid extraction has been found to be effective for decaffeinating coffee, and the process is showing promise as a means of lowering cholesterol in whole eggs, egg yolk, and butter.

TYPES OF EXTRACTION PROCESSES

Extraction processes may be classified as follows:

1. *Single-stage batch processing*. In this process, the solid is contacted with solute-free solvent until equilibrium is reached. The solvent may be pumped through the bed of solids and recirculated, or the solids may be soaked in the solvent with or without agitation. After equilibrium, the solvent phase is drained out of the solids. Examples are brewing of coffee or tea and water decaffeination of raw coffee beans.
2. *Multistage cross-flow extraction*. In this process, the solid is contacted repeatedly, each time with solute-free solvent. A good example is soxhlet extraction of fat in food analysis. This procedure requires a lot of solvent or, in the case of a soxhlet, a lot of energy in vaporizing and condensing the solvent for recycling; therefore, it is not used as an industrial separation process.
3. *Multistage countercurrent extraction*. This process utilizes a battery of extractors. Solute-free solvent enters the system at the opposite end from the point of entry of the unextracted solids. The solute-free solvent contacts the solids in the last extraction stage, resulting in the least concentration of solute in the solvent phase at equilibrium. Thus, the solute carried over by the solids after separation from the solvent phase at this stage is minimal. Solute-rich solvent, called the *extract*, emerges from the system at the first extraction stage after contacting the solids which have just entered the system. A stage-to-stage flow of the solids. The same solvent is used from stage to stage; therefore, the solute concentration in the solvent phase increases as the solvent moves from one stage to the next, while the solute concentration in the solids decreases as the solids move in the opposite direction. A good example of a multistage countercurrent extraction process is oil extraction from soybeans using a "Rotocell" (trademark of the Blaw-Knox Corporation). In this system, (Figure 14.1) one cylindrical tank is positioned over another. The top tank rotates while the lower tank is stationary. Both the top and bottom tanks are separated into equal sized sectors such that the contents of each sector are not allowed to mix. Each sector of the top tank is fitted with a swinging false bottom to retain the solids, while a pump is installed to draw out solvent from each of the sectors except one in the lower tank. A screw conveyor is installed in one of the sectors in the lower tank to remove the spent solids and convey them to a desolventization system. The false bottom swings out after the last extraction stage to drop the solids out of a sector

Fig. 14.1. The Rotocel solvent extractor. Source: Becker, K. W. *Chem. Eng. Prog. Symp. Ser.* 64(86):60, 1968. Reproduced by permission of the American Institute of Chemical Engineers © 1968 AIChE.

in the top tank into a bottom sector fitted with the screw conveyor. The movement of the sectors on the top tank is indexed such that, with each index, each sector is positioned directly over a corresponding one in the lower tank. Thus, solvent draining through the bed of solids in a sector in the top tank all goes into one sector in the lower tank. Solvent taken from the sector forward of the current one is pumped over the bed of solids, drains through the bed, and enters the receiving tank, from which another pump transfers this solvent to the top of the bed of solids in the preceding sector. After the last extraction stage, the swinging false bottom drops down, releasing the solids, the swinging false bottom is lifted into place, and the empty sector receives fresh solids to start the process over again. A similar system, although of a different design, is employed in the beet sugar industry.

4. *Continuous countercurrent extraction*. In this system, the physical appearance of an extraction stage is not well defined. In its most simple form, it may consist of an inclined screw conveyor. The conveyor is initially filled with the solvent to the overflow level at the lower end, and solids are introduced at the lower end. The screw moves the solids up through the solvent. Fresh solvent introduced at the highest end moves countercurrent to the flow of solids, picking up solute from the solids as the solvent moves down. Eventually, the solute-rich solvent collects at the lowermost end of the conveyor and is withdrawn through the overflow.

In this type of extraction system, the term *height of a transfer unit* (*HTU*) is used to represent the length of the conveyor, where the solute transfer from the solids to the solvent is equivalent to one equilibrium stage in a multistage system.

Discussion in this chapter is limited to solid-liquid extraction, the most common type used in the food industry.

GENERAL PRINCIPLES

The following are the physical phenomena involved in extraction processes.

Diffusion: Diffusion is the transport of molecules of a compound through a continuum in one phase or through an interface between phases.

In solid-liquid extraction (also known as *leaching*), the solvent must diffuse into the solid for the solute to dissolve in the solvent, and the solute must diffuse out of the solvent-saturated solid into the solvent phase. The rate of diffusion determines the length of time needed to achieve equilibrium between the phases. In the section "Mass Diffusion" in Chapter 12, the solution to the differential equation for diffusion through a slab is expressed as the average concentration against time in equation 41. The equation is valid when the quantity of solid is small relative to the solvent such that the concentration of solute in the solvent phase remains practically constant. Qualitatively, equation 41 shows that the time required for the diffusion process to reach equilibrium is inversely proportional to the square of the diffusion path. Thus, in solvent extraction, the smaller the particle size, the shorter the residence time needed for the solids to remain in the extraction stage.

Particle size, however, must be balanced by the need for the solvent to percolate through the bed of solids. Very small particles result in very slow movement of the solvent through the bed of solids. This increases the probability that fines will go with the solvent phase, interfering with subsequent solute and solvent recovery.

In soybean oil extraction, the soy beans are tempered to a certain moisture content so that they can be passed through flaking rolls to produce thin flakes without disintegrating into fine particles. The short diffusion path reduces equilibrium time in each extraction stage, while the relatively large flakes allow solvent introduced at the top of the bed of flakes to percolate unhindered through the bed.

In cottonseed and peanut oil extraction, a prepress is used to remove as much oil as can be mechanically expressed. The residue comes out of the expeller as small pellets which then go into the extractor.

A recent procedure for extraction of oil from rice bran involves the use of an extruder to heat the bran in order to inactivate lipoxygenase. The extruder produces small pellets, which facilitates the extraction process by minimizing the amount of fines that goes with the solvent phase.

In cane sugar diffusers, hammer mills are used to disintegrate the cane so that the thickness of each particle is not more than twice the size of the juice cells. Thus, equilibrium is almost instantaneous on contact of the particles with water. The cane may be prepressed through a roller mill to crush the cane and produce very finely shredded solids for the extraction battery.

Solubility. The highest possible solute concentration in the final extract leaving an extraction system is the saturation concentration. Thus, the solvent/solids ratio must be high enough so that when fresh solvent contacts fresh solids, the resulting solution on equilibrium is below the saturation concentration of the solute.

In systems where the solids are repeatedly extracted with recycled solvent (e.g., supercritical fluid extraction), a high solute solubility reduces the number of solvent recycles needed to obtain the desired degree of solute removal.

Equilibrium. When the solvent/solid ratio is adequate to satisfy the solubility of the solute, equilibrium is a condition in which the solute concentrations in the solid and solvent phases are equal. Thus, the solution adhering to the solids has the same solute concentration as the liquid or solvent phase. When the amount of solvent is inadequate to dissolve all the solute present, equilibrium is considered as a condition in which no further changes in solute concentration in either phase will occur with prolonged contact time. In order for equilibrium to occur, enough contact time must be allowed for the solid and solvent phases.

The extent to which the equilibrium concentration of the solute in the solvent phase is reached in an extraction stage is expressed as a stage efficiency. If equilibrium is reached in an extraction stage, the stage is 100% efficient and is designated as an *ideal stage*.

SOLID-LIQUID EXTRACTION: LEACHING

The Extraction Battery: Number of Extraction Stages. Figure 14.2 is a schematic diagram of an extraction battery with n stages. The liquid phase is designated the *overflow*, the quantity of which is represented by V. The solid phase is designated the *underflow*, the quantity of which is represented by L. The extraction stages are numbered 1 after the mixing stage where the fresh solids first contact the solute-laden extract from the other stages and n as the last stage where fresh solvent first enters the system and where the spent solids leave the system. The stage where extract from the first extraction stage contacts the fresh solids is called the *mixing stage*. It is different from the other stages in the extraction battery because at this stage the solids have to absorb a much larger amount of solvent than in the other stages.

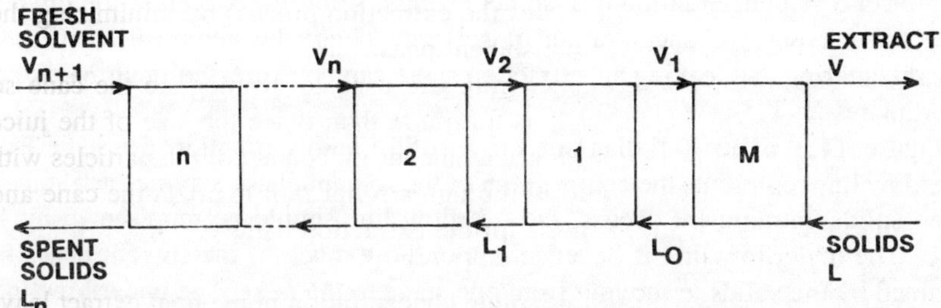

Fig. 14.2. Schematic diagram of an extraction battery for multistage countercurrent extraction.

Each of the stages is considered an ideal stage. Solute concentration in the underflow leaving stage *n* must be at the designated level considered for completeness of the extraction process. Residual solute in the solids fraction must be maintained at a low level. If it is a valuable solute, the efficiency of solute recovery is limited only by the cost of adding more extraction stages. On the other hand, if the spent solids is also valuable and the presence of solute in the spent solids affects its value, then the number of extraction stages must be adequate to reduce the solute level to a minimum desirable value. For example, in oilseed extraction, the residual amount of oil in the meal must be very low; otherwise, the meal will rapidly become rancid.

The number of extraction stages may be determined using a stage-by-stage material balance. Since the only conditions known are those at the entrance to and outlet from the extraction system, stage-by-stage material balancing involves setting up a system of equations which are solved simultaneously to determine the conditions of the solute concentration in the underflow and overflow leaving each stage. The procedure involves assuming a number of ideal stages, solving the equations, and determining if the calculated level of solute in the underflow from stage *n* matches the specified value. The process is very tedious. A graphical method is generally used.

Determination of the Number of Extraction Stages Using the Ponchon-Savarit Diagram. This graphical method for determining the number of extraction stages in a multistage extraction process involves the use of an *X-Y* diagram. The coordinates of this diagram are defined as follows:

$$Y = \frac{\text{solid}}{\text{solute} + \text{solvent}}$$

$$X = \frac{\text{solute}}{\text{solute} + \text{solvent}}$$

where solid = concentration of insoluble solids, solute = concentration of solute, and solvent = concentration of solvent. Thus, the composition of any stream entering and leaving an extraction stage can be expressed in terms of the coordinates X, Y.

Figure 14.3 is an X-Y diagram for a solid-liquid extraction process. The overflow line represents the composition of the solvent phase leaving each stage. If no solids entrainment occurs, the overflow line should be represented by $Y = 0$. The underflow line is dependent upon how much of the solvent phase is retained by the solids in moving from one stage to the next. The underflow line is linear if the solvent retained is constant and curved if the solvent retained by the solids varies with the concentration of the solute. Variable solvent retention in the underflow occurs when the presence of the solute significantly increases the viscosity of the solvent phase.

The Lever Rule in Plotting the Position of a Mixture of Two Streams in an X-Y Diagram. Let two streams with mass R and S, and with coordinates X_r, Y_r and X_s and Y_s, respectively, be mixed together to form T with coordinates X_t and Y_t. The diagram for the material balance is shown in Figure 14.4-I. If R and S consists of only solute and solvent, solute balance results in:

$$RX_r + SX_s = (R + S)X_t \tag{1}$$

A solids balance gives:

$$RY_r + SY_s = (R + S)Y_t \tag{2}$$

Solving for R in equation 1 and S in equation 2 and dividing:

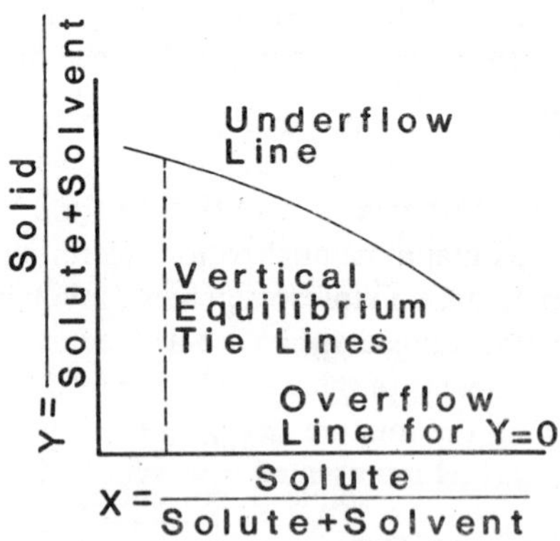

Fig. 14.3. The underflow and overflow curves and equilibrium tie lines on a Ponchon-Savarit diagram.

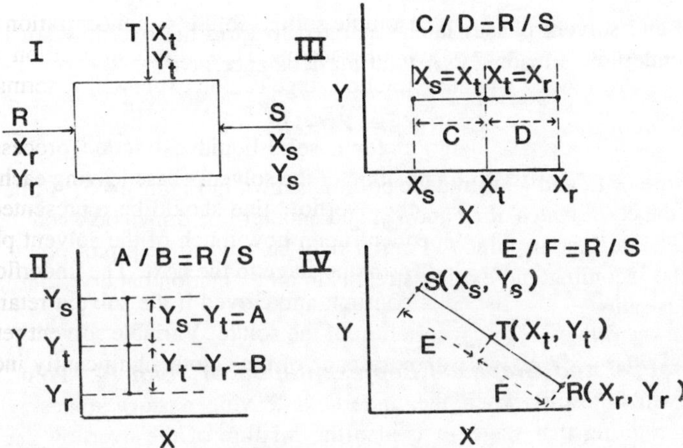

Fig. 14.4. Representation of a material balance on the *X-Y* diagram, and the lever rule for plotting the mass ratio of process streams as the ratio of the distances between points on the *X-Y* diagram.

$$\frac{R}{S} = \frac{X_s - X_t}{X_t - X_r} = \frac{Y_s - Y_t}{Y_t - Y_r} \tag{3}$$

Figure 14.4-II represents the term involving Y in equation 3. The coordinate for the mixture should always be between those of its components. The ratio of the distance between the line $Y = Y_s$ and the line $Y = Y_t$, represented on the diagram by A, and the distance between the line $Y = Y_t$ and the line $Y = Y_r$, represented on the diagram by B, equals the ratio of the mass of R and S.

Figure 14.4-III represents the term involving X in equation 3. Again, the ratio of the masses of R and X equals the ratio of distance C to distance D in the *X-Y* diagram. The composite of the material balance is drawn in Figure 14.4-IV. The coordinates of points S and R, when plotted on the *X-Y* diagram and joined together by a straight line, result in the point representing T being on the line between S and R. The ratio of the distance on the line between S and T, represented by E, and that between R and T, represented by F, is the ratio of the mass of R to S.

These principles show that a material balance can be shown on the *X-Y* diagram, with each process stream represented as a point on the diagram, and any mixture of streams can be represented by a point on the line drawn between the coordinates of the components of the mixture. Exact positioning of the location of the point representing the mixture can be done using the lever rule on distances between the points as represented in Figure 14.4-IV.

Mathematical and Graphical Representation of Point J in the Ponchon-Savarit Diagram. Consider underflow, L_b, mixing with overflow, V_a, and the solid and solvent phases later separated to form the overflow stream, V_b, and the underflow stream, L_a. A total mass balance gives:

$$L_b + V_a = V_b + L_a = J \tag{4}$$

The mixture of L_b and V_a forms the point J, shown in Figure 14.5. A line drawn between the coordinates of L_b and V_a, and another line drawn between the point V_b and L_a, intersect at point J. Point J may be used to help plot points representing the incoming and exiting streams in an extraction battery from the solvent solids ratio.

Mathematical and Graphical Representation of Point P. Point P is a mixture which results when the underflow leaving a stage mixes with the overflow entering that stage. It is also the mixture of the overflow leaving a stage and the underflow entering that stage. The mixture representing point P is shown schematically in Figure 14.6. A total mass balance around the system represented by the dotted line in Figure 14.5 gives:

$$-V_b + L_b = -V_a + L_a = -V_{n+1} + L_n = P \tag{5}$$

Thus, point P is an extrapolation of the line which joins V_b and L_b, the line which joins V_a and L_a, and the line which joins V_{n+1} and L_n. All lines which join the underflow stream leaving a stage and the overflow stream entering that stage all meet at a common point, P. This is a basic principle used to draw the successive stages in a Ponchon-Savarit diagram for stage-by-stage analysis of an extraction process.

Equation of the Operating Line and Representation on the X-Y Diagram. Figure 14.7 represents an extraction battery with n cells. Each stage in the extractor may also be called an *extraction cell*. Cell $n + 1$ is the cell to which underflow enters after cell n. The subscripts on the overflow stream, V, and the underflow stream, L, represent the cell from which the stream is leav-

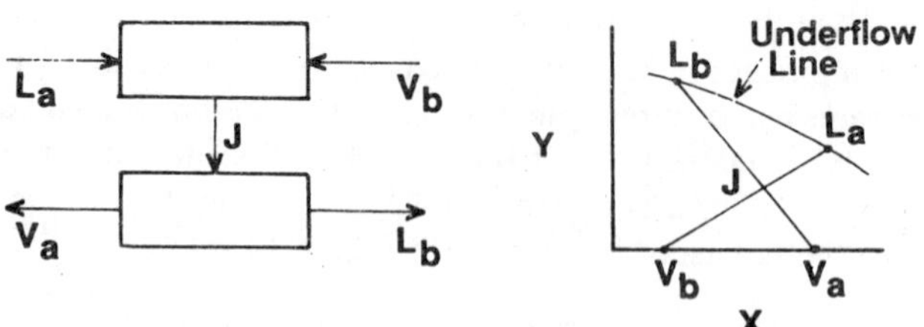

Fig. 14.5. Representation of point J as the intersection of lines connecting four process streams entering and leaving a system.

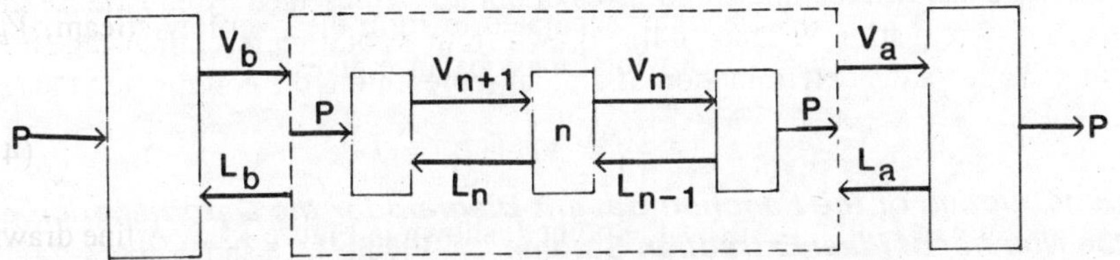

Fig. 14.6. Representation of point P as the common point through which all lines connecting the underflow leaving and the overflow entering an extraction stage must pass.

ing. Thus, the solvent phase entering cell n comes from cell $n + 1$ and is designated V_{n+1}, and the solid phase leaving cell n is designated L_n. If cell n is the last stage in the extraction battery, $V_{n+1} = V_b$ and $L_n = L_b$. Let V represent the mass of solute and solvent in an overflow stream, and let L represent the mass of solute and solvent in an underflow stream. A material balance around the battery of n cells is as follows:

Total mass balance:

$$V_{n+1} + L_a = L_n + V_a; \qquad V_{n+1} = L_n + V_a - L_a \tag{6}$$

Solute balance:

$$V_{n+1}X_{n+1} = V_aX_{V_a} + L_aX_{L_a} + L_nX_n \tag{7}$$

Substituting equation 6 in equation 7 and solving for X_{n+1}:

$$X_{n+1} + \frac{L_n}{L_n + V_a - L_a}X_n + \frac{V_aX_{V_a} - L_aX_{L_a}}{L_n + V_a - L_a} \tag{8}$$

Equation 8 is the equation of an operating line. It shows that the points representing overflow and underflow leaving stage $n + 1$, which have the X-coordinate X_{n+1}, are on the same line drawn through the points representing overflow and underflow leaving stage n, which have the X-coordinate X_n.

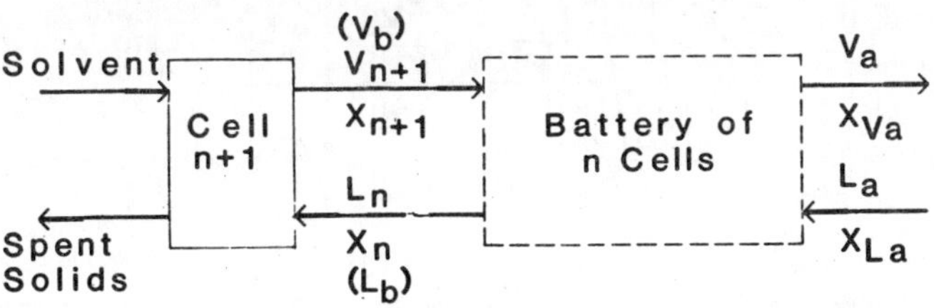

Fig. 14.7. Diagram of the system used to make a material balance for deriving the equation of the operating line in a Ponchon-Savarit diagram.

It can be shown that the coordinates of point P represented by equation 5 can also satisfy equation 8. Thus, a line drawn from a point representing stage n to point P allows the determination of a point representing the coordinate of stage $n + 1$.

Construction of the Ponchon-Savarit Diagram for the Determination of the Number of Ideal Extraction Stages. Figure 14.8 shows the X-Y diagram for determination of the number of ideal stages. Known points representing the coordinates of solvent-saturated solids from the mixing stage, L_a, the extract, V_a, the spent solids, L_b, and the fresh solvent entering the system, V_b, are plotted first on the X-Y diagram. A total material balance must first be made to determine the coordinates of L_b and V_a from the solvent/solids ratio.

Point P is established from the intersection of lines going through L_b and V_b and through L_a and V_a.

The condition of equilibrium is represented by a vertical line. Equilibrium means that X_{L_n} and X_{V_n} are equal. The equilibrium line is also known as a *tie line* and may not be vertical if equilibrium is not achieved. However, it is easier to assume that equilibrium will occur, determine the number of ideal extraction stages, and incorporate the fact that equilibrium does not occur in terms of stage efficiency.

The succeeding stages are established by drawing a line from the point representing stage n on the underflow line to point P. The intersection of this line with the overflow curve will determine X_{n+1}.

EXAMPLE 1: Draw a diagram for a single stage extraction process involving beef (64% water, 20% fat, 16% nonextractable solids) and isopropyl alcohol

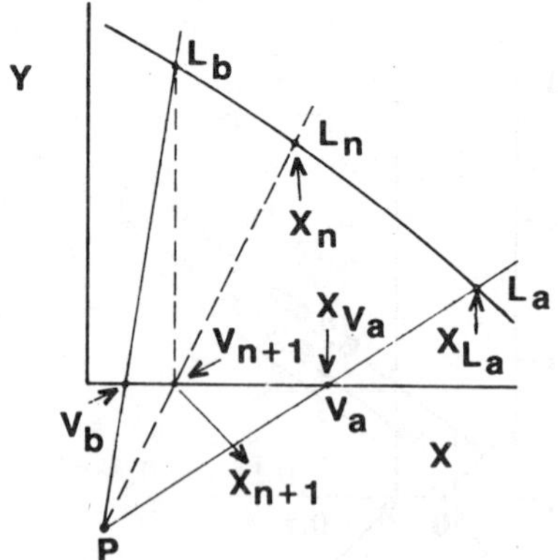

Fig. 14.8. Plotting of point P, the operating line, and the equilibrium tie-line on a Ponchon-Savarit diagram.

in a 1 to 5 ratio. Isopropyl alcohol and water are totally miscible, and the mixture is considered to be the total solvent. Assume all fat dissolves in this solvent. Following equilibrium, the solids fraction separated by filtration retained 10% by weight of the total solution of fat and solvent. Determine graphically the concentration of fat in the extract and calculate the fat content in the solvent free solids.

Solution: Basis: 100 kg beef. Since the *X-Y* diagram uses as a denominator for the *X* and *Y* coordinates only the mass of the solution phase rather than total mass, the distances on the diagram should be scaled on the basis of the mass of solute and solvent present in each stream. Let L_a represent the beef.

$$X_{L_a} = 20/(20 + 64) = 0.238$$

$$Y_{L_a} = 16/(20 + 64) = 0.190$$

The solvent, V_b, will have the coordinates 0,0 since it contains no solute nor solids. L_a and V_b are plotted in Figure 14.9.

Point *J*, the point representing the mixture before filtration is determined using the lever rule as follows: The ratio of the length of line segment L_aJ/length of line segment V_bL_a equals the ratio of the mass of solvent and solute in V_b to the total mass of solvent and solute in V_b and L_a.

$$\text{Let: } V_bL_a = 100 \text{ units}$$

The ratio $V_b/(L_a + V_b) = 500/(20 + 64 + 500) = 0.856$. Therefore, the distance V_bJ should be 85.6 units. For example, when plotting on a scale of

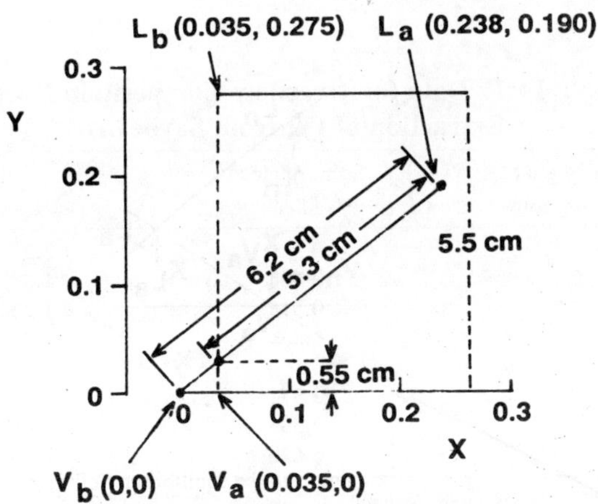

Fig. 14.9. *X-Y* diagram for a single stage extraction process.

1 cm = 0.5 units for X and Y, length of V_bL_a = 6.2 cm. Thus, L_aJ should be 6.2(0.856) = 5.31 cm long. These distances when plotted establish point J which has coordinates (0.035, 0.0275). The vertical line X = 0.035 is the equilibrium line. Assuming no insoluble solids in the extract, V_a has coordinates (0.035,0). Point L_b is determined by plotting the distance V_aL_b (5.5 cm) which is 10 times V_aJ (0.55 cm) since 10% of the solution is retained in the underflow. The ratio X_{L_b}/Y_{L_b} = solute/solids in L_b = mass fraction solute in the solvent free solids = 0.127.

EXAMPLE 2: Table 14.1 presents the amount of solution retained in soybean meal as a function of the oil concentration.

(a) Draw the underflow curve.

(b) If a solvent/soy ratio of 0.5/1 is used for extraction and the original seed contains 18% oil, determine the number of extraction stages needed so that the meal after final desolventization will have no more than 0.01 kg oil/kg oil free meal.

Solution: (a) The coordinates of the points representing the underflow curve are given in Table 14.1. The parameter Y is the reciprocal of the solution retained per kilogram of solids. The underflow curve is linear except for the last three points at high oil concentrations, where there is a slight deviation from linearity. Fig 14.10 shows the Ponchon-Savarit diagram for this problem. The underflow curve is plotted in the figure.

(b) The composition of the final extract and spent solids stream is calculated by performing a material balance around the whole system. Basis: 1 kg soybean; solvent = 0.5 kg. From the specified level of oil in the extracted meal:

Table 14.1. Data for Overflow Composition During Extraction of Oil from Soybeans

$X = \frac{\text{kg oil}}{\text{kg soln.}}$	Soln. retained kg/kg solids	$Y = \frac{\text{Solids}}{\text{kg soln.}}$
0	0.5	2.0
0.1	0.505	1.980
0.2	0.515	1.942
0.3	0.530	1.887
0.4	0.550	1.818
0.5	0.571	1.751
0.6	0.595	1.680
0.7	0.620	1.613

Solution represents the solvent phase, mass of solvent plus mass of dissolved solute.

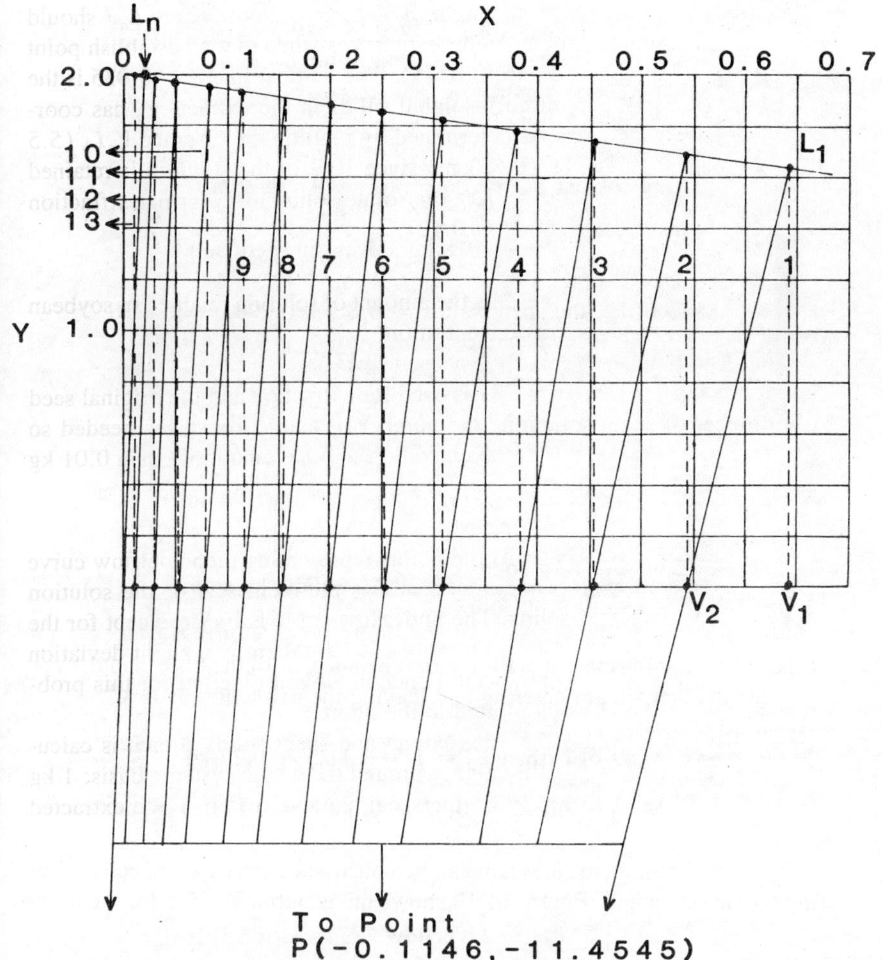

Fig. 14.10. The underflow curve and operating lines establishing the number of equilibrium stages needed to solve the example problem on continuous countercurrent oil extraction from soybeans.

$$\text{kg oil in spent solids} = \frac{0.01 \text{ kg oil}}{\text{kg solids}} 0.82 \text{ kg solids}$$

$$= 0.0082$$

From the data in Table 14.1, at very low oil contents, the amount of solution retained by the solids in 0.5 kg/kg solids.

$$\text{kg soln. in spent solids} = \frac{0.5 \text{ kg soln.}}{\text{kg solids}} 0.82 \text{ kg solids}$$

$$= 0.41$$

$$\text{kg solvent} = \frac{1 \text{ kg solvent}}{1.0082 \text{ kg soln.}} 0.41 \text{ kg soln}$$

$$= 0.4018 \text{ kg solvent in spent solids}$$

$$\text{Mass of } L_n \text{, the spent solids} = 0.82 + 0.0082 + 0.4018$$

$$= 1.23 \text{ kg}$$

Coordinates of spent solids, L_n:

$$X = \frac{0.0082}{0.41} = 0.02 \text{ kg solute/soln.}$$

$$Y = \frac{1 \text{ kg solids}}{0.5 \text{ kg soln.}} = 2.0 \text{ kg solids/kg soln.}$$

Figure 14.11 shows the overall material balance, with the mass and the components of each stream entering and leaving the extraction battery.

$$\text{Mass of extract } V_a = 1.5 - 1.23 = 0.27 \text{ kg}$$

$$\text{Mass of oil } V_a = 0.82 - 0.0082 = 0.1718 \text{ kg}$$

The Ponchon-Savarit diagram can be constructed only for the cells following the mixing stage. Figure 14.12 shows the material balance for the mixing

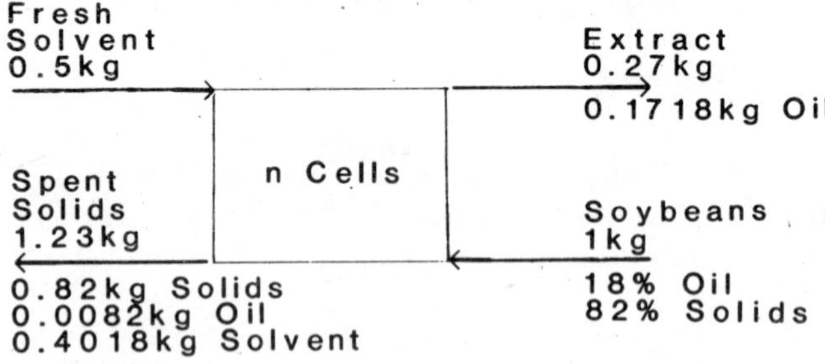

Fig. 14.11. Overall material balance to determine the extract mass and composition in the example problem on soybean extraction.

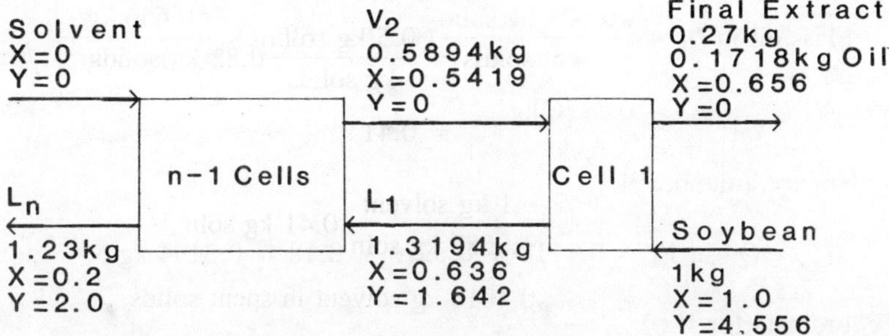

Fig. 14.12. Diagram showing the material balance around the mixing stage, Cell 1, to establish the composition of the overflow stream leaving and the underflow stream entering the extraction battery after the mixing stage.

stage which is considered as cell 1 in this case, and for the stream entering and leaving stage 2. Material balance around cell 1:

$$\text{Final extract, } V_a\text{: } X = \frac{0.1718}{0.27}$$

$$= 0.636 \text{ kg oil/kg soln.}$$

Since the last three data points deviate slightly from linearity, the amount of solution retained by the solids in stream L_1 is calculated by interpolation from the tabulated data instead of the regression equation. Since the final extract, V_a, is in equilibrium with the solids in stream, L_1, the X coordinate of stream L_1 must be the same as for the final extract. Thus, X for stream $L_1 = 0.636$ kg oil/kg soln. From Table 14.1 by interpolation, the solution retained by the solids if the solute concentration is 0.636 kg oil/kg soln. is:

$$= 0.595 + \frac{(0.620 - 0.595)(0.656 - 0.0.6)}{(0.7 - 0.6)}$$

$$= 0.609 \text{ kg soln/kg solids.}$$

$$\text{Total mass of } L_1 = 0.82 \text{ kg solids} + 0.609(0.82)$$

$$= 1.3194 \text{ kg}$$

Y for stream $L_1 = 1/0.609 = 1.642$ kg solids/kg soln. The mass and composition of stream V_1 can now be calculated.

$$\text{Mass of } V_2 = L_1 + V_a - 1$$

$$= 1.3194 + 0.27 - 1 = 0.5894 \text{ kg}$$

$$\text{Mass oil in } L_1 = \left[\frac{0.609 \text{ kg soln.}}{\text{kg solids}}(0.82 \text{ kg solids})\right]\left(\frac{0.656 \text{ kg oil}}{\text{kg soln.}}\right)$$

$$= 0.3276 \text{ kg}$$

Oil balance around cell 1:

$$\text{Oil in } V_2 = 0.1718 + 0.3276 - 0.18 = 0.3194 \text{ kg}$$

X coordinate for V_2:

$$X = \frac{0.3194}{0.5894} = 0.5419$$

The streams to be considered to start plotting the Ponchon-Savarit diagram are streams V_2 and L_1, and the solute-free solvent stream and the spent solids stream, L_n.

V_2 and L_1 are first plotted in Fig. 14.10. A line is drawn through the two points and extended beyond the graph. The equation of the line which connects the two points may also be determined. Points representing the solute-free solvent and the spent solids stream, L_n, are plotted next. The solute-free solvent is represented by the origin. A line is drawn through the two points and extended until it intersects the previous line drawn between L_1 and V_2. The equation for the line may also be determined and the point of intersection, representing point P, can be calculated by solving the two equations simultaneously. The coordinate of point P is -0.1146, -11.4545.

An equilibrium line drawn from V_2 to the underflow curve locates a point which represents L_3. Drawing a line from this point to connect with point P locates an intersection on the overflow curve for the point representing V_3. The process is repeated until the last line drawn from the underflow curve to point P intersects the overflow curve at a point which is equal to or less than the specified X for the spent solids, L_n. The number of extraction stages is the number of equilibrium tie-lines drawn in the diagram in Fig. 14.10. In this problem, 13 equilibrium stages are required.

SUPERCRITICAL FLUID EXTRACTION

Supercritical fluid extraction may be done on solids or liquids. The solvent is a gas at conditions of temperature and pressure at which the gas will not condense into a liquid phase. The gas has a density almost that of a liquid, but it is not a liquid. In addition, the solubility of solutes in a supercritical fluid approaches the solubility in a liquid. Thus the solubility of solutes from solids using a supercritical fluid is very similar to that for solid-liquid extractions.

Carbon dioxide is the fluid most often investigated for use on food products.

It is inexpensive, readily available, nonpolluting, nontoxic and has good solubility for solutes encountered in foods. Typical operating conditions are 300 atm pressure and 40°C.

Supercritical fluid extraction is done in a single-stage contractor, with or without recycling of the supercritical fluid. When recycling is used, the process involves a reduction of pressure to allow the supercritical fluid to lose its ability to dissolve the solute, after which the solid is allowed to separate by gravity and the gas at a slightly lower pressure is compressed back to the supercritical conditions and recycled. Temperature reduction may also be used to drop the solute and the solvent is reheated for recycling without the need for recompression.

Figure 14.13 shows a schematic diagram of a supercritical fluid extraction system. The basic components are an extractor tank and expansion tank. Supercritical fluid conditions are maintained in the extractor. Temperature is usually maintained under controlled conditions in both tanks. Charging and emptying the extractor is a batch operation. The supercritical fluid is allowed to expand in the expansion tank, then recompressed and recycled. Heat exchanges are needed to maintain temperatures and prevent excessive cooling at the expansion values due to the Joule-Kelvin effect.

Two of the major problems of supercritical fluid extraction are channeling of fluid flow through the bed of solids and entrainment of the nonextractable com-

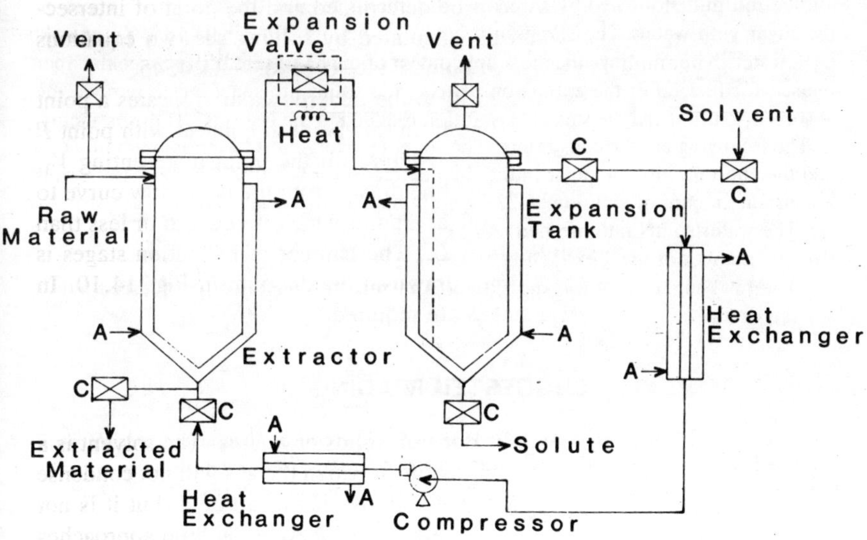

Fig. 14.13. Schematic diagram of a supercritical fluid extraction system.

ponent by the supercritical fluid. The time for contacting is related to the solubility of the solute in the supercritical fluid and the rate of flow of the fluid through the bed of solids. A large quantity of solute to be extracted requires a higher rate of fluid flow to carry out the extraction within a reasonable length of time. Supercritical fluid penetration into the interior of a solid is rapid, but solute diffusion from the solid into the supercritical fluid may be slow and may contribute to the prolonged contact time needed for extraction. Supercritical fluid extraction is one of many experimental separation processes currently under investigation which holds some promise for the future in the food industry.

PROBLEMS

1. Ground roasted coffee contains 5% soluble solids, 3% water, and 92% inert insoluble solids. In order to obtain an extract with a high soluble solids content without having to concentrate it for spray drying, a countercurrent extraction process is to be used to prepare the extract. It is desired that the final extract contain 0.1 kg solubles/kg water and that the solubles of the spent coffee grounds not exceed 0.005 kg/kg dry inert solids.
   (a) Determine the water/coffee ratio to be used in the extraction.
   (b) The coffee grounds carry 1 kg water/kg of solubles-free inert solids, and this quantity is constant with the solute concentration in the extract. Calculate the number of extraction stages needed for this process.
2. A process for extracting sugar from sweet sorghum involves pressing the cane through a three-roll mill, followed by shredding the fibrous residue (bagasse) and extracting the sugar with water. The sorghum originally contains 20% fiber, 16% sugar, and 64% water. After milling, the moisture content of the bagasse is 55%. Since the fiber is used for fuel, after the extraction battery, the solids are squeezed to remove the absorbed solution and the squeezed solution is added to the last stage of the extractor.

   The following are the constraints: The sugar recovery must be a minimum of 99%, and the concentration of sugar in the final extract must be 10%. The bagasse carries a constant amount of solution, 1.22 kg solution/kg fiber. Calculate:
   (a) The water/solids ratio needed.
   (b) The number of ideal extraction stages.
   (c) The final sugar content if the extract is mixed with the juice first pressed out of the cane.

SUGGESTED READING

Johnston, K. P., and Penninger, J. M. L. (eds.). 1989. *Supercritical Fluid Science and Technology*. ACS Symposium Series 406. American Chemical Society, Washington, D.C.

McCabe, W. L., Smith, J. C., and Harriott, P. 1985. *Unit Operations in Chemical Engineering*, 4th ed. McGraw-Hill Book Co., New York.

Perry, R. H., and Chilton, C. H. (eds.). 1973. *The Chemical Engineers Handbook*, 5th ed. McGraw-Hill Book Co., New York.

Prabhudesai, R. K., 1979. Leaching. In: *Handbook of Separation Techniques for Chemical Engineers*, Schweitzer, P.A. (ed.). McGraw-Hill Book Co., New York.

Appendixes

Appendix A.1. Conversion Factors Expressed as a Ratio

Denominator I	Numerators II	Numerators III
Acre	0.4046873	Hectare
Acre	43560.0	$(ft)^2$
Atm (std)	101325.0	Pascal
Atm (std)	14.696	lb_f/in^2
Atm (std)	29.921	in Hg
Atm (std)	76.0	cm Hg @ 0°C
Atm (std)	33.899	ft H_2O @ 32.2°F
BTU	1.0550*E + 10	Erg
BTU	778.3	ft · lb_f
BTU	3.9292*E − 4	Hp · h
BTU	1054.8	Joule
BTU	252	Cal
BTU	2.928*E − 4	kW · h
BTU/min	0.023575	Hp
BTU/min	0.01757	kW
BTU/min	17.5725	Watt
BTU/s	1054.35	Watt
BTU/h	0.29299	Watt
$\frac{BTU}{h\,(ft^2)(°F)}$	5.678263	$W/m^2 \cdot K$
$\frac{BTU}{h\,(ft)(°F)}$	1.730735	W/m · K
BTU/lb	2326.0	J/kg
$\frac{BTU}{lb\,(°F)}$	4186.8	J/kg · K
Bushel	1.2445	ft^3
Bushel	0.035239	m^3
Cal	4.1868	Joule
Cal	3.9684*E − 3	BTU
Cal	4.1868*E + 7	Erg
centimeter	0.3937	Inch
cm Hg @ 0°C	1333.33	Pascal
cm H_2O @ 4°C	98.0638	Pascal
cPoise	0.01	g/cm · s
cPoise	3.60	kg/m · h
cPoise	6.72*E − 4	lb/ft · s
Joule	10.000*E + 7	Erg
Joule	0.73756	ft · lb_f
cPoise	0.001	Pa · s
$(cm)^3$	3.531*E − 5	$(ft)^3$
$(cm)^3$	0.061023	$(in)^3$
$(cm)^3$	2.642*E − 4	Gal
$(ft)^3$	7.48052	Gal
$(ft)^3$	28.316	Liter
Cup	2.36588*E − 4	$(m)^3$
Dyne	1.000*E − 5	Newton
Dyne · cm	1.000*E − 7	Newton · m
Dyne	2.248*E − 6	lg_f
Erg	9.486*E − 11	BTU
Erg	2.389*E − 8	Cal
Erg	1.000*E − 7	Joule
Foot	0.3048	Meter
$(ft)^3$	2.83168*E − 2	$(m)^3$
$(ft)^2$	9.29030*E − 2	$(m)^2$
ft · lb_f	1.355818	Joule
Gal (U.S.)	0.13368	$(ft)^3$
Gal (U.S.)	3.78541	Liter
Gal (U.S.)	3.78541*E − 3	$(m)^3$
Gram	2.2046*E − 3	Pound
Hectare	2.471	Acre
Hp	42.44	BTU/min
Hp	33,000	ft · lb_f/min
Hp	0.7457	kW
Hp(boiler)	33,480	BTU/h
Inch	2.5400*E − 2	Meter
in Hg @ 0°C	3.38638*E + 3	Pascal
in Hg @ 0°C	0.4912	lb_f/in^2
Joule	9.48*E − 4	BTU
Joule	0.23889	Cal
Pound	453.5924	Gram
Pound	0.45359	kg

Appendix A.1. (*Continued*)

Denominator	Numerators		Denominator	Numerators	
I	II	III	I	II	III
Joule	2.77*E − 4	W · h	lb_f	4.44823	Newton
			$lb_f/in.^2$	0.068046	Atm (std)
kg	2.2046	Pound	$lb_f/in.^2$	68947	dynes/cm^2
km	3281	Foot	$lb_f/in.^2$	2.3066	ft H_2O @ 39.2°F
km	0.6214	Mile	$lb_f/in.^2$	2.035	in Hg @ 0°C
kW	3413	BTU/h	lb_f/ft^2	47.88026	Pascal
kW · h	3.6*E + 6	Joule	$lb_f/in.^2$	6894.757	Pascal
Liter	0.03532	$(ft)^3$	Qt (U.S.)	9.4635*E − 4	$(m)^3$
Liter	0.2642	Gal (U.S.)	Qt (U.S.)	946.358	$(cm)^3$
Liter	2.113	Pint	Qt (U.S.)	57.75	$(in.)^3$
			Qt (U.S.)	0.9463	Liter
Meter	3.281	Foot	Qt (U.S.)	0.25	Gallon
Meter	39.37	Inch			
			Ton (metric)	1000	kg
Newton	1.000*E + 5	Dyne	Ton (metric)	2204.6	Pound
			Ton (short)	2000	Pound
Oz (liq)	29.57373	$(cm)^3$	Ton (refri)	12,000	BTU/h
Oz (liq)	1.803	$(in)^3$	Torr (mmHg		
Oz (av)	28.3495	Gram	@ 0 °C)	133.322	Pascal
Oz (av)	0.0625	Pound			
			Watt	3.413	BTU/h
Pascal	1.4504*E − 4	$lb_f/in.^2$	Watt	44.27	ft · lb_f/min
Pascal	1.0197*E − 5	kg_f/cm^2	Watt	1.341*E − 3	Hp
Pint	28.87	$(in.)^3$	Watt · h	3.413	BTU
Poise	0.1	Pa · s	Watt · h	860.01	Cal
lb_f	444823	Dyne	Watt · h	3600	Joule

To use: multiply quantities having units under Column I with the factors under Column II to obtain quantities having the units under Column III. Also use as a ratio in a dimensional equation. Example: 10 acres = 10 × 0.4,046,873 hectares. The dimensional ratio is (0.4046873) hectare/acre. The symbol *E represents exponents of 10. 9.486*E − 11 = 9.486×10^{-11}

Appendix A.2. Properties of Superheated Steam

Temp. °F	1 psi, T_s = 101.74°F: v	h	5 psi, T_s = 162.24°F: v	h	10 psi, T_s = 193.21°F: v	h
200	392.5	1150.2	78.14	1148.6	38.84	1146.6
250	422.4	1172.9	84.21	1171.7	41.93	1170.2
300	452.3	1195.7	90.24	1194.8	44.98	1193.7
350	482.1	1218.7	96.25	1218.0	48.02	1217.1
400	511.9	1241.8	102.24	1241.3	51.03	1240.6
450	541.7	1265.1	108.23	1264.7	54.04	1264.1
500	571.5	1288.6	114.21	1288.2	57.04	1287.8
600	631.1	1336.1	126.15	1335.9	63.03	1335.5

Absolute Pressure lb_f/in^2 (psi)

Temp. °F	14.696 psi, T_s = 212.00°F: v	h	15 psi, T_s = 213.03°F: v	h	20 psi, T_s = 227.96°F: v	h
250	28.42	1168.8	27.837	1168.7	20.788	1167.1
300	30.52	1192.6	29.889	1192.5	22.356	1191.4
350	32.60	1216.3	31.939	1216.2	23.900	1215.4
400	34.67	1239.9	33.963	1239.9	25.428	1239.2
450	36.72	1263.6	35.977	1263.6	26.946	1263.0
500	38.77	1287.4	37.985	1287.3	28.457	1286.9
600	42.86	1335.2	41.986	1335.2	31.466	1334.9

Absolute Pressure lb_f/in^2 (psi)

Temp. °F	25 psi, T_s = 240.07°F: v	h	30 psi, T_s = 250.34°F: v	h	35 psi, T_s = 259.29°F: v	h
250	16.558	1165.6				
300	17.829	1190.2	14.810	1189.0	12.654	1187.8
350	19.076	1214.5	15.589	1213.6	12.562	1212.7
400	20.307	1238.5	16.892	1237.8	14.453	1237.1
450	21.527	1262.5	17.914	1261.9	15.334	1261.3
500	22.740	1286.4	18.929	1286.0	16.207	1285.5
600	25.153	1334.6	20.945	1334.2	17.939	1333.9

v = specific volume in ft^3/lb; h = enthalpy in BTU/lb.
T_s = saturation temperature at the designated pressure.
Source: Abridged from ASME. 1967. *Steam Tables. Properties of Saturated and Superheated Steam*—from 0.08865 to 15,500 lb per sq in. absolute pressure. American Society of Mechanical Engineers, NY. Used with permission.

Appendix A.3. Saturated Steam Tables—English Units

Temp. °F	Abs. pressure lb/in.^{2}	Specific Volume (ft^2/lb) Sat. liquid v_f	Evap. v_{fg}	Sat. vapor v_g	Enthalpy (BTU/lb) Sat. liquid h_f	Evap. h_{fg}	Sat. vapor h_g
32	0.08859	0.016022	3304.7	3304.7	−0.0179	1075.5	1075.5
35	0.09998	0.016020	2950.5	2950.5	3.002	1073.8	1076.8
40	0.12163	0.016019	2445.8	2445.8	8.027	1071.0	1079.0
45	0.14753	0.016020	2039.3	2039.3	13.044	1068.2	1081.2
50	0.17796	0.016023	1704.8	1704.8	18.054	1065.3	1083.4
55	0.21404	0.016027	1384.2	1384.2	23.059	1062.5	1085.6
60	0.25611	0.016033	1207.6	1207.6	28.060	1059.7	1087.7
65	0.30562	0.016041	1022.8	1022.8	33.057	1056.9	1089.9
70	0.36292	0.016050	868.3	868.4	38.052	1054.0	1092.1
75	0.42985	0.016061	740.8	740.8	43.045	1051.3	1094.3
80	0.50683	0.016072	633.3	633.3	48.037	1048.4	1096.4
85	0.59610	0.016085	543.9	543.9	53.028	1045.6	1098.6
90	0.69813	0.016099	468.1	468.1	58.018	1042.7	1100.0
95	0.81567	0.016114	404.6	404.6	63.008	1039.9	1102.9
100	0.94924	0.016130	350.4	350.4	67.999	1037.1	1105.1
105	1.10218	0.016148	304.6	304.6	72.991	1034.3	1107.2
110	1.2750	0.016165	265.4	265.4	77.98	1031.4	1109.3
115	1.4716	0.016184	232.03	232.0	82.97	1028.5	1111.5
120	1.6927	0.016204	203.25	203.26	87.97	1025.6	1113.6
125	1.9435	0.016225	178.66	178.67	92.96	1022.8	1115.7
130	2.2230	0.016247	157.32	157.33	97.96	1019.8	1117.8
135	2.5382	0.016270	138.98	138.99	102.95	1016.9	1119.9
140	2.8892	0.016293	122.98	123.00	107.95	1014.0	1122.0
145	3.2825	0.016317	109.16	109.18	112.95	1011.1	1124.1
150	3.7184	0.016343	97.05	97.07	117.95	1008.2	1126.1
155	4.2047	0.016369	86.53	86.55	122.95	1005.2	1128.2
160	4.7414	0.016395	77.27	77.29	127.96	1002.2	1130.2
165	5.3374	0.016423	69.19	69.20	132.97	999.2	1132.2
170	5.9926	0.016451	62.04	62.06	137.97	996.2	1134.2
175	6.7173	0.016480	55.77	55.79	142.99	993.2	1136.2
180	7.5110	0.016510	50.21	50.22	148.00	990.2	1138.2
185	8.3855	0.016543	45.31	45.33	153.02	987.2	1140.2
190	9.340	0.016572	40.941	40.957	158.04	984.1	1142.1
195	10.386	0.016605	37.078	37.094	163.06	981.0	1144.1
200	11.526	0.016637	33.622	33.639	168.09	977.9	1146.0
205	12.776	0.016707	30.567	30.583	173.12	974.8	1147.8
210	14.132	0.016705	27.822	27.839	178.16	971.6	1149.8
212	14.696	0.016719	26.782	26.799	180.17	970.3	1150.5
220	17.186	0.016775	23.131	23.148	188.23	965.2	1153.4
225	18.921	0.016812	21.161	21.177	193.28	961.9	1155.2
230	20.791	0.016849	19.379	19.396	198.33	958.7	1157.1
235	22.804	0.016887	17.766	17.783	203.39	956.5	1158.9
240	24.968	0.016926	16.304	16.321	208.45	952.1	1160.6
245	27.319	0.016966	14.998	15.015	213.52	948.8	1162.4
250	29.840	0.017006	13.811	13.828	218.59	945.5	1164.1

Appendix A.3. (*Continued*)

Temp. °F	Abs. pressure lb/in.^{2}	Specific Volume (ft^2/lb) Sat. liquid v_f	Evap. v_{fg}	Sat. vapor v_g	Enthalpy (BTU/lb) Sat. liquid h_f	Evap. h_{fg}	Sat. vapor h_g
255	32.539	0.017047	12.729	12.747	223.67	942.1	1165.8
260	35.427	0.017089	11.745	11.762	228.76	938.6	1167.4
265	38.546	0.017132	10.858	10.875	233.85	935.2	1169.0
270	41.875	0.017175	10.048	10.065	238.95	931.7	1170.7
275	45.423	0.017219	9.306	9.324	244.06	928.2	1172.2
280	49.200	0.017264	8.627	8.644	249.17	924.6	1173.8
285	53.259	0.017310	8.0118	8.0291	254.32	920.9	1175.3
290	57.752	0.017360	7.4468	7.4641	259.45	917.3	1176.8

Source: Abridged from: ASME 1967. *Steam Tables. Properties of Saturated and Superheated Steam.* American Society of Mechanical Engineers, NY. Used with permission.

Appendix A.4. Saturated Steam Tables—Metric Units

Temperature °C	Absolute pressure kPa	Enthalpy (MJ/kg) Saturated liquid h_f	(MJ/kg) Evaporation h_{fg}	Saturated vapor h_g
0	0.6108	−0.00004	2.5016	2.5016
2.5	0.7314	0.01049	2.4956	2.5061
5	0.8724	0.02100	2.4897	2.5108
7.5	1.0365	0.03151	2.4839	2.5153
10	1.2270	0.04204	2.4779	2.5200
12.5	1.4489	0.05253	2.4720	2.5245
15	1.7049	0.06292	2.4661	2.5291
17.5	2.0326	0.07453	2.4595	2.5342
20	2.3366	0.08386	2.4544	2.5381
22.5	2.7248	0.09780	2.4484	2.5428
25	3.1599	0.10477	2.4425	2.5473
27.5	3.6708	0.11522	2.4367	2.5518
30	4.2415	0.12566	2.4307	2.5563
32.5	4.8913	0.13611	2.4246	2.5609
35	5.6238	0.14656	2.4188	2.5653
37.5	6.4488	0.15701	2.4129	2.5699
40	7.3749	0.16745	2.4069	2.5744
42.5	8.4185	0.17789	2.4009	2.5788
45	9.5851	0.18834	2.3949	2.5832
47.5	10.8868	0.19880	2.3889	2.5877
50	12.3354	0.20925	2.3829	2.5921
52.5	13.9524	0.21971	2.3769	2.5966
55	15.7459	0.23017	2.3705	2.6000
57.5	17.7295	0.24062	2.3648	2.6054
60	19.9203	0.25109	2.3586	2.6098
62.5	22.3466	0.26155	2.3525	2.6140
65	25.0159	0.27202	2.3464	2.6184
67.5	27.9479	0.28249	2.3402	2.6226
70	31.1622	0.29298	2.3339	2.6270
72.5	34.6961	0.30345	2.3276	2.6312
75	38.5575	0.31394	2.3214	2.6354
77.5	42.7706	0.32442	2.3151	2.6395
80	47.3601	0.33492	2.30879	2.64373
82.5	52.5777	0.34542	2.30251	2.64792
85	57.8159	0.34659	2.29611	2.65199
87.5	63.7196	0.36643	2.28971	2.65606
90	70.1059	0.37693	2.28320	2.66025
92.5	77.0489	0.38747	2.27669	2.66420
95	84.5676	0.39799	2.27023	2.66821
97.5	92.6379	0.40853	2.26349	2.67214
100	101.3250	0.41908	2.25692	2.67606
102.5	110.7410	0.42962	2.25035	2.67996
105	120.8548	0.44017	2.24354	2.68368
107.5	131.7114	0.45074	2.23674	2.68752
110	143.3489	0.46132	2.22994	2.69129

Appendix A.4. (*Continued*)

		Enthalpy (MJ/kg)		
Temperature °C	Absolute pressure kPa	Saturated liquid h_f	(MJ/kg) Evaporation h_{fg}	Saturated vapor h_g
112.5	155.8051	0.47190	2.22313	2.69508
115	169.1284	0.48249	2.21615	2.69874
117.5	183.3574	0.49309	2.20929	2.70241
120	198.5414	0.50372	2.20225	2.70607
122.5	214.8337	0.51434	2.19519	2.70949
125	232.1809	0.52499	2.18807	2.71311
127.5	250.6391	0.53565	2.18083	2.71651
130	270.2538	0.54631	2.17365	2.71991
132.5	291.0837	0.55698	2.11632	2.72331
135	313.1771	0.56768	2.15899	2.72654

Source: Calculated from ASME 1967. *Steam Tables. Properties of Saturated and Superheated Steam*—from 0.08865 to 15,500 lb per sq in. absolute pressure. American Society of Mechanical Engineers, NY Used with permission.

Appendix A.5. Flow Properties of Food Fluids

Product	% Solids	Temperature °C	Flow Constants: n (dimensionless)	b dyne · s^n/cm^2
Applesauce	11	30	0.34	116
		82	0.34	90
Apple juice	50–65.5 Brix	30	0.65	—
	10.5–40 Brix	30	1.0	—
Apricot puree	16	30	0.30	68
		82	0.27	56
Apricot concentrate	26	4.5	0.26	860
		25	0.30	670
		60	0.32	400
Banana puree	—	24	0.458	65
Grape juice	64 Brix	30	0.9	—
	15–50 Brix	30	1.0	—
Orange juice concentrate	—	15	0.584	11.9
		0	0.542	18.0
Orange juice concentrate	30 Brix	30	0.85	—
	60 Brix	30	0.55	15.5
	65 Brix	30	0.91	2.6
Pear puree	18.3	32	0.486	22.5
		82	0.484	14.5
	26	32	0.450	62
		82	0.455	36
	31	32	0.450	109
		82	0.459	56
	37	32	0.479	355
		82	0.481	160
Peach puree	12	30	0.28	72
		82	0.27	58
Plum juice	14	30	0.34	22
		82	0.34	20
Tomato juice	12.8	32	0.43	20
		82	0.345	31.2
	25	32	0.405	129
		82	0.43	61
	30	32	0.40	187
		82	0.445	79
Tomato catsup (0.15 g tomato solids/g catsup)	36	30	0.441	81

Source: Holdsworth, S. D. Applicability of rheological models to the interpretation of flow and processing behavior of fluid food products. *J. Tex. Studies* 2(4): 393–418, 1971.

Appendix A.6. Psychrometric Chart—English Units

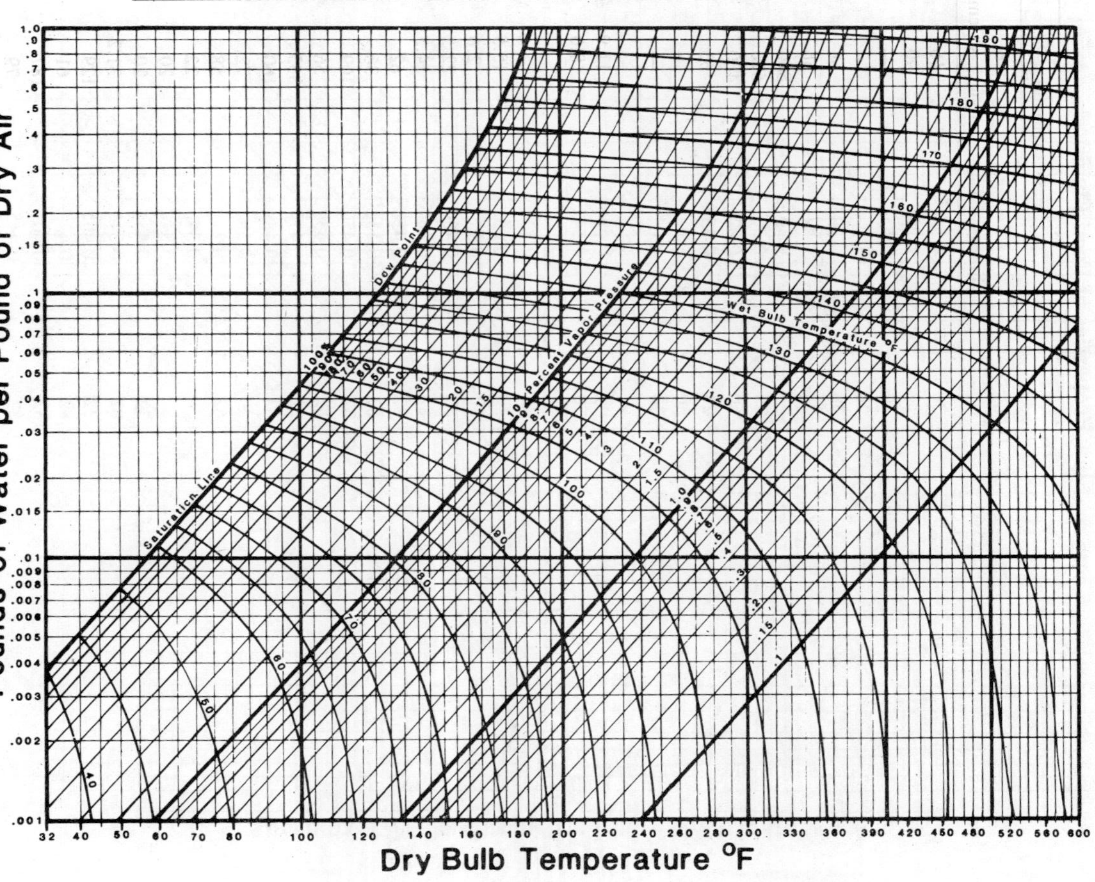

Appendix A.7. Psychrometric Chart—Metric Units

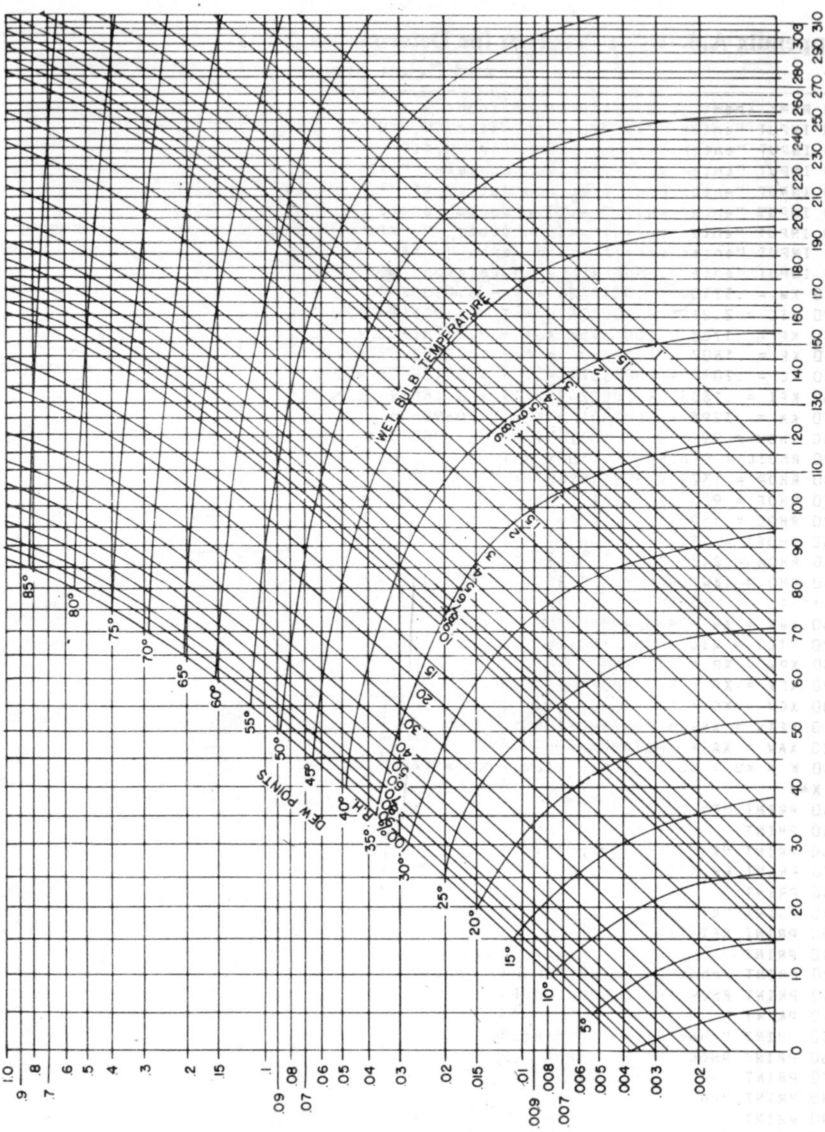

Appendix A.8. Basic Program for Determination of Thermal Conductivity Above and Below Freezing

```
10 REM: Themal conductivity of foods
20 INPUT "enter mass fraction water  ", XW
30 INPUT "enter mass fraction ice  ",XIC
40 INPUT "enter mass fraction protein  ",XP
50 INPUT "enter mass fraction fat  ",XF
60 INPUT "enter mass fraction carbohydrate  ", XC
70 INPUT "enter mass fraction fiber  ", XFI
80 INPUT "enter mass fraction ash  ", XA
90 INPUT "enter temperature in Celsius ",T
110 KW = .57109 + .0017625 * T - 6.7306E-06 * T^2
120 KIC = 2.2196 - .0062489 * T = 1.0154E-04 * T^2
130 KP = .1788 = .0011958 * T - 2.7178E-06 * T^2
140 KF = .1807 - .0027604 * T - 1.7749E-07 * T^2
150 KC = .2014 + .0013874 * T - 4.3312E-06 * T^2
160 KFI = .18331 + .0012497 * T - 3.1683E-06 * T^2
170 KA = .3296 + .001401 * T - 2.9069E-06 * T^2
180 RHOW = 997.18 + .0031439 * T - .0037574 * T^2
190 RHOIC = 916.89 - .13071 * T
200 RHOP = 1329.9 - .5181401 * T
210 RHOF = 925.59 - .41757 * T
220 RHOC = 1599.1 - .32056 * T
230 RHOFI = 1311.5 - .36589 * T
240 RHOA = 2423.8 - .28063 * T
250 RHO = (XW/RHOW + XIC/RHOIC + XP/RHOP + XF/RHOF + XC/RHOC + XFI/RHOFI + XA/RH
OA)^-1
260 XWV = XW * RHO/RHOW
270 XICV = XIC * RHO/RHOIC
280 XPV = XP * RHO/RHOP
290 XFV = XF * RHO/RHOF
300 XCV = XC * RHO/RHOC
310 XFIV = XFI * RHO/RHOFI
320 XAV = XA * RHO/RHOA
330 K = KW * XWV + KIC * XICV + KP * XPV + KF * XFV + KC * XCV + KFI * XFIV + KA
* XAV
340 PRINT "T= ", T ;"C"
350 PRINT
360 PRINT "kw","kic","kp","kf","kc"
370 PRINT KW,KIC,KP,KF,KC,
380 PRINT
390 PRINT "kfi", "ka"
400 PRINT KFI, KA
410 PRINT
420 PRINT "rhow","rhoic","rhop", "rhof"
430 PRINT RHOW,RHOIC,RHOP,RHOF
440 PRINT
450 PRINT "rhoc","rhofi","rhoa"
460 PRINT RHOC, RHOFI, RHOA
470 PRINT
480 PRINT "rho= ", RHO
490 PRINT
500 PRINT "xwv","xicv","xpv","xfv"
510 PRINT XWV,XICV,XPV,XFV
520 PRINT
530 PRINT "xcv", "xfiv", "xav"
540 PRINT XCV,XFIV,XAV
550 PRINT
560 PRINT "k=  ",K
570 INPUT "If you want k at another temperature enter 1, else press return  ",L
580 IF L=1 THEN GOTO 90 ELSE END
```

Appendix A.9. Thermal Conductivity of Materials of Construction and Insulating Materials

	BTU/h(ft)(°F)	W/m(K)
Building materials		
Asbestos cement boards	0.43	0.74
Building brick	0.40	0.69
Building plaster	0.25	0.43
Concrete	0.54	0.93
Concrete blocks		
Two oval core, 8 in. thick	0.60	1.04
Two rectangular core, 8 in. thick	0.64	1.11
Corkboard	0.025	0.043
Felt (wool)	0.03	0.052
Glass	0.3–0.61	0.52–1.06
Gypsum or plasterboard	0.33	0.57
Wood (laminated board)	0.045	0.078
Wood (across grain, dry)		
Maple	0.11	0.19
Oak	0.12	0.21
Pine	0.087	0.15
Wood (plywood)	0.067	0.12
Rubber (hard)	0.087	0.15
Insulating materials		
Air		
32°F (0°C)	0.014	0.024
212°F (100°C)	0.0183	0.032
392°F (200°C)	0.0226	0.039
Fiberglass (9 lb/ft density)	0.02	0.035
Polystyrene		
2.4 lb/ft density	0.019	0.032
2.9 lb/ft density	0.015	0.026
1.6 lb/ft density	0.023	0.040
Polyurethane (5–8.5 lb/ft density)	0.019	0.033
Hog hair with asphalt binder (8.5 lb/ft density)	0.028	0.048
Mineral wool with binder	0.025	0.043
Metals		
Aluminum		
32°F (0°C)	117	202
212°F (100°C)	119	205
572°F (300°C)	133	230
Cast iron		
32°F (0°C)	32	55
212°F (100°C)	30	52
572°F (300°C)	26	45

Appendix A.9. (*Continued*)

	$\frac{BTU}{h(ft)(°F)}$	$\frac{W}{m(K)}$
Copper		
32°F (0°C)	294	509
212°F (100°C)	218	377
572°F (300°C)	212	367
Steel (carbon)		
212°F (100°C)	26	45
572°F (300°C)	25	43
Steel, stainless type 304 or 302	10	17
Steel, stainless type 316	9	15

Appendix A.10. Thermal Conductivity of Foods

Food	Temp. °C	Thermal conductivity W/(m · K)	Food	Temp. °C	Thermal conductivity W/(m · K)
Apple juice	80	0.6317	Lemon	—	1.817
Applesauce	29	0.5846	Limes		
Avocado	—	0.4292	Peeled	—	0.4900
Banana	—	0.4811	Margarine	—	0.2340
Beef	5	0.5106	Milk		
	10	0.5227	3% fat	—	0.5296
Beets	28	0.6006	2.5% fat	20	0.5054
Broccoli	−6.6	0.3808	Oatmeal, dry	—	0.6404
Butter	—	0.1972	Olive oil	5.6	0.1887
Butterfat	−10.6			100	0.1627
	to 10	0.1679	Onions	8.6	0.5746
Cantaloupe	—	0.5711	Oranges		
Carrots			peeled	28	0.5800
Fresh	—	0.6058	Orange juice	−18	2.3880
Puree	—	1.263	Peaches	28	0.5815
Corn			Peanut oil	3.9	0.1679
Yellow	—	0.1405	Pear	8.7	0.5954
Dent	—	0.577	Pear juice	20	0.4760
Egg white	—	0.338		80	0.5365
Egg yolk	2.8	0.5435	Peas	2.8–	
Fish	−10	1.497	Blackeye	16.7	0.3115
Cod	3.9	0.5019	Pineapple	—	0.5486
Salmon	−2.5	1.2980	Plums	—	0.5504
			Pork	6	0.4881
Gooseberries	—	0.2769		59.3	0.5400
Dry	—	0.3288	Potato, raw	—	0.554
Wet	—	0.0277	Poultry,	—	0.4119
Frozen			broiler		
Grapefruit	—	1.3500	Sesame oil	—	0.1755
Mashed			Strawberries	13.3	0.6750
Honey		0.5019		−12.2	1.0970
80% water	2	0.4154	Tomato	—	0.5279
80% water	69	0.6230	Turkey	2.8	0.5019
14.8% water	69	2.4230		−10	1.461
Ice	−25	0.4500	Turnips	—	0.5625
Lamb	5.5	0.4777			
	61.1				

Source: Excerpted from *Food Technol.* 34(11):76–94, 1980.

Appendix A.11. Basic Program for Thermophysical Properties of Foods from Their Composition

```
0 REM: Physical Properties Estimation based on a paper by Choi and Okos, 1986.
  Food Engineering and Process Applications, M. LeMaguer and P. Jelen eds. Else
  vier Science Pub. N. Y.
0 DIM COMP(6)
0 INPUT "COMPONENT 1 IS WATER, ENTER WT. FRACTION =  ",COMP(1)
0 INPUT  "COMPONENT 2 IS PROTEIN, ENTER WT. FRACTION =  ", COMP(2)
0 INPUT "COMPONENT 3 IS FAT, ENTER WT. FRACTION =  ", COMP(3)
0 INPUT "COMPONENT 4 IS CARBOHYDRATES, ENTER WT. FRACTION=  ", COMP(4)
0 INPUT "COMPONENT 5 IS FIBER; ENTER WT. FRACTION =  ", COMP(5)
0 INPUT "COMPONENT 6 IS ASH; ENTER WT. FRACTION=  ",COMP(6)
0 INPUT "ENTER TEMPERATURE IN DEGREE CELSIUS  ", T
00 KW = .57109 + .0017625 * T - 6.7036E-06 * T^2
10 KP = .17881 + .0011958 * T - 2.7178E-06 * T^2
20 KF = .18071 - .0027604 * T - 1.7749E-07 * T^2
21 KC = .20141 + .0013874 * T - 4.3312E-06 * T^2
30 KFIB= .18331 + .0012497 * T - 3.1683E-06 * T^2
40 KASH= .32962 + .0014011 * T - 2.9069E-06 * T^2
50 RHOW= 997.18 + .0031429 * T - .0037574 * T^2
60 RHOP= 1329.9 - .5185 * T
70 RHOF = 925.59 - .41757 * T
80 RHOC = 1599.1 - .31046 * T
90 RHOFI= 1311.5 - .36589 * T
00 RHOA = 2423.8 - .28063 * T
10 CPW= 4176.2 -.090864 * T + .0054731 * T^2
20 CPP = 2008.2 + 1.2089 * T - .0013129 * T^2
30 CPF = 1984.2 + 1.4733 * T - .0048008 * T^2
40 CPC=  1548.8 + 1.9625 * T - .0059399 * T^2
50 CPFI = 1845.9 + 1.8306 * T - .0046509 * T^2
60 CPA = 1092.6 + 1.8896 * T - .0036817 * T^2
69  RHO = (COMP(1)/RHOW + COMP(2)/RHOP + COMP(3)/RHOF + COMP(4)/RHOC + COMP(5)/R
OFI + COMP(6)/RHOA)^-1
71 XVW = (COMP(1)/RHOW) * RHO
80 XVP = (COMP(2)/RHOP) * RHO
90 XVF = (COMP(3)/RHOF) * RHO
00 XVC = (COMP(4)/RHOC) * RHO
10 XVFI = (COMP(5)/RHOFI) * RHO
20 XVA = (COMP(6)/RHOA) * RHO
30 K = XVW * KW + XVP * KP + XVF * KF + XVC * KC + XVFI * KFIB + XVA * KASH
40 PRINT "K=  ", K
60 PRINT "RHO= ", RHO
65 CP = CPW * COMP(1) + CPP * COMP(2) + CPF * COMP(3) + CPC * COMP(4) + CPFI *
OMP(5) + CFA * COMP(6)
70 PRINT "CP= ", CP
80 ALPHA = XVW * KW/(RHOW * CPW) + XVP * KP/(RHOP * CPP) + XVF * KF/
RHOF * CPF) + XVC
 KC/(RHOC * CPC) + XVFI * KFIB/(RHOFI * CPFI) + XVA * KASH/(RHOA * CPA)
90 PRINT "ALPHA = ", ALPHA
00 END
0 REM: Physical Properties Estimation based on a paper by Choi and Okos, 1986.
  Food Engineering and Process Applications, M. LeMaguer and P. Jelen eds. Else
  vier Science Pub. N. Y.
0 DIM COMP(6)
0 INPUT "COMPONENT 1 IS WATER, ENTER WT. FRACTION =  ",COMP(1)
0 INPUT  "COMPONENT 2 IS PROTEIN, ENTER WT. FRACTION =  ", COMP(2)
0 INPUT "COMPONENT 3 IS FAT, ENTER WT. FRACTION =  ", COMP(3)
0 INPUT "COMPONENT 4 IS CARBOHYDRATES, ENTER WT. FRACTION=  ", COMP(4)
0 INPUT "COMPONENT 5 IS FIBER; ENTER WT. FRACTION =  ", COMP(5)
0 INPUT "COMPONENT 6 IS ASH; ENTER WT. FRACTION=  ",COMP(6)
0 INPUT "INPUT "ENTER TEMPERATURE IN DEGREE CELSIUS   ", T
00 KW = .57109 + .0017625 * T - 6.7036E-06 * T^2
10 KP = .17881 + .0011958 * T - 2.7178E-06 * T^2
20 KF = .18071 - .0027604 * T - 1.7749E-07 * T^2
```

Appendix A.11. (*Continued*)

```
121 KC = .20141 + .0013874 * T - 4.3312E-06 * T^2
130 KFIB= .18331 + .0012497 * T - 3.1683E-06 * T^2
140 KASH= .32962 + .0014011 * T - 2.9069E-06 * T^2
150 RHOW= 997.18 + .0031429 * T - .0037574 * T^2
160 RHOP= 1329.9 - .5185 * T
170 RHOF = 925.59 - .41757 * T
180 RHOC = 1599.1 - .31046 * T
190 RHOFI= 1311.5 - .36589 * T
200 RHOA = 2423.8 - .28063 * T
210 CPW= 4176.2 -.090864 * T + .0054731 * T^2
220 CPP = 2008.2 + 1.2089 * T - .0013129 * T^2
230 CPF = 1984.2 + 1.4733 * T - .0048008 * T^2
240 CPC=  1548.8 + 1.9625 * T - .0059399 * T^2
250 CPFI = 1845.9 + 1.8306 * T - .0046509 * T^2
260 CPA = 1092.6 + 1.8896 * T - .0036817 * T^2
269  RHO = (COMP(1)/RHOW + COMP(2)/RHOP + COMP(3)/RHOF + COMP(4)/RHOC + COMP(5)/R
HOFI + COMP(6)/RHOA)^-1
271 XVW = (COMP(1)/RHOW) * RHO
280 XVP = (COMP(2)/RHOP) * RHO
290 XVF = (COMP(3)/RHOF) * RHO
300 XVC = (COMP(4)/RHOC) * RHO
310 XVFI = (COMP(5)/RHOFI) * RHO
320 XVA = (COMP(6)/RHOA) * RHO
330 K = XVW * KW + XVP * KP + XVF * KF + XVC * KC + XVFI * KFIB + XVA * KASH
340 PRINT "K=  ", K
360 PRINT "RHO= ", RHO
365 CP = CPW * COMP(1) + CPP * COMP(2) + CPF * COMP(3) + CPC * COMP(4) + CPFI *
COMP(5) + CFA * COMP(6)
370 PRINT "CP= ", CP
380 ALPHA = XVW * KW/(RHOW * CPW) + XVP * KP/(RHOP * CPP) + XVF * KF/
(RHOF * CPF) + XVC
* KC/(RHOC * CPC) + XVFI * KFIB/(RHOFI * CPFI) + XVA * KASH/(RHOA * CPA)
390 PRINT "ALPHA = ", ALPHA
400 END
10 REM: Physical Properties Estimation based on a paper by Choi and Okos, 1986.
   Food Engineering and Process Applications, M. LeMaguer and P. Jelen eds. Else
   vier Science Pub. N. Y.
20 DIM COMP(6)
30 INPUT "COMPONENT 1 IS WATER, ENTER WT. FRACTION =  ",COMP(1)
40 INPUT  "COMPONENT 2 IS PROTEIN, ENTER WT. FRACTION =  ", COMP(2)
50 INPUT "COMPONENT 3 IS FAT, ENTER WT. FRACTION =  ", COMP(3)
60 INPUT "COMPONENT 4 IS CARBOHYDRATES, ENTER WT. FRACTION=  ", COMP(4)
70 INPUT "COMPONENT 5 IS FIBER; ENTER WT. FRACTION =  ", COMP(5)
80 INPUT "COMPONENT 6 IS ASH; ENTER WT. FRACTION=  ",COMP(6)
90 INPUT "ENTER TEMPERATURE IN DEGREE CELSIUS   ", T
100 KW = .57109 + .0017625 * T - 6.7036E-06 * T^2
110 KP = .17881 + .0011958 * T - 2.7178E-06 * T^2
120 KF = .18071 - .0027604 * T - 1.7749E-07 * T^2
121 KC = .20141 + .0013874 * T - 4.3312E-06 * T^2
130 KFIB= .18331 + .0012497 * T - 3.1683E-06 * T^2
140 KASH= .32962 + .0014011 * T - 2.9069E-06 * T^2
150 RHOW= 997.18+ .0031429 * T - .0037574 * T^2
160 RHOP= 1329.9 - .5185 * T
170 RHOF = 925.59 - .41757 * T
180 RHOC = 1599.1 - .31046 * T
190 RHOFI= 1311.5 - .36589 * T
200 RHOA = 2423.8 - .28063 * T
210 CPW= 4176.2 -.090864 * T + .0054731 * T^2
220 CPP = 2008.2 + 1.2089 * T - .0013129 * T^2
230 CPF = 1984.2 + 1.4733 * T - .0048008 * T^2
240 CPC=  1548.8 + 1.9625 * T - .0059399 * T^2
250 CPFI = 1845.9 + 1.8306 * T - .0046509 * T^2
260 CPA = 1092.6 + 1.8896 * T - .0036817 * T^2
269  RHO = (COMP(1)/RHOW + COMP(2)/RHOP + COMP(3)/RHOF + COMP(4)/RHOC + COMP(5)/R
HOFI + COMP(6)/RHOA)^-1
```

Appendix A.11. (*Continued*)

```
271 XVW = (COMP(1)/RHOW) * RHO
280 XVP = (COMP(2)/RHOP) * RHO
290 XVF = (COMP(3)/RHOF) * RHO
300 XVC = (COMP(4)/RHOC) * RHO
310 XVFI = (COMP(5)/RHOFI) * RHO
320 XVA = (COMP(6)/RHOA) * RHO
330 K = XVW * KW + XVP * KP + XVF * KF + XVC * KC + XVFI * KFIB + XVA * KASH
340 PRINT "K=  ", K
360 PRINT "RHO= ", RHO
365 CP = CPW * COMP(1) + CPP * COMP(2) + CPF * COMP(3) + CPC * COMP(4) + CPFI *
COMP(5) + CFA * COMP(6)
370 PRINT "CP= ", CP
380 ALPHA = XVW * KW/(RHOW * CPW) + XVP * KP/(RHOP * CPP) + XVF * KF/
(RHOF * CPF) + XVC
* KC/(RHOC * CPC) + XVFI * KFIB/(RHOFI * CPFI) + XVA * KASH/(RHOA * CPA)
390 PRINT "ALPHA = ", ALPHA
400 END
10 REM: Physical Properties Estimation based on a paper by Choi and Okos, 1986.
   Food Engineering and Process Applications, M. LeMaguer and P. Jelen eds. Else
   vier Science Pub. N. Y.
20 DIM COMP(6)
30 INPUT "COMPONENT 1 IS WATER, ENTER WT. FRACTION =  ",COMP(1)
40 INPUT  "COMPONENT 2 IS PROTEIN, ENTER WT. FRACTION =  ", COMP(2)
50 INPUT "COMPONENT 3 IS FAT, ENTER WT. FRACTION =  ", COMP(3)
60 INPUT "COMPONENT 4 IS CARBOHYDRATES, ENTER WT. FRACTION=  ", COMP(4)
70 INPUT "COMPONENT 5 IS FIBER; ENTER WT. FRACTION =  ", COMP(5)
80 INPUT "COMPONENT 6 IS ASH; ENTER WT. FRACTION=  ",COMP(6)
90 INPUT "ENTER TEMPERATURE IN DEGREE CELSIUS   ", T
100 KW = .57109 + .0017625 * T - 6.7036E-06 * T^2
110 KP = .17881 + .0011958 * T - 2.7178E-06 * T^2
120 KF = .18071 - .0027604 * T - 1.7749E-07 * T^2
121 KC = .20141 + .0013874 * T - 4.3312E-06 * T^2
130 KFIB= .18331 + .0012497 * T - 3.1683E-06 * T^2
140 KASH= .32962 + .0014011 * T - 2.9069E-06 * T^2
150 RHOW= 997.18 + .0031429 * T - .0037574 * T^2
160 RHOP= 1329.9 - .5185 * T
170 RHOF = 925.59 - .41757 * T
180 RHOC = 1599.1 - .31046 * T
190 RHOFI= 1311.5 - .36589 * T
200 RHOA = 2423.8 - .28063 * T
210 CPW= 4176.2 -.090864 * T + .0054731 * T^2
220 CPP = 2008.2 + 1.2089 * T - .0013129 * T^2
230 CPF = 1984.2 + 1.4733 * T - .0048008 * T^2
240 CPC=  1548.8 + 1.9625 * T - .0059399 * T^2
250 CPFI = 1845.9 + 1.8306 * T - .0046509 * T^2
260 CPA = 1092.6 + 1.8896 * T - .0036817 * T^2
269  RHO = (COMP(1)/RHOW + COMP(2)/RHOP + COMP(3)/RHOF + COMP(4)/RHOC + COMP(5)/R
OFI + COMP(6)/RHOA)^-1
271 XVW = (COMP(1)/RHOW) * RHO
280 XVP = (COMP(2)/RHOP) * RHO
290 XVF = (COMP(3)/RHOF) * RHO
300 XVC = (COMP(4)/RHOC) * RHO
310 XVFI = (COMP(5)/RHOFI) * RHO
320 XVA = (COMP(6)/RHOA) * RHO
330 K = XVW * KW + XVP * KP + XVF * KF + XVC * KC + XVFI * KFIB + XVA * KASH
340 PRINT "K=  ", K
360 PRINT "RHO= ", RHO
365 CP = CPW * COMP(1) + CPP * COMP(2) + CPF * COMP(3) + CPC * COMP(4) + CPFI *
OMP(5) + CFA * COMP(6)
370 PRINT "CP= ", CP
380 ALPHA = XVW * KW/(RHOW * CPW) + XVP * KP/(RHOP * CPP) + XVF * KF/
RHOF * CPF) + XVC
 KC/(RHOC * CPC) + XVFI * KFIB/(RHOFI * CPFI) + XVA * KASH/(RHOA * CPA)
390 PRINT "ALPHA = ", ALPHA
400 END
```

Appendix A.12. Correlation Equations for Heat Transfer Coefficients

Correlation equations for heat transfer coefficients:

Very viscious fluids flowing inside horizontal tubes

$$Nu = 1.62\left[Pr\ Re\ \frac{d}{L}\right]^{0.33}[1 + 0.015\ (Gr)^{0.33}]\left[\frac{\mu_f}{\mu_w}\right]^{0.33}$$

Very viscous fluids flowing inside vertical tubes

$$\text{Nu} = 0.255 Gr^{0.25}\ Re^{0.07}\ Pr^{0.37}$$

Fluids in laminar flow inside bent tubes

$$h = h_r\left[\frac{1 + 21}{Re^{0.14}}\right]\left[\frac{d}{D}\right]$$

$h_r = h$ in straight tube, $d =$ tube diameter, $D =$ diameter of curvature of bend.

Evaporation from heat exchange surfaces

$$\frac{q}{A} = 15.6P^{1.156}(T_w - T_s)^{2.30/P^{0.0324}}$$

$q/A =$ heat flux, $T_w =$ wall temperature, $T_s =$ saturated temperature of vapor at pressure P.

Condensing vapors outside vertical tubes

$$Nu_L = 0.925\left[\frac{L^3\rho^2 g\lambda}{\mu k\Delta T}\right]^{0.25}$$

Condensation outside horizontal tubes

$$Nu_{Do} = 0.73\left[\frac{Do^3\rho^2 g\lambda}{k\mu\Delta T}\right]^{0.25}$$

Condensation inside horizontal tubes

$$Nu_{Di} = 0.612\left[\frac{Di^3\rho_1(\rho_1 - \rho_v)g}{k\mu\Delta T}\right]^{0.25}$$

Condensation inside horizontal tubes

$$Nu_{Di} = 0.024(Re)^{0.8}(Pr)^{0.43}\rho_{corr}$$

Re is based on the total mass of steam entering the pipe. $\rho_{corr} = [0.5/\rho_v](\rho_1 - \rho_v)(x_i - x_o)$. $x =$ steam quality, subscripts i and o refer to inlet and exit, and 1 and v refer to condensate and vapor.

Fluids in cross-flow to a bank of tubes

Re is based on fluid velocity at the entrance to the tube bank.

Appendix A.12. (*Continued*)

Tubes in line: a = diameter to diameter distance between tubes.

$$Nu = [1.517 + 205Re^{0.38}]^2 \left[\frac{4a}{(4a - \pi)}\right]$$

Tubes staggered, hexagonal centers; a = distance between tube rows.

$$Nu = [1.878 + 0.256Re^{0.36}]^2 \left[\frac{4a}{(4a - \pi)}\right]$$

Laminar flow in annuli

$$Nu = 1.02Re^{0.45}Pr^{0.5}\left(\frac{De}{L}\right)^{0.4}\left(\frac{D_2}{D_1}\right)^{0.8} Gr^{0.05}\left(\frac{\mu}{\mu_1}\right)^{0.14}$$

De = hydraulic diameter; subscripts 1 and 2 refer to outside diameter of inner cylinder and inside diameter of outer cylinder, respectively.

Turbulent flow in annuli

$$Nu = 0.02Re^{0.8}Pr^{0.33}\left(\frac{D_2}{D_1}\right)^{0.53}$$

Finned tubes

Nu_d, Re_d, and A_0 use the outside diameter of the bare tube. A = total area of tube wall and fin.

$$\text{Tubes in line:} \quad Nu_d = 0.3\ Re_d^{0.625}\left(\frac{A_0}{A}\right)^{0.375} Pr^{0.33}$$

$$\text{Staggered tubes:} \quad Hu_d = 0.45\ Re_d^{0.625}\left(\frac{A_0}{A}\right)^{0.375} Pr^{0.33}$$

Swept surface heat exchangers

N = rotational speed of blades, D = inside diameter of heat exchanger, V = average fluid velocity, L = swept surface length.

$$Nu = 4.9Re^{0.57}Pr^{0.47}\left(D\frac{N}{V}\right)^{0.17}\left(\frac{D}{L}\right)^{0.37}$$

Individual particles

The Nusselt number and the Reynolds number are based on the particle characteristic diameter and the fluid velocity over the particle.

$$\text{Particles in a packed bed:} \quad Nu = 0.015Re^{1.6}Pr^{0.67}$$

$$\text{Particles in a gas stream:} \quad Nu = 2 + 0.6Re^{0.5}Pr^{0.33}$$

Sources: Perry and Chilton, *Chemical Engineers Handbook*, 5th ed., McGraw-Hill Book Co., New York; Rohsenow and Hartnett, *Handbook of Heat Transfer*, McGraw-Hill Book Co., New York; Hausen, *Heat Transfer in Counterflow, Parallel Flow and Cross Flow*, McGraw-Hill Book Co., New York; Schmidt, *Kaltechn.* 15:98, 1963 and 15:370, 1963; Ranz and Marshall, *Chem. Eng. Prog.* 48(3):141 1952.

Appendix A.13. Basic Program for Evaluating Temperature Response of a Brick Shaped Solid

```
10 REM "surface and center temperature in a slab"
20 DIM BI(3), DELTA1(6), DELTA2(6), DELTA3(6)
30 TM = 100:TO = 4
40 L1 = .0245:L2 = .0508: L3 = .1016
50 H1 = 125: H2 = 125: H3 = 125
60 K = .455: RHO = 1085: CP = 4100
70 ALPHA = K/(RHO * CP)
80 FOR TIME = 0 TO 600 STEP 60
90 BI(1) = H1 * L1/K
100 GOSUB 420
110 YX = YXI + YX2 + YX3 + YX4 + YX5
120 IF TIME = 0 GOTO 130 ELSE PRINT "YX = ";YX, "YX1 = ";YX1
130 IF YX>1 THEN YX = 1
140 YCX = YCX1 + YCX2 + YCX3 + YCX4 + YCX5 + YCX6
150 IF TIME = 0 GOTO 160 ELSE PRINT "YCX = "; YCX, "YCX1 = ";YCX1
160 IF YCX>1 THEN YCX = 1
170 BI(2) = H2 * L2/K
180 GOSUB 940
190 YY = YY1 + YY2 + YY3 + YY4 + YY5 + YY6
200 IF TIME = 0 GOTO 210 ELSE PRINT "YY = "; YY,"YY1 = ";YY1
210 IF YY>1 THEN YY = 1
220 YCY = YCY1 + YCY2 + YCY3 + YCY4 + YCY5 + YCY6
230 IF TIME = 0 GOTO 240 ELSE PRINT "YCY = "; YCY, "YCY1 = "; YCY1
240 IF YCY>1 THEN YCY = 1
250 BI(3) = H3 * L3/K
260 GOSUB 1460
270 YZ = YZ1 + YZ2 + YZ3 + YZ4 + YZ5 + YZ6
280 IF TIME = 0 GOTO 290 ELSE PRINT "YZ = ";YZ, "YZ1 = ";YZ1
290 IF YZ>1 THEN YZ = 1
300 YCZ = YCZ1 + YCZ2 + YCZ3 + YCZ4 + YCZ5 + YCZ6
310 IF TIME = 0 GOTO 320 ELSE PRINT "YCZ = ";YCZ, "YCZ1 = ";YCZ1
320 IF YCZ>1 THEN YCZ = 1
330 YS = YX * YY * YZ
340 YC = YCX * YCY * YCZ
350 TS = TM - YS * (TM-TO)
360 TC = TM - YC * (TM-TO)
370 IF TIME = 0 GOTO 380 ELSE PRINT "TIME = ";TIME, "TS = ";TS, "TC = ";TC
380 NEXT TIME
390 FH = 1/(ALPHA * .4343 * ((DELTA1(1)^2/L1^2) + (DELTA2(1)^2/L2^2) +
(DELTA3(1)^2/L3^2)))
400 PRINT "FH = ",FH
410 END
420 REM SUBROUTINE FOR CALCULATING DELTA(N)
430 REM DETERMINE ROOTS OF DELTA1
440 LO = 0: HI = 1.57
450 GOSUB 870
460 DELTA1(1) = X
470 LO = 3.14: HI = 4.71
480 GOSUB 870
490 DELTA1(2) = X
500 LO = 6.28; HI = 7.85
510 GOSUB 870
520 DELTA1(3) = X
530 LO = 9.42 : HI = 11
540 GOSUB 870
550 DELTA1(4) = X
560 LO = 12.5 : HI = 17.3
570 GOSUB 870
580 DELTA1(5) = X
590 LO = 15.7: HI = 17.3
600 GOSUB 870
```

Appendix A.13. (*Continued*)

```
610 DELTA1(6) = X
620 NUMX1 = 2 * SIN(DELTA1(1)) * COS(DELTA1(1))
630 DENUMX1 = DELTA1(1) + SIN(DELTA1(1)) * COS(DELTA1(1))
640 YX1 = (NUMX1/DENUMX1) * EXP(-ALPHA * TIME * DELTA1(1)^2/L1^2)
650 YCX1 = YX1/COS(DELTA1(1))
660 NUMX2 = 2 * SIN(DELTA1(2)) * COS(DELTA1(2))
670 DENUMX2 = DELTA1(2) + SIN(DELTA1(2)) * COS(DELTA1(2))
680 YX2 = (NUMX2/DENUMX2) * EXP(-ALPHA * TIME * DELTA1(2)^2/L1^2)
690 YCX2 = YX2/COS(DELTA1(2))
700 NUMX3 = 2 * SIN(DELTA1(3)) * COS(DELTA1(3))
710 DENUMX3 = DELTA1(3) + SIN(DELTA1(3)) * COS(DELTA1(3))
720 YX3 = (NUMX3/DENUMX3) * EXP(-ALPHA * TIME * DELTA1(3)^2/L1^2)
730 YCX3 = YX3/COS(DELTA1(3))
740 NUMX4 = 2 * SIN(DELTA1(4)) * COS(DELTA1(4))
750 DENUMX4 = DELTA1(4) + SIN(DELTA1(4)) * COS(DELTA1(4))
760 YX4 = (NUMX4/DENUMX4) * EXP(-ALPHA * TIME * DELTA1(4)^2/L1^2)
770 YCX4 = YX4/COS(DELTA1(4))
780 NUMX5 = 2 * SIN(DELTA1(5)) * COS(DELTA1(5))
790 DENUMX5 = DELTA1(5) + SIN(DELTA1(5)) * COS(DELTA1(5))
800 YX5 = (NUMX5/DENUMX5) * EXP(-ALPHA * TIME * DELTA1(5)^2/L1^2)
810 YCX5 = YX5/COX(DELTA1(5))
820 NUMX6 = 2 * SIN(DELTA1(6)) * COS(DELTA1(6))
830 DENUMX6 = DELTA1(6) + SIN(DELTA1(6)) * COS(DELTA1(6))
840 YX6 = (NUMX6/DENUMX6) * EXP(-ALPHA * TIME * DELTA1(6)^2/L1^2)
850 YCX6 = YX6/COS(DELTA1(6))
860 RETURN
870 REM SUBROUTINE FOR CALCULATING ROOTS OF DNTANDN = BI
880 X = .5 * (LO + HI)
890 TEST = X * TAN(X) - BI(1)
900 IF TEST>0 THEN HI = X ELSE LO = X
910 IF ABS(HI - LO) <.00001 THEN GOTO 930
920 IF ABS(TEST)> .001 THEN GOTO 880
930 RETURN
940 REM SUBROUTINE FOR CALCULATING DELTA(N)
950 REM DETERMINE ROOTS OF DELTA2
960 LO = 0: HI = 1.57
970 GOSUB 1390
980 DELTA2(1) = X
990 LO = 3.14: HI = 4.71
000 GOSUB 1390
010 DELTA2(2) = X
020 LO = 6.28: HI = 7.85
030 GOSUB 1390
040 DELTA2(3) = X
050 LO = 9.42 : HI = 14.1
060 GOSUB 1390
070 DELTA2(4) = X
080 LO = 12.5 : HI = 14.1
090 GOSUB 1390
100 DELTA2(5) = X
110 LO = 15.7: HI = 17.3
120 GOSUB 1390
130 DELTA2(6) = X
140 NUMY1 = 2 * SIN(DELTA2(1)) * COS(DELTA2(1))
150 DENUMY = DELTA2(1) + SIN(DELTA2(1)) * COS(DELTA2(1))
160 YY1 = (NUMY1/DENUMY1) * EXP(-ALPHA * TIME * DELTA2(1)^2/L2^2)
170 YCY1 = YY1/COS(DELTA2(1))
180 NUMY2 = 2 * SIN(DELTA2(2)) * COS(DELTA2(2))
190 DENUMY2 = DELTA2(2) + SIN(DELTA2(2)) * COS(DELTA2(2))
200 YY2 = (NUMY2/DENUMY2) * EXP(-ALPHA * TIME * DELTA2(2)^2/L2^2)
210 YCY2 = YY2/COS(DELTA2(2))
220 NUMY3 = 2 * SIN(DELTA2(3)) * COS(DELTA2(3))
230 DENUMY2 = DELTA2(3) + SIN(DELTA2(3)) * COS(DELTA2(3))
```

Appendix A.13. (*Continued*)

```
1240 YY3 = (NUMY3/DENUMY3) * EXP(-ALPHA * TIME * DELTA(3)^2/L2^2)
1250 YCY3 = YY3/COS(DELTA2(3))
1260 NUMY4 = 2 * SIN(DELTA2(4)) * COS(DELTA2(4))
1270 DENUMY4 = DELTA2(4) + SIN(DELTA2(4)) * COS(DELTA2(4))
1280 YY4 = (NUMY4/DENUMY4) * EXP(-ALPHA * TIME * DELTA2(4)^2/L2^2)
1290 YCY4 = YY4/COS(DELTA2(4))
1300 NUMY5 = 2 * SIN(DELTA2(5)) * COS(DELTA2(5))
1310 DENUMY5 = DELTA2(5) + SIN(DELTA2(5)) * COS(DELTA2(5))
1320 YY5 = (NUMY5/DENUMY5) * EXP(-ALPHA * TIME * DELTA2(5)^2/L2^2)
1330 YCY5 = YY5/COS(DELTA2(5))
1340 NUMY6 = 2 * SIN(DELTA2(6)) * COS(DELTA2(6))
1350 DENUMY6 = DELTA2(6) + SIN(DELTA2(6)) * COS(DELTA2(6))
1360 YY6 = (NUMY6/DENUMY6) * EXP(-ALPHA * TIME * DELTA2(6)^2/L2^2)
1370 YCY6 = YY6/COS(DELTA2(6))
1380 RETURN
1390 REM SUBROUTINE FOR CALCULATING ROOTS OF DNTANDN = BI
1400 X = .5 * (LO + HI)
1410 TEST = X * TAN(X) - BI(2)
1420 IF TEST > 0 THEN HI = X ELSE LO = X
1430 IF ABS(HI - LO) < .00001 THEN GOTO 1450
1440 IF ABS(TEST) > .001 THEN GOTO 1400
1450 RETURN
1460 REM SUBROUTINE FOR CALCULATING DELTA(N)
1470 REM DETERMINE ROOTS OF DELTA3
1480 LO = 0: HI = 1.57
1490 GOSUB 1910
1500 DELTA3(1) = X
1510 LO = 3.14: HI = 4.71
1520 GOSUB 1910
1530 DELTA3(2) = X
1540 LO = 6.28: HI = 7.85
1550 GOSUB 1910
1560 DELTA3(3) = X
1570 LO = 9.42: HI = 11
1580 GOSUB 1910
1590 DELTA3(4) = X
1600 LO = 12.5: HI = 14.1
1610 GOSUB 1910
1620 DELTA3(5) = X
1630 LO = 15.7: HI = 17.3
1640 GOSUB 1910
1650 DELTA3(6) = X
1660 NUMZ1 = 2 * SIN(DELTA3(1)) * COS(DELTA3(1))
1670 DENUMZ1 = DELTA3(1) + SIN(DELTA3(1)) * COS(DELTA3(1))
1680 YZ1 = (NUMZ1/DENUMZ1) * EXP(-ALPHA * TIME * DELTA3(1)^2/L3^2)
1690 YCZ1 = YZ1/COS(DELTA3(1))
1700 NUMZ2 = 2 * SIN(DELTA3(2)) * COS(DELTA3(2))
1710 DENUMZ2 = DELTA3(2)( + SIN(DELTA3(2)) * COS(DELTA3(2))
1720 YZ2 = (NUMZ2/DENUMZ2) * EXP(-ALPHA * TIME * DELTA3(2)^2/L3^2)
1730 YCZ2 = YZ2/COS(DELTA3(2))
1740 NUMZ3 = 2 * SIN(DELTA3(3)) * COS(DELTA3(3))
1750 DENUMZ3 = DELTA3(3) + SIN(DELTA3(3)) * COS(DELTA3(3))
1760 YZ3 = (NUMZ3/DENUMZ3) * EXP(-ALPHA * TIME * DELTA3(3)^2/L3^2)
1770 YCZ3 = YZ3/COS(DELTA3(3))
1780 NUMZ4 = 2 * SIN(DELTA3(4)) * COS(DELTA3(4))
1790 DENUMZ4 = DELTA3(4) + SIN(DELTA3(4)) * COS(DELTA3(4))
1800 YZ4 = (NUMZ4/DENUMZ4) * EXP(-ALPHA * TIME * DELTA3(4)^2/L3^2)
1810 YCZ4 = YZ4/COS(DELTA3(4))
1820 NUMZ5 = 2 * SIN(DELTA3(5)) * COS(DELTA3(5))
1830 DENUMZ5 = DELTA3(5) + SIN(DELTA3(5)) * COS(DELTA3(5))
1840 YZ5 = (NUMZ5/DENUMZ5) * EXP(-ALPHA * TIME * DELTA3(5)^2/L3^2)
1850 YCZ5 = YZ5/COS(DELTA3(5))
```

Appendix A.13. (*Continued*)

```
860 NUMZ6 = 2 * SIN(DELTA3(6)) * COS(DELTA3(6))
870 DENUMZ6 = DELTA3(6) + SIN(DELTA3(6)) * COS(DELTA3(6))
880 YZ6 = (NUMZ6/DENUMZ6) * EXP(-ALPHA * TIME * DELTA3(6)^2/L3^2)
890 YCZ6 = YZ6/COS(DELTA3(6))
900 RETURN
910 REM SUBROUTINE FOR CALCULATING ROOTS OF DNTANDN = BI
920 X = .5 * (LO + HI)
930 TEST = X * TAN(X) - BI(3)
940 IF TEST > 0 THEN HI = X ELSE LO = X
950 IF ABS(HI - LO) < .00001 THEN GOTO 1970
960 IF ABS(TEST) > .001 THEN GOTO 1400
970 RETURN
```

Appendix A.14. Basic Program for Evaluating Local Heat Transfer Coefficient from Temperature Response of a Brick-Shaped Solid

```
10 REM DETERMINATION OF HEAT TRANSFER COEFFICIENT FROM FH
20 FH = 33596!
30 L1 = .0254: L2 = .0508: L3 = .1016
40 K = .455: RHO = 1085: CP = 4100
50 ALPHA = K/(RHO * CP)
60 H = 5
80 BI1 = H * L1/K
90 GOSUB 200
100 BI2 = H * L2/K
110 GOSUB 290
120 BI3 = H * L3/K
130 GOSUB 380
140 TEST = 1-(FH * ALPHA * .4343 * ((DELTA1^2/L1^2) + (DELTA2^2/L2^2) + (DELTA3^2/
L3^2)))
141 IF TEST > 0 THEN DH = .01 ELSE DH = -.01
180 PRINT "HEAT TRANSFER COEFFICIENT = "; H, "test = ";TEST
181 H = H + DH
182 IF ABS(TEST) > .01 THEN GOTO 80
190 END
200 REM DETERMINE ROOTS OF DELTA
210 LO = 0: HI = 2
220 X = .5 * (LO + HI)
230 TEST = X * TAN(X) - BI1
240 IF TEST > 0 THEN HI = X ELSE LO = X
250 IF ABS (HI - LO) < .00001 THEN GOTO 270
260 IF ABS(TEST) > .001 THEN GOTO 220
270 DELTA1 = X
280 RETURN
290 REM DETERMINE ROOTS OF DELTA
300 LO = 0: HI = 2
310 X = .5 * (LO + HI)
320 TEST = X * TAN(X) - BI2
330 IF TEST > 0 THEN HI = X ELSE LO = X
340 IF ABS (HI - LO) < .00001 THEN GOTO 360
350 IF ABS(TEST) > .001 THEN GOTO 310
360 DELTA2 = X
370 RETURN
380 REM DETERMINE ROOTS OF DELTA
390 LO = 0: HI = 2
400 X = .5 * (LO + HI)
410 TEST = X * TAN(X) - BI3
420 IF TEST > 0 THEN HI = X ELSE LO = X
430 IF ABS (HI - LO) < .00001 THEN GOTO 450
440 IF ABS(TEST) > .001 THEN GOTO 400
450 DELTA3 = X
460 RETURN
```


```
10 REM DETERMINATION OF HEAT TRANSFER COEFFICIENT FROM FH
20 FH = 33596!
30 LI = .0254: L2 = .0508: L3 = .1016
40 K = .455: RHO = 1085: CP = 4100
50 ALPHA = K/(RHO * CP)
60 H = 5
80 BI1 = H * L1/K
90 GOSUB 200
100 BI2 = H * L2/K
110 GOSUB 290
120 BI3 = H * L3/K
130 GOSUB 380
140 TEST = 1-(FH * ALPHA * .4343 * ((DELTA1^2/L1^2) + (DELTA2^2/L2^2) + (DELTA3^2/
L3^2)))
141 IF TEST > 0 THEN DH = .01 ELSE DH = -.01
```

Appendix A.14. (*Continued*)

```
80 PRINT "HEAT TRANSFER COEFFICIENT = "; H, "test = ";TEST
81 H = H + DH
82 IF ABS(TEST) > .01 THEN GOTO 80
90 END
00 REM DETERMINE ROOTS OF DELTA
10 LO = 0: HI = 2
20 X = .5 * (LO + HI)
30 TEST = X * TAN(X) - BI1
40 IF TEST > 0 THEN HI = X ELSE LO = X
50 IF ABS (HI - LO) < .00001 THEN GOTO 270
60 IF ABS(TEST) > .001 THEN GOTO 220
70 DELTA1 = X
80 RETURN
90 REM DETERMINE ROOTS OF DELTA
00 LO = 0: HI = 2
10 X = .5 * (LO + HI)
20 TEST = X * TAN(X) - BI2
30 IF TEST > 0 THEN HI = X ELSE LO = X
40 IF ABS (HI - LO) < .00001 THEN GOTO 360
50 IF ABS(TEST) > .001 THEN GOTO 310
60 DELTA2 = X
70 RETURN
80 REM DETERMINE ROOTS OF DELTA
90 LO = 0: HI = 2
00 X = .5 * (LO + HI)
10 TEST = X * TAN(X) - BI3
20 IF TEST > 0 THEN HI = X ELSE LO = X
30 IF ABS (HI - LO) < .00001 THEN GOTO 450
40 IF ABS(TEST) > .001 THEN GOTO 400
50 DELTA3 = X
60 RETURN
```

Appendix A.15. Thermal Conductivity of Water as a Function of Temperature

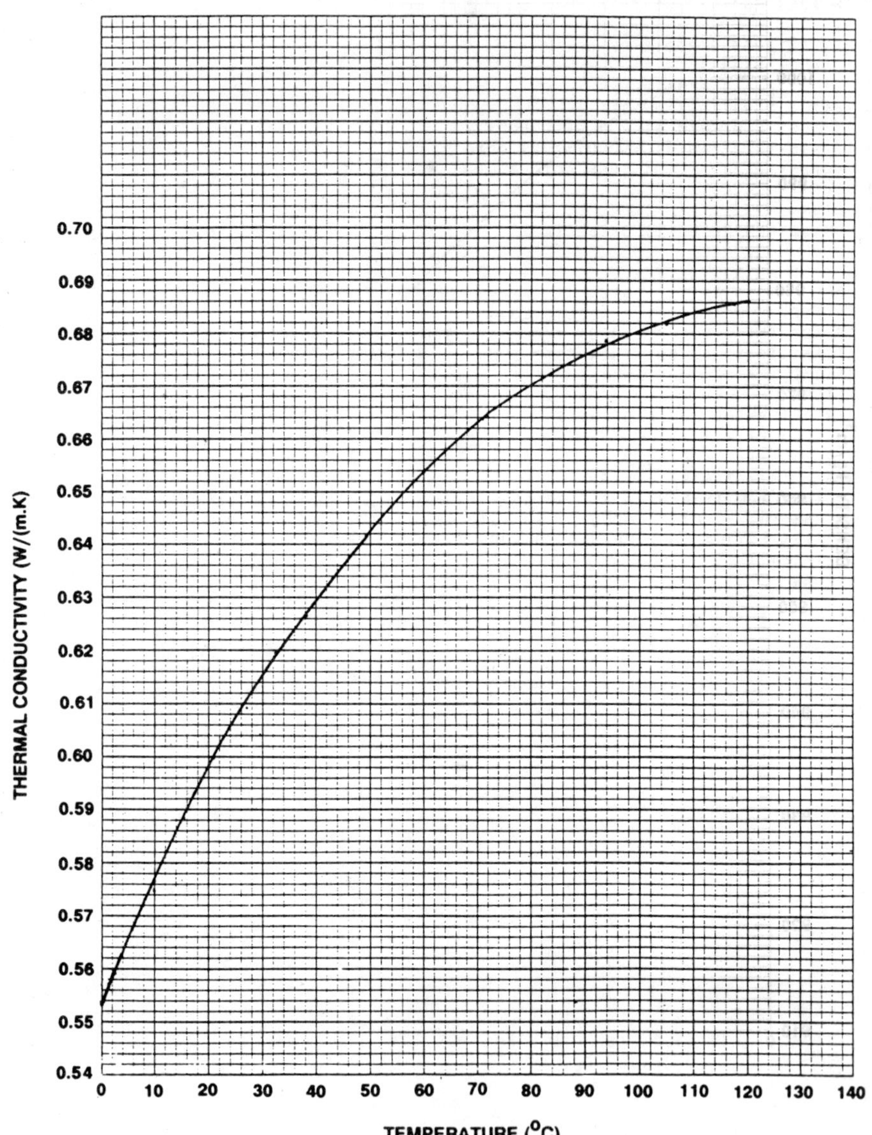

Appendix A.16. Density of Water as a Function of Temperature

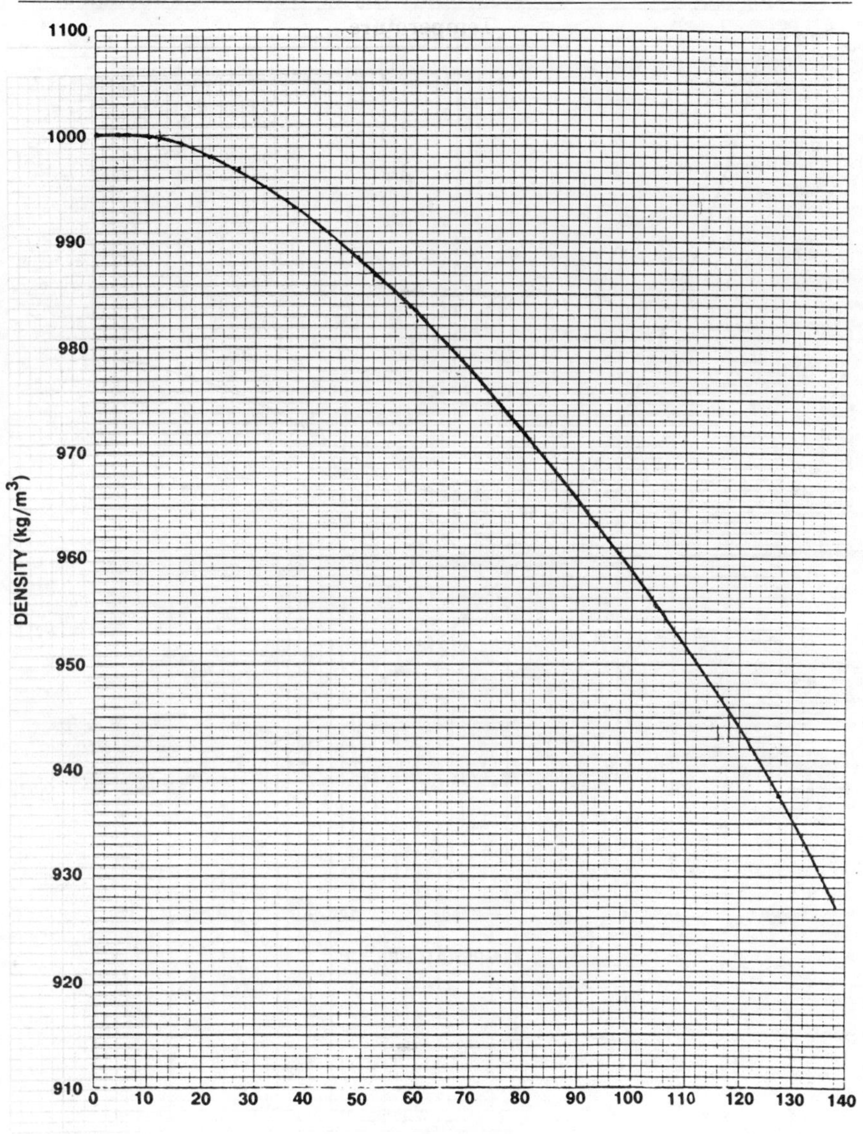

Appendix A.17. Viscosity of Water as a Function of Temperature

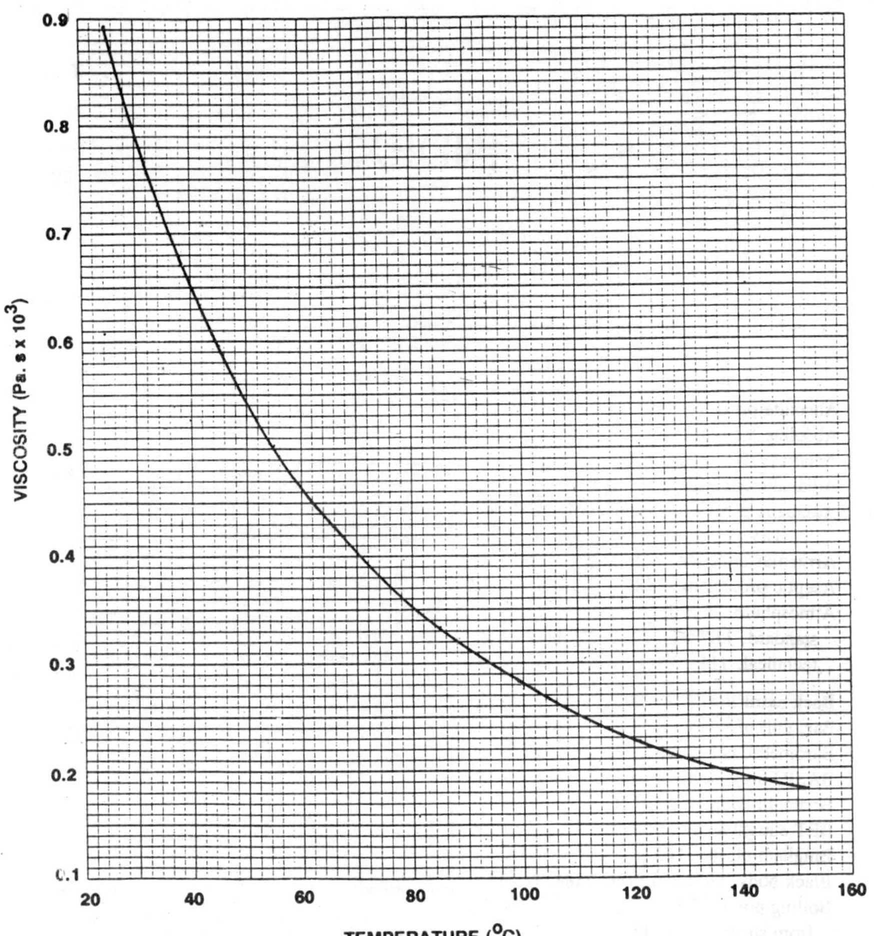

INDEX